Notes on Units and References

Emanating from the work of both scientists and engineers, the published literature on coastal processes uses various systems of units, including the English system (feet, inches, etc.) and the full range of metric units. Although there has been a trend toward the common use of *Système International* (SI) units, this is far from being universal and many workers object to the constraints of SI units, which are not always convenient. This text reflects the diverse usage of measurement systems, although in almost all cases metric units are employed. Since I am drawing on material from the existing literature, it often was not possible or would have been extremely difficult to convert to SI units or even from English to metric units. In a sense, it is of educational value that this text retains the variety of measurement units used in coastal research, reflecting what is still found in the published literature.

A few publications, the *Proceedings of the Coastal Engineering Conferences* being an example, have changed their names over the years, in some cases with each successive conference. In citing these sources I have used the same name throughout the text. In the case of conferences, generally the conference itself is held in one year, while the proceedings are not published until the following year. The dates of references given in the text are the years of the conferences.

Beach Processes and Sedimentation

Second Edition

PAUL D. KOMAR

College of Oceanic & Atmospheric Sciences
Oregon State University

Prentice Hall
Upper Saddle River, New Jersey 07458

Library of Congress Cataloging-in-Publication Data

Komar, Paul D.
    Beach processes and sedimentation / Paul D. Komar.—2nd ed.
        p.   cm.
    Includes bibliographical references and indexes.
    ISBN 0-13-754938-5
    1. Coast changes.   2. Marine sediments.   3. Ocean waves.
    4. Tides.   I. Title.
    GB451.2.K65   1998
    551.45′7—dc21   97-22893
                                                         CIP

Executive Editor: Robert A. McConnin
Total Concept Coordinator: Kimberly P. Karpovich
Cover Designer: Karen Salzbach
Manufacturing Manager: Trudy Pisciotti
Production Supervision/Composition: BookMasters, Inc.

© 1998, 1976 by Prentice-Hall, Inc.
Simon & Schuster/A Viacom Company
Upper Saddle River, New Jersey 07458

Printed in the United States of America

10 9 8 7 6 5 4 3 2 1

ISBN 0-13-754938-5

ISBN 0-13-754938-5

Prentice-Hall International (UK) Limited, *London*
Prentice-Hall of Australia Pty, Limited, *Sydney*
Prentice-Hall Canada Inc., *Toronto*
Prentice-Hall Hispanoamericana, S.A., *Mexico*
Prentice-Hall of India Private Limited, *New Delhi*
Prentice-Hall Japan, Inc., *Tokyo*
Simon & Schuster Asia Pte. Ltd., *Singapore*
Editora Prentice-Hall do Brasil, Ltda., *Rio de Janeiro*

To my parents,
who took me to the beach as a child
and instilled in me the desire to understand.

Contents

Preface

There has been a remarkable increase in the scientific and engineering literature dealing with the coastal zone during the 20 years since publication of the first edition of this text. To suggest that the present edition is a "revision" is therefore something of an understatement. In 1976 when the first edition was published, little was known concerning the patterns of wave transformations and dissipation within the surf zone and how these water motions produce cross-shore movements of sediment resulting in beach-profile variations. Profile responses to storms had been documented, but there was little understanding of the underlying causes. By 1976, suggestions had been made that edge waves might be an important form of energy in the nearshore, responsible for the generation of rip currents and for the rearrangement of beach sediments into crescentic bars. At that time, however, no direct measurements had been made of edge waves on ocean beaches to verify that they actually exist outside of the laboratory, and this uncertainty resulted in much debate as to their relevance. This debate has been muted by the clear documentation of the presence of edge waves in the nearshore and that they often contain substantial amounts of energy. We continue to explore their roles in contributing to the transport of sediment in the nearshore and their effects on the beach morphology. These are only a few examples of research advances during the past 20 years, and many others could be cited touching on almost every aspect of beach processes and sedimentation.

In writing this text I was faced with a large volume of literature from scientific and engineering journals and conference proceedings, more than I could satisfactorily summarize in the text while maintaining the total number of pages reasonable and the product readable. Of course, I read many more papers than could be cited and in the end had to select for inclusion only representative publications covering each of the topics. These have tended to be the historic publications (the "first" contribution), those that made significant and lasting contributions, and finally I have tried to cite at least one recent paper that establishes the "state of the art" on the topic and from which the reader can derive an up-to-date list of relevant references. My choices of papers to be summarized were often subjective, so I apologize to the researchers whose important publications have been left out.

The primary intent of this book is to serve as a textbook in college courses. At Oregon State University I participate in the teaching of two courses that cover this material. One is taken by advanced undergraduates (juniors and seniors) having various backgrounds and graduate students who come mainly from Engineering and our Marine Resource Management Program. The objective of that course is to present an introductory overview of beach processes, relying almost entirely on the contents of this book with only a few additional reading assignments. Some students who enter the course are concerned about whether they have sufficient mathematical skills. Much to their relief, this book is not especially mathematical, generally requiring only algebra and trigonometry. This is because derivations are not included, my goal being to present the results of the derivations and discuss what they mean physically and how they actually agree with the real world.

The second course I teach at Oregon State University presents a more advanced treatment of beach processes and is designed for science and engineering students who intend to specialize in coastal research or its applications. Here the students are expected not only to cover the introductory material contained in the text but also are assigned original journal papers where they are expected to absorb more of the details of the research. Comparatively few original publications can be reviewed during a one-term course, so the expectation is that this approach will initiate a continuation of self-motivated reading, guided by the summaries presented in the text. In that respect, this new edition should also be of service to established scientists, engineers, and individuals concerned with coastal-zone management. With the growth in the volume of literature, increasing each year, we have tended to become more specialized, giving attention only to that literature immediately relevant to our own research. My hope is that this text will serve to provide an updated yet comprehensive review of the total subject for even the most established researchers.

I am indebted to many people who have assisted me during the several years involved in preparing this second edition. My colleagues Robert Holman, William McDougal, and Reggie Beach have been of immense help, first in discussing the topics, clearing up my questions regarding specific points in the literature, and finally in reviewing chapters. I especially want to thank the many students who gave this book the acid test and quickly demonstrated any weaknesses in presentation. David Reinert undertook nearly all of the figure drafting and photography required in the preparation of this edition. Thanks also to the following reviewers: George P. Burbanck, Hampton University; Robert G. Dean, University of Florida; Anthony J. Bowen, Dalhousie University; and Christopher T. Baldwin, Sam Houston State University. Finally, I would like to thank the people who generously supplied me with original photographs and diagrams.

P. D. K.
Corvallis, Oregon

1

An Introduction to the Study of Beaches

The shore is an ancient world, for as long as there has been an earth and sea there has been this place of the meeting of land and water. Yet it is a world that keeps alive the sense of continuing creation and of the relentless drive of life. Each time that I enter it, I gain some new awareness of its beauty and its deeper meanings, sensing that intricate fabric of life by which one creature is linked with another, and each with its surroundings.

Rachel Carson
The Edge of the Sea (1959)

The above quotation by Rachel Carson articulates the sense of heightened awareness derived by many people when they visit a beach. This sense of awareness is certainly what inspired the nineteenth-century artist Martin Johnson Heade to paint the view near Newport, Rhode Island, reproduced in Figure 1-1. The painting captures the surf breaking on the beach among the rocks and the atmosphere of an approaching storm. Other people go to the beach simply to soak up the sun, to improve their tans, to surf on the waves, or to throw a Frisbee. Whatever the reason, people are drawn to the shore, initially perhaps only during vacations, but then to own a home overlooking the sea with access to a beach. Unfortunately, this attraction to the coast has led to "people pollution" (Fig. 1-2).

About two-thirds of the world's population lives within a narrow belt directly landward from the ocean's edge. The thirty coastal states of the United States, including those bordering the Great Lakes, contain 62 percent of the total population and twelve of the thirteen largest cities; 53 percent of the U.S. population lives within 50 miles (80 km) of the shore (Edwards, 1989). Australia is even more oriented toward the sea—83 percent of its population lives near the coast, 25 percent within 3 km, and all of its major cities are found on the coast. It has been estimated that if everyone in the world decided to visit the 440,000 km of the world's shoreline (including the Arctic and Antarctic), each person would have less than 13 cm of shore (Inman and Brush, 1973, p. 26). Such population pressure can be seen at crowded public beaches and in the proliferation of seaside condominiums, hotels/motels, recreational-vehicle parks, and miniature-golf courses—developments, which, in many cases, destroy the esthetic value that originally drew people to the coast.

There are inherent dangers in living on the coast. Steep cliffs are undercut by the dynamics of waves, at times sliding away so that homes and sections of highway fall into the surf (Fig. 1-3). Beaches are inherently unstable, as sand is constantly shifted about by waves, nearshore currents, and the wind. At times the shoreline migrates landward, destroying homes built too close to the sea. Hurricanes have destroyed property worth millions of

Figure 1-1 *Approaching Storm: Beach Near Newport* by the nineteenth-century American artist Martin Johnson Heade. [Courtesy of the Museum of Fine Arts, Boston: Karolik Collection]

Figure 1-2 Spring break on Mustang Island, Texas, the ultimate beach "experience" and an example of "people pollution." [Courtesy of Marine Advisory Service, Texas A&M Sea Grant College]

Figure 1-3   (a) An episode of rapid sea-cliff erosion on the Oregon coast, which undermined the foundations of these two homes. The right house was moved back from the cliff edge. (b) The second house was not moved but was supported by a structure consisting of I-beams and railroad ties.

dollars on the Atlantic and Gulf coasts of the United States (Fig. 1-4). An assessment of shoreline changes along the U.S. coast has been made by investigators at the University of Virginia (Dolan, Hayden, and May, 1983; May, Dolan, and Hayden, 1983; Dolan, Trossbach, and Buckley, 1990). A state-by-state summary of shoreline-change rates, based on their results, is given in Table 1-1. The barrier islands of the Atlantic coast are retreating at an average rate of 0.8 m/year, while the mean rate for the Gulf coast barriers is 1.6 m/year, with some islands having rates as high as 15 m/year.

The shorelines of other nations also are retreating. Probably the most extreme rates found are those along the Nile Delta of Egypt, where the shoreline recession in places averages 50–100 m/year, with maximum rates of 200 m/year (Frihy et al., 1991). Erosion there is due in large part to the construction of the Aswan High Dam on the Nile River, which has cut off sand delivery from the river to the delta beaches (Chapter 3).

The destruction of homes has a visual and human impact, and therefore much interest and news coverage has focused on coastal erosion. This interest has intensified in recent years as our concern grows about the prospects of global warming resulting from the greenhouse effect. Average global temperatures are predicted to increase 1.5–4.5°C by the year 2050 (National Research Council, 1983). Such prophecies of global warming in turn have led to estimates for an accelerated rise in sea level, caused by increased glacial melting and the thermal

A

B

Figure 1-4 Hurricanes and other severe storms represent the main threat to coastal developments. (a) The destruction of Galveston, Texas, by a hurricane during September 8, 1900. It has been estimated that 6,000 people were killed, making this event the greatest natural catastrophe that has been experienced in the United States. [Courtesy of Rosenberg Library, Galveston, TX] (b) Damage to South Carolina's Sullivans Island by Hurricane Hugo during September 1989. [Reprinted from S. J. Williams et al., *Coasts in Crisis*, U.S. Geological Survey Circular No. 175, 1991.]

TABLE 1-1 SHORELINE CHANGE RATES [Used with permission of American Geophysical Union, from S. K. May, R. Dolan and B. P. Hayden, Erosion of U.S. Shorelines, *EOS*. Copyright © 1983 American Geophysical Union.]

Region	Average Shoreline Change Rate (m/year)	Range of Extreme Accretion or Erosion Rate (m/year)		
Atlantic Coast	−0.8	25.5	to	−24.6
Maine	−0.4	1.9	to	−0.5
New Hampshire	−0.5	−0.5	to	−0.5
Massachusetts	−0.9	4.5	to	−4.5
Rhode Island	−0.5	−0.3	to	−0.7
New York	0.1	18.8	to	−2.2
New Jersey	−1.0	25.5	to	−15.0
Delaware	0.1	5.0	to	−2.3
Maryland	−1.5	1.3	to	−8.8
Virginia	−4.2	0.9	to	−24.6
North Carolina	−0.6	9.4	to	−6.0
South Carolina	−2.0	5.9	to	−17.7
Georgia	0.7	5.0	to	−4.0
Florida	−0.1	5.0	to	−2.9
Gulf of Mexico	−1.8	8.8	to	−15.3
Florida	−0.4	8.8	to	−4.5
Alabama	−1.1	0.8	to	−3.1
Mississippi	−0.6	0.6	to	−6.4
Louisiana	−4.2	3.4	to	−15.3
Texas	−1.2	0.8	to	−5.0
Pacific Coast	0.0	10.1	to	−5.0
California	−0.1	10.1	to	−4.2
Oregon	0.1	5.0	to	−5.0
Washington	0.5	5.0	to	−3.9
Alaska	−2.4	2.9	to	−6.0

Negative values indicate erosion, positive values accretion.

expansion of sea water (Chapter 4). For example, Gornitz, Lebedeff, and Hansen (1982) have predicted a rise of 20–30 cm by the year 2050. Although there are many uncertainties in such predictions, the potential impacts need to be given serious consideration in coastal-management decisions, since there would be a near-catastrophic increase in coastal-erosion problems.

The American naturalist William Beebe wrote that the beach is "the battleground of the shore." Waves and currents constantly shift sand about, removing it from one area and depositing it in another. Sand added to the beach during calm weather is removed to the offshore during storms, only to return with the next calm period. The natural beach therefore shows short-term fluctuations within a longer-term dynamic equilibrium. Over hundreds of years a particular stretch of coast may slowly erode and retreat or build outward into the sea. The development of the coast often interferes with this natural variability in that once a house or road is built, human reaction is to maintain the property in the face of the natural processes of shoreline erosion.

> The real conflict of the beach is not between sea and shore, for theirs is only a lover's quarrel, but between man and nature. On the beach, nature has achieved a dynamic equilibrium that is alien to man and his static sense of equilibrium. Once a line has been established, whether it be a shoreline or a property line, man unreasonably expects it to stay put (Soucie, 1973, p. 56).

To maintain property lines, humans have pitted themselves against nature, constructing sea-walls and other shoreline stabilization structures. Not only is this a futile attempt at battling nature, it is expensive, destroys the esthetic value of the coast, and disrupts the natural equilibrium of the system.

An additional factor in the population pressure on the coast is the impact of human waste discharge into the shallow waters adjacent to the beaches and in the offshore, including industrial wastes and municipal sewage. Many beaches have been closed for extended periods and swimming has been prohibited because of the danger to the public. As a result of the extensive news coverage during the summer of 1988 when hypodermic syringes and other dangerous wastes washed onto the beaches of New York and New Jersey, many tourists decided to spend the summer elsewhere, resulting in major economic impacts to the coastal communities that rely on summer visitors. More recently, the Natural Resources Defense Council reported that during 1995 over 3,500 beaches in the United States were closed for a day or longer due to pollution problems.

Because of growing pressures for development and its associated problems, many nations have adopted plans for the management of their coastal zones. While these programs vary in their emphasis, most reflect a concern for the preservation of coastal environments in the light of increasing developmental pressures and multiple uses. The main initiative in the United States has been the Coastal Zone Management Act passed by Congress in 1972. It has four main objectives:

1. to protect fragile coasts;
2. to minimize life and property losses from coastal hazards;
3. to create better conditions for coastal resource use, including improved access for recreation; and
4. to promote intergovernmental cooperation in dealing with coastal-zone issues.

All of the states with ocean shorelines, except for Georgia and Texas, have adopted coastal-management programs. The approaches and goals of these states differ, as does the resulting effectiveness of their programs. Some states have established set-back lines that prohibit the construction of houses and large buildings close to the first line of dunes or to the edge of a sea cliff and outlaw the construction of permanent seawalls and other "hard" protection structures. In contrast, some states have little or no restrictions on where developers can build—if property is threatened, a permit to erect a seawall can be easily obtained. In some instances the state has passed the task of establishing management programs to its coastal counties, so no uniformity of approach exists even within the state. Although the Coastal Zone Management Act has for the most part led to the improved management of the U.S. coast, in many states the term "coastal management" is still an oxymoron.

USES OF THE BEACH

When one considers a vacation at a beach, activities such as swimming, surfing, sunbathing, beach combing, walking, jogging, fishing, and picnicking come to mind (Fig. 1-5). Good beaches attract thousands of visitors each year and are economically important to adjacent communities. The particular activities carried out at a beach are dependent on factors

Figure 1-5   A wide variety of recreational activities occur on beaches, making them
a natural playground for people of all ages. [The photo of the woman reprinted with
permission, Co Rentmeester, *Life* © Times Inc.]

such as the temperature of the water, the size of the waves, and the coarseness of the beach
sediment. Beaches in Florida and California are used for swimming, whereas only the hardi-
est individuals venture into the water off the coasts of Maine and Washington. Surfers travel
long distances to find the beach with the "right kind of waves."

There are other groups interested in beaches. In certain parts of the world, valuable
gems and minerals have been mined from beach sands, including diamonds in South Africa
and gold in Nome, Alaska. Economically important minerals such as magnetite, ilmenite, zircon, rutile, monazite, and platinum are also found on some beaches. The main components of
beaches—sand, gravel, and cobbles—also have been removed for use as aggregate, at times
to the detriment of the beach and to its recreational uses.

In addition to the direct uses of the beach already mentioned, there is an indirect use
that one sometimes forgets: The beach serves as a natural buffer between the ocean and land.
Waves arriving at the coast often contain a tremendous amount of energy. These waves break
on the beach and expend most of their energy before reaching the shoreline and coastal properties. During a storm, the beach is able to modify its slope and overall morphology to dissipate the waves, while not being destroyed itself. However, human activities, such as sand
mining or the interruption of sediment movement by the construction of jetties, can eliminate
the buffering capacity of the beach, leading to an intensified attack by the waves on coastal
properties. Attempts then are made to use a variety of seawalls, and even junk cars and tires,

Figure 1-5   (*Continued*)

to prevent the resulting erosion, a job done formerly and much more effectively by the natural beach.

It is apparent that the potential uses of the beach and adjacent waters are in many cases incompatible. As beaches become even more intensely utilized because of increasing populations, we must establish priorities for their use. The beach is one of our most important natural resources. Once destroyed, its repair is difficult if not impossible and is always costly.

THE STUDY OF BEACHES AND COASTAL PROCESSES

The preservation of beaches and the protection of coastal properties require an understanding of nearshore processes: the motions of waves, the generation of nearshore currents, the movement of beach sediment, and the resulting variability in the beach morphology. Workers in several disciplines have participated in the study of beaches, research that has led to our present understanding of those processes. Historically, the first professionals who seriously studied beaches and coasts were the geographers and geo-morphologists, groups that continue to make significant contributions. Their interests lie in recognizing and understanding the evolution of various coastal forms. They focus on studying changes in particular stretches of coast by comparing existing configurations with old surveys and photographs, describing various beach forms, and attempting to establish classifications of different coastal types. The more recent contributions by geographers and geomorphologists have been in coastal planning, particularly in drawing up guidelines for the wise utilization of our coastal resources.

Geologists (sedimentologists) also have made important contributions to the study of beaches, especially of sediments. In their investigations they have concentrated on various properties of the sediments, such as the statistics of grain-size distributions, including size variations across and along the beach, the degree of roundness and shapes of sediment grains, and sedimentary structures found within beach deposits. Geologists also have made important contributions to the study of beach profiles, in utilizing heavy minerals as natural tracers to determine sediment sources and transport paths and in determining the origin and developmental history of barrier islands and related coastal environments.

The coastal engineer has the unenviable task of protecting our coasts from erosion, maintaining our beaches, and at the same time building jetties, breakwaters, and other forms of boat harbors. These tasks often are contradictory; the construction of jetties or other structures commonly results in the destruction of the adjacent beaches and may even affect shorelines many kilometers away. The coastal engineer must anticipate this and try to prevent the destruction through an understanding of the basic processes involved in the nearshore. This interest has led engineers to undertake fundamental studies that have resulted in important contributions to the understanding of the coastal environment.

World War II found the military with insufficient knowledge to conduct the landing operations required in the Pacific war and in Europe, such as occurred during the D-Day Normandy invasion. If such operations were to avoid disaster, knowledge about the generation and propagation of waves on the ocean and their breaking characteristics in shallow water, about the formation of longshore currents in the nearshore, about the movement of beach sediment, and about factors affecting the morphology of the beach was needed. A crash program resulted, leading to significant advances in the knowledge of the nearshore zone. Although much of this effort was conducted by coastal engineers, a group of university scientists interested in the problems of the nearshore environment evolved from this program. Today, these scientists are concentrated mainly in the oceanography departments and special coastal-studies programs of universities. These scientists examine problems of a fundamental nature and participate in studies of tides and waves that require the use of facilities more readily available at such institutions.

The study of beaches has given rise to unusual laboratory and field facilities. The researcher at times turns to laboratory experiments that can provide more control over the waves and beach conditions than can be found on ocean shores where the environment is constantly changing. Experiments are performed in wave channels or basins, where an oscillating paddle replaces the wind in generating waves (Fig. 1-6). The equipment is often built on a large scale, so waves can be generated that are similar to those found on natural beaches. For example, the wave channel at Oregon State University (OSU) is 104-m long, with a water depth of 4.6 m (Fig. 1-6). Waves with heights up to 1.5 m have been used in experiments, larger than common on many ocean beaches—the OSU wave channel has even been used for surfboard riding. Laboratory wave basins are more three dimensional than wave channels, providing a longer stretch of shore, which makes it possible to study wave-induced longshore currents and sand transport along the length of the beach. Most unusual is the circular wave basin where the beach is not interrupted by side walls and therefore permits laboratory studies on an "infinitely" long beach; the waves are generated by an oscillating cylinder located at the center of the circular tank (Dalrymple and Dean, 1972).

Ultimately, what we know about beaches and coasts must come from studies within the natural environment. This has required the development of wave sensors, current meters, optical sensors of suspended sediments, and a variety of other equipment that can survive in the harsh conditions experienced on ocean beaches. In most cases, the equipment is reasonably small and portable, so it can be transported from one site to another. In recent years, coastal-research centers devoted to the field study of beach processes have been established, a prime example being the Field Research Facility (FRF) of the U.S. Army Corps of Engineers located in Duck, North Carolina (Fig. 1-7). These centers provide the basic tools to measure waves, tides, and beach profiles, as well as computers that can process the data (Birkemeier et al., 1981). The coastal-research centers have become the sites for major field experiments, where scientists and engineers from universities and governmental agencies come together for many weeks at a time. These intensive studies have been extremely important in contributing to the understanding of the processes of nearshore waves and currents, the movement of sediment, and beach-morphology changes.

LITERATURE SOURCES

One of the principal difficulties in the study of beaches and coasts is searching through the large number of journals, technical reports, and other publications that contain pertinent information. Table 1-2 contains a list of important journals and proceedings of conferences that are held regularly. These could be classified roughly as geological journals, engineering publications, and oceanographic literature. However, in recent years the division between these disciplines has been breaking down as more cooperative research is undertaken, and this is reflected in the published literature. Each discipline makes its contribution to the study of beach processes, and although differences in approach and philosophy remain, there is now a more unified attack on the problems of the nearshore environment.

Figure 1-6 The 104-m long wave channel at Oregon State University, in use during the 1991 SUPERTANK experiments undertaken to investigate waves in the nearshore and the resulting beach-profile responses. [Courtesy of N. Kraus]

A

B

Figure 1-7 The Field Research Facility of the U.S. Army Corps of Engineers, Coastal Engineering Research Center, at Duck, North Carolina. (a) This photograph shows the CRAB, used to measure beach profiles. (b) This photograph shows the retrieval of an instrument array containing electromagnetic current meters. [Courtesy of R. Holman]

A partial list of books pertinent to the study of nearshore processes is given in Table 1-3. Here I have distinguished between engineering texts, books for geologists and geographers, those concerned with coastal-zone management, and books suitable for the layman. Books in the "Living With" series can serve as guides to the U.S. coast.

Literature dealing with the coastal zone is growing at an astonishing pace. The challenge of today's scientists is to maintain a broad perspective of the problems and related research, while at the same time becoming specialized in their own research efforts.

TABLE 1-2 JOURNALS AND CONFERENCE PROCEEDINGS DEALING WITH COASTAL STUDIES

Coastal Engineering
Coastal Engineering in Japan
Coastal Management Journal
Coastal Sediments (Conferences of the American Society of Civil Engineers)
Coastal Zone Management Journal
Continental Shelf Research
Geographical Journal
Geography
International Journal of Estuarine and Coastal Law
Journal of Coastal Conservation
Journal of Coastal Research
Journal of Fluid Mechanics
Journal of Geology
Journal of Geomorphology
Journal of Geophysical Research
Journal of the Institute of Civil Engineers
Journal of Sedimentary Geology
Journal of Sedimentary Petrology
Journal of Shoreline Management
Journal of Waterway, Harbors and Coastal Engineering
Marine Geology
Ocean & Shoreline Management
Proceedings of the Conferences on Coastal Engineering (Amer. Soc. Civil Engrs.)
Sedimentology
Shore and Beach
Zeitschrift für Geomorphologie (Journal of Geomorphology)

TABLE 1-3 BOOK SOURCES IN COASTAL STUDIES

Geological and Geographic
Bird, E. C. F. (1970). *Coasts.* Cambridge, MA: MIT Press.
Bird, E. C. F. (1985). *Coastline Changes: A Global Review.* New York: John Wiley & Sons.
Bird, E. C. F. and M. L. Schwartz [editors] (1985). *The World's Coastline.* New York: Van Nostrand Reinhold.
Carter, W. G. (1988). *Coastal Environments.* New York: Academic Press.
Davies, J. L. (1979). *Geographical Variation in Coastal Development* (2nd ed.). New York: Hafner.
Davis, R. A. [editor] (1985). *Coastal Sedimentary Environments* (2nd ed.). New York: Springer-Verlag.
Hardisty, J. (1990). *Beaches: Form and Process.* New York: Unwin Hyman Inc.
King, C. A. M. (1972). *Beaches and Coasts* (2nd ed.). New York: St. Martin Press.
Kinsman, B. (1965). *Wind Waves.* Englewood Cliffs, NJ: Prentice-Hall.
Komar, P. D. [editor] (1983). *CRC Handbook of Coastal Processes and Erosion.* Boca Raton, FL: CRC Press.
Pethnick, J. (1984). *An Introduction to Coastal Geomorphology.* London, England: Edward Arnold.
Shepard, F. P. and H. R. Wanless (1971). *Our Changing Coastlines.* New York: McGraw-Hill.

TABLE 1-3   (CONTINUED)

Steers, J. A. (1948). *The Coastlines of England and Wales.* Cambridge, England: Cambridge University Press.

Zenkovich, V. P. (1967). *Processes of Coastal Development.* Edinburgh, Scotland: Oliver and Boyd.

Engineering

CERC (1984). *Shore Protection Manual* (2 volumes). Coastal Engreering Research Center, U.S. Army Corps of Engineers.

Horikawa, K. (1978). *Coastal Engineering.* New York: John Wiley & Sons.

Ippen, A. T. [editor] (1966). *Estuary and Coastline Hydrodynamics.* New York: McGraw-Hall.

Muir Wood, A. M. (1969). *Coastal Hydraulics.* New York: Macmillan.

Sawaragi, T. (1995). *Coastal Engineering—Waves, Beaches, Wave-Structure Interations.* New York: Elsevier.Silvester, R. (1974). *Coastal Engineering* (2 volumes). New York: Elsevier.

Sorensen, R. M. (1978). *Basic Coastal Engineering.* New York: John Wiley & Sons.

Wiegel, R. L. (1964). *Oceanographical Engineering.* Englewood Cliffs, NJ: Prentice-Hall.

Coastal-Zone Management

Clark, J. R. (1977). *Coastal Ecosystem Management.* New York: John Wiley & Sons.

Ketchum, B. H. [editor] (1972). *Critical Problems of the Coastal Zone.* Cambridge, MA: MIT Press.

National Academy Press (1990). *Managing Coastal Erosion.* Washington, D.C.

Layman-Level Presentations

Bascom, W. (1980). *Waves and Beaches* (2nd ed.). Garden City, NJ: Anchor-Doubleday.

Davis, R. A. (1994). *The Evolving Coast.* New York: Scientific American Library.

Fox, W. T. (1983). *At the Sea's Edge.* Englewood Cliffs, NJ: Prentice-Hall.

Kaufman, W. and O. Pilkey (1979). *The Beaches Are Moving.* Garden City, NJ: Anchor-Doubleday.

Kuhn, G. G. and F. P. Shepard (1984). *Sea Cliffs, Beaches and Coastal Valleys of San Diego County.* Berkeley: University of California Press.

Manley, S. and R. Manley (1968). *Beaches: Their Lives, Legends and Lore.* Radnor, PA: Chilton Book Co.

Pilkey, O. H. and K. L. Dixon (1996). *The Corps and the Shore.* Washington, D.C.: Island Press.

"Living With" Series [Durham, NC: Duke University Press]

Living with the Alabama/Mississippi Shore

Living with the California Coast

Living with Chesapeake Bay

A Moveable Shore (Connecticut)

From Currituck to Calabash

Living with the East Florida Shore

Living with the West Florida Shore

Living with the Lake Erie Shore

Living with Long Island's South Shore

Living with the Louisiana Shore

Living with the Coast of Maine

Living with the New Jersey Shore

Living with the Coast of Puget Sound

Living with the South Carolina Shore

Living with the Texas Shore

The Northwest Coast: Living with the Shores of Washington and Oregon

REFERENCES

BIRKEMEIER, W. A., A. E. DEWALL, C. S. GORBICS, and H. C. MILLER (1981). *A User's Guide to CERC's Field Research Facility.* U.S. Army Corps of Engineers, Coastal Engineering Research Center, Misc. Report No. 81-7, p. 118.

DALRYMPLE, R. A. and R. G. DEAN (1972). The Spiral Wavemaker for Littoral Drift Studies. *Proceedings of the 13th Conference on Coastal Engineering. Amer. Soc. Civil Engrs.,* pp. 689–705.

DOLAN, R., B. HAYDEN, and S. MAY (1983). Erosion of U.S. Shorelines. In *Handbook of Coastal Processes and Erosion*, P. D. Komar (editor), pp. 285–299. Boca Raton, FL: CRC Press.

DOLAN, R., S. TROSSBACH, and M. BUCKLEY (1990). New Shoreline Erosion Data for the Mid-Atlantic Coast. *Journal of Coastal Research* 6: 471–477.

EDWARDS, S. F. (1989). Estimates of Future Demographic Changes in the Coastal Zone. *Coastal Management* 17: 229–240.

FRIHY, O. E., A. M. FANOS, A. A. KHAFAGY, and P. D. KOMAR (1991). Patterns of Nearshore Sediment Transport Along the Nile Delta, Egypt. *Coastal Engineering* 15: 409–429.

GORNITZ, V., S. LEBEDEFF, and J. HANSEN (1982). Global Sea Level Trend in the Past Century. *Science* 215: 1611–1614.

INMAN, D. L. and B. M. BRUSH (1973). The Coastal Challenge. *Science* 181: 20–32.

MAY, S. K., R. DOLAN, and B. P. HAYDEN (1983). Erosion of U.S. Shorelines. *EOS* 64: 551–553.

NATIONAL RESEARCH COUNCIL (1983). *Changing Climate.* Report of the Carbon Dioxide Assessment Committee. Washington, D.C.: The National Academy Press.

SOUCIE, G. (1973). Where Beaches Have Been Going: Into the Ocean. *Smithsonian* 4(3): 55–61.

WILLIAMS, S. J. et al. (1991). *Coasts in Crisis.* U.S. Geological Survey Circular No. 175.

2

The Geomorphology of Eroding and Accreting Coasts

If the great geological labours of the oceans, such as the erosion of cliffs, the demolition of promontories, and the construction of new shores, astonish the mind of man by their grandeur, on the other hand, the thousand details of the strands and beaches charm by their infinite grace and marvelous variety.

Elisee Reclus
The Ocean, Atmosphere, and Life (1873)

The coastal landscapes of Earth are made up of a great variety of forms: the barrier islands on the eastern shore of the United States contrast with a predominance of beaches backed by sea cliffs on the west coast; the rapidly evolving shorelines of the Mississippi and Nile River deltas; the mangroves of Florida and the north coast of Australia; the reef-fringed islands of the Caribbean and South Pacific; the glaciated coasts of western Canada and Norway. In this chapter we examine the range of erosional and accretional coastal forms and consider classification schemes that have been developed to organize coastal types into recognizable patterns and settings.

COASTAL MORPHOLOGY

Erosional Coastal Features

Sea cliffs are common along many of the world's coastlines, with the waves either breaking directly against the rocks or on a beach separating the cliff and surf (Fig. 2-1). The retreat of cliffs under the attack of waves may be rapid, with the changes being readily apparent within one's lifetime. Homes and other structures unwisely built too close to the cliff edge may be undermined and destroyed. Entire medieval villages in England have disappeared, their former locations now being well out to sea (Fig. 2-2).

Associated with the retreating rocky coast is a diversity of characteristic erosional features, some of which are diagrammed in Figure 2-3. The erosion of the cliff leaves a gently sloping rock platform called the *wave-cut platform*, where the rock layers have been sliced off by the surging motion of the waves. The waves may find local weaknesses in the cliff, which they excavate to form *sea caves*. Commonly observed are *sea stacks*, where more resistant portions of the rock remain within the surf, separated from the retreating cliff. More spectacular but rarer is the natural *sea arch*, where wave erosion has hollowed out a line of weakness in an otherwise resistant promontory creating an opening.

Figure 2-1   Cliff erosion along the Victoria coast of Australia, leaving a group of remnant sea stacks known as the Twelve Apostles.

Marine erosion produces a general retreat of the coast, but the recession is usually not uniform because of variations in rock resistance. The more resistant rocks retreat slowly and remain as headlands, stacks, and offshore islands, while the weaker formations are cut back to form embayments between the headlands. Solid and massive igneous rocks, most metamorphic rocks, and some limestones are particularly resistant to wave attack and hence form headlands. In contrast, poorly cemented sandstone, shale, and rocks that have numerous bedding planes or are weakened by closely spaced joints and faults are easily eroded by waves. The most rapid erosion occurs in unconsolidated sediments such as glacial deposits. One of the more famous examples of sea-cliff erosion by wave action is found on the coast of Holderness, England, where the glacial deposits are retreating at rates up to 1.75 m/year (Valentin, 1954).

The orientation of rock formations is an important factor in the resulting configuration of the eroding coast. Where the beds trend normal to the coast and stand nearly vertical (Fig. 2-4), the result is pronounced headlands formed by the resistant strata, with sheltered embayments where the weaker formations are located. Because of wave refraction, the bending of the wave crests, that changes the direction of wave movement (Chapter 5), the energy of the waves is concentrated on the resistant headlands and weakened within the bays. This

Figure 2-2   Dunwich, the site of a once important seaport in medieval England, containing eight churches, two hospitals, and a monastery, which have been destroyed by erosion. The photographs show the progressive loss of All Saints' Church, the last of the eight churches, whose axis was at right angles to the retreating cliff. Today the church is entirely gone. The grave stone of John Brinkley Easy remained in the disappearing churchyard (*lower photo*), until it too was lost during a storm in December 1990. [Old photographs by F. Jenkins, Southwold, England]

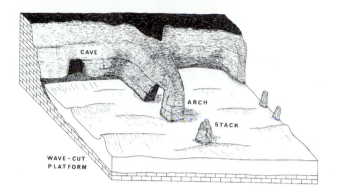

Figure 2-3 Erosional features associated with a retreating rocky coast.

pattern may prevent the continued higher rates of erosion within the bays, so there exists a maximum shore relief to which the coast is likely to erode.

When the layers of rock strata extend parallel to the coast, the erosion proceeds somewhat differently. The classic example of this situation is the Dorset coast of Purbeck in southern England (Fig. 2-5). Here several stages in the development of coastal erosion can be seen. Stair Hole demonstrates the earliest phase where the waves have cut out a series of caves and arches into the resistant Jurassic limestone, one cave having collapsed to produce a gap. The waves are beginning to scour out the weak Cretaceous sands and clays of the Wealdon Beds behind the resistant limestone. The next stage of advancing erosion is Lulworth Cove. At Lulworth Cove the gap in the Jurassic limestone is wider, and the weak Wealdon Beds have been carved into an almost circular bay. Erosion further landward is being hindered by the high ridge of resistant Cretaceous chalk backing Lulworth Cove. An even more advanced stage of erosion is demonstrated by Mupe-Worbarrow Bay farther to the east (Fig. 2-5). A broad embayment has developed, and the waves are now concentrating on the erosion of the high chalk ridge.

Where the rock strata trend nearly parallel to the coast and dip seaward (Fig. 2-6), landsliding and rock falls occur, since following wave undercutting, the cliff material readily slides down the bedding-plane surfaces into the sea. A well-known example is the Portuguese Bend landslide in Palos Verdes, California, near Los Angeles (Merriam, 1960), which covers some 4 million square meters and is moving seaward at rates of 1–3 cm/day, carrying homes and roads with it. The Jump-Off Joe landslide on the Oregon coast (Fig. 2-7) suddenly became active in 1942, destroying more than a dozen homes (Sayre and Komar, 1988). In spite of continued slumping over the years, in 1982 a condominium was built on the remaining bluff.

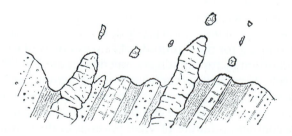

Figure 2-4 The configuration that results from erosion of rock strata trending normal to the coastline, where the more resistant layers form headlands and sea stacks while the weaker beds develop into embayments. [Reprinted from James Geikie, *Outlines of Geology* (London, Edward Stanford, 1896)]

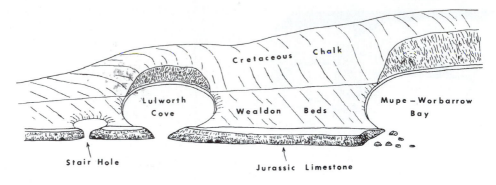

A

B

Figure 2-5   (a) The erosion of the Purbeck coast of Dorset, England. The bays are formed by waves eroding the weak Wealdon Beds (shales and sandstones) between the resistant Jurassic limestone and the ridge of Cretaceous chalk. (b) Stair Hole, Dorset, the initial stage of bay formation where the Jurassic limestone has been breached, and the weak Wealdon Beds are now being eroded. (c) Lulworth Cove, a more advanced stage of erosion where the entire width of the Wealdon Beds has been eroded away. [Photo courtesy of Aerofilms Ltd.]

Within three years, slope retreat had caused the foundation to fail and the structure had to be abandoned.

Considering the extent and importance of sea-cliff erosion and associated landsliding, it is surprising how few studies have focused on this problem, at least in comparison with beach-erosion problems and processes. Part of the reason for the limited study of sea-cliff erosion is the inherent difficulty in accounting for the multitude of factors involved (Trenhaile, 1987; Sunamura, 1992); the more important factors are diagrammed in Figure 2-8. One of the most difficult aspects is the cliff itself, its material composition and structure, the latter including bedding stratification (horizontal or dipping) and the presence of joints and faults. These factors are important in determining whether the cliff retreat takes the form of abrupt large-scale landsliding or the more continuous failure of small portions of the cliff face. The

C

Figure 2-5   (*Continued*)

Figure 2-6   When rock strata dip seaward, landslides and rock falls occur, leaving the bedding-plane surface as the coastal slope. [Reprinted from James Geikie, *Outlines of Geology* (London, Edward Stanford, 1896)]

processes of cliff attack are also complex. The retreat may be caused primarily by groundwater seepage and direct rain wash, with the ocean waves acting only to remove the accumulated talus at the base of the cliff. In other locations the waves play a more active role, directly attacking the cliff and cutting away its base.

An interesting series of laboratory experiments have been conducted by Sunamura (1976, 1982a, 1983) that simulate cliff erosion, experiments that illustrate the role of the fronting beach and its sediments. Artificial cliffs were constructed of loosely cemented sand. Initially there was no fronting beach, so the cliff was under the direct attack of waves generated by a paddle in the laboratory wave basin. The cliff retreated as the waves cut away its base (Fig. 2-9). The recession released a supply of sand that accumulated at the foot of the cliff and began to form a beach. This initially produced a higher rate of erosion of the cliff because the waves used this released sand as a "blasting" agent. However, at a later stage when still more sand had accumulated, the wider beach acted as a buffer that caused the waves to

Figure 2-7 The Jump-Off Joe landslide on the Oregon coast, which suddenly slumped during 1942, destroying several homes. The lower photograph shows the slide in the 1960s, when homes that had survived the initial movement were about to succumb to erosion. An attempt to develop the site in the 1980s led to further property losses. [From W. O. Sayre and P. D. Komar, The Jump-Off Joe Landslide at Newport, Oregon: History of Erosion, Development and Destruction, *Shore & Beach 56,* 1988.]

break offshore away from the cliff. At that stage the cliff erosion was reduced due to the presence of the expanding beach.

Similar factors and processes have been shown to be important to the erosion of cliffs backing ocean beaches. On the Yorkshire coast of northeast England, Robinson (1977) documented erosion at the base of a sea cliff by using a micro-erosion meter to measure the retreat of bare, solid rock surfaces. He found that cliff-base erosion rates were 15–20 times higher in places where there was a beach at the foot of the cliff, compared to where there was no beach. Processes of "sediment blasting" (corrasion) and wedging were important where a sand/gravel beach was present, whereas sporadic quarrying of blocks and more continuous micro-quarrying of shale fragments occurred where there was no beach. The zone of most intense erosion, centered at the junction of the beach and cliff, was limited to below a height of 10 cm up the face of the cliff. Maximum erosion occurred 14 cm below the beach face, due to mobilization of the sediment by wave action during storms and the resulting intense corrasion of the cliff face.

During the summer months when the beach is wide, it is clear that it can act as a buffer since the waves seldom if ever reach the cliff. However, in the winter beaches are cut back by storms, and the wave swash is then able to reach and attack the sea cliffs. The beach sand might then act as an agent to enhance cliff erosion. Important is the balance between the elevation of the beach versus the elevations of wave swash run-up superimposed on the mean

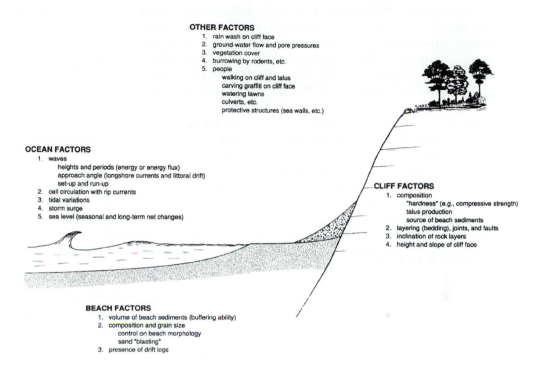

OTHER FACTORS
1. rain wash on cliff face
2. ground-water flow and pore pressures
3. vegetation cover
4. burrowing by rodents, etc.
5. people
   walking on cliff and talus
   carving graffiti on cliff face
   watering lawns
   culverts, etc.
   protective structures (sea walls, etc.)

OCEAN FACTORS
1. waves
   heights and periods (energy or energy flux)
   approach angle (longshore currents and littoral drift)
   set-up and run-up
2. cell circulation with rip currents
3. tidal variations
4. storm surge
5. sea level (seasonal and long-term net changes)

CLIFF FACTORS
1. composition
   "hardness" (e.g., compressive strength)
   talus production
   source of beach sediments
2. layering (bedding), joints, and faults
3. inclination of rock layers
4. height and slope of cliff face

BEACH FACTORS
1. volume of beach sediments (buffering ability)
2. composition and grain size
   control on beach morphology
   sand "blasting"
3. presence of drift logs

Figure 2-8   The many factors that are involved in sea-cliff erosion.

water level due to tides, etc. Shih et al. (1994) have developed a process-based model that accounts for the water-elevation factors of mean sea level, tidal stage, and wave run-up that are important to sea-cliff erosion. They applied their model to analyses of the frequency of wave attack and the resulting cliff retreat along the coast of Oregon.

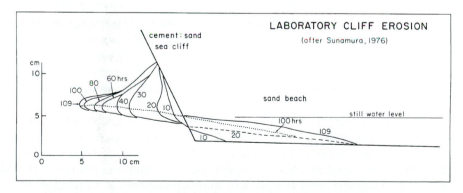

Figure 2-9   Erosion by waves of a cliff composed of cement plus sand in a laboratory experiment. The eroded notch extends obliquely upward, while the released sand has formed a beach at the front of the cliff. [Adapted with permission of The University of Chicago Press, from T. Sunamura, Feedback Relationship in Wave Erosion of Laboratory Rocky Coast, *Journal of Geology* 84, p. 429. Copyright © 1976 The University of Chicago Press.]

Sunamura (1977, 1982b, 1983) has analyzed cliff erosion in terms of the forces of wave attack versus the resistive strength of the cliff material. In a series of laboratory wave-channel experiments like those described above, cliffs having a range of compressive strengths were formed by varying the proportions of cement, fine sand, and water. Measurements also were obtained by Sunamura comparing erosion rates and the compressive strength of rocks composing cliffs on the east coast of Japan, where the wave forces are roughly constant along the cliff length. The laboratory and field studies support the existence of a critical wave height necessary to initiate erosion, the value of which depends on the compressive strength of the rock. Sunamura's field measurements established quantitatively that the larger but rarer waves are responsible for cliff erosion, as compared with smaller but more frequent waves. The result is that cliff recession is extremely episodic, and this can affect estimates based on comparisons of sequences of aerial photographs, maps, or direct surveys.

The episodic nature of cliff retreat is illustrated further by the field measurements of McGreal (1979) on the eastern coast of Northern Ireland. He found that, during 2 years of observations, erosion was associated with only 40 events, that is, with approximately 3 percent of all high tides. He showed that back-shore and cliff-base erosion requires the coincidence of well-defined tidal and meteorological conditions, the latter affecting the mean water level through storm surge as well as in the wave run-up.

The spatial and temporal variations in cliff erosion often go hand-in-hand and can be caused by a number of factors. On the East Anglia coast of England, Robinson (1980) demonstrated the significance of an offshore shoal (Sizewell Bank) in the North Sea to the long-term erosion of the adjacent coast. The slow northward migration of the shoal has caused periods of erosion of beaches and cliffs to alternate with periods of beach accretion; there is a temporal variation at a specific location, involving decades to centuries, as well as spatially along the length of coast at any given time. This is attributed mainly to the effects of the shoal on wave refraction, which focuses the wave energy (Chapter 5), rather than to a sediment exchange between the beach and offshore. The ancient town of Dunwich within the study area has undergone massive erosion over the centuries (Fig. 2-2), but cliff retreat during the past 50 years has been much smaller. Robinson showed that this change coincided with a reduction in wave energy on the Dunwich shore as Sizewell Bank grew northward and altered the pattern of wave refraction.

Farther to the north on the Holderness coast, Pringle (1985) has shown that a localized reduction in the level of the beach at the base of the cliff is important to along-coast variations in cliff-erosion rates. The beach lowering is part of what is known locally as an "ord," a shore-oblique trough that cuts down through the beach sediment to the shore platform. The longshore migration of an ord, averaging 500 m/year, produces a slow shift in the zone of maximum cliff erosion.

There is considerable spatial and temporal variability in cliff retreat along the Oregon coast that results from its tectonic setting as well as from local beach processes (Komar and Shih, 1993). Every few hundred years an earthquake associated with plate subduction causes major parts of the coast to abruptly subside by 1–2 m and initiates coast-wide cliff erosion. However, in the intervening centuries of subduction-stress accumulation, the uplift of the coast reduces the erosional impact. In areas rising faster than the global rise of sea level (Chapter 4), sea-cliff erosion has effectively ceased, while in other areas the rate of sea-level rise exceeds the tectonic uplift and cliff erosion continues. Superimposed on this coast-wide tectonic control of sea-cliff retreat are more local factors, including the buffering capacity of the beach to protect the cliff (Shih and Komar, 1994). This buffer of protection is determined by the volume of sand on the beach and the coarseness of the beach sediment that controls the annual cycle in the beach morphology (including the elevation of the profile; Chapter 7) and the development of erosive rip-current embayments (Chapters 8 and 11). The rip embayments in large part determine where the wave swash can reach the cliff, resulting in localized erosion. The longshore positions of rip embayments can be particularly important in controlling the spatial variability in cliff recession during any single storm, while the temporal variability of episodic erosion at a specific cliff site may be associated with shifting positions of rip currents as they move about from year to year.

Kuhn and Shepard (1983) documented that cycles of weather conditions, particularly the intensity of storms, have resulted in long-term changes in sea-cliff erosion along the southern California coast. Records for the past 2 centuries show that during the 3 or more recent decades, both rainfall and high storm winds have been less severe than in preceding decades. Features of coastal erosion that appear to show some relation to this cycle are the arches and stacks that were well developed along the coast in the past century but now have largely disappeared. The formation of arches and stacks appears to correspond with past periods of more intense sea-cliff erosion, while their destruction has occurred during the present period of more gradual cliff retreat.

Kuhn and Shepard (1983) also concentrated on documenting the human-induced effects on sea-cliff erosion in southern California. These effects include increased occurrences of landslides due to enhanced groundwater resulting from irrigation and septic tanks, the development of gullies and canyons where culverts have been constructed under highways and railways, increased slope failures due to the grading of bluff tops, and erosion induced by the construction of sea walls and revetments (Chapter 12).

Although generally not as significant as wave action, the activities of organisms that burrow into the rock or scrape vegetation from its surface may in some instances be important in eroding sea cliffs. Burrowing clams such as *Pholadidae* (Fig. 2-10) may weaken an otherwise resistant sea cliff, so that it succumbs more quickly to wave attack. Jehu (1918) reported localities on the coast of England where pholads have lowered the chalk at rates of 1.8–3.3 cm/year. Vita-Finzi and Cornelius (1973) found cliff-erosion rates of 0.25 cm/year in Oman, caused by the burrowing date-stone mollusk *Lithophaga*. By studying the ingestion rate and stomach contents of periwinkles, a small intertidal snail, North (1954) estimated that, with their scraping radulae, 100 snails could remove 86 cm^3 of rock material yearly, or almost 1 liter in a decade. In addition to mollusks, other organisms may aid in rock erosion—marine worms, barnacles that bore into the rock (Ahr and Stanton, 1973), and sea urchins. As well as physically boring into the rock or scraping its surface, these organisms can enhance chemical solution. Emery (1946) has demonstrated that at night organisms locally increase the acidity of the water, and this may result in dissolution of the rock, producing high-tide rock basins. The small pock marks observed near the water's edge that commonly house periwinkles, chitons, and limpets may be due to biochemical leaching. Finally, humans can also be an important agent in the direct erosion of sea cliffs, particularly children who carve graffiti onto the cliff or excavate caves.

Accretional Coastal Features

Sand, gravel, and cobbles eroded from sea cliffs or delivered to the coast from rivers are worked into a variety of accretional shoreline forms (Fig. 2-11). A common feature is the *spit* (Fig. 2-12): a beach and associated backshore and dunes that are tied to the coast at one end,

Figure 2-10 The rock burrowing clam *Pholadidae*. "Their ceaseless, and as yet inexplicable toil enables them to pierce deeply into the most compact stones. We are astonished, in splitting marble, to find living shells in the midst of blocks which the chisel of the sculptor cuts with difficulty." [Etching and quotation from Pouchet (1884)]

while free at the other. Most spits grow in the direction of predominant longshore sediment transport (Fig. 2-13) and are often a continuation of the beach that is adjacent to the coast. Other spits depart from the trend of the coast and align themselves nearly at right angles to the prevailing wave direction. The free end of a spit may terminate in a hook or recurve as seen on Hurst Castle Spit (Fig. 2-14), formed either by wave refraction around the end (Evans, 1942) or by the interplay of wave trains arriving from different directions (King and McCullagh, 1971). Older hooks can also be seen on Hurst Castle Spit, trailing landward with almost the same configuration as the active hook. Spits are most common on irregular coasts where they grow across bay mouths, confining the bay except for the tidal inlet at the end of the spit. The growth of spits in this way is an important factor in smoothing an initially irregular coastline (Johnson, 1919).

A growing spit can deflect the mouth of a river or entrance to a bay, prolonging it in the direction of longshore sediment transport. The mouth of the River Alde, Suffolk, England (Fig. 2-15), has been deflected for about 18 km by a gravel spit (Steers, 1962). Records indi-

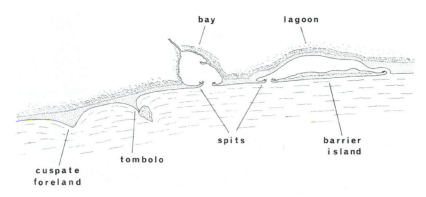

bay   lagoon

spits   barrier
                  island

tombolo

cuspate
foreland

Figure 2-11   Features associated with accreting coasts.

cate that at the time Orford Castle was built in 1165, it stood at the mouth of the river. Maps dating back to the sixteenth century show the spit still ending opposite Orford Castle, so most of the spit growth has occurred during the last 400 years. Old maps and written records are very useful in this way to examine long-term changes. Comparisons of dated aerial

Figure 2-12   Netarts Spit on the Oregon Coast. The inlet leading to Netarts Bay is in the foreground. [Courtesy of the Oregon Department of Transportation]

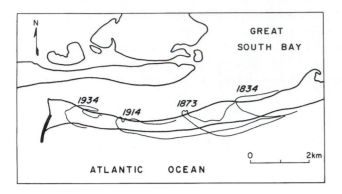

Figure 2-13   The long-term extension of Fire Island, a sand spit on the south shore of Long Island, New York. [Adapted from Sand Transfer, Beach Control, and Inlet Improvements, Fire Island Inlet to Jones Beach, T. Saville, *Proceedings of the 7th Conference on Coastal Engineering,* 1960. Reproduced with permission from the American Society of Civil Engineers.]

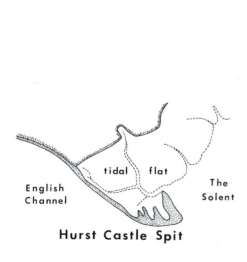

Hurst Castle Spit

Figure 2-14   Hurst Castle Spit on the south coast of England, composed of gravel (shingle). Recurves have developed on the end of the spit where it enters deep water, formed by the interplay of waves having different directions of approach.

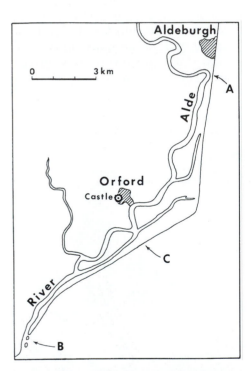

Figure 2-15   The River Alde, Suffolk, England, whose mouth has been deflected for about 18 km by the growth of the gravel (shingle) spit of Orford Ness. The river comes within 45 m of the sea at point A. The present river mouth is at B, while its approximate position in 1165 at the time of the building of Orford Castle was at C. [Adapted with permission of J. A. Steers from *The Sea Coast, 3rd Ed.* Copyright © 1962.]

photographs can be used to trace the shorter-term evolution of spits and other coastal features [see Shepard and Wanless (1971) for numerous examples].

When a spit links the mainland to an offshore island, it becomes a *tombolo* (Fig. 2-16). A tombolo can also develop from a cusp-shaped body growing in the lee and eventually connecting with an island. The term comes from Italy, where these features are particularly well developed.

A variety of cuspate shoreline forms may be observed in the coastal zone. The largest, the *cuspate forelands,* are sometimes associated with offshore islands or shoals, the shape of the cusp being governed by wave refraction around the island (Russell, 1958). Other cuspate forelands clearly are independent of such a mode of formation. Dungeness on the south coast of England (Fig. 2-17) is thought to have been built out at a point of convergence of the longshore transport, the sediment having accumulated from two directions. Because of the position of France across the English Channel, waves arriving diagonally at that portion of the coast are stronger and more frequent than those coming directly from offshore, accounting for the growth of Dungeness outward into the channel. The development of cuspate forelands such as Cape Hatteras and Cape Lookout (Fig. 2-18) on the southeastern coast of the United States has received considerable attention. Suggestions of origin have included eddies associated with the Gulf Stream, structural uplift, wave refraction by offshore shoals, and the reworking of river delta sediments (Hoyt and Henry, 1971). There is also a variety of cuspate shoreline features on a smaller scale than cuspate forelands—sand waves, giant cusps, and beach cusps. The origins of these features are considered in Chapter 11.

A major depositional landform having particular significance to coastal development and associated management problems is the *barrier island.* A barrier island can be defined as

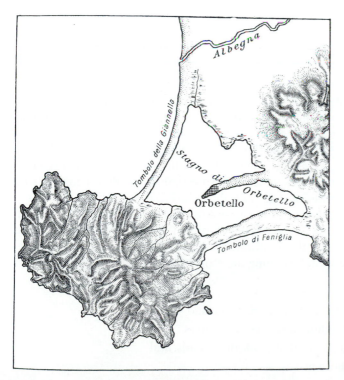

Figure 2-16 A double tombolo on the west coast of Italy, where two spits have become connected with an offshore island. [Reprinted with permission of John Wiley & Sons Inc. from D. W. Johnson, *Shore Processes and Shoreline Development.* Copyright © 1919 John Wiley & Sons Inc., New York.]

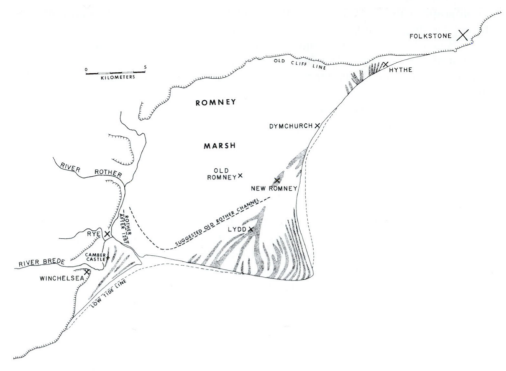

Figure 2-17   Dungeness, a type of cuspate foreland on the coast of England. The shaded areas represent old beach ridges composed of gravel (shingle), revealing how the foreland has built out with time and also has migrated toward the northeast. The port of Romney had to be moved as the coast accreted, but the Rother River changed course during the thirteenth century so that Rye became the principal port.

an unconsolidated elongate body of sand or gravel lying above the high-tide level and separated from the mainland by a lagoon or marsh. This feature is illustrated by the Outer Banks of North Carolina (Fig. 2-18). Barrier islands dominate most of the east and Gulf coasts of the United States, are present along the Arctic coast of Alaska and within the Gulf of Alaska, and are locally found in eastern Canada. As will be discussed later, this distribution in large part is determined by patterns of global tectonics, which in turn establish whether a coast is subsiding or rising, and determine quantities of sediments delivered to the coast from rivers.

   An excellent review of barrier islands has been prepared by Nummedal (1983), who examined the physical processes that have led to the development of our present barriers and the natural forces of sediment dispersal that continue to modify them. The thesis of his review is that the most important factors determining the morphology and evolution of barrier islands are

- the history of sea-level rise during the last few thousand years,
- the longshore distribution of sediment sources and sinks,
- water exchange through tidal passes that flank the islands, and
- the energy level of the nearshore waves.

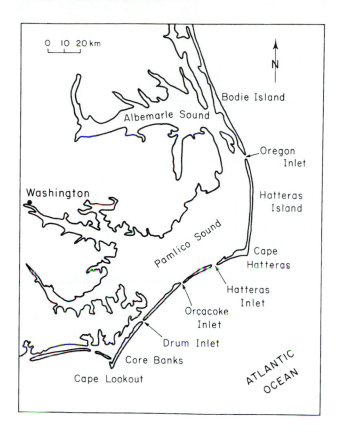

Figure 2-18 The Outer Banks of North Carolina, which consist of a series of barrier islands separated by inlets.

Most commonly recognized is the importance of the Holocene era rise in sea level, a rise that continues today (Chapter 4). As a result, many barrier islands have migrated landward across the coastal plain, a movement that also raises their elevations so they do not become submerged. The migration is episodic, primarily taking place during major storms and hurricanes when a storm surge (Chapter 4) elevates the water level to the point where much of the island becomes submerged. Water currents associated with the storm surge combine with hurricane-generated waves to erode sediment from the fronting beach and transport it as overwash across the island. This sediment is typically deposited atop the barrier island, raising its elevation, and is also carried across the island to form an overwash fan along the landward edge of the island. The combined effect is to move the island landward and upward. The result is well illustrated by the layering found by Godfrey and Godfrey (1973) within the Core Banks portion of the Outer Banks of North Carolina (Fig. 2-19), documented by a series of cores obtained at intervals across the width of the island. The layering of the sand together with carbon-14 dating of shell and marsh materials allowed them to decipher the history of island migration. Also shown in Figure 2-19 is the profile change produced by Hurricane Ginger, which occurred in 1971 during the study of Godfrey and Godfrey, further demonstrating how sand is eroded from the ocean beach and is deposited atop the island by overwash processes.

The landward migration of a barrier island is primarily a response to a rising sea level—not just a global (eustatic) rise but also that contributed locally by land subsidence. The other

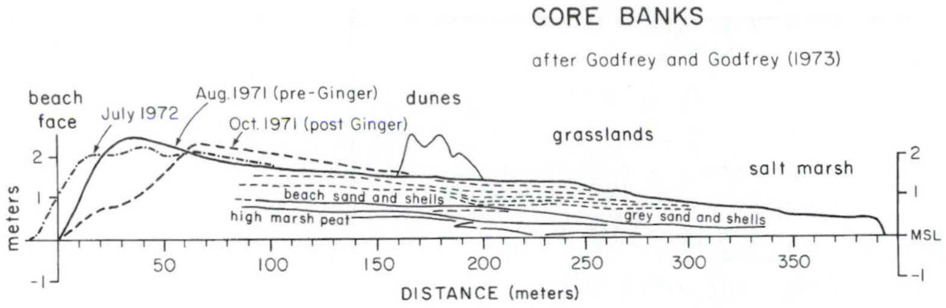

CORE BANKS

after Godfrey and Godfrey (1973)

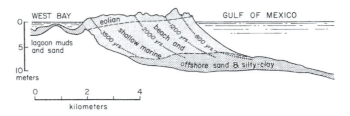

Figure 2-19 A cross section of Core Banks, a barrier island on the Outer Banks of North Carolina. The island consists of layers of sand deposited when the sea washed over the island during extreme storms. Also shown are profile changes that occurred during Hurricane Ginger in 1971, including the overwash deposit laid down by that storm. [Adapted with permission of State University of New York, from P. J. Godfrey and M. M. Godfrey, Comparison of Ecological and Geomorphic Interactions Between Altered and Unaltered Barrier Island Systems in North Carolina, *Coastal Geomorphology*, p. 246. Copyright © 1973 State University of New York.]

important factor is the sediment supply to the island. If the sediment supply is sufficient, then the island can build upward and even seaward, maintaining itself above the rising sea without having to migrate landward. This situation was found to be the case for Galveston Island, Texas, studied in detail by Bernard, LeBlanc, and Major (1962). A diagrammatic cross section of the island is shown in Figure 2-20, demonstrating that from about 3,500 years ago to the present, the barrier island has prograded in the seaward direction and become considerably wider. The barrier island itself is flanked both seaward and landward by silts and clays, respectively, representing offshore and lagoonal deposits. The vertical sequence of sediments through the barrier is the same as the offshore sequence across the barrier surface and out onto the shelf. The barrier is capped by aeolian dune sand, below which is beach sand, while at still greater depths within the barrier island is sand that accumulated in progressively deeper water, finally giving way to silt and clay that had been deposited on what was formerly the outer continental shelf. This is a *regressive sequence*, one where landward sediments are deposited on top of more seaward ones and reflect a seaward shift of the shoreline. In the case of Galveston Island, the supply of sand has been sufficient to overcome the rise in sea level, so the island has shown a net accretion with the shoreline migrating seaward.

GALVESTON ISLAND, TEXAS

Figure 2-20 Diagrammatic cross section of Galveston Island, Texas, showing the principal environments of sediment deposition and the vertical sequence of the deposits, which indicate that the island has grown upward and seaward through time. [Adapted with permission of Geological Society of America, from H. A. Bernard, R. J. LeBlanc and C. F. Major, Recent and Pleistocene Geology of Southwest Texas, Field Excursion No. 3, *Geology of the Gulf Coast and Central Texas.* Copyright © 1962 Geological Society of America.]

A *transgressive sequence* is one where the shoreline has moved landward, such that the sediments deposited in seaward environments stratigraphically end up on top of sediments that originated in more landward environments. This is the case for barrier islands that have migrated landward in response to the rise in sea level, the sand of the barrier having been deposited over the top of lagoonal mud. This can be seen in the layering within Core Banks (Fig. 2-19), where Godfrey and Godfrey (1973) found lagoonal peat and mud below the sand of the island. In some instances, the lagoonal mud, marsh deposits, and even trees become exposed on the ocean side of the barrier as the beach face retreated (Fig. 2-21). The stratigraphy of a transgressive barrier-island system on the coast of Delaware (Fig. 2-22) has been

Figure 2-21 Lagoonal marsh sediments and peat that have become exposed on the beach-face along Cape Hatteras, North Carolina, providing evidence for the landward migration of the barrier island. [From R. Dolan and H. Lins, *The Outer Banks of North Carolina*, Geological Survey Prof. Paper #1177-B.]

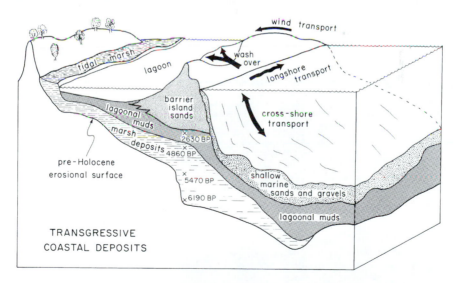

Figure 2-22 Representation of the sedimentary units in the barrier islands of Delaware on the U.S. east coast. The sequence and layering indicates that the island has migrated landward as sea level rose during the past several thousand years. [Used with permission of the State University of New York, from J. C. Kraft, R. B. Biggs and S. D. Halsey, Morphology and Vertical Sedimentary Sequence Models in Holocene Transgressive Barrier Systems, *Coastal Geomorphology*, p. 329. Copyright © 1973 State University of New York.]

studied in detail by Kraft, Biggs, and Halsey (1973). Long cores through the barrier island first passed through the sand of the island, then lagoonal mud (tidal marsh deposits that had originated on the landward side of the lagoon), and finally into the pre-Holocene "bedrock" of the landmass. Carbon-14 dates further establish the ages of the respective sediments, reaching about 6,000 years BP (before the present) at the base of the deposits representing the Holocene transgression. Also diagrammed in Figure 2-22 is the movement of sediment on the barrier island and in the shallow offshore, including overwashes that produce the landward migration of the island, cross-shore sediment transport associated with storms, and a longshore transport to the north. The transgressive stratigraphy of Figure 2-22 shows that Delaware's barrier islands have migrated landward from a more seaward position; carbon-14 dating of the sediments indicate that this movement has been rapid.

A clear indication of the landward migration of barrier islands is provided by the "permanent" structures of humankind, in particular, lighthouses built along the coasts. A good example is the Morris Island lighthouse (Fig. 2-23) on the coast of South Carolina. According to Leatherman and Møller (1991), this lighthouse was constructed in 1876, at which time it was firmly on land. But shoreline erosion along Morris Island has averaged nearly 8 m/year as the entire island migrated landward. The island eventually moved beyond the base of the lighthouse, so for the last 50 years, this beacon has stood in progressively deeper water of the Atlantic Ocean.

Figure 2-23 The Morris Island lighthouse, South Carolina, that a century ago was constructed on the barrier island but is now well offshore due to the landward migration of the island. [Courtesy of S. Leatherman]

The rise in sea level was very rapid throughout most of the Holocene era, starting about 20,000 years ago, but the rate has slowed considerably during the last 5,000 years (Chapter 4). This reduction in the rate of sea-level rise has allowed some coasts to become accretional (regressive), while having been erosional (transgressive) throughout the early Holocene era. This is shown by the stratigraphy of the coast of Nayarit, Mexico (Fig. 2-24), investigated by Curray, Emmel, and Crampton (1969). The old alluvium deposited by a river is overlain by a sheet-sand layer several meters thick, deposited by the last transgression from the rising sea. Above that layer is a regressive sand deposit, formed when the coast began to prograde about 4,500 years ago. The prograding sand is extending out over the shelf mud, and on the landward side, the coastal sand is in turn being covered by lagoonal and marsh deposits.

Transgressive and regressive barrier islands can be found in close proximity, at times even in separate parts of the same barrier island, depending largely on the relative sediment supplies. The contrast is also reflected in the morphologies of the islands and has been used to define "high-profile" and "low-profile" barrier islands, based on observations along the Gulf Coast of the United States (Nummedal, 1983). The low-profile type occurs where sediment supplies are small, so the island is low in relief. Hurricanes cause massive overwash and the landward migration of low-profile islands. The high-profile islands are closer to sediment sources, with the abundance of sand allowing the development of high dune ridges within the islands and vegetated foredunes immediately behind the beach. The relief of the island minimizes overwash during hurricanes. The erosion instead involves the waves and storm surge cutting a scarp into the dunes, which reform in the months and years following the storm.

A similar contrast in barrier-island morphologies is found on the Outer Banks of North Carolina but where the high-profile type is the result of decades of dune stabilization accompanying extensive public and private development (Godfrey and Godfrey, 1973; Dolan, 1973). The difference between natural and stabilized barrier islands is illustrated in Figure 2-25. The natural island is of the low-profile type and is characterized by overwash processes and landward migration, seen earlier for Core Banks (Fig. 2-19), an undeveloped portion of the Outer Banks. In contrast, the stabilized barrier island is characterized by a pronounced dune immediately backing the beach. The formation of this continuous dune ridge

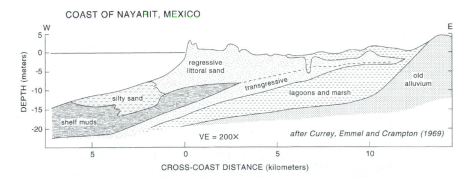

Figure 2-24   Sediments on the coast of Nayarit, Mexico, formed first by the transgression of the sea over the land when the Holocene rate of sea-level rise was rapid, followed by a regression when the rate of rise was slower so that the supply of sand from nearby rivers could build out the shore. [Adapted from J. R. Curray, F. J. Emmel and P. J. S. Crampton, Holocene History of a Strand Plain, Lagoonal Coast, Nayarit, Mexico, *Lagunas Conseras un Simposio,* 1969, UNESCO.]

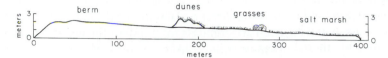

A   Natual barrier island profile

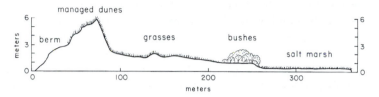

B   Modified profile of developed island

Figure 2-25 (a) Profile across a natural barrier island of the Outer Banks, North Carolina, compared with the (b) profile of an island modified by the artificial formation of a large foredune. [Used with permission of the State University of New York, from R. Dolan, Barrier Islands: Natural and Controlled, *Coastal Geomorphology*, p. 269. Copyright © 1973 State University of New York.]

resulted from the erection of dune fences and the planting of dune grasses between 1936 and 1940. This in effect has created a high-profile island, but one that is artificial rather than natural as found on the Gulf Coast. Its origin is not due to a greater abundance of sand but instead represents a redistribution of a fixed quantity of sand on the island—the natural profile of the island has been reshaped into one having a large foredune. This has reduced overwash and in the short term has helped to shelter developments on these altered islands. However, the beaches on the stabilized islands are significantly narrower than those on the natural islands (Fig. 2-25). Rather than permitting overwash, storm erosion results in the protective dune being cut back with the sand transported offshore. This is illustrated by the contrasting profiles of Figure 2-26, showing the natural and stabilized barrier islands' response

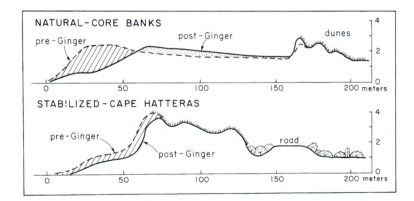

Figure 2-26 The response to Hurricane Ginger in 1971 by the natural and stabilized barrier islands of the Outer Banks, North Carolina. [Used with permission of the State University of New York, from R. Dolan, Barrier Islands: Natural and Controlled, *Coastal Geomorphology*, p. 269. Copyright © 1973 State University of New York.]

to Hurricane Ginger in 1971 (Dolan, 1973). On the natural barrier island, sand eroded from the beach is mainly transported over the island by overwash, so the response is one that maintains the island in the face of a rising sea. In contrast, on the stabilized island the eroded sand moves offshore, and it is uncertain whether all of that sand returns to the beach following the storm or whether there has been a net loss. In any case, the island is unable to modify its shape and position in response to the rising sea, and in the long term, this may result in the destruction of the island as well as the homes and other structures it contains.

A storm can pile up gravel and shells into a *beach ridge* well above the normal high-tide level at the back of the beach, while at the same time transport the finer sand to offshore bars (Chapter 7). On a prograding beach, a series of beach ridges may develop from successive storms, each of which forms a ridge parallel to the shoreline (Lewis and Balchin, 1940). The relief of each ridge is small, so they are best viewed from the air. This is seen in Figure 2-27, a series of some 250 beach ridges on the coast of Nayarit, Mexico, which have been studied by Curray and Moore (1964). They calculated that 12–20 years were required for the formation of individual ridges. Most of the observed series of beach ridges were built since sea level reached its present position some 5,000 years ago. Each ridge was formed individually as a shore deposit; the present beach generally representing the most recent ridge of the series. As such, they constitute lines of growth of the shoreline, and when preserved, enable one to decipher the history of shoreline evolution. An example is the sequence of beach ridges that are found on the Gulf coast of Texas and Louisiana (Fig. 2-28), locally termed "cheniers"

Figure 2-27 Beach ridges and the modern shoreline of the coast of Nayarit, Mexico. [Courtesy of J. Curray]

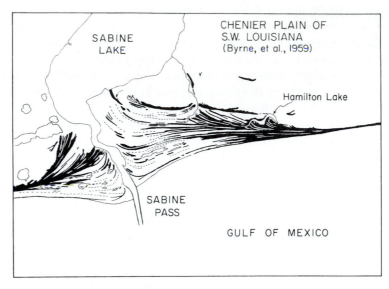

Figure 2-28 Cheniers or beach ridges on the coast of Louisiana, providing evidence of how the shoreline built out with time. [used with permission of Gulf Coast Association of the Geologic Society, from J.V. Byrne, D.O. Le Roy and C.M. Ritey, The Chenier Plain and Its Stratigraphy, Southwestern Louisiana, *Proceedings of the 9th Annual Meeting Gulf Coast Association of the Geologic Society 9.* Copyright © 1959 Gulf Coast Association of the Geologic Society.]

(Russell and Howe, 1935). These beach ridges are up to 4.5 m thick, 200 m in width, and up to 50 km long. From analyses of cores, Byrne, LeRoy, and Riley (1959) were able to identify marsh, bay, mud flat, open Gulf, and beach sediments within the chenier plains. The ages of these cheniers range from 300 to 2,800 years (Gould and McFarlan, 1959). Another example of the use of beach ridges to decipher the changing configuration of the coast is illustrated by Cape Canaveral (Fig. 2-29) on the Atlantic coast of Florida. The cape has slowly migrated southward, adding new ridges to its south side while erosion along the north margin has truncated older beach ridges. The pattern of beach ridges on Dungeness, the cuspate foreland in southern England (Fig. 2-17) reveals a similar alongshore migration that has been analyzed by Lewis (1932) and Lewis and Balchin (1940).

COASTAL CLASSIFICATIONS

A variety of coastal features have been considered and generally referred to as erosional or accretional. It is worthwhile to further categorize coastal morphology in order to bring about a better understanding of the factors that are significant in controlling the morphology. This has led to the development of coastal-classification schemes. Coastal morphology reflects the complex imprint of the tectonic setting, modified by the combined actions of more local agents and processes. The resulting classifications have included purely descriptive schemes, as well as systems that relate to physical processes important to the morphology.

The classification scheme developed by Inman and Nordstrom (1971) is modern in its outlook in that it is based on the concepts of global plate tectonics discovered in the 1960s. Inman and Nordstrom recognized that the gross aspects of the topography of a coast are related to its position on the moving tectonic plates. For example, this can explain the observed contrast in the west versus the east coasts of the United States, shorelines dominated, re-

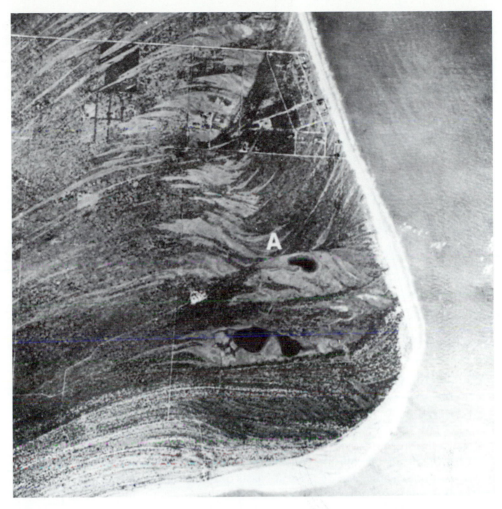

A

Figure 2-29   (a) Aerial photograph of Cape Canaveral, Florida, showing a complex pattern of beach ridges formed during its growth, including the former location of the cape's end at A. (b) An interpretation derived from the beach ridge patterns, indicating that the Cape has migrated southward as it grew seaward. [Adapted with permission of McGraw-Hill Inc., from F. P. Shepard and H. R. Wanless, *Our Changing Coastlines*, p. 149. Copyright © 1971 McGraw-Hill Inc.]

spectively, by erosion (i.e., sea cliffs, rocky headlands, etc.) and deposition (barrier islands). Along the northwest coast of the United States, portions of the tectonic plate in the ocean basin are moving eastward and colliding with the continental plate of North America. This collision has resulted in earthquakes, coastal uplift, and the formation of mountains immediately inland from the coast. In contrast, the east and Gulf coasts of the United States lie within the interior of the North American plate, and therefore experience little tectonic activity and are either stable or subside rather than being uplifted. Furthermore, larger quantities of sediments are delivered by rivers to the east and Gulf coasts due to the overall asymmetry of North America, with the largest mountains and divides located closer to the west coast.

Inman and Nordstrom (1971) have classified coasts into three categories according to their positions within the moving tectonic plates:

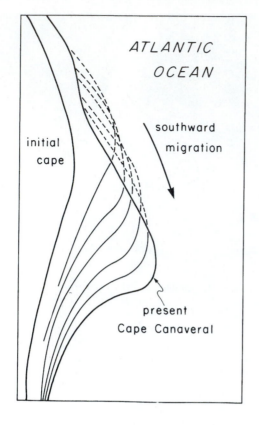

B

Figure 2-29   *(Continued)*

1. Collision coasts (convergent margins)
    a. Continental collision coast: the margin of a thick continental plate colliding with a thin oceanic plate (e.g., west coasts of North and South America)
    b. Island arc collision coasts: along island arcs where thin oceanic plates collide (e.g., the Aleutian island arc)
2. Trailing-edge coasts (divergent margins)
    a. Neo-trailing-edge coasts: new trailing-edge coasts formed near beginning spreading centers and rifts (e.g., the Red Sea and Gulf of California)
    b. Amero-trailing-edge coasts: the trailing edge of a continent having a collision coast on its opposite side (e.g., east coasts of the Americas)
    c. Afro-trailing-edge coasts: the coast on the opposite side of the continent is also trailing (e.g., the east and west coasts of Africa)
3. Marginal sea coasts: coasts fronting on marginal seas and protected from the open ocean by island arcs (e.g., Korea)

Figure 2-30 shows the worldwide distribution of coastal and shelf types as determined by Inman and Nordstrom. They distinguish between Afro-trailing-edge coasts and Amero-trailing-edge coasts due to differences in the resulting sediment supplies. As was pointed out previously, the east and Gulf coasts of the United States receive large quantities of sediments since many of the rivers originate in mountainous regions and have very large drainage basins.

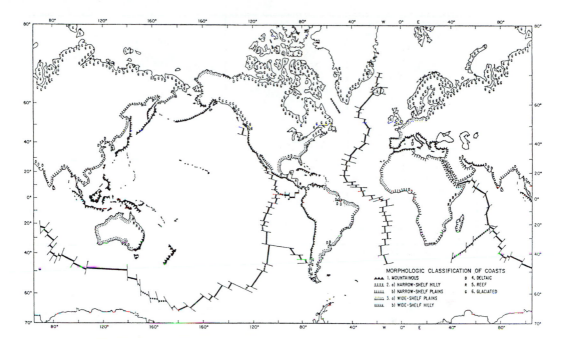

Figure 2-30 A tectonic-based classification of coasts. [Used with permission of the University of Chicago Press, from D. L. Inman and C. E. Nordstrom, On the Tectonic and Morphological Classification of Coasts, *Journal of Geology 79,* p. 10. Copyright © 1971 The University of Chicago Press.]

This contrasts with Africa, which lacks significant mountains due to the absence of plate collisions, and accordingly smaller quantities of sediment are delivered to the coastal zone.

As noted by Inman and Nordstrom, the categories within their classification correspond at least grossly to the physical nature of the coast. The collision coasts are all relatively straight and mountainous and generally are characterized by sea cliffs, raised terraces, and narrow continental shelves. The trailing-edge types of coasts are more variable. The "Amero" types have low-lying depositional coastal forms such as barrier islands and the widest continental shelves. The neo-trailing-edge coasts only recently have come into being, as spreading between two newly formed plates has split the continental crust apart. The coast is typically steep with beaches backed by sea cliffs, so in many respects these new-trailing-edge coasts are similar to collision coasts. The marginal-sea coasts have the greatest diversity of form. The land may be low lying or hilly, and the form of the coast can be dominated by local processes such as the formation of river deltas.

Inman and Nordstrom refer to the above-mentioned tectonic classification as the control on first-order features, the gross morphology of the coast. They further recognize the importance of second-order features superimposed on the first-order tectonic control. They have in mind the formation of river deltas, the development of barrier islands, the erosion of sea cliffs, special processes such as glaciation that produces fjords, and organic action that forms coral reefs and mangrove coastlines. This level of smaller-scale coastal features and processes has been the primary focus of most classifications. For example, Figure 2-31 is derived from the classification of Shepard (1976), which relates the groupings of such features to the level of coastal stability.

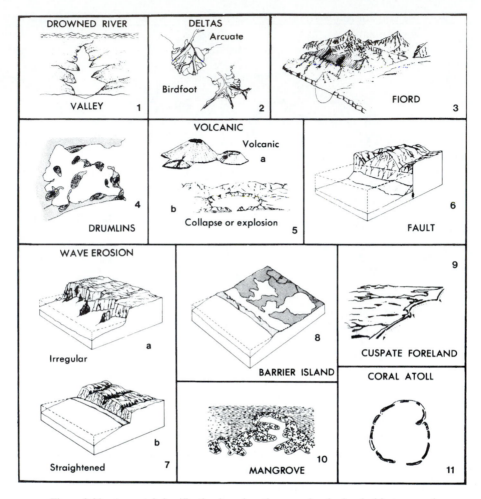

Figure 2-31   A coastal classification based on the second-order level of features and processes, the tectonic settling being the first-order control. [Used with permission of Geoscience Publications, from F. P. Shepard, Coastal Classification and Changing Coastlines, *Geoscience and Man 14*, p. 55. Copyright © 1976 Geoscience Publications.]

     Such classifications can be useful in helping one to focus on the tectonic controls, geologic history, and the many processes that have shaped a specific area of shore or have given rise to a certain coastal landform. For example, Glaeser (1978) has examined the global distribution of barrier islands in terms of the classification of Inman and Nordstrom (1971), leading to a better understanding of the physical setting required for the formation of that coastal landform.

SUMMARY

     In this chapter we have examined the variety of erosional and depositional coastal forms. An extensive nomenclature has been introduced to describe the features of coasts. Terms have been introduced (denoted by italics) that will be needed in subsequent chapters.

We also have considered various classification schemes that are useful in organizing coastal types into recognizable patterns. In the next chapter we will consider the details of beaches found within coasts of different types, including the compositions of their sediments and variety of morphologies.

REFERENCES

AHR, W. M. and R. J. STANTON (1973). The Sedimentologic and Paleoecologic Significance of Lithotyra, a Rock-Boring Barnacle. *Journal of Sedimentary Petrology* 43: 20–23.

BERNARD, H. A., R. J. LeBLANC, and C. F. MAJOR (1962). Recent and Pleistocene Geology of Southwest Texas, Field Excurison No. 3. In *Geology of the Gulf Coast and Central Texas,* Geological Society Annual Meeting Guidebook, pp. 175–224.

BYRNE, J. V., D. O. LeRoy, and C. M. RILEY (1959). The Chenier Plain and Its Stratigraphy, Southwestern Louisiana. *Proceedings of the 9th Annual Meeting Gulf Coast Association of the Geologic Society* 9: 237–260.

CURRAY, J. R. and D. G. MOORE (1964), Holocene Regressive Littoral Sand, Costa de Nayarit, Mexico. In *Deltaic and Shallow Marine Deposits,* L.M.J.U. van Straaten (editor), pp. 76–82. Amsterdam: Elsevier.

CURRAY, J. R., F. J. EMMEL, and P. J. S. CRAMPTON (1969). Holocene History of a Strand Plain, Lagoonal Coast, Nayarit, Mexico. In *Lagunas Conseras, un Simposio,* A. A. Castanares and F. B. Phleger (editors), pp. 63–100, UNESCO.

DOLAN, R. (1973). Barrier Islands: Natural and Controlled. In *Coastal Geomorphology,* D. R. Coates (editor), pp. 263–278. Binghampton, NY: State University of New York.

DOLAN, R. and H. LINS (1986). *The Outer Banks of North Carolina.* Geological Survey Prof. Paper # 1177-B.

EMERY, K. O. (1946). Marine Solution Basins. *Journal of Geology* 54: 209–228.

EVANS, O. F. (1942). The Origin of Spits, Bars, and Related Structures. *Journal of Geology* 50: 846–863.

GEIKIE, J. (1896). *Outlines of Geology.* London, England: Edward Stanford.

GLAESER, J. D. (1978). Global Distribution of Barrier Islands in Terms of Tectonic Setting. *Journal of Geology* 86: 283–297.

GODREY, P. J. and M. M. GODFREY (1973). Comparison of Ecological and Geomorphic Interactions Between Altered and Unaltered Barrier Island Systems in North Carolina. In *Coastal Geomorphology,* D. R. Coates (editor), pp. 239–258. Binghampton, NY: State University of New York.

GOULD, H. R. and E. McFARLAN (1959). Geologic History of the Chenier Plain, Southwestern Louisiana. *Trans. Gulf Coast Assoc. Geol. Soc. 9th Annual Meeting* 9: 261–269.

HOYT, J. H. and V. J. HENRY (1971). Origin of Capes and Shoals Along the Southeastern Coast of the United States. *Geological Society of America Bulletin* 82: 59–66.

INMAN, D. L. and C. E. NORDSTROM (1971). On the Tectonic and Morphological Classification of Coasts. *Journal of Geology* 79: 1–21.

JEHU, T. J. (1918). Rock Boring Organisms As Agents in Coast Erosion. *Scottish Geographic Magazine* 34: 1–11.

JOHNSON, D. W. (1919). *Shore Processes and Shoreline Development.* New York: John Wiley & Sons [facsimile edition: Hafner, New York (1965)].

KING, C. A. M. and M. J. McCULLAGH (1971). A Simulation Model of a Complex Recurved Spit. *Journal of Geology* 79: 22–37.

KOMAR, P. D. and S.-M. SHIH (1993). Cliff Erosion Along the Oregon Coast: A Tectonic-Sea Level Imprint Plus Local Controls by Beach Processes. *Journal of Coastal Research* 9: 747–765.

KRAFT, J. C., R. B. BIGGS, and S. D. HALSEY (1973). Morphology and Vertical Sedimentary Sequence Models in Holocene Transgressive Barrier Systems. In *Coastal Geomorphology,* D. R. Coates (editor), pp. 321–354. Binghampton, NY: State University of New York.

KUHN, G. G. and F. P. SHEPARD (1983). Beach Processes and Sea Cliff Erosion in San Diego County, California. In *Handbook of Coastal Processes and Erosion,* P. D. Komar (editor). Boca Raton, FL: pp. 267–284. CRC Press.

LEATHERMAN, S. P. and J. J. MØLLER (1991). Morris Island Lighthouse: A "Survivor" of Hurricane Hugo. *Shore & Beach* 59: 11–15.

LEWIS, W. V. (1932). The Formation of Dungeness Foreland. *Geographic Journal* 80: 309–324.

LEWIS, W. V. and W. G. V. BALCHIN (1940). Past Sea Levels at Dungeness. *Geographic Journal* 96: 258–285

McGREAL, W. S. (1979). Marine Erosion of Glacial Sediments from a Low-Energy Cliff Line Environment near Kilkeel, Northern Ireland. *Marine Geology* 32: 89–103.

MERRIAM, R. (1960). Portuguese Bend Landslide, Palos Verdes Hills, California. *Journal of Geology* 68: 140–153.

NORTH, W. J. (1954). Size, Distribution, Erosion Activities, and Gross Metabolic Efficiency of the Marine Intertidal Snails, *Littorina planaxis* and *L. scutulata. Biological Bulletin* 106: 185–197.

NUMMEDAL, D. (1983). Barrier Islands: In *Handbook of Coastal Processes and Erosion,* P. D. Komar (editor), pp. 77–121. Boca Raton, FL: CRC Press.

POUCHET, (1884). *The Universe.* Edinburgh, Scotland: Blackie and Sons.

PRINGLE, A. W. (1985). Holderness Coast Erosion and the Significance of Ords. *Earth Surface Processes and Landforms* 10: 107–124.

ROBINSON, A. H. W. (1980). Erosion and Accretion Along Part of the Suffolk Coast of East Anglia, England. *Marine Geology* 37: 133–146.

ROBINSON, L. A. (1977). Marine Erosive Processes at the Cliff Foot. *Marine Geology* 23: 257–271.

RUSSELL, R. J. (1958). Long Straight Beaches. *Ecol. Geol. Helv.* 51: 591–598.

RUSSELL, R. J. and H. V. HOWE (1935). Cheniers of Southwestern Louisiana. *Geographical Review* 25: 449–461.

SAVILLE, T. (1960). Sand Transfer, Beach Control, and Inlet Improvements, Fire Island Inlet to Jones Beach. *Proceedings of the 7th Conference on Coastal Engineering, Amer. Soc. Civil Engrs.,* pp. 785–807.

SAYRE, W. O. and P. D. KOMAR (1988). The Jump-Off Joe Landslide at Newport, Oregon: History of Erosion, Development and Destruction. *Shore & Beach* 56: 15–22.

SHEPARD, F. P. (1976). Coastal Classification and Changing Coastlines. *Geoscience and Man* 14: 53–64.

SHEPARD, F. P. and H. R. WANLESS (1971). *Our Changing Coastlines.* New York: McGraw-Hill. P. 579.

SHIH, S.-M., and P. D. KOMAR (1994). Sediments, Beach Morphology and Sea Cliff Erosion Within an Oregon Coast Littoral Cell. *Journal of Coastal Research* 10: 144–157.

SHIH, S.-M., P. D. KOMAR, K. TILLOTSON, W. G. McDOUGAL, and P. RUGGIERO (1994). Wave Run-Up and Sea-Cliff Erosion. *Proceedings of the 24th Coastal Engineering Conference, Amer. Soc. Civil Engrs.,* pp. 2170–2184.

STEERS, J. A. (1962). *The Sea Coast* (3rd ed.). London, England.

SUNAMURA, T. (1976). Feedback Relationship in Wave Erosion of Laboratory Rocky Coast. *Journal of Geology* 84: 427–437.

SUNAMURA, T. (1977). A Relationship Between Wave-Induced Cliff Erosion and Erosive Rate of Waves. *Journal of Geology* 85: 613–618.

SUNAMURA, T. (1982a). A Wave Tank Experiment on the Erosional Mechanism at a Cliff Base. *Earth Surface Processes and Landforms* 7: 333–343.

SUNAMURA, T. (1982b). A Predictive Model for Wave-Induced Cliff Erosion, with Application to Pacific Coasts of Japan. *Journal of Geology* 90: 167–178.

SUNAMURA, T. (1983). Processes of Sea Cliff and Platform Erosion. In *Handbook of Coastal Processes and Erosion,* P. D. Komar (editor), pp. 233–265. Boca Raton, FL: CRC Press.

SUNAMURA, T. (1992). *Geomorphology of Rocky Coasts.* New York: John Wiley & Sons.

TRENHAILE, A. S. (1987). *The Geomorphology of Rock Coasts.* Oxford, England: Clarendon Press.

VALENTIN, H. (1954). Der Landverlust in Holderness, Ostengland, von 1852 bis 1952. *Die Erde* 3(4): 296–315.

VITA-FINZI, C. and P. F. S. CORNELIUS (1973). Cliff Sapping by Molluscs in Oman. *Journal of Sedimentary Petrology* 43: 31–32.

3

Beach Morphology and Sediments

The Walrus and the Carpenter were walking close at hand:
They wept like anything to see such quantities of sand:
"If this were only cleared away," they said, "It would be grand!"
"If seven maids with seven mops swept it for half a year,
Do you suppose," the Walrus said, "That they could get it clear?"
"I doubt it," said the Carpenter, and shed a bitter tear.

Lewis Carroll
Through the Looking-Glass (1871)

Most beaches are composed of sediments derived from the disintegration of the land—sand and gravel eroded from terrestrial rocks. The composition of beach sediments reflects the nature of the source rocks and often can be used to assess relative contributions and transport paths from the sources to the beaches. Waves and nearshore currents continuously rework the accumulated beach sediment, rounding the particles and sorting them by size, shape, and density. The beach takes on a form that reflects the totality of water and sediment movements.

GENERAL BEACH MORPHOLOGY AND NOMENCLATURE

It is best to begin with the final product, the overall morphology of the beach, which reflects the composition of its sediments and the physical processes of waves, currents, and sediment transport. The *beach* is usually defined as an accumulation of unconsolidated sediment (sand, gravel, cobbles, and boulders) extending from the mean low-tide line to some physiographic change such as a sea cliff or dune field or to the point where permanent vegetation is established. This definition may satisfy the sunbather, as it focuses on the dry portion of this environment, but it is unsatisfactory from our standpoint since this limited view of the beach does not include any portion that is permanently underwater. We require a more inclusive term, one that encompasses the underwater portion of the environment, since that is where the more important processes of waves and sediment transport occur, which are ultimately responsible for the formation of the beach. The term *littoral* will be used to denote the entire environment—the zone extending across the exposed beach into the water to a depth at which the sediment is less actively transported by surface waves. This is a rather imprecise definition since waves occasionally transport sediments at considerable water depths, but in general we can take the depth limit for the littoral zone to be 10–20 m. In actual practice, the term beach is used more loosely than indicated above and is commonly taken as synonymous with our definition of the littoral zone. The term *coastal* is even more inclusive in that this region extends inland to include sea cliffs, dune fields, and estuaries. The *nearshore zone* extends seaward from the shoreline to just beyond the region in which the waves break; this term is particularly useful when discussing waves, currents, and other physical processes within this environment.

Figure 3-1 illustrates the terminology used to describe the action of waves and currents in the nearshore, while Figure 3-2 diagrams the terms for the beach-profile morphology. An alphabetical list of standard nomenclature and their definitions are given in Table 3-1.

There is a close correspondence between the zones of Figures 3-1 and 3-2, respectively, used to describe the waves and currents and the resulting nearshore morphology. In Chapter 8 it will be seen that longshore currents and water movements associated with nearshore circulation cells (including rip currents) are generated within the surf zone (Fig. 3-1), and these currents interact with the waves to rearrange the bottom sediments to form longshore bars and troughs observed in beach profiles (Fig. 3-2). This contrasts with the swash zone and foreshore where true longshore currents do not develop, and the water motions instead consist of a sequence of swash run-up and run-down. When there is a longshore component to the movement, water and sediment particles follow a zigzag or sawtooth longshore motion along the steeper foreshore slope of the profile. Later chapters will detail the continuous interplay or feedback between the beach morphology and patterns of nearshore waves and currents, to a degree that it is seldom possible to establish which is the controlling factor or which came first.

The presence and width of the surf zone is primarily a function of the beach slope and secondarily depends on the tidal stage. Beaches of low slope, those normally composed of fine sand, are characterized by wide surf zones. In contrast, steeply sloping gravel and cobble beaches seldom possess a surf zone; the waves break close to shore and develop directly into an intense swash that runs up and down the beach face. Moderately sloping beaches commonly lack a surf zone at high tide, when the waves break close to shore over the steeper

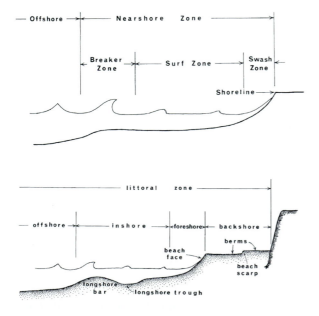

Figure 3-1   The terminology used to describe processes of waves and currents in the nearshore. See Table 3-1 for definitions.

Figure 3-2   The terminology used to describe the beach profile. See Table 3-1 for definitions.

TABLE 3-1 NOMENCLATURE USED TO DESCRIBE THE LITTORAL ZONE AND BEACH PROFILES DIAGRAMMED IN FIGURES 3-1 AND 3-2

Backshore: The zone of the beach profile extending landward from the sloping foreshore to the point of development of vegetation or change in physiography (sea cliff, dune field, etc.)

Beach face: The sloping nearly planar section of the beach profile below the berm, which is normally exposed to the swash of waves.

Berm (beach berm): The nearly horizontal portion of the beach or backshore formed by the deposition of sediments by waves. Some beaches have more than one berm at slightly different levels, separated by a scarp.

Breaker zone: The portion of the nearshore region in which the waves arriving from offshore become unstable and break. With uniform waves, such as generated in a laboratory wave tank, the zone may be reduced to a breaker line. On a wide, flat beach, secondary breaker zones may occur where reformed waves break for a second time.

Foreshore: The sloping portion of the beach profile lying between a berm crest (or in the absence of a berm crest, the upper limit of wave swash at high tide) and the low-water mark of the run-down of the wave swash at low tide. This term is often synonymous with the beach face but is commonly more inclusive, also containing some of the flat portion of the beach profile below the beach face.

Inshore: The zone of the beach profile extending seaward from the foreshore to just beyond the breaker zone.

Longshore bar: An underwater ridge of sand running roughly parallel to the shore, sometimes continuous over large distances, at other times having roughly even breaks along its length. It may become exposed at low tide. Often there is a series of such ridges parallel to one another but at different water depths, separated by longshore troughs.

Longshore trough: An elongated depression extending parallel to the shoreline and any longshore bars that are present, often representing the low point in the profile between successive bars.

Offshore: The comparatively flat portion of the beach profile extending seaward from beyond the breaker zone (the inshore) to the edge of the continental shelf. This term is also used to refer to the water and waves seaward of the nearshore zone.

Scarp: A nearly vertical escarpment cut into the beach profile by wave erosion. Its height is generally less than a meter, although higher examples are found. The scarp may be at the top of the beach face when erosion is occurring, but older scarps can be found on the berm as shown in Figure 3-2 due to former episodes of erosion.

Shoreline: The line of demarcation between the water and the exposed beach. The water's edge.

Surf zone: The portion of the nearshore region in which bore-like waves occur following wave breaking. This portion extends from the inner breakers shoreward to the swash zone.

Swash zone: The portion of the nearshore region where the beach face is alternately covered by the run-up of the wave swash and exposed by the backwash.

beach face, but develop a surf zone at low tide when the wave action is over the flatter portion of the beach profile. This contrast in beach types has given rise to the classification scheme diagrammed in Figure 3-3, developed by Wright and Short (1983). The *dissipative beach* [Fig. 3-3(a)] is the type having a low-sloping profile, such that the waves first break well offshore and continuously lose energy when they travel as breaking bores across the wide surf zone. If the breaker heights increase during a storm, the waves simply break farther offshore with minimal increase in the incident-wave energy at the shoreline (Chapter 6). The morphology of the dissipative beach, therefore, acts to dissipate the energy of the wind-generated waves. In contrast, on the *reflective beach* of Figure 3-3(c), the incident waves break close to shore with little prior loss of energy. The *intermediate beach* [Fig. 3-3(b)] incorporates a series of morphological types that are more three dimensional, in that some involve complex water-circulation patterns and bar-trough systems (Wright and Short, 1983), which will be considered in detail in Chapter 11.

A. Dissipative Beach

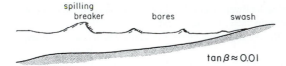

B. Intermediate Beach

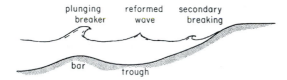

C. Reflective Beach

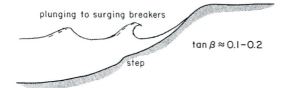

Figure 3-3 Dissipative, intermediate, and reflective beaches, a system developed by Wright and Short (1983) to classify beach morphologies and the accompanying patterns of nearshore waves and currents. The classification depends on the angle of the beach slope β and on the wave conditions.

COMPOSITION OF BEACH SEDIMENTS

Sediments in the littoral zone may be composed of any material that is available in significant quantities and is of a suitable grain size to remain on the beach. The composition therefore closely reflects the various sources and their relative importance. Most of the beaches of the temperate regions of the earth are composed principally of quartz and feldspar grains, derived ultimately from the weathering of granites and metamorphic gneisses and schists, rocks that are abundant on the continents.

In addition to the quartz and feldspar grains, beach sediments generally also contain small amounts of heavy minerals such as hornblende, garnet, epidote, tourmaline, zircon, and magnetite. These accessory minerals are much denser than quartz and feldspar and generally are darker in color. The heavy-mineral grains usually appear in the otherwise tan-colored, quartz-rich beach sand as a few dark specks. However, because of their different hydraulic behavior from the quartz and feldspar grains, resulting from their greater densities and generally smaller diameters, the heavy minerals often are concentrated by waves and currents to form dark laminae within the beach sand (Fig. 3-4). The specific heavy minerals present in the beach sand reflect the types of rocks from which they are derived. Granite may provide different heavy minerals than does schist, gneiss, or volcanic rock, and one type of granite could

Figure 3-4 Laminations and a heavy-mineral rich "black sand" seen within a trench dug into the beach face of Silver Strand Beach, California. Shovel at right for scale.

provide heavy minerals not supplied by another type of granite. The presence of a particular heavy mineral or series of heavy minerals in a beach sand can point to the erosion of a certain rock type or recycled sediment deposit. Sometimes proportions of those heavy minerals can yield information as to the relative importance of several rock types as sediment sources. Furthermore, the distribution of those heavy minerals along the coast can indicate the direction of sediment transport—the heavy minerals can be traced back to their sources.

One of the first and best studies of the sources of beach sediments was that of Trask (1952), who was interested in the problems that arose at Santa Barbara, California, following construction of a breakwater that blocked the natural longshore transport of beach sand (Chapter 8). Trask demonstrated that a significant portion of the sand filling the Santa Barbara harbor came from a distance of more than 160 km up the coast. He established this by using the mineral augite as a tracer, a black ferromagnesian silicate mineral commonly found in volcanic rocks. In that portion of the California coast, the augite is derived from rocks near Morro Bay to the north of Santa Barbara. This mineral together with the lighter quartz and feldspar, plus other heavy minerals, are transported to the south along the beaches for 160 km before being trapped by the Santa Barbara breakwater. The streams between Morro Bay and Santa Barbara do not supply any additional augite, so its proportion among the heavy minerals progressively decreases. Using the 215,000 m^3/year value for the rate of sediment trapped at the Santa Barbara breakwater and applying the data of Trask (1952, 1955) on the contents of augite in beach sands along the coast, Bowen and Inman (1966) were able to compute the longshore sediment transport rates at Surf and Gato to the north of Santa Barbara. The basis for the computation was the progressive dilution of augite in the beach sand as more sediment (without augite) is contributed to the beach, increasing the total quantity of transported sand but decreasing the average concentration of augite.

The heavy-mineral contents of beach and estuarine sands have been particularly useful in studies of sediment sources and transport paths along the Oregon coast [see review in Clemens and Komar (1988a)]. The Columbia River is potentially the largest source of sediment (Fig. 3-5), and its sand contains a diverse mineral content due to the many rock types within its immense drainage basin. A number of small rivers drain the Coast Range of Oregon, but the sand of those rivers is limited in content to augite and a small amount of brown hornblende, derived from the volcanic rocks found in their drainage basins. The south coast

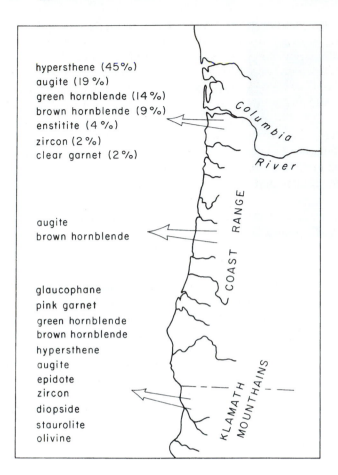

hypersthene (45%)
augite (19%)
green hornblende (14%)
brown hornblende (9%)
enstitite (4%)
zircon (2%)
clear garnet (2%)

augite
brown hornblende

glaucophane
pink garnet
green hornblende
brown hornblende
hypersthene
augite
epidote
zircon
diopside
staurolite
olivine

Figure 3-5   Minerals within beach sands on the Oregon coast are derived from three principal sources, the Columbia River on the north, the Coast Range, and the Klamath Mountain metamorphic rocks in the south. These sources contain different suites of minerals, permitting one to trace their movement along the coast. [From Tracers of Sand Movement on the Oregon Coast, K. E. Clemens and P. D. Komar, *Proceedings of the 21st Coastal Engineering Conference,* 1988. Reproduced with permission from the American Society of Civil Engineers.]

of Oregon and northern California is backed by the Klamath Mountains, containing various metamorphic rocks as well as granite. These rocks yield a wide spectrum of heavy minerals, including glaucophane and pink garnet, which are particularly useful in tracing the paths of sediments derived from this metamorphic source. The metamorphic minerals, originating in the far south, are found in Oregon beach sands along the full length of the coast (Clemens and Komar, 1988b). This is initially surprising in that there is a series of large headlands and stretches of rocky coast that should block any longshore transport of sand. These headlands segment the coast into a series of self-contained pockets or *littoral cells.* The conclusion is that the metamorphic minerals derived from the Klamath Mountains reached these northerly beaches by having first moved north thousands of years ago at the time of lowered sea levels during the ice ages (Chapter 4), a time when headlands did not interrupt such longshore movements. The sand then moved onshore with the progressive rise in sea level at the end of the ice age. The sand within any littoral cell of the Oregon coast is a mixture of minerals derived from the Klamath Mountains, the Coast Range, and the Columbia River, recording a longshore mixing of sand derived from those sources thousands of years ago.

The heavy minerals of beach sands may become concentrated by the waves, particularly within the swash zone, because their hydraulic properties differ from quartz and feldspar; at times the heavy minerals approach 100 percent of the beach-sand content. Such deposits are termed *black sands* in reference to their color. These concentrated black sands or *placers* may contain valuable minerals and have been mined for gold, platinum, chromite, magnetite (iron), and the titanium minerals ilmenite, monazite, and rutile (Komar, 1989). A typical black-sand deposit is seen in Figure 3-4, covered by a layer of quartz-rich sand of lighter color. The black-sand layer becomes exposed on the beach face during the high wave conditions of winter, when the overlying layer of quartz-rich sand is preferentially transported offshore. The heavy minerals have higher densities than quartz and feldspar and generally also have smaller grain diameters. This combination of factors is important to their selective entrainment by flowing water as illustrated in Figure 3-6. The large low-density quartz and feldspar grains project higher above the sediment bed and therefore are more exposed to the flow and are easily picked up by the moving water. In contrast, the small heavy minerals are more difficult to entrain by the flowing water and are left behind to form a lag when the light minerals are eroded and transported away (Komar and Wang, 1984).

Local pocket beaches with limited source areas may be dominated by exotic minerals within their sediments, even on the continents where quartz-feldspar sands otherwise predominate. For example, Sand Dollar Beach south of Point Sur, California, displays a bright green sand derived from the serpentine of the local Franciscan metamorphic rocks. The beautiful pocket beach at Tintagel, Cornwall, in England, below the castle of the legendary King Arthur, is made up of flat pebbles composed of very fine-grained mica schist.

The littoral sediments of volcanic islands may consist almost entirely of fragments of andesitic and basaltic lavas or individual minerals derived from the lavas. Bennett and Martin-Kaye (1951) describe the beach at Black Bay, Grenada, in the West Indies, which consists of nearly 100 percent augite grains that were eroded from local volcanic rocks. Hawaii's green-sand and black-sand beaches are well-known, the green-sand beaches consist of olivine

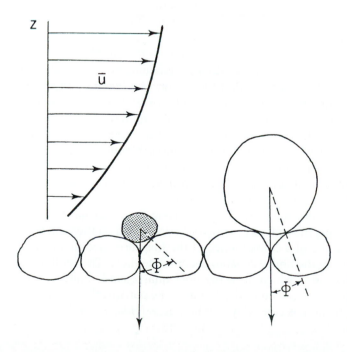

Figure 3-6 Schematic diagram illustrating the entrainment by flowing water of a large grain of quartz or feldspar versus a small, dense heavy-mineral particle. The flowing water forms a profile of increasing velocities $\bar{u}$ above the bottom. The pivoting angle Φ is important to grain entrainment and depends on the size of the grain relative to the underlying particles. [Used with permission of CRC Press, from P. D. Komar, *Physical Processes of Waves and Currents and the Formation of Marine Placers, Reviews in Aquatic Sciences,* p. 401. Copyright © 1989 CRC Press.]

that originated as crystals within the surrounding volcanic rocks, while the black-sand beaches consist of fresh microcrystalline lava and volcanic glass formed where hot lavas flowed into the sea, the lava exploding when it came into contact with the cold water (Wentworth and MacDonald, 1953; Moberly, Baver, and Morrison, 1965). Other beaches in Hawaii range from black volcanic sands to white sands composed entirely of calcium carbonate grains, with the relative proportions depending on how recently there has been volcanic activity in the area, the intensity of weathering of the rocks, and the vigor of growth of marine organisms and coral reefs.

Shells and shell fragments are important in many beach sediments, especially those in the tropics where biologic productivity is high and chemical weathering of the inland rocks tends to be intense. Moberly, Baver, and Morrison (1965) found that, for the Hawaiian Islands as a whole, foraminiferans form the chief carbonate fraction in most beaches, followed by mollusks, red algae, and echinoid fragments. Coral was a distant fifth, in spite of the fact that the sand is commonly referred to as "coral sand." In most cases the remains of these organisms are broken into small fragments and worn smooth by wave action into sand and larger particles of white calcium carbonate. Shell material may also be abundant in beaches at high latitudes because the supply of terrigenous sand is either very low or of the wrong grain size for the particular beach as determined by the wave energy. Where other sources are negligible, shells may be the primary contributor of beach material. This explains the unusual shell beach at the far north location of John O'Groats, Scotland (58.5°N latitude), which consists of 97 percent shells (Raymond and Hutchins, 1932).

The calcium carbonate content of the beaches along the southeastern Atlantic coast of the United States shows a general increase from north to south because of the increasing productivity of offshore mollusks and the decreasing supply of quartz-feldspar sand to the south (MacCarthy, 1933; Gorsline, 1963; Giles and Pilkey, 1965). To the north of the Carolinas the average beach contains very little calcareous material, usually less than one-half percent. This contrasts with Florida, where there is minimal continental sand supply to the Atlantic coast beaches and broken shells of marine organisms may form nearly the entire beach sediment. However, Giles and Pilkey found that this pattern is not a simple progressive increase in carbonate content from north to south—the content is lower in Georgia than in Florida and increases slightly to the north to a maximum in North Carolina, and then decreases rapidly still further to the north. Giles and Pilkey hypothesized that there is less carbonate on Georgia beaches because the lower wave energy there is insufficient to break and grind shells down to the proper size for littoral sediments. The grain size of sand is also a minimum on the Georgia coast, increasing both to the north and south; therefore, an alternative possibility is that the decrease is brought about by processes of selective sorting. The carbonate grains are generally large in comparison with quartz sand derived from the continent, so they may be selectively transported and concentrated in the coarser-grained beaches of higher energy, both to the north and south of Georgia.

An example of the concentration of shells by selective-sorting processes has been described by Watson (1971) as found along the central portion of Padre Island on the Gulf coast of Texas. Central Padre Island is the site of a convergence of the longshore sediment transport, which causes shells and sand from the entire coast to accumulate there. The shell material is concentrated on the beach by winds, which blow the finer quartz sand inland to form dunes. The beaches of central Padre Island are composed of up to 80 percent shells and shell fragments, in spite of the fact that the area is generally rich in quartz-feldspar sand.

Some beaches in the tropics or in arid regions are composed of oolites, nearly spherical sand-sized carbonate grains. These are formed by direct precipitation from salt water and therefore require warm, shallow, agitated waters supersaturated with calcium carbonate. Eardley (1939) has studied oolite formation on beaches on the Great Salt Lake in Utah, and Rusnak (1960) has described an occurrence in the highly saline Laguna Madre, the lagoon landward from Padre Island, Texas. Oolites also are found on many beaches of the Mediterranean Sea coast of North Africa. Fabricius, Berdau, and Munnich (1970) have concluded that the oolitic beaches of Tunisia are apparently not forming at present but rather have been eroded from fossil oolitic rocks that outcrop in the vicinity. However, the oolitic beaches along the Egyptian coast west of Alexandria are certainly forming under present-day conditions. Some of the most impressive oolitic beaches are found in the Caribbean, and particularly in the Bahamas where extensive deposits occur both on the beaches and on the offshore shoals of the Bahaman Banks.

A few unusual beaches are the inadvertent by-products of human activities. Bascom (1960) reported the existence of a beach at Fort Bragg, California, composed of old tin cans washed in from the nearby city dump, the cans having been neatly arranged by the waves (the dump is no longer in use, and the beach now consists of a nearly solid mass of cans rusted together). In the coal-mining districts of England, a number of beaches have high proportions of sand-size coal fragments. Bird (1968) describes a beach on Brownsea Island in Poole Harbour, Dorset, England, that is composed of broken fragments of earthenware, derived from a former pipe works.

BEACH SEDIMENT GRAIN SIZES

The nomenclature used to classify sediment particles according to their grain diameters is summarized in Table 3-2, with the divisions being based on multiples of 2. The distributions in diameters of grains found within a beach sand generally are determined by sieving, while the diameters of pebbles and larger particles are measured individually with a caliper or similar devise (Blatt, Middleton, and Murray, 1972). Statistics such as the mean or median grain

TABLE 3-2 CLASSIFICATION OF SEDIMENT GRAIN SIZES BY THEIR DIAMETERS

	Diameter	
Size Nomenclature	Millimeters	ϕ Units*
Boulders	> 256	> −8
Cobbles	64 to 256	−6 to −8
Pebbles	4 to 64	−2 to −6
Granule	2 to 4	−1 to −2
Very Coarse Sand	1 to 2	0 to −1
Coarse Sand	0.5 to 1	1 to 0
Medium Sand	0.25 to 0.5	2 to 1
Fine Sand	0.125 to 0.25	3 to 2
Very Fine Sand	0.0625 to 0.125	4 to 3
Silt	0.0039 to 0.0626	8 to 4

*The ϕ scale is related to the diameter D in millimeters by $D = 1/2^{\phi}$.

size and the standard deviation are calculated from the measured distributions of particle diameters. Alternately, the sizes of particles may be expressed in terms of their settling velocities, usually measured directly in the laboratory by having them settle through a column of water. In many applications, the measured settling velocity is to be preferred over the diameter, as the settling velocity accounts for the particle density and shape as well as its diameter and better represents the hydraulic behavior of the grains while being transported by waves and currents.

Particles in beach sediments vary from more than a meter for boulders to less than 0.1 mm for very-fine sand grains. In parts of Labrador, Alaska, Scotland, and Argentina, beaches commonly consist of large cobbles and boulders. Such coarse material can also be found in many pocket beaches throughout the world, especially those in the vicinity of rocky headlands. Beaches of flat pebbles and cobbles, called *shingle*, are particularly common in England and may even be said to be typical for that area; indeed, the term "beach" is derived from an Anglo-Saxon word that referred to shingle. More familiar to most beach visitors are the long stretches of shore consisting of sand grains in the size range 0.1–2 mm.

There are three dominant factors that control the mean grain size of beach sediments:

(1) the sediment source,

(2) the wave-energy level, and

(3) the general offshore slope upon which the beach is constructed.

The importance of the source is obvious. The beach environment will select out the grain sizes that are appropriate for its particular conditions. If it so happens that the sources provide no appropriate material of the right sizes, then there would be no beach. Assuming the sources do provide the proper grain sizes, then there remains a complex relationship between the energy level of the nearshore waves and currents, the general offshore slope, and the resulting grain sizes of the beach deposit. There is a general tendency for the highest-energy beaches with the largest waves to have the coarsest grains. This is particularly apparent if only one small geographical area or only one stretch of beach is considered. For example, the more exposed beaches of an island are generally made up of the coarsest grains, other factors being constant (Marshall 1929; Moberly, Baver, and Morrison, 1965; Dobkins and Folk, 1970). A stretch of beach may be coarser at one end than at the other because of the greater wave heights at the coarser-grained end. Such relationships are responsible for many (but not all) examples of longshore sorting by grain size. However, a simple correlation between grain size and wave-energy level cannot be made. This is readily apparent when one realizes that medium-sand beaches may be found in lakes with waves of heights less than 5 cm, as well as on ocean beaches having an extreme level of wave energy. Headlands often have small pocket beaches composed of cobbles and boulders, while the long stretches of nearby beaches consist of sand. While this may in part be due to higher wave-energy levels on the headlands, also important is the offshore slope upon which the beaches are formed, the slope being considerably steeper off from the headlands. Apparently because of the high slope, as well as greater wave energies, even coarse sand within headland beaches can be maintained in suspension so those particle sizes move offshore, leaving only gravel and cobbles. Initially the bedrock slope of the nearshore would govern which grain sizes remain to form a beach. Once the beach is established, however, its slope would govern the lower limit of grain sizes that could remain within the nearshore.

Intuition suggests that there must be some quantitative relationship between slope, energy level of the waves, and the smallest sizes of grains that remain on the beach. Bagnold (1963) suggested that *autosuspension* might be the controlling factor—that grains of a certain size and finer become permanently raised above the bottom into suspension and therefore move offshore without redeposition. For autosuspension, the settling velocity w_s of the grains is related to the beach slope S and to the horizontal orbital water motions under the waves, u, by

$$w_s < uS \qquad (3.1a)$$

In Chapter 5 it will be shown that the orbital velocity depends on the height of the waves, H, and is inversely proportional to the wave period, T, the time interval between successive waves. Equation (3.1a) can be modified to

$$w_s < \frac{\pi H S}{T} \qquad (3.1b)$$

According to this relationship, the greater the wave height or bottom slope, the larger the maximum settling velocity of the sediment grains that would remain permanently in suspension and be lost to the offshore. Dean (1973, 1983) has derived a similar relationship by considering the interval between successive waves, the wave period T, in comparison with the distance the suspended grains settle through the water during that time. The hypothesis is that if the fall time of suspended material is greater than the wave period, then that material will remain suspended and probably will be carried seaward beyond the surf zone. If the sediment is suspended to some fraction δ of the water depth h, then according to Dean, this criterion would yield

$$w_s < \frac{\delta h}{T} \qquad (3.2a)$$

In the shallow water of the beach, the local wave height H is controlled by the water depth h according to the relationship $H/h = \gamma$, where γ has values on the order of 0.8 to 1.2 when the waves are breaking and 0.4 when they have developed into bores (Chapter 6). Accordingly, Equation (3.2a) becomes

$$w_s < \frac{\delta H}{\gamma T} \qquad (3.2b)$$

which has the same form as the autosuspension criterion of Equation (3.1b), except for its lack of a dependence on the slope S. These relationships appear to provide reasonable results. For example, taking an average beach slope of $3°$ ($S = 0.05$) and an orbital velocity $u = 25$ cm/sec as a reasonable average for nearshore wave conditions, Equation (3.1a) yields $w_s < 1.25$ cm/sec, which is the settling velocity of 0.15-mm diameter quartz sand. Similarly, for a suspension height $\delta h = 10$ cm and a wave period $T = 10$ sec, Equation (3.2a) yields $w_s < 1$ cm/sec, which corresponds to a grain diameter of 0.1 mm. Finer grain sizes than this should tend to remain permanently in suspension, not returning to the bed between waves, and therefore would drift offshore. Although protected low-energy beaches sometimes contain finer material, open-ocean shores rarely include grains less than this calculated 0.1–0.15 mm diameter, so there appears to be a basic validity to the concept that permanently suspended grains cannot remain on the beach and some validity to the above relationships derived to predict that limit.

However, there have been no detailed investigations completed to firmly establish these criteria, so they must be used with caution.

GRAIN-SIZE SORTING ON BEACHES

The sorting of sediments along a beach profile produces cross-shore variations in sediment grain sizes that are readily apparent and accordingly have attracted a number of investigators (Krumbein and Griffith, 1938; Krumbein, 1938; Evans, 1939; Bascom, 1951; Inman, 1953; Miller and Zeigler, 1958, 1964; Fox, Ladd, and Martin, 1966; Greenwood and Davidson-Arnott, 1972). The coarseness of the sediment reflects the bottom topography and the local intensity of turbulence and wave-energy dissipation. The largest sediment particles generally are located in the zone of most intense wave breaking, with a decrease in grain sizes both toward deeper water and shoreward across the surf and swash zones. This variation is illustrated by the nearly tideless Lake Michigan in the study of Fox, Ladd, and Martin (1966), who measured sediment grain-size parameters across the nearshore at South Haven, Michigan (Fig. 3-7). The principal sediment modes are a medium sand of approximately 0.3 mm diameter and a very-coarse sand to granules centered at 2 mm. The different locations on the profile are represented by varying proportions of those two modes. The mean grain size is greatest within the wave plunge point at the base of the beach face, decreasing up the foreshore slope as well as offshore within the trough of the beach profile. There is a second slight coarsening of the sediment over the offshore bar. The incoming waves first break over the offshore bar but without much dissipation of energy and generation of turbulence. The waves then reform over the trough and break for a second time, plunging at the base of the beach face where they expend most of their energy. Thus the mean grain sizes closely reflect the energy level of the wave processes. The intensity of the swash decreases up the beach face, and this produces a parallel decrease in grain size. The sediment sorting, as reflected in the stan-

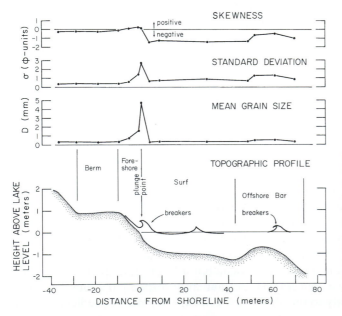

Figure 3-7 Grain-size parameters across a Lake Michigan beach, corresponding to the local wave-energy level and turbulence. Skewness refers to the asymmetry of the grain-size distribution, and standard deviation is related to the degree of sorting. [Adapted with permission of Society for Sedimentary Geology, from W. T. Fox, J. W. Ladd and M. K. Martin, A Profile of the Four Moment Measures Perpendicular to a Shore Line, South Haven, Michigan, *Journal of Sedimentary Petrology 36*, p. 1129. Copyright © 1966 Society for Sedimentary Geology.]

dard deviation (Fig. 3-7), is poorest at the plunge point and over the offshore bar since at those locations the sediment consists of mixtures of the two modes, medium sand and granules, and this mixing increases the overall range of sizes. Elsewhere the medium-sand mode exists alone, so the sorting is correspondingly better.

Similar distributions of mean grain sizes across beach profiles have been found by other investigations, including studies of gravel and cobble beaches (Krumbein and Griffith, 1938) and of beaches where tides are important (Bascom, 1951; Inman, 1953; Miller and Zeigler, 1958, 1964; Birkemeier et al., 1985). Figure 3-8 shows grain-size distributions of samples collected at intervals across the beach profile at the Field Research Facility, Duck, North Carolina. Wide ranges of sizes are present, with many of the individual samples being multimodal. Otherwise, the pattern of changing sizes across the profile is much like that seen in Figure 3-7 for the Lake Michigan beach. The coarsest grain sizes are found in the swash zone at and above mean sea level, with a tendency for decreasing grain sizes up the beach face in the landward direction and also toward the offshore.

Miller and Zeigler (1964) studied an area of irregular bottom topography with transverse spit-like bars extending out from the shore at the tip of Cape Cod, Massachusetts. In spite of the irregularity in morphology, there was a fairly regular trend of median sediment sizes, with the coarsest material being in the breaker zone and finer sediments both shoreward and in the offshore. There was some rhythmic longshore periodicity in the median grain sizes, but the presence of the shore-attached bars was not reflected in the cross-shore change in grain sizes. On California beaches, Bascom (1951) found the coarsest mean grain size at the plunge point of breaking waves and also found coarse material on the exposed summer berm. He hypothesized that the coarse berm deposit represented material carried up and over the berm crest by maximum wave uprush. Alternatively, in some cases it may represent a lag

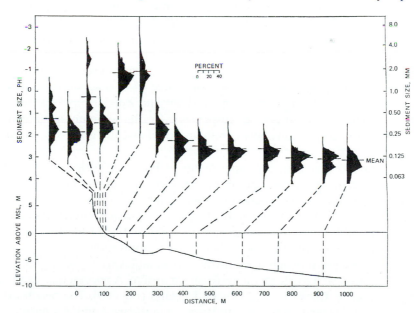

Figure 3-8 Grain-size distribution of samples collected across a beach profile at the Field Research Facility of the U.S. Army Corps of Engineers, Duck, North Carolina. [From W. A. Birkemeier et al., *A User's Guide to the Coastal Engineering Research Center's (CERC's) Field Research Facility*, 1985, U.S. Army Corps of Engineers.]

deposit, the finer sand grains having been selectively removed by the wind and carried inland to form dunes.

Relatively few attempts have been made to quantify or to understand the processes responsible for the observed grain sorting across the beach profile. Evans (1939) described qualitatively the relationship between the swash energy (velocity) and the resulting pattern of grain sizes across the foreshore zone. As the swash of a wave ascends the sloping beach, more and more of its energy is lost due to friction and percolation into the beach face, and the velocity gradually decreases to zero. The energy of the swash run-up has its source in the arriving waves, but the velocity of the backwash as the flow returns down the beach face is due to gravity; the flow starts from rest and accelerates until it reaches its maximum velocity at the base of the slope where it collides with the next incoming wave. Because of these variations in velocity, the sediment is sorted as it is moved about on the beach, the smaller grains accumulating in the slower moving water near the top of the swash and the coarsest material near the plunge line where the energy and turbulence are greatest.

Miller and Zeigler (1958) presented a more advanced model for cross-shore sediment sorting. In the breaker zone they envisioned a net vertical movement of water and sediment near the base of the turbulent breakers such that the finer sizes are lifted farthest above the bottom to be subsequently swept up the beach face or drift offshore. Only the coarsest sizes remain close to the bottom under this vertical motion and therefore remain within the breaker zone. This vertical sorting by grain size within the breakers was partially substantiated by measurements of the suspended sediments. In the foreshore region, Miller and Zeigler made an assumption that as the onshore wave swash decreases in velocity, its sediment load is dropped along the way, the coarsest material first, followed by progressively finer grains up the slope. As a first approximation they assumed that the backwash velocity increases linearly as it starts from zero at the top of the swash run-up. Using considerations of the threshold of sediment motion under turbulent flow, they then determined the final sediment distribution across the swash zone following the backwash. Because of the increase in velocity downslope, they concluded that there should be a band of graded small sizes near the top of the foreshore, with the grain size increasing progressively down the slope as the breakers are approached. This pattern is of course what is actually found within the swash and breaker zones. Although the study of Miller and Zeigler represented an advance in attempts to understand the processes responsible for the sediment sorting, they provided no real quantitative comparisons between their model and the grain-size sorting on some particular beach.

The nearshore sand tracer experiments of Ingle (1966) and others have in part confirmed the expected patterns of grain sorting. Tagging different grain sizes with various colors of fluorescent dyes, Ingle found that there is a general tendency for the coarser gains to move seaward out of the surf zone to the breaker zone, while at the same time the finer grains remained within the surf zone. The results indicate that a true sorting does take place, with each grain size seeking out the zone where it is in equilibrium.

Some stretches of beach demonstrate systematic variations in grain sizes in the longshore direction as well as along the cross-shore profile. A spectacular example of longshore sorting is found on Chesil Beach, Dorset, England (Fig. 3-9). The beach extends for 28 km from the Isle of Portland on the east to Bridport in the west and is separated from the mainland over most of its length by a shallow lagoon known as The Fleet. Chesil Beach appears to be virtually a closed system, with little new material being added to the beach under present conditions. The shingle consists of 98.5 percent flint and chert, the remainder being almost entirely quartzite pebbles (Carr, 1969). There is no immediately obvious local source for the pebbles having these compositions. Bray (1997) has concluded that they are derived from sea-cliff erosion to the west beyond Bridport. That area of cliffs has experienced massive landsliding, and the debris contains a small amount of shingle like that found on Chesil Beach. Bray further suggested that the large landslides temporarily act like groynes in that they extend well out into the water, blocking the shingle that accumulates locally on beaches. As a landslide erodes back and can no longer act as a groyne, it eventually releases the shingle which then moves eastward to Chesil Beach. Thus, according to the hypothesis offered by Bray, the contribution of shingle to Chesil Beach is episodic, and it has been several hundred years since the last addition of material. This would allow the waves sufficient time to sort the shingle into the observed ranges of sizes along the length of Chesil Beach, even though all sizes are derived from the western end of the beach.

Grain sorting along the length of Chesil Beach is striking, as can be seen in the series of photographs of Figure 3-9. At Portland, in the east, the median shingle size is 6 cm, with individual cobbles over 10 cm diameter. Midway along the beach opposite Abbotsbury, the size has decreased to about 1.25 cm, and at Bridport at its western end the beach is composed of pea-size grains with a large portion of sand. Stories are told that during past centuries, when smugglers reached the shore at night, they could judge exactly where they were just by taking up a handful of pebbles. The pattern of pebble-size variations apparently has remained constant during this century, as the measurements of Cornish (1898) and Carr (1969) agree very closely. Being a closed system, Chesil Beach has aligned itself to face approximately the direction of dominant waves arriving from the Atlantic (Lewis, 1938). Although the shingle may alternately be transported in opposite directions along the beach, the overall net movement or littoral transport of shingle along Chesil Beach must be zero.

Such a pronounced pattern of grain sorting has led to considerable speculation as to its cause. Early thoughts on the matter are summarized by Lord Avebury (1902):

> There has been much difference of opinion whether the shingle travels from east to west or from west to east. The latter view was advocated by Sir John Coode, Sir John Rennie, Sir John Hawkshaw, and Mr. Gregory. Sir J. Prestwich and Sir G. Airy, on the contrary, maintained that the shingle travels from east to west. This view is also prevalent among the fishermen of the locality, and is based on the idea that the stones are gradually worn down, and consequently that the smaller ones are those that have travelled farthest.

The local fishermen seem to have developed the most logical explanation for the sorting—the progressive abrasion of the particles as they move away from their supposed source in the Isle of Portland. However, their explanation is wrong, since the flint and quartzite pebbles are not derived from the limestones of the Isle of Portland, abrasion appears to be minor, and there is no westward net transport.

Can the increase in pebble sizes to the east be accounted for by a parallel increase in the wave energy? This is certainly the cause for observed longshore variations in grain sizes on some beaches. For example, Halfmoon Bay on the central California coast is protected on its north by a rocky headland, while the southern end is fully exposed to the waves (Bascom, 1951). This variation in wave energy is reflected in a progressive change in particle sizes, the finer sand having accumulated in the low-energy sheltered area behind the headland, and the coarsest sand being found on the fully exposed southern beach. According to the measurements of Bascom, the grain size is 0.17 mm at its sheltered north end and progressively

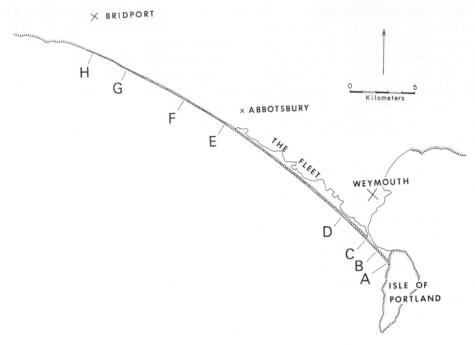

Figure 3-9 Longshore sorting of beach sediments along the length of Chesil Beach, Dorset, England. The map gives locations of the sediments shown in the series of photographs, ranging from cobbles (A) to the east to pea-size gravel mixed with sand (H) in the west.

increases to 0.65 mm at the exposed south end. There is some evidence that variations in wave energies along the length of Chesil Beach might similarly be a factor in producing grain-size changes. King (1972, pp. 308–311) appears to support this hypothesis, based on the fact that the coarsest material is found where the local offshore slope is the steepest. The reasoning is that the steeper offshore slopes allow more of the initial wave energy to be concentrated on the beach. Wave measurements obtained by Hardcastle and King (1972) from the beach opposite Wyke Regis and West Bexington, a span of about three-quarters of the total length of the beach, yielded average wave heights that were a factor 1.1 larger at Wyke Regis at the east end near Portland. On the other hand, Carr (1971) reported that their wave measurements showed no evidence of wave-height variations along the length of the beach. It is questionable, therefore, whether longshore variations in the wave energy along Chesil Beach are sufficient to account for the observed grain-size sorting, even though this does offer a satisfactory explanation at other coastal sites.

Lewis (1938) explained the sorting along Chesil Beach in terms of different wave energies reaching the beach from various directions. The beach is approximately perpendicular to the approach of the dominant waves of high energy from the southwest. The next largest waves come from the west or southwest and would move pebbles of all sizes to the east toward Portland where they would be blocked. Smaller waves generated in the English Channel arrive from the south to southeast, and according to Lewis would be capable of moving only the smaller shingle westward. A long-term interplay between larger waves from the west and

Figure 3-9 (*Continued*)

smaller waves from the southeast would cause a progressive shift of the coarsest shingle to the east and the finest to the west. The overall net transport could still be zero under this sorting scheme. This model appears to provide the best explanation for the longshore grain-size sorting observed on Chesil Beach.

Other examples of longshore variations in grain sizes do not involve closed systems such as Chesil Beach and cannot be explained in terms of longshore changes in the wave energy. Instead, they appear to be the product of selective transport processes, with some grain sizes moving alongshore at faster rates than others, or with the progressive loss of certain grain sizes into the offshore. For example, Pettijohn and Ridge (1932) found a progressive decrease in grain size along the 12-km length of Cedar Point, Ohio, a sand spit that extends out into Lake Erie. The coarsest grain size (approximately 0.5 mm) is found on the beach next to the glacial-deposit sediment source, and the finest sediment (0.15 mm) is located at the terminal end of the spit. Progressive abrasion of the sediment in the direction of transport was ruled out, as was a relationship to a systematic decrease in wave energy along the spit. Instead, the grain-size decrease was attributed to selective longshore transport, the smallest sizes moving alongshore at the greatest rate and the coarsest material remaining close to the original source area.

Schalk (1938) found just the opposite trend in grain sizes away from the source area along the outer beach of Cape Cod, Massachusetts. The finest sediment is found in the mid-Cape region, the area of the eroding cliffs that supply sediments to the beach. Both north and south from this cliff source (the sediment transport directions diverge from this area), the median grain size of the beach sediment increases. Shalk demonstrated that this pattern results from the offshore loss of the finer grains within the material supplied to the beach by cliff erosion, lost as the sediment is transported alongshore away from the source.

Thus it appears that longshore variations in grain sizes can be produced in at least four ways: (1) by parallel variations in the wave energy; (2) by selective rates of transport, with the finer grains generally outdistancing the coarser; (3) by selective removal of the finer grain sizes from the beach (carried onshore by winds or offshore by waves), leaving the remaining beach sediment coarser in median size; and (4) the interplay of waves reaching the beach from different directions with contrasting energy levels as suggested for Chesil Beach. Many examples of longshore variations in grain sizes likely result from combinations of these mechanisms.

GRAIN ABRASION AND SORTING BY SHAPE

The intense water motions of the surf zone can lead to the abrasion of sediment grains as they collide with one another. This abrasion is reflected in their *roundness* or *angularity*, the degree to which their sharp edges and corners have been worn smooth. Another important attribute of the sediment grains is their shape as measured by *sphericity*, the degree to which the shape departs from a sphere. For example, beach pebbles and shingles are commonly disc-shaped to rodlike, shapes that affect their movement on the beach.

Sand-size sediment grains are not significantly abraded while transported in rivers (Kuenen, 1959) and generally reach the coast retaining their original angularity. Once in the surf, however, grain abrasion occurs leading to a progressive increase in roundness, but the rate of change for sand-size grains is apparently very slow, involving thousands of years. An indication of this is provided by the studies of Clemens and Komar (1988a, 1988b) on the Oregon coast. As discussed earlier, the large headlands now prevent the longshore movement of beach sand and form limits to a series of isolated littoral cells. Tillamook Head (Fig. 3-10) is the northern-most headland. The Columbia River to the north supplies large volumes of sand to the adjacent ocean beaches, which have built seaward during historic times. South of Tillamook Head the composition of the beach sand indicates that it has been derived from rivers draining the Coast Range and the Klamath Mountains in southern Oregon (Fig. 3-5), with some contribution as well from the Columbia River. As a result, there is a distinct compositional difference of the beach sands to the north and south of this headland, reflecting the different sources. Furthermore, north of Tillamook Head the mineral grains are highly angular and many delicate crystals are present, while to the south the grains are noticeably more rounded. This difference is quantified by the histograms of grain roundness in Figure 3-10, ranging from VA = very angular to WR = well rounded, established using the photo-comparison roundness scale of Shepard and Young (1961). The interpretation is that the beach sand to the north of Tillamook Head has been derived from the Columbia River within

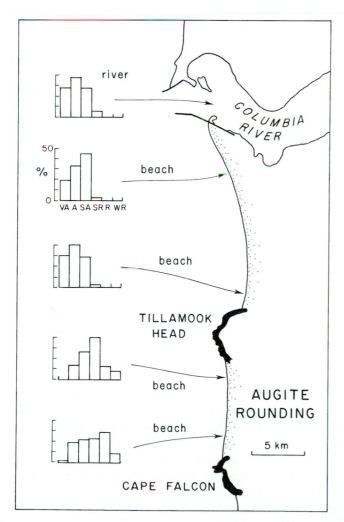

Figure 3-10 Histograms of rounding for augite sand grains in river and beach sands on the north coast of Oregon: VA = very angular, A = angular, SA = subangular, SR = subrounded, R = rounded, and WR = well rounded. The sand north of Tillamook Head is systematically more angular, as it only reached the littoral zone during recent centuries from the Columbia River, and the surf has not had sufficient time to abrade its edges. In contrast, the sand to the south of Tillamook Head has been on the beach for thousands of years and has been abraded by grain collisions. [From Tracers of Sand Movement on the Oregon Coast, K. E. Clemens and P. D. Komar, *Proceedings of the 21st Coastal Engineering Conference,* 1988. Reproduced with permission from the American Society of Civil Engineers.]

the last few hundred years and therefore still retains its angularity; the sand to the south of the headland reached the pocket beach with the rise in sea level at the end of the ice ages and has been on that beach for thousands of years during which time the grains have been abraded to higher levels of roundness.

The photo-comparison scale of grain roundness as used by Clemens and Komar (1988b) is a rather crude approach and is effective only if there are substantial differences in the degrees of grain rounding. More sophisticated analysis techniques are available. For example, Osborne and Yeh (1991) examined the degree of roundness and sphericity of sand grains using Fourier grain-shape analyses, a mathematical analysis that divides the shapes of grains into their component parts consisting of multiple-leafed clover shapes; their application was specifically to document sediment sources to beaches in southern California. Williams and Morgan (1993) employed a scanning electron microscope to examine small-scale features (pits, grooves, etc.) on the surfaces of sand grains to determine the sources of sediment found on the beaches of Fire Island, New York, establishing the importance of an offshore source. Such approaches can therefore use grain shapes and surface textures to determine sediment sources, while at the same time provide information concerning the processes of transport leading to grain abrasion.

Gravel is readily abraded and can achieve a high degree of rounding much faster than sand. This is shown by the study of Grogan (1945) on a Lake Superior beach. He found that rhyolite (volcanic) pebbles are progressively rounded as they move alongshore away from their source. The parent rock outcrops at one end of a long stretch of beach, supplying angular blocks controlled by the jointing of the rhyolite. As the pebbles move away from the source, they progressively develop a high degree of roundness. The rounding initially develops rapidly as the rough edges of the blocks are worn away, but with a subsequent decrease in the rate of additional rounding at greater distances from the source. The sphericity of the pebbles also increases in the direction of transport but not markedly so. Therefore, there is relatively little modification of the overall shapes of the pebbles as their edges are rounded. This was further established by Sames (1966) who compared beach and river pebbles on the coast of Japan, pebbles composed of resistant chert and quartzite. He could find no tendency for the beach pebbles to change their overall shapes and sphericity; they instead retained the flatness inherited from the bedrock source. However, in a similar study of pebbles in the rivers and beaches of Tahiti, Dobkins and Folk (1970) concluded that abrasion yields more disc-like forms as well as higher degrees of roundness. It was found that the size of pebbles that achieves the most nearly disc-like shape depends on the wave energy and character of the beach surface. For each grain size there is an optimum intensity of wave action that best produces a sliding motion, and this is the wave intensity that develops the best discs. Any pebbles that are larger tend to remain stationary and are not abraded by sliding to form discs, while smaller pebbles are rolled and tossed randomly by the waves and thus are abraded on all sides and the shapes are not altered.

The movement of pebbles on beaches is particularly affected by their shapes, since their overall departure from a spherical form governs how well they can be rolled about by the waves and currents. Landon (1930) showed that spherical pebbles are less stable on a beach than are flat disc-like forms, the spherical pebbles tending to roll offshore to deeper water while the flat ones preferentially remain on the beach. Excellent examples of this cross-shore sorting of pebbles by shape and size are found in the study by Bluck (1967), who investigated pebble beaches in southern Wales. As shown in Figure 3-11, Bluck found that beaches can be subdivided into four zones on the basis of pebble shape and size. Farthest shoreward on the gravel ridge is a large-disc zone, typified by cobble-size discs. Next seaward is a zone

Figure 3-11 Sorting of pebbles by size and shape across a gravel beach at Sker Point in South Wales, with large disc-shaped pebbles high on the beach and the more spherical and rod shaped pebbles at the base of the profile. Photograph B is of the Large Disc Zone, C is of the Imbricate Zone, D is of the Outer Frame Zone, and E is of the Infill Zone. [Reprinted with permission of Society for Sedimentary Geology from B. J. Bluck, Sedimentation of Beach Gravels: Examples from South Wales, *Journal of Sedimentary Petrology 37,* p. 153. Copyright © 1967 Society for Sedimentary Geology.]

composed mainly of imbricated disc-shaped pebbles, the discs being stacked on edge but dipping seaward. This offshore-dipping imbrication is enhanced on beaches where the backwash is weak. Where there is strong backwash or an impermeable layer below the discs, the imbrication may be completely destroyed or even dip landward. Sometimes there is a zone of sandy bottom seaward of the imbricated zone, over which the pebbles move quickly. These pebbles accumulate on the seaward side of the sandy zone to form a band composed of spherical and rod-shaped pebbles. The seaward-most zone consists of a framework of large cobbles containing an infilling of rod-shaped pebbles. Thus the examples studied by Bluck demonstrate a pronounced cross-shore sorting of pebbles by shape. He attributed this sorting in part to the ability of the backwash of the waves to roll the more spherical grains down the beachface, as demonstrated by Landon (1930). Bluck further suggested that the sorting is partly produced during storm conditions when fragments of all shapes are thrown forward by the waves, with the discoidal particles being most easily lifted above the sea floor and tending to have lower settling velocities, so they are thrown further up the beach than are the other shapes.

THE BUDGET OF LITTORAL SEDIMENTS

The use of compositions of beach sediments to identify their sources is often one component in the development of a *budget of littoral sediments*. The budget of sediments is simply an application of the principle of conservation of mass to the littoral sediments—the time rate of change of sand within the system is dependent upon the rate at which sand is brought into the system versus the rate at which sand leaves the system. An analysis, therefore, involves evaluations of the relative importance of various sediment sources and losses to the nearshore zone and a comparison of the net gain or loss with the observed rate of beach erosion or accretion. The development of a sediment budget can sometimes offer an explanation for the initiation of an erosion problem and may provide an avenue for remedial action. Applications often involve determining what effects humans have on the nearshore sediment system when altered by development. For example, damming a river might cut off a significant source of sand to the beach, which then begins to erode since the processes involved in sand losses continue to operate. Construction of a seawall to prevent cliff erosion may similarly lead to beach erosion if that cliff had been an important source of sand, the seawall in effect acting like the dam on the river in halting sediment delivery to the littoral zone.

The budget of littoral sediments involves making assessments of the sedimentary contributions (credits) and losses (debits) and equating these to the net gain or loss (balance of sediments) in a given sedimentary compartment or littoral cell (Bowen and Inman, 1966; Komar, 1996). The balance of sediments between the credits and debits should be approximately equivalent to the local beach erosion or accretion. Table 3-3 summarizes the possible credits and debits of sand for a littoral sedimentary budget, while some of the more important components are diagrammed in Figure 3-12. In general, the longshore movement of sand into a littoral compartment, river transport, and sea-cliff erosion provide the major natural credits; longshore movement out of the compartment, offshore transport (especially through submarine canyons), and wind transport shoreward to form sand dunes are the major debits. Included in Table 3-3 are the major human-induced credits and debits, respectively, beach nourishment that is increasingly used to rebuild lost beaches (Chapter 12), and mining that directly removes sediment from the nearshore.

TABLE 3-3   THE BUDGET OF LITTORAL SEDIMENTS [Adapted from A. J. Bowen and D. L. Inman, *Budget of Littoral Sands in the Vicinity of Point Arguello, California,* U.S. Army Coastal Engineering Research Center Technical Memo No. 19.]

Credit	Debit	Balance
Longshore transport into area	Longshore transport out of area	Beach accretion or erosion
River transport	Wind transport away from the beach	
Sea cliff erosion	Offshore transport	
Onshore transport	Deposition in submarine canyons	
Biogenous deposition	Solution and abrasion	
Hydrogenous deposition	Mining	
Wind transport onto beach		
Beach nourishment		

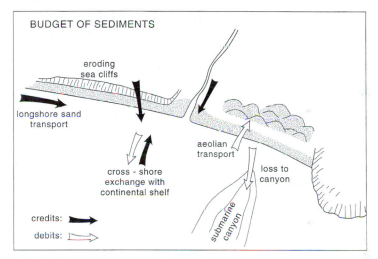

Figure 3-12   Schematic of the principal components that are involved in the development of a budget of littoral sediments. [From P. D. Komar, The Budget of Littoral Sediments—Concepts and Applications, *Shore & Beach* 64, 1996.]

The main challenge in developing a budget of sediments is to accurately assess the contributions and losses. Contributions from rivers can be evaluated using a variety of techniques. Gauging stations are found on some rivers where measurements of the flow discharge are made daily, and in some cases, data also are obtained on the quantities of suspended and bedload sediments. Best and Griggs (1991) provide an example of the use of such data in their development of a sediment budget for the Santa Cruz littoral cell on the central-California coast. They established rating curves for sediment-transport loads as functions of the daily river discharge and then used probabilities of discharge events to evaluate annual sediment yields from the rivers. If sediment-transport data are not available for the river-gauging site, engineering formulae relating the sediment transport rate to the river's discharge, bed stress, or the flow power (Graf, 1971) can be used to calculate expected sediment yields. In the absence of direct measurements of river-flow discharges and accompanying sediment transport, the sediment yield from a river can be estimated based on rainfall data, the size of the river's drainage basin, and its degree of vegetation cover (Langbein and Schumm, 1958; Scott,

Ritter, and Knott, 1968). Whichever approach is used to calculate the total sediment yield from the river, this volume must be revised to yield the quantities of sand and gravel that are sufficiently coarse to remain on the beach, eliminating the faction that is too fine and is quickly lost offshore. Usually, an analysis of the grain-size distribution of the beach sediment itself can establish which sizes are likely to remain (Best and Griggs, 1991).

Contributions of sediment to the beach from sea-cliff erosion require measurements of the long-term recession rates of the cliff. This retreat rate is converted into an annual sediment yield by multiplying the rate by the surface area of the eroding cliff. As with the river-sediment yield, this estimate of the total amount of sediment released during sea-cliff erosion must be reduced to the percentage that actually remains on the beach. For example, Valentin (1954) has shown that only about 3 percent of the sediment eroded from the boulder-clay cliffs of the Holderness, England, reaches the sandy spit of Spurn Head to the south.

A review is provided by Komar (1996) of the techniques that have been used to evaluate sediment quantities derived from other sources listed in Table 3-3 and also to estimate the debits within the sediment budget. It is difficult to make reasonable assessments for some of these components, particularly the exchange with the offshore continental shelf where the sediment transfer can represent either a credit or debit. When the various sediment credits and debits have been quantitatively evaluated, the two are compared to arrive at a balance for the sediment budget. If the debits are greater than the credits, there will be a net deficit that should be reflected as a decrease in the total volume of beach sediment—beach erosion will occur. If the credits outweigh the debits, beach accretion should prevail. If there is beach erosion or accretion, measurements of the volumes of sand involved yield a direct evaluation of the balance within the budget, an assessment that is independent of the sum of the credits and debits. Indeed, in most sediment budgets this balance determined by direct measurement of the erosion or accretion is one of the best-established parts of the budget and is often known before estimates are made of the individual credits and debits. The art of formulating a budget entails evaluating the various credits and debits such that their balance agrees reasonably well with the measured erosion or accretion.

Sediment budgets have been developed for a variety of coastal settings. Some coasts can be naturally divided into a series of compartments or littoral cells, which are semi-contained entities where one can better develop a budget. An extreme case would be a pocket beach, isolated between headlands. The sediment budget would then attempt to balance the sediment credits and debits (if any), with the balance equaling the rate of accretion or erosion of the beach within the isolated pocket. For example, Sunamura and Horikawa (1977) have developed a sediment budget for Kujykuri Beach (Fig. 3-13), located east of Tokyo, Japan. The principal sources of sediment to the central pocket beach are two rivers to the south; their contributions were estimated to be 112,000 and 156,000 m^3/year based on calculations with river-transport formulae. The erosion of bluffs at the two ends of the pocket beach contributed 82,000 and 6,000 m^3/year (Fig. 3-13). Furthermore, the beaches at the ends of the pocket have eroded over the long term with the sand moving toward the center, so the beach has become straighter with time. This erosion of the north and south ends of the beach has contributed, respectively, 62,000 and 17,000 m^3/year to the central pocket. All of these inputs to the pocket beach are diagrammed in Figure 3-13; there were no significant losses. The sum of the sediment contributions is 469,000 m^3/year, which is essentially equal to the value (494,000 m^3/year) of beach accretion within the pocket determined by comparisons of old (1903) and recent maps, the accretion having yielded shoreline advance rates of 2–3 m/year (Sunamura and Horikawa, 1977).

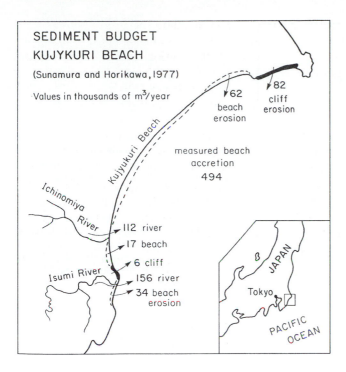

SEDIMENT BUDGET
KUJYKURI BEACH
(Sunamura and Horikawa, 1977)

·Values in thousands of m³/year

Kujyukuri Beach

62
beach
erosion

82
cliff
erosion

measured beach
accretion
494

Ichinomiya
River

112 river

17 beach

6 cliff

156 river

34 beach
erosion

Isumi River

JAPAN

Tokyo

PACIFIC
OCEAN

Figure 3-13 The Kujykuri Beach in Japan, with rivers, cliff erosion, and beach recession at the ends of the pocket supplying sediment to the central beach, which has experienced significant accretion. [Adapted from Sediment Budget in Kujkuri Coastal Area, Japan, *Coastal Sediment '77*, 1977. Reproduced with permission from American Society of Civil Engineers.]

Although the development of a budget of sediments is aided by having a littoral cell defined by headlands, budgets also have been formulated for open stretches of coast. Pierce (1969), Stapor (1971, 1973), Inman and Dolan (1989), and Kana (1995) have developed sediment budgets for stretches of barrier islands. This requires assessments of sand washed or blown inland over the island, as well as accurate evaluations of net longshore and cross-shore sediment transport rates. Particularly important are sand movements through the tidal inlets and long-term migrations of the inlets. In their study of Oregon Inlet on the Outer Banks of North Carolina, Inman and Dolan found that, as the inlet migrates landward and alongshore, the ebb-tide shoals in the offshore lag behind the migration and their sediments are only gradually returned to the beach by onshore transport under waves.

The coast of southern California consists of a series of littoral cells (Inman and Frautschy, 1966) (Fig. 3-14), each of which contains a complete system of sand sources (rivers and cliff erosion), longshore transport to the southeast, and losses of sediment from the littoral zone (blown inland to form dunes or carried into the deep ocean through submarine canyons). The typical cell begins with a promontory or stretch of rocky coast where the supply of sand is restricted. The longshore transport of sand is downcoast to the southeast, determined by the prevailing wave approach from the northwest. Downdrift from the rocky coast, the beach gradually widens as sand is supplied by rivers and from sea-cliff erosion. Each littoral cell is terminated in the south by a submarine canyon that captures the beach sand and funnels it into the deep ocean where it is permanently lost from the littoral zone. This causes the next littoral cell to begin anew with a rocky coast devoid of beach sand, and the system is repeated.

The littoral-sediment budget developed by Bowen and Inman (1966) focused on the stretch of the California coast from Pismo Beach at the north to Santa Barbara at the south (Fig. 3-15), a portion of the Santa Barbara littoral cell. Johnson (1959) had earlier examined

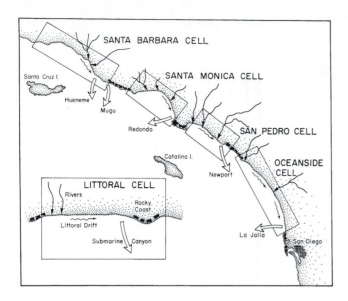

Figure 3-14 The southern California coast can be divided into a series of littoral cells, with each being a system where rivers and cliff erosion supply sand to the beach. The sand then moves southeast as littoral drift where it eventually is trapped within submarine canyons and is lost to the offshore. [Adapted from Littoral Processes and the Development of Shorelines, D. L. Inman and J. D. Frautschy, *Proceedings Coastal Engineering Specialty Conference,* 1966. Reproduced with permission from the American Society of Civil Engineers.]

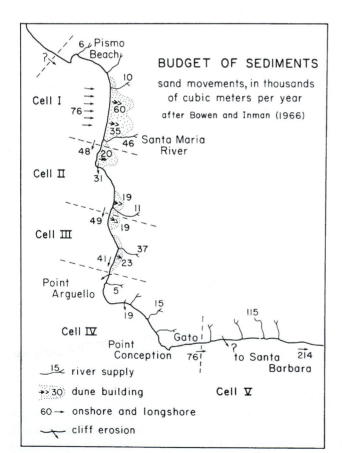

Figure 3-15 The budget of littoral sediments for the southern California coast to the immediate north of Santa Barbara. Values given in the figure represent thousands of cubic meters of sand per day. [From A. J. Bowen and D. L. Inman, *Budget of Littoral Sand in the Vicinity of Point Arguello, California,* U.S. Army Coastal Engineering Research Center Technical Memo No. 19.]

the supply and losses of sand to the beaches as far north as Point Lobos and Monterey Bay, focusing particularly on the sediment supply from rivers, the principal source of sand to these beaches. Originally the sand contributed to this cell moved south as littoral drift until it was trapped in the Hueneme and Mugu submarine canyons (Fig. 3-14). Construction of the Santa Barbara breakwater in 1929 (Chapter 9) interrupted that longshore sand transport and for a time acted as the southern limit to sand movement (dredging operations now bypass the sand to the downdrift side of the breakwater). The budget developed by Bowen and Inman (1966) for the central portion of this cell is shown in Figure 3-15. They considered five sub-cells defined by the arcuate beaches upcoast from rocky points that act like groynes (Chapter 12) but are too small to completely interrupt the longshore sand movement. The main contribution to subcell I comes from offshore sources (76,000 m^3/year) where Bowen and Inman (1966) found a large sand reservoir. Another sediment contribution to subcell I is the Santa Maria River (46,000 m^3/year), and there may be some longshore transport arriving from the north. A large volume of sand is blown inland, evaluated from migration rates of sand dunes and their sizes. Approximately 48,000 m^3/year is estimated to remain on the beach and move south as littoral drift into subcell II. There are various river sources, some cliff erosion, and losses to inland sand dunes in the remaining subcells (Fig. 3-15). The input to subcell V from longshore transport (76,000 m^3/year) and from streams (115,000 m^3/year) approximately balances the rate of sand accumulation due to blockage at the Santa Barbara breakwater (214,000 m^3/year). Although many of these quantities represent rough approximations, the budget of sediments still provides a useful framework for balancing gains versus losses and considering the impacts of humans on the system. For example, the budget developed by Bowen and Inman for the Santa Barbara littoral cell represents the modern condition where dam construction on the major rivers already has resulted in decreased supplies of sand to the littoral zone. They briefly discuss what the system must have been like prior to the dams and describe likely responses of the beaches due to dam construction and sediment losses.

The damming of a river, cutting off its sediment supply to the coast, can produce a significant imbalance in the budget of littoral sediments, leading to major beach erosion. The most extreme example of this has been the construction of the Aswan High Dam in 1964 on the Nile River, which has resulted in shoreline recession rates on the order of 100–200 m/year on the delta near the river mouth [see Frihy et al. (1991) and the references therein]. Another example is the construction of the Akosombo Dam on the Volta River of West Africa, which blocked approximately 99.5 percent of the river's drainage area, again resulting in large-scale shoreline erosion (Ly, 1980).

In addition to the construction of dams on rivers, other human activities have altered the sediment processes of beaches and otherwise have affected the budget of sediments. In particular, the construction of breakwaters, jetties, or groynes can interrupt the longshore sediment movement, reducing or completely eliminating the natural delivery of littoral sediments to the downdrift beaches (Chapter 9 and 12). A more direct human impact is sand mining, particularly of the beach sediment but also in the common practice of mining sand and gravel from rivers and streams that reduces contributions to the coastal beaches. Less obvious is the dredging of sand from inlets and harbors, sediment that originated on nearby beaches. Dean (1988) has reported that on the Florida coast, poor sand management practices at channels modified for navigation have accounted for 80–85 percent of the beach erosion during the last 40–50 years. During that time, more than 40 million cubic meters of high quality sand from Florida's east coast inlets have been dredged and disposed of offshore in water that is too deep to assure its return to the active nearshore system. In the opposite sense, beach nourishment has become common on eroding beaches, the practice of importing sand into the littoral zone, sediment that has been dredged and pumped from the offshore or trucked in from an inland source (Chapter 12). With beach nourishment, humans provide a credit within the overall budget of sediments, in many cases to offset a deficit resulting from activities such as dam construction or the building of jetties.

One of the chief advantages in the development of a budget of sediments is that it compels one to take a broad view of a coastal area, examining the potential sources of sediments, their transport paths, and finally the mechanisms of sediment loss. Even though it is often difficult, or even impossible, to make quantitative assessments of all the credits and debits and to achieve a balance that is in agreement with the observed rate of erosion or accretion, undertaking the exercise can be fruitful, as it allows one to recognize that the construction of a dam on a river hundreds of kilometers away may ultimately have a destructive impact on the beach. The proliferation of shore-protection structures along an eroding sea cliff may cut off the beach's primary source of sand. The formulation of a sediment budget can serve to organize what is known about a coastal area or littoral cell and identify where gaps in understanding exist, thereby providing a guide for future research. The development of a budget of sediments should be the first order of business in any investigation of erosion impacts and in considerations of possible remedial measures. The development of sediment budgets should be the primary goal of all coastal-management programs.

SUMMARY

Beaches throughout the world show wide variations in sediment compositions and grain sizes. On the continents we are most familiar with tan-colored, quartz-rich beach sand, but in some places the dark heavy minerals become concentrated into black-sand deposits. The black sands of volcanic islands differ in that their grains consist of small fragments of basalt or volcanic glass. Olivine-rich green sands also are found on the beaches of volcanic islands such as Hawaii. White-colored calcium carbonate grains derived from the fragmentation and abrasion of shells and coral debris are important on many beaches, particularly in the tropics where biologic productivity is high.

The grain sizes of beach sediments range from boulders down to fine sand a fraction of a millimeter in diameter. On many beaches there are interesting grain-size sorting patterns along beach profiles as well as in the longshore direction. An important concept is the budget of littoral sediments, an attempt to assess the contributions of sediment to the nearshore from rivers, cliff erosion, etc., versus losses to the offshore or blown inland to form dunes. The balance of these contributions and losses is reflected in beach accretion or erosion, and accordingly, sediment budgets can be important in analyzing the impact of humans when they dam rivers, build seawalls to halt cliff erosion, or otherwise alter the natural budget of sediments leading to beach erosion.

REFERENCES

AVEBURY, LORD. (1902). *The Scenery of England.* London, England: Macmillan.

BAGNOLD, R. A. (1963). Mechanics of Marine Sedimentation. In *The Sea,* M. N. Hill (editor). Volume 3, pp. 507–528. New York: Interscience.

BASCOM, W. N. (1951). The Relationship Between Sand-Size and Beach-Face Slope. *Transactions American Geophysical Union* 32: 866–874.

BASCOM, W. N. (1960). Beaches. *Scientific American* 203: 80–94.

BENNETT, H. S. and P. MARTIN-KAYE (1951). The Occurrence and Derivation of an Augite-Rich Beach Sand, Grenada, B.W.I. *Journal of Sedimentary Petrology* 21: 200–204.

BEST, T. C. and G. B. GRIGGS (1991). A Sediment Budget for the Santa Cruz Littoral Cell, California. In *From Shoreline to Abyss,* R. Osborne (editor). SEPM Special Publication No. 46, pp. 35–50.

BIRD, E. C. F. (1968). *Coasts.* Cambridge, MA: MIT Press.

BIRKEMEIER, W. A., H. C. MILLER, S. D. WILHELM, A. E. DEWALL, and C. S. GORBICS (1985). *A User's Guide to the Coastal Engineering Research Center's (CERC'S) Field Research Facility.* U.S. Army Corps of Engineers, Waterways Experiment Station, Vicksburg, Mississippi.

BLATT, H., G. MIDDLETON, and R. MURRAY (1972). *Origin of Sedimentary Rocks.* Englewood Cliffs, NJ: Prentice-Hall.

BLUCK, B. J. (1967). Sedimentation of Beach Gravels: Examples from South Wales. *Journal of Sedimentary Petrology* 37: 128–156.

BOWEN, A. J. and D. L. INMAN (1966). *Budget of Littoral Sands in the Vicinity of Point Arguello, California.* U.S. Army Coastal Engineering Research Center Technical Memo No. 19.

BRAY, M. J. (1997). Littoral Cells and Budget Analysis for Sediment Management in West Dorset, England. *Journal of Coastal Research* (in press).

CARR, A. P. (1969). Size Grading Along a Pebble Beach: Chesil Beach, England. *Journal of Sedimentary Petrology* 39: 297–311.

CARR, A. P. (1971). Experiments on Longshore Transport and Sorting of Pebbles: Chesil Beach, England. *Journal of Sedimentary Petrology* 41: 1084–1104.

CLEMENS, K. E. and P. D. KOMAR (1988a). Tracers of Sand Movement on the Oregon Coast. *Proceedings of the 21st Coastal Engineering Conference. Amer. Soc. Civil Engrs.,* pp. 1338–1351.

CLEMENS, K. E. and P. D. KOMAR (1988b). Oregon Beach-Sand Compositions Produced by the Mixing of Sediments under a Transgressing Sea. *Journal of Sedimentary Petrology* 58: 519–529.

CORNISH, V. (1898). On Sea Beaches and Sand Banks. *Geographical Journal* 2: 628–647.

DEAN, R. G. (1973). Heuristic Models of Sand Transport in the Surf Zone. *Proceedings of the 1st Australian Conference on Coastal Engineering. Engineering Dynamics in the Surf Zone, Sydney,* pp. 209–214.

DEAN, R. G. (1983). Principles of Beach Nourishment. In *Handbook of Coastal Processes and Erosion,* P. D. Komar (editor), pp. 217–231, Boca Raton, FL: CRC Press.

DEAN, R. G. (1988). Managing Sand and Preserving Shorelines. *Oceanus* 31: 49–55.

DOBKINS, J. E. and R. L. FOLK (1970). Shape Development on Tahiti-Nui. *Journal of Sedimentary Petrology* 40: 1167–1203.

EARDLEY, A. J. (1939). Sediments of Great Salt Lake, Utah. *American Association of Petroleum Geologists Bulletin* 22: 1359–1387.

EVANS, O. F. (1939). Sorting and Transportation of Material in the Swash and Backwash. *Journal of Sedimentary Petrology* 9: 28–31.

FABRICIUS, F. H., D. BERDAU, and K. O. MUNNICH (1970). Early Holocene Ooids in Modern Littoral Sands Reworked from a Coastal Terrace, Southern Tunisia. *Science* 169: 757–760.

FOX, W. T., J. W. LADD, and M. K. MARTIN (1966). A profile of the Four Movement Measures Perpendicular to a Shore Line, South Haven, Michigan. *Journal of Sedimentary Petrology* 36: 1126–1130.

FRIHY, O. E., A. M. FANOS, A. A. KHAFAGY, and P. D. KOMAR (1991). Patterns of Nearshore Sediment Transport Along the Nile Delta, Egypt. *Coastal Engineering* 15: 409–429.

GILES, R T. and O. H. PILKEY (1965). Atlantic Beach and Dune Sediments of the Southern United States. *Journal of Sedimentary Petrology* 35: 900–910.

GORSLINE, D. S. (1963). Bottom Sediments of the Atlantic Shelf and Slope Off the Southern United States. *Journal of Geology* 71: 422–440.

GRAF, W. H. (1971). *Hydraulics of Sediment Transport:* New York: McGraw-Hill.

GREENWOOD, B. and R. G. D. DAVIDSON-ARNOTT (1972). Textural Variation in the Subenvironments of the Shallow-Water Wave Zone, Kouchibouguas Bay, New Brunswick. *Canadian Journal of Earth Sciences* 9: 679–688.

GROGAN, R. M. (1945). Shape Variation of Some Lake Superior Beach Pebbles. *Journal of Sedimentary Petrology* 15: 3–10.

HARDCASTLE, R. J. and C. A. M. KING (1972). Chesil Beach Sea Wave Records. *Civil Engineering and Public Works Review* 67(788): 299–300.

INGLE, J. C. (1966). *The Movement of Beach Sand.* Amsterdam: Elsevier.

INMAN, D. L. (1953). *Areal and Seasonal Variations in Beach and Nearshore Sediments at La Jolla, California.* U.S. Army Corps of Engineers Beach Erosion Board Technical Memo No. 39.

INMAN, D. L. and J. D. FRAUTSCHY (1966). Littoral Processes and the Development of Shorelines. *Proceedings of the Coastal Engineering Speciality Conference*, Santa Barbara, CA: *Amer. Soc. Civil Engrs.*, pp. 511–536.

INMAN, D. L. and R. DOLAN (1989). The Outer Banks of North Carolina: Budget of Sediment and Inlet Dynamics Along a Migrating Barrier System. *Journal of Coastal Research* 5: 193–237.

JOHNSON, J. W. (1959). The Supply and Loss of Sand to the Coast: *Journal of Waterways and Harbors Division, Amer. Soc. Civil Engrs.* 85: 227–251.

KANA, T. W. (1995). A Mesoscale Sediment Budget for Long Island, New York. *Marine Geology* 126: 87–110.

KING, C. A. M. (1972). *Beaches and Coasts* (2nd ed.). New York: St. Martin's Press.

KOMAR, P. D. (1989). Physical Processes of Waves and Currents and the Formation of Marine Placers. *Reviews in Aquatic Sciences* 1: 393–423.

KOMAR, P. D. (1996). The Budget of Littoral Sediments—Concepts and Applications. *Shore & Beach* 64: 18–26.

KOMAR, P. D. and C. WANG (1984). Processes of Selective Grain Transport and the Formation of Placers on Beaches. *Journal of Geology* 92: 637–655.

KRUMBEIN, W. C. (1938). Local Areal Variation of Beach Sands. *Geological Society of America Bulletin* 49: 653–658.

KRUMBEIN, W. C. and J. S. GRIFFITH (1938). Beach Environment in Little Sister Bay, Wisconsin. *Geological Society of America Bulletin* 49: 629–652.

KUENEN, Ph.H. (1959). Experimental Abrasion, 3: Fluviatile Action on Sand. *American Journal of Science* 257: 172–190.

LANDON, R. E. (1930). An Analysis of Beach Pebble Abrasion and Transportation. *Journal of Geology* 38: 437–446.

LANGBEIN, W. B. and S. A. SCHUMM (1958). Yield of Sediment in Relation to Mean Annual Precipitation. *Transactions of the American Geophysical Union* 39: 1076–1084.

LEWIS, W. V. (1938). Evolution of Shoreline Curves. *Proceedings of the Geological Association* 49: 107–127.

LY, C. K. (1980). The Role of the Akosombo Dam on the Volta River in Causing Coastal Erosion in Central and Eastern Ghana (West Africa). *Marine Geology* 37: 323–332.

MACCARTHY, G. R. (1933). Calcium Carbonate in Beach Sands. *Journal of Sedimentary Petrology* 3: 61–67.

MARSHALL, P. (1929). Beach Gravels and Sands. *Transactions of the New Zealand Institute* 60: 324–365.

MILLER, R. L. and J. M. ZEIGLER (1958). A Model Relating Sediment Pattern in the Region of Shoaling Waves, Breaker Zone, and Foreshore. *Journal of Geology* 66: 417–441.

MILLER, R. L. and J. M. ZEIGLER (1964). A Study of Sediment Distribution in the Zone of Shoaling Waves Over Complicated Bottom Topography: In *Papers in Marine Geology,* R. L. Miller (editor), pp. 133–153, New York: Macmillan.

MOBERLY, R., L. D. BAVER, and A. MORRISON (1965). Source and Variation of Hawaiian Littoral Sand. *Journal of Sedimentary Petrology* 35: 589–598.

OSBORNE, R. H. and C.-C. YEH (1991). Fourier Grain-Shape Analysis of Coastal and Inner Continental-Shelf Sand Samples: Oceanside Littoral Cell, Southern Orange and San Diego Counties, Southern

California. In *From Shoreline to Abyss,* R. H. Osborne (editor), pp. 51–66, Special Publication No. 46, Society for Sedimentary Geology, Tulsa, Oklahoma.

PETTIJOHN, F. J. and J. D. RIDGE (1932). A Textural Variation Series of Beach Sands from Cedar Point, Ohio. *Journal of Sedimentary Petrology* 2: 76–88.

PIERCE, J. W. (1969). Sediment Budget Along a Barrier Island Chain. *Sedimentary Geology* 3: 5–16.

RAYMOND, P. E. and F. HUTCHINS (1932). A Calcareous Beach at John O'Groats, Scotland. *Journal of Sedimentary Petrology* 2: 63–67.

RUSNAK, G. A. (1960). Some Observations of Recent Oolites. *Journal of Sedimentary Petrology* 30: 471–480.

SAMES, C. W. (1966). Morphometric Data of Some Recent Pebble Associations and Their Application to Ancient Deposits. *Journal of Sedimentary Petrology* 36: 126–142.

SCHALK, M. (1938). A Textural Study of the Outer Beach of Cape Cod, Massachusetts. *Journal of Sedimentary Petrology* 8: 41–54.

SCOTT, K. M., J. R. RITTER, and J. M. KNOTT (1968). *Sedimentation in the Piru Creek Watershed, Southern California.* U.S. Geological Survey Water Supply Paper No. 1, 798-E, Washington, D.C.: U.S. Government Printing Office.

SHEPARD, F. P. and R. YOUNG (1961). Distinguishing Between Beach and Dune Sands. *Journal of Sedimentary Petrology* 31: 196–214.

STAPOR, F. W. (1971). Sediment Budgets on a Compartmented Low-to-Moderate Energy Coast in North-West Florida. *Marine Geology* 10: M1–M7.

STAPOR, F. W. (1973). History and Sand Budgets of the Barrier Island System in the Panama City, Florida, Region. *Marie Geology* 14: 227–286.

SUNAMURA, T. and K. HORIKAWA (1977). Sediment Budget in Kujykuri Coastal Area, Japan. *Coastal Sediments '77, Amer. Soc. Civil Engrs.,* pp. 475–487.

TRASK, P. D. (1952). *Sources of Beach Sand at Santa Barbara, California, As Indicated by Mineral Grain Studies.* U.S. Army Corps of Engineers. Beach Erosion Board Technical Memo No. 28.

TRASK, P. D. (1955). *Movement of Sand around Southern California Promontories.* U.S. Army Corps of Engineers. Beach Erosion Board Technical Memo No. 76.

VALENTIN, H. (1954). Der Landverlust in Holderness, Ostengland, von 1852 bis 1952. *Die Erde,* 3(4): 296–315.

WATSON, R. L. (1971). Origin of Shell Beaches, Padre Island, Texas. *Journal of Sedimentary Petrology* 41: 1105–1111.

WENTWORTH, C. K. and G. A. MacDONALD (1953). *Structures and Forms of Basaltic Rocks in Hawaii.* U.S. Geological Survey Bulletin No. 944, p. 98 Washington, D.C.

WILLIAMS, A. T. and P. MORGAN (1993). Scanning Electron Microscope Evidence for Offshore-Onshore Sand Transport at Fire Island, New York, USA. *Sedimentology* 40: 63–77.

WRIGHT, L. D. and A. D. SHORT (1983). Morphodynamics of Beaches and Surf Zones in Australia. In *Handbook of Coastal Processes and Erosion,* P. D. Komar (editor), pp. 35–64. Boca Raton, FL: CRC Press.

4

The Changing Level of the Sea

Schleiden relates, that an old monk from a convent in the neighbourhood stated that in his youth he had gathered grapes near the monument, in a spot where now the boats of the fisherman are rocked by the waves.

F. A. Pouchet
The Universe (1884)
[The monument mentioned is the ruins of
the Temple of Jupiter, Serapis, Italy]

The level of the sea is everchanging. The principal fluctuations are the tides. This daily rise and fall of water has been described romantically by Defant (1958) as "the heartbeat of the ocean, a pulse that can be felt all over the world." Smaller, but still important, are seasonal variations in mean sea levels produced by changing water temperatures, wind patterns, and ocean currents. During the most recent millions of years of geologic history, the level of the sea has undergone profound cycles in response to the growth and retreat of glaciers. There is a growing concern that the long-term increase in sea level, which has been associated with the melting of glaciers during the past several thousand years, will accelerate in the near future as temperatures of the ocean's water increase due to greenhouse warming. These and other water-level variations are graphed in Figure 4-1 in terms of the amplitude of the change versus the time involved. Included is the decadal change in water levels in Lake Michigan, representative of the Great Lakes, illustrating that processes of water-level variations are important in lakes as well as in the sea.

Our interest in the varying water levels of the sea and in lakes results from the responses of coasts to those changes. We already have seen in Chapter 2 that the origin and migration of barrier islands are dependent on the progressive rise in sea level. This is the case for coastal erosion, as many problems are brought about by elevated water levels, even though the daily erosion and beach response are produced by storm waves. For example, extensive erosion occurred along the coasts of California and Oregon during the 1982–1983 El Niño, when mean water levels were raised by tens of centimeters for several months, and more intense storms occurred. Beaches also respond to the daily tides, continuously undergoing profile shifts due to the varying water levels, achieving average profiles that depend on the overall range of tides.

The objective of this chapter is to examine the many processes that affect sea levels, processes ranging from the daily tides to changes in mean sea level over thousands of years. Some discussion will be included concerning the parallel responses of the coastal zone, but much of that consideration will be left to later chapters where we will examine the beach morphology and in particular the profile responses to various factors.

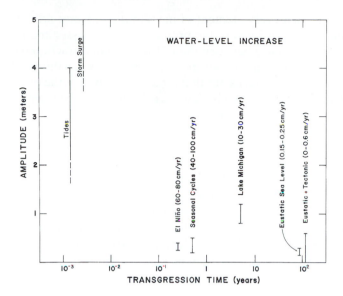

Figure 4-1 Average amplitudes of water-level changes due to various processes versus their time scales of development. The long-term global rise associated with glacial melting is given as a change over 100 years, even though it has persisted for thousands of years. Water-level variations are also important in lakes, indicated here by the decadal cycles that have been experienced in Lake Michigan. [Used with permission of Journal of Coastal Research from P. D. Komar et al., The Response of Beaches to Sea-level Changes: A Review of Predictive Models, *Journal of Coastal Research 7*, p. 900. Copyright © 1991 Journal of Coastal Research.]

TIDES—THE HEARTBEAT OF THE OCEAN

The most easily recognizable change in the level of the sea is that associated with tides, a variation that can involve several meters of change in water levels during a single day. This extreme variation can be important to processes acting within the nearshore. On a tideless sea, the area of the beach coming under wave attack is governed principally by the sizes of the waves and their run-up. In contrast, on beaches where there is a marked tidal range, the positions of wave action and the shoreline migrate continuously so a much wider zone is affected. This has an important influence on the resulting morphology of the beach (Chapter 7) and also affects the cross-shore distribution of sediments. Tides give rise to currents that are strongest within the entrances to bays and lagoons and are often the main process responsible for keeping those inlets open.

Observations over the centuries of water-level changes associated with tides have inspired a number of hypotheses as to their origin. In his geography, written in AD 902, Ibn al-Fakih retold some of the legends concerning the cause of tides. One legend relates that "verily the Angel, who is set over the seas, places his foot in the sea and thence comes the flow; then he raises it and thence comes the ebb" (Clancy, 1969). Another legend had it that tides are due to the breathing cycle of a giant whale. Although imaginative, these hypotheses have proved to be without foundation.

It is to Sir Isaac Newton (1642–1727) and his publication of *Philosophiae Naturalis Principia Mathematica* in 1686 that we owe our first real understanding of why tides exist and behave as they do. Newton's ability to explain tides was strong supporting evidence for his general theory of gravitation. He established that the gravitational pull of the moon and sun on the water of the oceans is responsible for the tides. Newton's analysis, now termed the *Equilibrium Theory,* is over simplistic in that it envisions an Earth covered with water, a simplification that allowed him to focus mainly on the tide-generating forces.

Newton's law of gravitation states that every element of mass in the universe attracts every other element with a force that is proportional to the product of their masses and inversely proportional to the square of the distance between them. The force of attraction between Earth and the Moon is the vector sum of a great many pairs of forces between the small elements of mass that make up the two bodies. It turns out that for spherical bodies the overall net attraction is the same as if all the mass of the two bodies were concentrated at their respective centers. For the Earth-Moon system the net force of attraction then becomes

$$F = G\frac{m_e m_m}{R^2} \tag{4.1}$$

where m_e and m_m are, respectively, the total masses of Earth and the Moon, R is the distance between their centers, and $G = 6.6 \times 10^{-11}$ Newton $\cdot$ m^2/kg^2 is the universal gravitational constant. Although the net force of attraction between Earth and the Moon is given by Equation (4.1), individual elements of Earth will be attracted by the Moon with a slightly different force, due to their varying distances from the Moon. A chunk of Earth or water on the surface facing the Moon is attracted more strongly than a similar chunk on the opposite side, away from the Moon. It is these small departures from the mean net force of attraction that are responsible for the tides.

Because of this mean force of attraction between Earth and the Moon, they orbit one another, each moving approximately in a circle about a common center of mass (Fig. 4-2). Since the mass of Earth is much greater than that of the Moon, this center of mass is actually located within the body of Earth, so that while the Moon sweeps freely around this point, Earth's motion is more of a wobble. The force of attraction then provides the centripetal force that produces the circular motion—there is a mutual acceleration toward one another. Some writers introduce a centrifugal force (fleeing a center) to oppose the force of Equation (4.1) in order to "keep the Moon out there," that is, to maintain it in its circular path. This is faulty reasoning, as a body in equilibrium moves in a straight line through space at a constant speed. The circular motions of Earth and the Moon show that their motions are not one of a balance of forces but rather that there is a net force and acceleration toward one another. The centrifugal force is a bogus nonexistent force, for which there is no need, including accounting for the generation of tides.

Nowhere on Earth's surface will the force of attraction by the Moon be exactly equal in magnitude and direction to the centripetal force, equal to the mean force given by Equation (4.1); the forces that govern the motions of Earth and the Moon about one another. The difference between these forces at any point provides the net force that is responsible for tide generation on Earth. This is illustrated in Figure 4-3 where f_a is the local attraction of the ele-

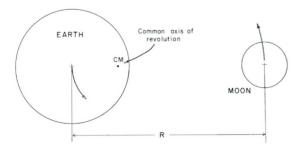

Figure 4-2   Rotation of Earth and the Moon about a common center of mass. Because of the greater mass of Earth, the center is located within Earth, so in their mutual rotation the Moon swings freely while Earth's motion is more of a wobble.

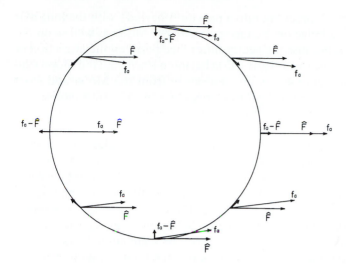

Figure 4-3 The vector subtraction of $\hat{F}$, the total force of attraction between Earth and the Moon and the local attraction f_a by the Moon of an element of water on Earth's surface.

ment of water by the Moon, from which the mean force of balance $\hat{F}$ is subtracted vectorally. It is the pattern of the net force $f_a - \hat{F}$ diagrammed in Figure 4-3 that draws the water out into tidal bulges as depicted in Figure 4-4. Strictly speaking it is the component of $f_a - \hat{F}$ tangent to the surface of Earth that is important. The component normal to Earth's surface is in line with the full force of gravity and is negligible in comparison; a person with the Moon overhead would have his weight decreased by only a ten-millionth of his original weight. In contrast, although the component of $f_a - \hat{F}$ that is tangent to the surface is also small, it acts at right angles to Earth's gravitational attraction. Although small, it has no opposing force so it can produce appreciable effects. It is this tangential component, depicted in Figure 4-4, that draws the water into bulges. The bulges would continue to grow until the pressure gradient associated with the sloping water surface of each bulge offsets and balances the tangential tide-producing force component. If Earth were completely covered with water, it would be

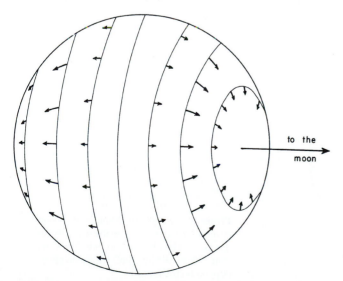

to the
moon

Figure 4-4 The tangential component of $f_a - \hat{F}$ resulting from the Moon's attraction of the surface water, the component that draws the water out into tidal bulges.

drawn out into an egg-like shape (more precisely, into a prolate spheroid), with the long axis pointing toward the Moon. There is a bulge on the face opposite the Moon as well as on the side directly facing the Moon. This is because the vector difference between the attraction of each water element by the Moon and the overall attraction between Earth and the Moon (the centripetal acceleration), shown in Figure 4-3, is directed away from the Moon and from Earth's center. This bulge is not due to a centrifugal force resulting from Earth's motions.

How would these tides appear to an observer on Earth? Consider an observer at point A on the equator in Figure 4-5. As Earth rotates daily the observer would see two high tides of equal height as he/she passes under the bulges, separated by two low tides. Another observer at point B would similarly see two high tides of equal height, but the height would be somewhat lower than seen by A. Our observers would find that the periodicity of the occurrence of the high tides is not 12 hours, as one might at first expect. Rather, each high tide occurs about 12 hours and 25 minutes after the preceding one. The observer would pass under the same tidal bulge 24 hours and 50.47 minutes following his/her previous passage under that particular bulge. This departure from an exact 12 and 24 hours is brought about by the orbital motion of the Moon around Earth in the same direction as Earth's rotation. While Earth is turning on its axis, the Moon is also revolving around Earth.

If tide observers A and B continue on the job for an extended period of time, they would begin to see some variation in the heights of the tides. One variation would be due to the eccentricity of the Moon's orbit around Earth. As depicted in Figure 4-6, the moon's path around Earth is elliptical rather than circular and is closest to the Earth at *perigee* (about 357,000 km) and farthest away at *apogee* (407,000 km). Although the departure from a true circle is not great, the varying distance will have an observable effect on the tides, since the tidal force is strongly dependent on this distance. The periodicity of this variation is 27.55 days. There is a similar variation in the tide associated with the eccentricity of Earth's orbit around the Sun. The periodicity of this variation is one anomalistic year, which is slightly longer than the "true" or sidereal year, because the line connecting Earth and Sun at perigee processes around Earth's orbit by an average of 11 seconds of arc per year.

Now that we are considering the Sun, let us examine how the presence of the egg-shaped tidal bulges, due to the sun's attraction, affects our picture of tides on an Earth cov-

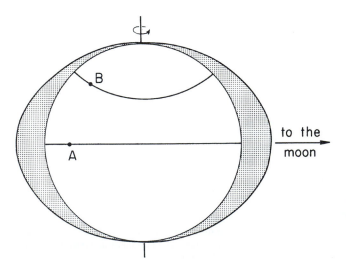

Figure 4-5 Simple tidal bulges of an Earth-covered ocean due to the attraction by the moon, seen by observers A and B as a succession of high and low tides whose heights depend on the latitude.

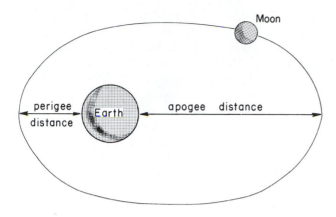

Figure 4-6 The elliptical path of the Moon around Earth. The varying distance has a significant effect on the tide-generating forces.

ered with water. The Sun's mass is about 27 million times greater than the mass of the Moon, so one might expect that it would be far more important in producing tides. But the sun is also much farther away, and this reduces the tide-generating force. The greater distance of the Sun reduces its force on Earth by a factor of 59 million in comparison with the Moon. Combining the offsetting effects of mass and distance, the Sun's tide-generating force is roughly 27/59 that of the Moon, a little less than half. The tidal bulge produced by the gravitational attraction of the Sun would not be as large as that caused by the Moon.

In Figure 4-7 Earth, the Moon, and the Sun are approximately in line, the conditions of full and new Moon, an orientation that is termed *syzygy*. It is apparent that the tidal bulges of the Moon and the Sun are additive in these two cases, so the high tides produced will be extreme, the so-called *spring tides*. In Figure 4-8 the Sun and the Moon are in quadrature, that is, the tide-generating forces are operating at right angles to one another. This produces the

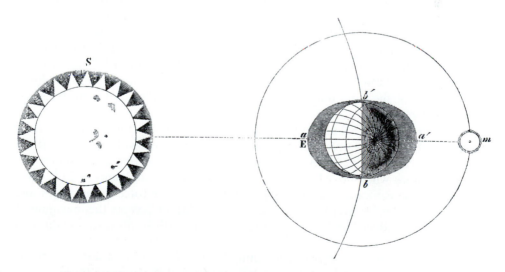

Figure 4-7 Spring tides produced by the alignment of the Sun, the Moon, and Earth. [From Rev. Lewis Tomlinson, *Recreations in Astronomy*, 1858.]

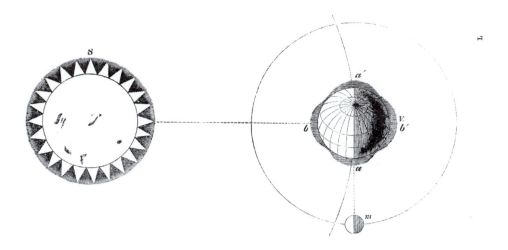

Figure 4-8 Neap tides produced by the Sun and Moon acting in opposition. [From Rev. Lewis Tomlinson, Recreations in Astronomy, 1858.]

minimum tidal range, the *neap tides*. This variation from spring to neap tides occurs fortnightly (2 weeks). On an earth covered with water, spring tides would be about 20 percent greater than the average tidal range; neap tides would be about 20 percent lower than average (on the real Earth, the difference depends on the location).

So far we have assumed that the Moon's orbit about Earth, Earth's orbit around the Sun, and Earth's equator all lie in the same plane. In fact they do not, but rather are tilted with respect to one another. The greatest departure is the tilting of Earth's equatorial plane and hence its axis, with respect to the plane of Earth's orbit around the Sun, the two planes making an angle of 23.5° (Fig. 4-9). The direction of Earth's axis in space remains nearly constant, as shown. The plane of the Moon's orbit is nearly the same as the plane of Earth's orbit about the Sun, the angle between the planes being only 5°, and to a first approximation, it can be considered the same. Because of the tilting of Earth's equatorial plane, the line of the solar tidal force varies by 47° (2 × 23.5°) north and south on Earth's surface throughout 1 year. What effect will this tilting have on tidal observers on Earth? Figure 4-10 shows that as Earth rotates, both observers A and B will see two high tides, but the heights of the tides will no longer be the same; the declination or tilting of Earth's axis has produced a daily variation in the heights of successive tides.

The 5° tilt of the Moon's orbit with respect to Earth's orbit produces a long-term tidal variation. The intersection of the plane of the Moon's orbit with the plane of Earth's orbit forms a line. This line of intersection slowly rotates, completing a revolution in 18.6 years. There are times when the Moon's declination reaches a maximum of 23.5 + 5 = 28.5°. At these times the center line of the Moon's tide-generating force varies north and south on Earth's surface by 57° over the course of a month. After 9.3 years the maximum declination of the Moon is reduced to 23.5 − 5 = 18.5°, and the north-south variation during the month is only 37°.

This results in a number of generating forces that can be observed in the tides; these forces are summarized in Table 4-1. Symbols are given that represent these various forces: M and S, respectively, represent lunar and solar constituents, while other letters signify various

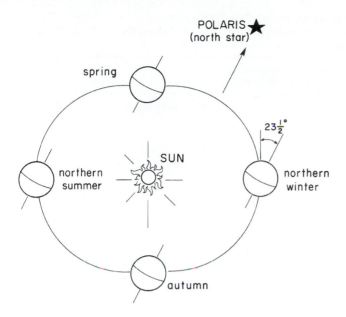

Figure 4-9 The 23.5° eccentricity of Earth's axis with respect to the plane of its orbit around the Sun.

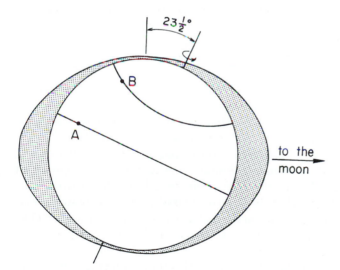

Figure 4-10 The effect of the 23.5° eccentricity on the tides as seen by observers A and B.

solar-lunar interactions. Their relative magnitudes in comparison with the M_2 main lunar semidiurnal constituent, the strongest force, also are listed in Table 4-1. The nature of the tide at a given location is governed by which of these constituents are most important (those having the greatest amplitudes).

The highest tides of the month are the spring tides, produced by the alignment of Earth, the Moon, and the Sun (syzygy). Unusually high spring tides occur when the Moon happens to be at perigee in its orbit while simultaneously in line with Earth and the Sun. These are termed *perigean spring tides*. This combined occurrence of perigee and syzygy adds about

TABLE 4-1 THE MOST IMPORTANT TIDE-GENERATING CONSTITUENTS
[Adapted with permission of The University of Michigan Press, from A. Defant, *Ebb and Flow.* Copyright © 1958 The University of Michigan Press.]

	Symbol	Period (solar hrs)	Amplitude ($M_2 = 100$)	Description
Semidiurnal tides (two tides per day)	M_2	12.42	100.0	Main lunar semidiurnal constituent
	S_2	12.00	46.6	Main solar semidiurnal constituent
	N_2	12.66	19.1	Lunar constituent due to monthly variation in the Moon's distance
	K_2	11.97	12.7	Solar-lunar constituent due to changes in declination of the sun and the moon throughout their orbital cycle
Diurnal tides (one tide per day)	K_1	23.93	58.4	Solar-lunar constituent
	O_1	25.82	41.5	Main lunar diurnal constituent
	P_1	24.07	19.3	Main solar diurnal constituent
Long-period tides	M_f	327.86	17.2	Moon's fortnightly constituent

40 percent to the total range, thereby significantly affecting water levels (again, on a water-covered Earth, differing locally on the real world). Wood (1977) has characterized the occurrence of a perigean spring tide as "a window for a potential flood." In themselves the unusual water levels may not produce appreciable coastal flooding and erosion but do increase the potential for those conditions. Wood has demonstrated in a comprehensive fashion that many of our major coastal-flooding events in the past have occurred when perigean spring tides combined with strong onshore winds that raised water levels higher as well as generated large waves. He found that more than 100 cases of major flooding associated with these conditions occurred on the North American coast during the 341 years between 1635 and 1976.

The greatest tide-generating forces occur in combination to produce the largest tides when, at the same time, the Sun is in perigee, the Sun and the Moon are in conjunction or opposition (spring tides), and both the Sun and the Moon have zero declination. This combination of circumstances happens about every 1,600 years, the last occurrence having been in about the year 1400 AD. The tides are now progressively decreasing and will reach a minimum in the year 2300.

Although the above Equilibrium Theory of tides developed by Newton adequately accounts for the principal causative forces, it fails in applications when one wants to describe the tides at some specific site. It has been found that high tides occur at the wrong times, when the site is not in line with the Moon's or the Sun's tide-generating forces. The range of the tide is not usually as predicted by the Equilibrium Theory, and the observed diurnal inequality often bears little resemblance to the theory. The Equilibrium Theory does predict the observed periodicities, or nearly so, and therefore should not be abandoned entirely. Rather, it has had to be modified to account for several complicating factors. The most obvious are the irregular shapes and varying depths of the oceans—the world is not uniformly covered with water as assumed by Newton. The rotation of Earth introduces the Coriolis force, which alters water movements associated with the tides. The assumption that water has no inertia and therefore responds immediately to gravitational forces is a particularly serious shortcoming of the Equilibrium Theory. The modern theory of tides, the *Dynamic Theory,* was first developed by the

French mathematician and scientist Pierre-Simon Laplace (1749–1827), who used the same tide-generating forces as Newton, but who accounted for these complicating factors.

In the Dynamic Theory, one thinks in terms of tidal waves rather than bulges. The periods of the waves correspond to the periods of the tide-generating forces and therefore have very long wave lengths. Let us consider the tidal waves generated in a rectangular ocean basin in the Northern Hemisphere. The tidal wave produced by a tide-generating force would advance from east to west until it meets the western boundary of the ocean. There it would be reflected and move in the opposite direction, to the east. The reflected wave will combine with other tidal waves moving toward the west in such a way that a *standing wave* is produced. This type of wave can also be formed in an enclosed basin such as a bath tub, in which the water has been set to oscillating by tilting the basin or otherwise disturbing the water. The motion of a standing wave is shown in Figure 4-11; the water oscillates about a *nodal line*, where no amplitude changes are observed but strong periodic oscillating currents are felt. The largest water-level changes, the amplitude of the standing wave, are observed at the limiting walls. The period of oscillation of such a basin, having one central nodal line, is given by

$$T_n = \frac{2L_b}{\sqrt{gh}} \tag{4.2}$$

where L_b is the east-west length of the basin, and h is the water depth ($\sqrt{gh}$ is the phase velocity of the wave; see Chapter 5). The natural period of oscillation of the water therefore depends on the geometry of the basin—its length and depth.

The tide-generating force does not simply produce such a standing wave and then cease to be effective. Rather, the force continues to act on the water in a rather complicated manner since it is always varying in direction and in magnitude. Tidal waves are therefore forced waves, as opposed to free waves. The situation is analogous to the swinging of a pendulum. If you give the pendulum a single push, it will swing in its natural period, which depends chiefly on its length. If you continuously apply forces to the swinging pendulum, it has no choice but to respond—it becomes a forced oscillation. If your applied force happens to correspond to and support the natural period of oscillation, you have a *resonant condition* and the amplitude of swing of the pendulum increases. In a similar fashion, if the natural period T_n given by Equation (4.2) corresponds to the periodicity of the tide-generating force, there will be a resonant condition and the amplitude of the standing wave will increase. It is in this way that the geometry of the ocean basin governs which of the tide-generating forces and periodicities will be most effective in producing tides.

Because of the rotation of Earth, the Coriolis force acts to modify the standing-wave oscillation. The Coriolis force in the Northern Hemisphere causes the currents associated

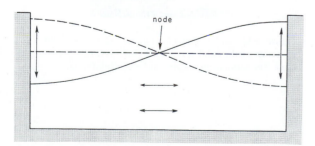

Figure 4-11 A standing wave within an enclosed basin, oscillating with a single node. The vertical motions are greatest at the end walls, whereas the horizontal current is strongest beneath the node.

with the oscillation to veer slightly to the right rather than proceed simply in an east-west direction. As the current flows toward the west, it swings to the north and piles up along the northern boundary of our rectangular ocean. Similarly, the eastward flow causes a pile-up of water along the southern boundary. The overall effect is that the tidal wave, instead of oscillating about a nodal line, now moves around a nodal point called the *amphidromic point* (from the Greek amphi, "around," and dromas, "running"). The high water of the tidal wave now progresses around the basin in a counterclockwise direction in the Northern Hemisphere as depicted in Figure 4-12. Such a rotating wave is known as a *Kelvin wave*. The range of the tide will be greatest around the edges of the ocean at the coasts and smallest in the center near the amphidromic point where the range is zero.

A series of Kelvin wave tidal systems is found around the coasts of the British Isles (Fig. 4-13), because the island is surrounded by a series of relatively small water bodies—the North Sea, two segments of the English Channel, and the Irish Sea. Three amphidromic systems are

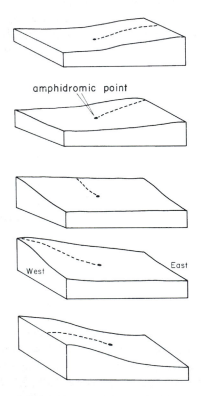

Figure 4-12 The counterclockwise motion of a Kelvin wave about an amphidromic point, shown by the progression of the dashed line with time, the form of a tidal wave in the Northern Hemisphere (the motion is clockwise in the Southern Hemisphere).

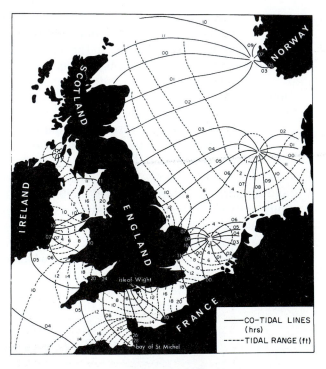

Figure 4-13 Tides around England. The cotidal lines give the positions of the tidal wave crest (high tide) throughout the daily cycle. The dashed curves give the mean tidal ranges.

found in the North Sea. In the system farthest to the north, the amphidromic point is very close to the coast of Norway. The dimensions of the North Sea are such that it resonates with the tidal movements in the open Atlantic to the north. The tidal waves can be viewed as sweeping in from the north, from the open Atlantic, and then undergoing a counterclockwise rotation about the amphidromic point. The North Sea is relatively shallow and results in bottom friction acting on the tidal wave. Friction or other losses of energy cause the amphidromic point to move away from the source of the tidal energy. In the case of the North Sea, the energy approaches from the Atlantic, and friction therefore shifts the amphidromic point eastward toward Norway. One result of this eastward shift is that the east coast of England has sizable tides, since it is distant from the amphidromic point, while the coast of Norway has smaller tides.

The English Channel demonstrates an even greater shift of the amphidromic point due to frictional effects. As seen in Figure 4-13, the point has shifted entirely onto land and is situated in the south of England. This results in the tidal waves moving in the same direction along both the southern English coast and northern French coast, on opposite sides of the Channel. The tidal range in the vicinity of the Isle of Wight, off the south coast of England, is small since it is near the amphidromic point; in contrast, the tidal range on the northern French coast is the highest in Europe.

The amphidromic systems around the British Isles, therefore, show some effects of shallow water on the motions of tides, as well as the influence of the irregularities of the coastal outline. Such effects are common and are generally important in understanding the tides at a specific coastal location. The heights of tides, and the associated tidal currents, may be considerably increased within gulfs and embayments on the coast due to combinations of convergence and resonant effects. These phenomena are illustrated by the Bay of Fundy, Canada, which achieves the highest range of tides found anywhere. The bay is long with a wide mouth (Fig. 4-14), progressively narrowing along its length and developing into two baylets at its landward termination. The tidal range at the bay mouth is large but not exceptional (about 3 m at spring tides). A considerable increase in range occurs as one proceeds up the bay. At Saint John, the range is increased to 7.6 m; near the end of Chignecto Bay, the range is 14.0 m; and at the end of the Minas Basin, the tidal range has increased to 15.6 m, the largest spring tidal range anywhere in the world. As the tidal front approaches the narrowing indentation of the bay, the enveloping shores constrict its movement and wedge the water together, thereby increasing the height of the tidal wave. Furthermore, the length of the bay in the

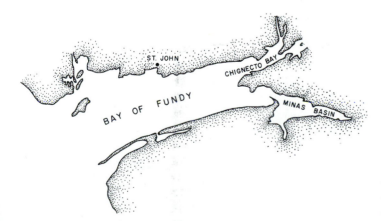

Figure 4-14 The Bay of Fundy, Canada, where the highest tides in the world occur.

direction of tidal-wave advance is such that it approximates an integral number of half-wave lengths for the appropriate depth of water, a condition where standing oscillations can develop within the bay that resonate with the tide-generating forces. This combination of effects explains the extreme tidal ranges found within the Bay of Fundy.

Another shallow-water effect on tidal waves is the development of asymmetry in the flood versus the ebb. Considerations of a tidal wave traveling in water of depth h leads to the velocity relationship

$$C = \left(1 + \frac{3}{2}\frac{\eta}{h}\right)\sqrt{gh} \qquad (4.3)$$

where η is the local height of the tidal wave above the still-water level (Fig. 4-15). Over the deep ocean η/h is small so that $C \approx \sqrt{gh}$. As the tidal wave approaches the coast, η/h increases as the water depth decreases. Since η is greater at the crest than at the troughs of the tidal wave, the crest will travel somewhat faster than the troughs. The crest will tend to overtake the preceding trough, causing the tidal wave to become asymmetrical with a steeper front. At a particular site of tidal observations, this shallow-water effect will be seen as a reduction in the period of flood (tidal rise) and an increase in the time of tidal ebb. Taken to its extreme, if the tidal wave proceeds up an estuary and river, the tidal wave may eventually oversteepen and "break," producing a *tidal bore* traveling up the river. Bores are found throughout the world, but the largest occur on the Amazon River in Brazil and on the Chientang-kiang in China. When the Amazon River bore moves upstream, it looks like a waterfall with a height of some 5 m, traveling at a rate of 10 m/sec. The bore on the Chien-tang-kiang, said to reach a height of up to 7.5 m, has been described by the ancients in graphic terms: "The surge thereof rises like a hill, and the wave like a house; it roars like thunder, and as it comes on it appears to swallow the heavens and bathe the sun" (Clancy, 1969).

Tides along the coasts of the contiguous United States are not particularly exciting with regard to their heights. However, they do provide good examples of how the natural period of oscillation of a water basin selects out the particular tide-generating force that corresponds most closely to that period, and by resonant interactions magnifies the significance of that force over the others. Because of their lengths and depths, the basins within the Atlantic

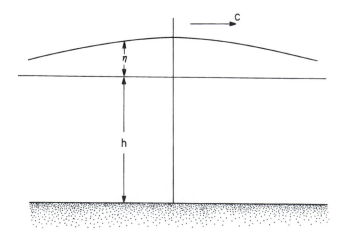

Figure 4-15 The tidal wave traveling in shallow water of depth h, with its local amplitude denoted by η.

Ocean have natural periods of oscillation that correspond roughly to the 12-hour tide-generating forces, so that *semidiurnal tides* prevail there, tides having two high and two low waters of approximately the same height during each day. Figure 4-16 shows a typical variation at New York Harbor, which illustrates a semidiurnal tide with only a slight tendency for mixed tides to occur. The resonant period for the Gulf of Mexico is closer to 24 hours, so it responds to and resonates with the tide-producing forces of that period (K_1, O_1, and P_1 of Table 4-1). The result is *diurnal tides,* with only a single high tide per day. The natural period of oscillation of the Pacific basin nearest the United States does not correspond particularly well to either the daily or semidaily tide-generating forces. The result is *mixed tides;* the example in Figure 4-16 from Seattle, Washington, illustrates this form. With mixed tides there are two high tides and two low tides per day, with strong inequalities in the heights of successive tides. During one part of the month there is a tendency for the diurnal tides to prevail, with the second tide of the day being a mere dimple on the water-level cycle. At another time in the same month, the tides are essentially semidiurnal, with two tides a day of nearly equal height. Such mixed tides prevail over the entire west coast of the United States.

Defant (1958) has shown that the form of the tide can be characterized by the relative magnitudes of the tidal constituents $M_2, S_2, K_1,$ and $O_1,$ since these are the major components. He proposed a form number

$$N_f = \frac{K_1 + O_1}{M_2 + S_2} \qquad (4.4)$$

which divides the tides roughly as follows:

$N_f = 0{-}0.25$ — semidiurnal form
$N_f = 0.25{-}1.5$ — mixed, predominately semidiurnal
$N_f = 1.5{-}3.0$ — mixed, predominately diurnal
$N_f =$ greater than 3.0 — diurnal form

Figure 4-17 illustrates tides having form numbers in these different ranges. This constitutes a classification of tides based on the periods of their principal generating forces, but does not relate to the absolute magnitudes of the tidal ranges. Davies (1964) has developed a classification based on the spring tidal ranges, employing the following categories:

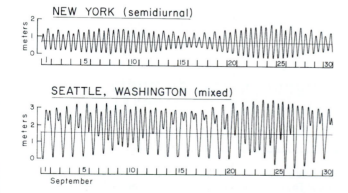

Figure 4-16 Semidiurnal tides at New York and mixed tides at Seattle, Washington. [Used with permission of The University of Michigan Press, from A. Defant, *Ebb and Flow*, p. 57. Copyright © 1958 The University of Michigan Press.]

Microtidal: less than 2 m
Mesotidal: 2–4 m
Macrotidal: greater than 4 m

Microtidal and mesotidal ranges are found generally on the open coasts of the world's oceans, as well as in the virtually landlocked seas such as the Mediterranean, Black, and Red Seas (Fig. 4-18). The higher macrotidal ranges are found locally in gulfs and embayments

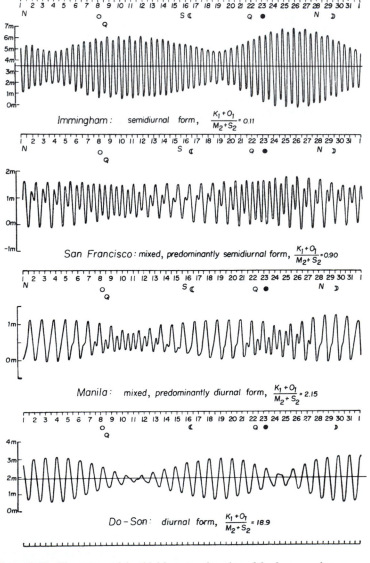

Figure 4-17 The nature of the tidal form as a function of the form number proposed by Defant. [Used with permission of The University of Michigan Press, from A. Defant, *Ebb and Flow*, p. 138. Copyright © 1958 The University of Michigan Press.]

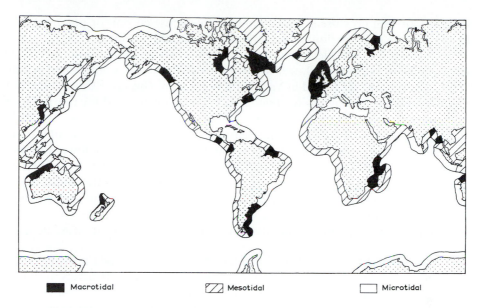

Macrotidal   Mesotidal   Microtidal

Figure 4-18   Areas of occurrence of microtides (range less than 2 m), mesotides (range 2–4 m) and macrotides (range greater than 4 m). [Used with permission of Gebruder Barntraeger, from J. L. Davies, A Morphogenic Approach to World Shorelines, *Zeitschrift für Geomorphologie*, p. 138. Copyright © 1964 Gebruder Barntraeger.]

along the coast since the shallow-water effects described above are required to produce such extreme tides.

The time-honored fashion of actually measuring tidal fluctuations is illustrated in Figure 4-19. This involves digging a well that is connected to the open sea by a pipe so the water level in the well and ocean will be the same. The pipe has a valve or other form of constriction that damps out short-period water oscillations like those due to ordinary ocean waves. This still-water level in the well could be continuously recorded by a float whose vertical movement turns a wire and pulley system; this is the old-fashioned way. Tides are now more commonly measured electronically using a pressure-sensitive device that can be placed on

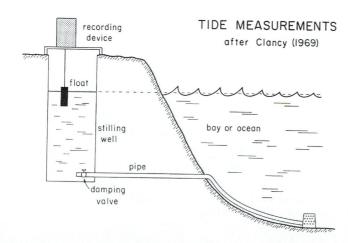

TIDE MEASUREMENTS
after Clancy (1969)

recording device

float

stilling well

pipe

bay or ocean

damping valve

Figure 4-19   The traditional station for measuring tides, consisting of a well connected to the sea by a pipe that has a valve to damp out short-period, water-level changes such as those due to waves.

the sea floor. The pressure is converted into an electric current or potential difference and damping of the short-period wave motions is accomplished electronically.

The recorded tidal elevations have to be referenced to some datum level. In the United States this is the mean of the lower-low water levels (MLLW), as illustrated in Figure 4-20 for a typical mixed tide measured in Yaquina Bay, Oregon. This cycle shows "higher-high tides," a "lower-high tide," a "higher-low tide," and a "lower-low tide." Many years of measurements of tides within the bay have permitted evaluations of averages for these four highs and lows; these average values are noted on the right-hand side of Figure 4-20. All of the elevations are given in reference to the mean lower-low water, the average of the lower-low daily tidal elevations, which is arbitrarily set at 0. In this example, the mean higher-high water is then at the elevation +8.38 ft MLLW, that distance above mean lower-low water. Since the 0 MLLW datum is placed at a relatively low water elevation, most tidal levels have positive values, while only the very lowest tides have negative elevations.

Some of the more important tidal elevations noted in Figure 4-20 are defined as follows:

Extreme high tide: The sum of the highest predicted tide and the highest recorded storm surge. The conjunction of such unusual and independent processes would be extremely unusual and accordingly would be expected to have a very long recurrence interval. This extreme tide level is sometimes used by engineers for the design of harbors.

Highest measured tide: The highest tide that has actually been measured by the tide gauge. Being a measured level, some of the elevation may be attributed to storm surge, seasonal fluctuations in water levels, and other non-tidal factors.

Highest predicted tide: The highest tide predicted by tide tables, the elevation attributable solely to tide-generating processes.

Mean higher-high water: The average of the measured higher-high tides.

Mean high water: The average of all measured high tides, including both the higher-high and lower-high recorded tides.

Local mean sea level: The mean water level at the gauge, determined by averaging the hourly height readings over the entire year. This may change from year to year, generally by a small amount, due to subsidence of the land and the global rise in sea level.

Mean sea level: The datum based on observations taken over a number of years at various tide stations along the coast of the United States and Canada. It officially is known as the *Sea Level Datum,* or the *National Geodetic Vertical Datum* (NGVD) of 1929, and is the most common datum used for land surveys.

Mean lower-low water: The average height of the lower-low tides. It serves as a reference datum for tidal elevations and for offshore water depths.

Lowest measured tide: The lowest tide level that has been measured by the gauge.

As noted above, the local mean sea level established at a specific tide gauge will vary from year to year, due to the global rise in sea level associated with the melting of glaciers and also due to vertical movements of the land upon which the tide gauge rests. The latter vary from place to place, so changes in the recorded mean sea level all vary. Using mean sea level as a datum for land surveys was found to be inconvenient, so it was decided to use the mean sea level as recorded in the 1929 NGVD, in effect ignoring any future changes in mean sea levels. The measured changes in mean sea level over the years are relatively small, so the difference with the NGVD 1929 level is generally less than 10 cm.

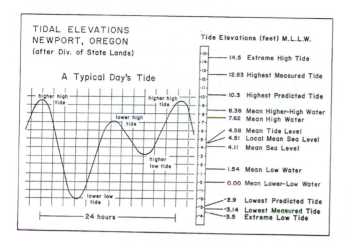

TIDAL ELEVATIONS
NEWPORT, OREGON
(after Div. of State Lands)

A Typical Day's Tide

Tide Elevations (feet) M.L.L.W.

14.5 Extreme High Tide

12.63 Highest Measured Tide

10.3 Highest Predicted Tide

8.38 Mean Higher–High Water
7.62 Mean High Water

4.58 Mean Tide Level
4.51 Local Mean Sea Level
4.11 Mean Sea Level

1.54 Mean Low Water

0.00 Mean Lower–Low Water

-2.9 Lowest Predicted Tide
-3.14 Lowest Measured Tide
-3.5 Extreme Low Tide

24 hours

Figure 4-20 A typical mixed tide as measured in Yaquina Bay, Oregon, with the tide-level elevations referenced to mean lower-low water (MLLW).

The measured tidal variation at some location is the complex summation of the several constituents caused by the various tide-generating forces (Table 4-1). Every tidal force fluctuates periodically, but when they are all added together they yield the observed tide. The analysis of measured tides is done by computer, a harmonic analysis or spectral analysis that determines the periods of the tidal constituents, their respective amplitudes, and the phase differences between the several constituents (the times when their maxima occur). These three parameters uniquely define each tidal constituent and can be used to reconstruct the total tidal cycle. This is illustrated in Figure 4-21, showing seven of the major constituents summed together to yield a computed tidal fluctuation (solid curve), which compares favorably with the tide actually observed (dashed curve). In general, the first seven to ten largest tidal components are sufficient to provide an accurately predicted tidal variation. Figure 4-22 shows the

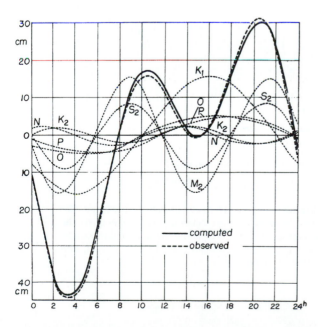

Figure 4-21 Summation of the components of the tidal cycle for 1 day to yield the computed tide, compared with the observed tide. [Used with permission of The University of Michigan Press, from A. Defant, *Ebb and Flow*, p. 50. Copyright © 1958 The University of Michigan Press.]

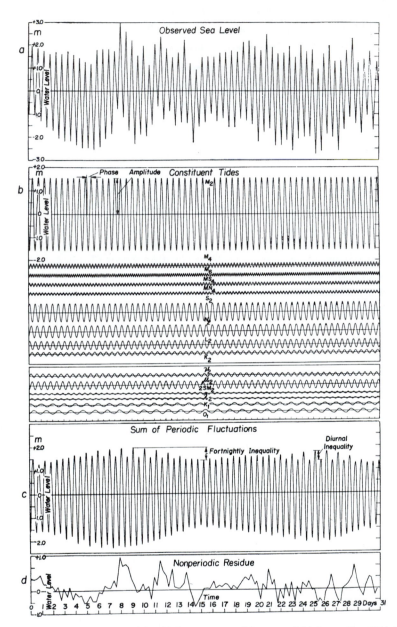

Figure 4-22 Summation of the tidal components (b) etc., to yield the predicted tidal variation (c). The observed tide (a) minus the predicted tide yields a nonperiodic residue (d) caused by winds, atmospheric pressures, and other non-tidal processes. [Used with permission of The University of Michigan Press, from A. Defant, *Ebb and Flow,* p. 51. Copyright © 1958 The University of Michigan Press.]

addition of fifteen components [Fig. 4-22(b)] used to obtain a predicted tide [Fig. 4-22(c)]. This is the basis for all predicted tides, no matter how far into the future, since the tide-generating forces and ocean-basin responses have been accounted for in the analysis. The difference between the observed and computed water levels [Fig. 4-22(a) minus Fig. 4-22(c)] yields a non-periodic residue [Fig. 4-22(d)], which could result from non-tidal factors such as winds and atmospheric pressure fluctuations that raise or depress the local water level.

MEAN SEA LEVEL DETERMINED FROM TIDE GAUGES

In addition to measuring the hourly fluctuations of tides, the many tide gauges along our coasts have been immensely important sources of information concerning daily to decadal variations in the mean level of the sea and also of long term land-level changes. If one takes the record from a tide gauge at some coastal station and averages it for a day, a mean sea level is determined for that day at that location. If the process is repeated on the following day, it is likely that the mean level will have changed, perhaps by as much as a meter or more. This is because the daily mean sea level is highly susceptible to local weather conditions. To eliminate these daily fluctuations, we could extend our averaging period to 1 month. This would yield a longer-term average sea level, but we would still find changes from 1 month to the next, although the differences would not be as great as those on a daily basis. Within a year the lowest and highest monthly values of sea level might differ by as much as 50 cm but are more typically on the order of 10–30 cm. Of importance here are seasonal variations in water temperatures, the strengths of offshore currents, and other factors that can affect the local level of the sea. Finally, if we average the tide-gauge record for the entire year, the mean sea level for that year is obtained. This level will change from year to year, sometimes irregularly, but at most locations reveals the presence of a long-term rise that supports the existence of a progressive worldwide increase in the mean level of the sea. The rate of this rise is on the order of 1–2 mm/year, 10–20 cm per century, and is interpreted as resulting from water being added to the oceans by the continued melting of glaciers and perhaps due to some thermal expansion as the water in the ocean progressively warms. These various aspects of sea-level change determined from tide gauges, either as fluctuations or net changes, are the topics of this section. These processes are immensely important to our coasts, as water levels are often a primary factor in governing beach and property erosion. A daily fluctuation in the mean water level may be important to episodic erosion and coastal flooding, while the long-term increase over the decades has been fundamental to the landward migration of barrier islands.

The most dramatic change in the mean water level on a daily basis is the *storm surge,* characterized by a rise in the water level in addition to the normal tidal variations. A storm surge is meteorological in origin and can be induced by the passage of a low atmospheric pressure system or by strong winds piling water against the coast. Fluctuations up to 1 m are common in the North Sea, with the extreme storm surge of January 31 to February 1, 1953, having produced a water-level increase of 3.3 m above the normal 1-m high tides at the Hooke of Holland, the Netherlands (Fig. 4-23). Large waves superimposed on the storm surge breached the system of dikes along the Dutch coast, with the North Sea water inundating 5 percent of the land area with salt or brackish water; 1,783 people drowned and the damage exceeded $250 million (Wemelsfelder, 1953). Storm surges generally accompany hurricanes and typhoons and can be as destructive as the waves generated by the storm. The most devastating storm surges have been experienced in the Bay of Bengal, along the coasts of India and

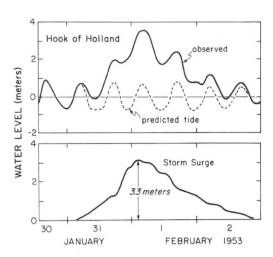

Figure 4-23 An extreme storm surge experienced on the coast of the Netherlands during 1953, measured by the tide gauge at the Hook of Holland. The lower graph represents the predicted tide subtracted from the measured water level, showing the contribution due to the storm surge. [Adapted from The Disaster in the Netherlands Caused by the Storm Flood of February 1, 1953, P. J. Wemelsfelder, *Proceedings 4th Coastal Engineering Conference*, 1953. Reproduced with permission from the American Society of Civil Engineers.]

Bangladesh, due to the low-lying topography of that area. In 1970, a storm surge associated with a typhoon killed an estimated 500,000 people. The most dramatic example of storm surge destruction along the U.S. coast was that at Galveston, Texas, where the "flood" of 1900 associated with winds of 120 miles per hour (about 50 m/sec) was 15 feet (4.6 m) above the normal 2-foot tides. Most of the city was destroyed (Fig. 1-4), and it is estimated that about 6,000 people were killed. More recently, when Hurricane Hugo made its landfall along the coast of South Carolina on September 21, 1989, the strong winds raised water levels by some 4–6 m above the predicted tide in the Charleston area (Leatherman and Møller, 1990).

Because of the cyclonic nature of hurricanes, there can be considerable alongcoast variations in the onshore component of the wind, in the local atmospheric pressure, and thus with the generated storm surge. This is illustrated in Figure 4-24 for Hurricane Carla, which occurred in 1961 on the northwest Gulf Coast. The distribution shows an asymmetry in storm-tide elevations along the coast with respect to the eye of the storm—the peak surge is some 10 km to the right of the point of landfall of the eye of the hurricane. Furthermore, the water elevations decline more rapidly to the left of the point of landfall than to the right, due to the distribution of the wind. The recorded storm surge elevations fall within a broad range because of local effects at the tide-gauge stations.

Monthly averages of sea levels determined from tide gauges eliminate most of the daily fluctuations, such as those associated with individual storms. However, a systematic pattern remains, revealing seasonal variations as illustrated by the examples in Figure 4-25. The most common pattern is one where mean-water levels are highest during the summer and lowest

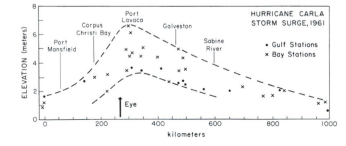

Figure 4-24 The distribution of storm surge associated with Hurricane Carla in 1961 along the northwest coast of the Gulf of Mexico. [Adapted with permission of CRC Press, from D. Nummedal, Barrier Islands, *Handbook of Coastal Processes and Erosion*, p. 111. Copyright © 1983 CRC Press.]

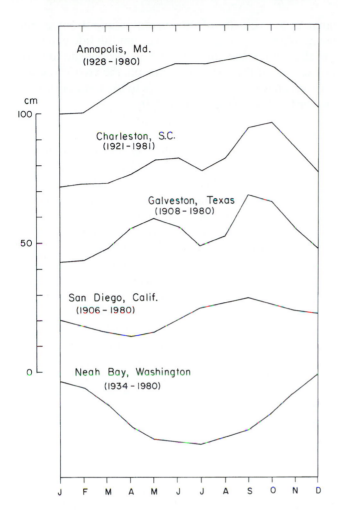

Figure 4-25 Examples of variations in monthly mean sea levels, showing contrasting seasonal cycles. [Used with permission of Society for Sedimentary Geology from P. D. Komar and D. B. Enfield, Short-term Sea-level Changes and Coastal Erosion, *Sea-level Fluctuations and Coastal Evolution,* p. 18. Copyright © 1987 Society for Sedimentary Geology.]

during the winter, illustrated most clearly by the example from Annapolis, Maryland. This trend results mainly from parallel variations in water temperatures produced by the seasonal solar radiation. During the summer, the local nearshore water heats up, and the thermal expansion produces temporary high-water levels; the cold water of the winter is denser, and this locally depresses the sea level. An inverse relationship is seen in Figure 4-25 for the monthly sea-level records from Neah Bay, Washington. This is due to upwelling during the summer, the rise of deep ocean water to the surface, so that the coastal water is colder and denser during the summer than in the winter.

Ocean currents can also cause sea-level changes through geostrophic effects, not just by producing upwelling. As discussed earlier, in the Northern Hemisphere the Coriolis force acts to turn currents toward the right (to the left in the Southern Hemisphere). This tends to raise water levels to the right of the flow until the cross-current pressure gradient of the sloping water surface counteracts the Coriolis force. The greater the velocity of the flow, the greater the cross-flow surface slope. Accordingly, the presence of the current can affect tide gauges to either side of the current. Any change in the velocity of the current is important, as this modifies the cross-flow slope and thus the recorded mean-water levels on the tide gauges. It is apparent that the effects on gauges to the right of the current will be just opposite to those experienced on gauges to the left. For example, if the current speeds up, the cross-flow surface slope will increase, and this will raise mean water levels on gauges to the right of the current while at the same time it lowers water levels on gauges to the left.

A good example of this pattern is that associated with the Florida Current (Fig. 4-26), a component of the Gulf Stream off the east coast of Florida. This current flows between the Florida peninsula on its left with Cuba and the Bahamas on its right. Sea levels can thereby be monitored on both sides of this major ocean current and the results utilized to infer fluctuations in the flow. Comparisons by Stommel (1953) and Wunsch, Hansen, and Zetler (1969) of tide-gauge records at Key West and Miami on the mainland with those at Havana, Cuba, and Cat Cay, the Bahamas, document the effects of the variable currents on the coastal water levels. The comparison between Miami and Cat Cay (Fig. 4-26) reveals that the seasonal variations at both locations show maxima toward the fall season, that at Cat Cay being slightly earlier than at Miami. These individual cycles may be influenced by a number of factors, but their difference indicates changes in the water-surface slope across the Strait, which mainly reflect variations in the strength of the Florida Current. The results show a maximum in July to August, with a minimum in October to November. These slopes imply a maximum current flow in mid-summer, the lowest flow being in the fall, results that are confirmed by direct measurements of the current. Blaha (1984) and Maul et al. (1985) provide more recent analyses of fluctuations in the Florida Current based on tide-gauge records of sea level, including more detailed considerations of other factors such as monthly changes in coastal winds and water temperatures.

Another excellent example is the water-level changes associated with the Pacific equatorial Currents investigated by Wyrtki (1973). This is a complex system of currents flowing in opposite directions but nearly parallel to the equator. The directions and strengths of the currents affect monthly mean sea levels recorded at tide gauges on scattered islands throughout the equatorial Pacific. Similar to the analyses of the tide gauges on opposite sides of the Florida Current, Wyrtki was able to decipher the more complex equatorial current system and compute the monthly mean discharges of the countercurrent for a 21-year period from the tide-gauge records.

Dramatic monthly changes in mean sea levels have periodically occurred on the equatorial islands studied by Wyrtki and along the Pacific coasts of North, Central, and South America. These extreme changes have been associated with the occurrence of an El Niño, which is triggered by the breakdown of the westward directed trade winds along the equator. One result of that cessation of winds toward the west Pacific is the release of a set-up of the sea level, which slopes upward from east to west, normally held in its inclination by the trade winds (Wyrtki, 1975). The released water moves eastward as a bulge in the sea level, held close to the equator by the Coriolis force, which acts in opposite directions north and south of the equator, turning the water back toward the equator rather than allowing the sea-level high to expand and dissipate. The movement of these sea-level "waves" eastward along the equator has been documented in tide-gauge records from islands near the equator (Wyrtki, 1977, 1983) and has been shown to represent sea-level variations up to 40–50 cm within a few months. This is illustrated by the monthly sea-level series in Figure 4-27 during the 1982–1983 El Niño (Wyrtki, 1983, 1984). Sea level at Rabaul in the western Pacific reached a peak in March or April and then began to drop. In the mid-Pacific at Fanning

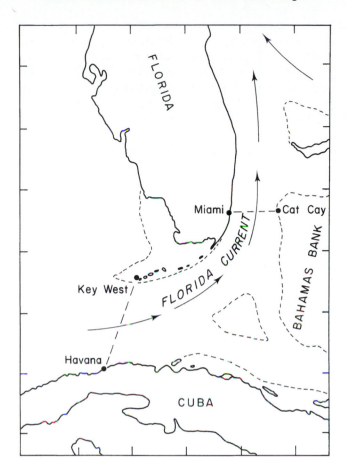

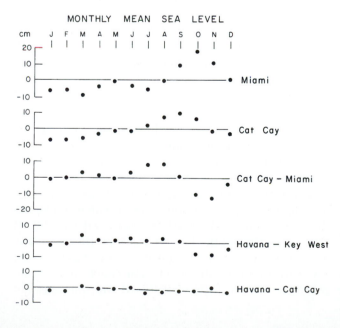

Figure 4-26 The passage of the Florida Current between the coast of Florida on its left and Cuba and the Bahamas on its right. The magnitude of the current affects sea levels measured by tide gauges and is reflected in differences in levels on opposite sides of the current. [Used with permission of Society for Sedimentary Geology from P. D. Komar and D. B. Enfield, Short-term Sea-level Changes and Coastal Erosion, *Sea-level Fluctuations and Coastal Evolution*, p. 20. Copyright © 1987 Society for Sedimentary Geology.]

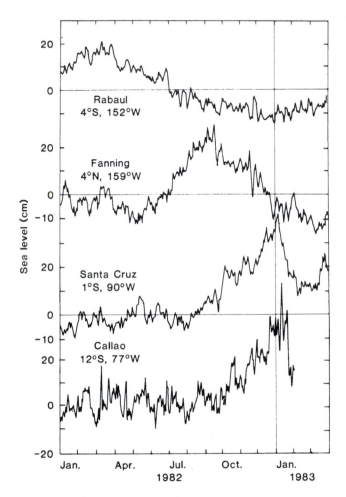

Figure 4-27 The sea-level "wave" associated with the 1982–1983 El Niño, measured by tide gauges at a series of islands from west to east along the equator within the Pacific Ocean. [Adapted with permission of American Geophysical Union, from K. Wyrtki, The Slope of Sea Level Along the Equator During the 1982/1983 El Niño, *Journal of Geophysical Research* 89. Copyright © 1983 American Geophysical Union.]

Island, due south of Hawaii, sea level was normal into June but then began to rise rapidly, reaching a maximum in September, which was some 30–40 cm higher than 6 months earlier. The peak of the rise passed Santa Cruz in the Galapagos during December and finally reached Callao on the coast of Peru during January, raising water levels by more than 50 cm. On reaching the west coast of South America, the sea-level bulge split, with part of it moving south along the coast of Chile and part moving north. The northward-moving bulges associated with El Niños have been followed in tide gauges along the coasts of Mexico, the continental United States, and have even been detected on the coast of Alaska (Enfield and Allen, 1980, 1983). The sea-level waves are held to the slope of the continental margin by refraction, much like the movement of edge waves along sloping beaches (Chapter 6). Although some energy is lost offshore, the amplitude at the shoreline itself is enhanced by the increasing Coriolis force at higher latitudes. Such sea-level waves associated with an El Niño typically raise water levels along the west coast of the United States by 10–20 cm. The 1982–1983 El Niño was exceptional in magnitude—on the California coast, the increased water levels combined with storm waves to produce extreme erosion (Flick and Cayan, 1985). On the coast of Oregon the generated sea-level wave combined with the seasonal variations in water levels to produce a 60-cm rise within 12 months, 35-cm higher than average (Huyer, Gilbert, and Pittock, 1983; Komar, 1986).

The combined effects of several causes of sea-level changes are illustrated by the study of Namias and Huang (1972), changes that spanned a decade and were associated with two large-scale, long-term climatic regimes covering the entire United States. The periods compared were 1948–1957 and 1958–1969, a rather rapid change having occurred during 1957–1958. In the latter period, the sea level off southern California was observed to rise by 5.6 cm over what it had maintained during the earlier interval. Investigating the possible causes, Namias and Huang found that the rise was mainly due to a warming of the surface water by 1 °C, causing a 3.7-cm rise in sea level, due to the thermal expansion of the water. This long-term increase in mean temperature resulted from a change in the prevailing wind patterns between the two decades. Changes in the mean atmospheric pressure accounted for a 0.6-cm rise, dynamic effects of ocean currents caused a 1.0-cm change, and the normal long-term sea level rise due to melting of glaciers accounted for another 1.0 cm. The net calculated change amounted to 6.3 cm, which compares favorably with the observed 5.6-cm rise.

Tide-gauge records are particularly useful in examining the year-by-year changes in the level of the sea compared with the level of the land. This is termed the *relative sea level,* since it reflects the changing level of the land upon which the tide gauge is mounted, as well as any worldwide (eustatic) rise in sea level produced by the melting of glaciers and the thermal expansion of ocean water due to a temperature increase. It is, of course, this local relative sea-level change that is important to erosion at the coastal site of the tide gauge, reflecting the increase or decrease in the water level in comparison with the land. Examples of records up to 80 years in length are shown in Figure 4-28. The curve from New York is typical of those for much of the east coast of the United States, indicating an average rise of approximately 3 mm/year. Roughly half of that rate can be ascribed to the global eustatic rise, the other half to land subsidence that prevails along much of the east coast of the United States. The result from Galveston, Texas, indicates a much higher relative sea level rise, averaging 6 mm/year, produced by the substantial sinking of that portion of the Gulf Coast resulting from the extraction of oil and groundwater. Data from the tide gauge at Astoria, Oregon, yield a nearly constant mean sea level (Fig. 4-28), caused by the rise of the land of that tectonically active continental margin at the same rate as the eustatic sea-level rise. An extreme case is that of Juneau, Alaska, where the land is rising at such a rate that there is a net lowering of the relative sea level and a seaward retreat of the shoreline. This is produced by the high rate of uplift of much of Alaska, in response to the unloading of the crust by the melting of glaciers that formerly covered the land. Similarly, the broad area of Scandinavia is rising due to rebound, with the middle of the area around the Gulf of Bothnia rising at a rate in excess of 5 mm/year relative to sea level (Sauramo, 1958). The hinge line for the uplift passes south of the coast of Norway, through southern Sweden, and off southern Finland.

It is apparent that the changing levels of the land can be as important, or more important, than the global eustatic rise in sea level. The land-level changes can also be documented by geodetic leveling, the periodic releveling of topographic survey stations or benchmarks. This resurveying is undertaken at intervals of 20–30 years and thus provides data regarding decadal elevation changes. Since none of the benchmarks can be fixed in elevation with certainty, releveling provides information on elevation changes within the series of benchmarks but not in absolute terms. One example is shown in Figure 4-29, based on the original data of Fairbridge and Newman (1968) but reanalyzed by Dean et al. (1987). The geodetic releveling roughly follows a north-south line along the Hudson River from the coasts of New York and

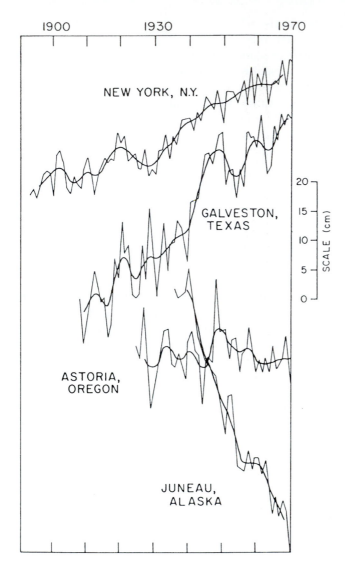

Figure 4-28 Yearly averaged relative sea levels as determined from tide-gauge records at various coastal sites, given by the thin jagged lines, while the heavier lines represent smoothed trends. The results illustrate the effects of a slow global rise in the water level within the oceans, plus local changes in the levels of the land masses. [From S. D. Hicks, On the Classification and Trends of Long Period Sea Level Series, *Shore & Beach,* 1972.]

New Jersey to the Canadian border. As analyzed by Fairbridge and Newman, the geodetic data were referenced to Rouses Point, New York, in the north; the rates were relative to that location rather than being actual rates of uplift or subsidence. The results as graphed in Figure 4-29 are from the reanalysis of Dean et al., which took the value of 2.7 mm/year relative sea-level rise determined from the tide gauge in New York City, accepted a value of 1.2 mm/year as the global rise in sea level (see Table 4-2), with the 1.5 mm/year difference representing the rate of subsidence of New York City. With that subsidence value for New York City, the remainder of the data from the geodetic releveling can be taken as reflecting land-level changes alone, and this is what is graphed in Figure 4-29. The results show a tilting of Earth's crust, with the northern section rising at about 1–2 mm/year, while New York City and the coastal area in general are subsiding. In this example, the uplift of the land to the north is, for the most part, due to continued rebound following the removal of the immense weight

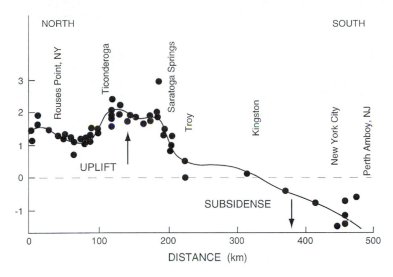

Figure 4-29 Geodetic releveling profile from northern New York south to the coasts of New York and New Jersey, showing the patterns of uplift versus coastal subsidence. [Used with permission of National Academy Press, from R. G. Dean et al., *Responding to Changes in Sea Level*, p. 22. Copyright © 1987 National Academy Press.]

of continental glaciers that formerly covered this area. Much of the subsidence of the coastal zone in this area can also be attributed to the former presence of glaciers. While the glaciers depressed the level of the inland area due to their weight, the adjacent ice-free area bulged upward in compensation. The removal of the ice has reversed the crustal movement, in part causing the subsidence of the uplifted area near the coast. Another factor is the weight of the added water to the ocean following glacial melting. That water now covers areas of the coast that were dry land 10–20 thousand years ago, and the weight of this water is depressing the level of the inundated continental shelf.

Another example of land-level changes deduced from geodetic surveys is shown in Figure 4-30 for a north-south transit along the length of the Oregon coast approximately parallel to the shore. The original analyses of the geodetic data were undertaken by Vincent (1989) and Mitchell et al. (1994), while Komar and Shih (1993) investigated the implications to alongcoast patterns of erosion. The land-level changes in this area are due to tectonic activity, this being a convergent margin where the oceanic Juan de Fuca plate collides with the continental North American plate. Figure 4-30 shows the existence of a systematic north to south variation, with the uplift of the land exceeding the global rise in sea level in the far north near Astoria and along the south coast toward the California border, while along the north-central portion of the coast the rise in sea level has exceeded the land-level rise. This pattern of land-level versus sea-level variations is paralleled by a corresponding intensity of sea cliff erosion, with the erosion being greatest along the section where the sea-level rise exceeds the land-level increase.

It has been seen from these examples that the relative sea-level variations recorded on tide gauges differ from site to site primarily due to land-level changes associated with tectonism, rebound, or subsidence and, in some areas, due to the extraction of oil or groundwater. Potentially there could also be variations due to oceanographic factors such as long-term

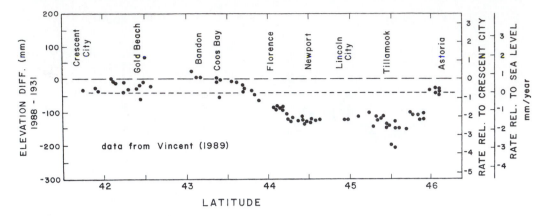

Figure 4-30   Geodetic releveling data along the length of the Oregon coast. The scale on the left gives elevation changes in reference to Crescent City, California, while the scales on the right place the changes in terms of rates, the last scale being a comparison with the eustatic rise in sea level. [Used with permission of Journal of Coastal Research, from P. D. Komar and S. M. Shih, Cliff Erosion Along the Oregon Coast: A Tectonic-Sea Level Imprint Plus Local Controls by Beach Processes, *Journal of Coastal Research 9*, p. 750. Copyright © 1993 Journal of Coastal Research.]

changes in offshore currents. This makes it difficult to decipher the tide-gauge records to establish the trends and processes responsible for the changes. In a series of papers covering various coastal areas of the world, Emery and Aubrey have used eigenfunction analysis to clarify the time and spatial variations of data from tide gauges; they have summarized their results in the text *Sea Levels, Land Levels, and Tide Gauges* (Emery and Aubrey, 1991). Eigenfunction analysis is an objective statistical technique for determining the dominant modes of variation within a data set; in the application to yearly sea levels measured on tide gauges, the approach focuses on the time variations but in particular on patterns of spatial variations throughout the area under study. An example of their results is shown in Figure 4-31 from analyses of tide gauge records along the coasts of Japan and Korea (Aubrey and Emery, 1986). Parts of Japan are sinking as fast as 24 mm/year, while others are rising at rates up to 6 mm/year. The pattern revealed by the analysis indicates that the island is tilting in response to tectonic and volcanic processes. Another result from their analyses is given in Figure 4-32, showing the pattern of annual relative sea-level change in gauges along the east coast of the United States (Aubrey and Emery, 1983). The eigenfunction analysis identified three coastal compartments where sea-level behavior is distinct, readily identified by the regression lines in Figure 4-32. The highest rates of relative sea-level rise are found in the Cape Hatteras area of North Carolina, systematically decreasing to both the north and south. The cause of this alongcoast pattern is uncertain. It does not appear to be tectonic in origin as it has not affected the depth of the shelf break. A third segment is identified north of Cape Cod, one that is interpreted to be associated with the rebound and subsidence produced by glaciers as discussed above for this area.

A number of studies, a few of which are listed in Table 4-2, have attempted to utilize the tide-gauge data to estimate the global rise in sea level, resulting from the continued melting of glaciers and thermal expansion of the ocean's water. These analyses do not utilize geodetic data to establish rates of land-level change, since such data are seldom available, but instead combine the tide-gauge data from many locations in the expectation that the land-level

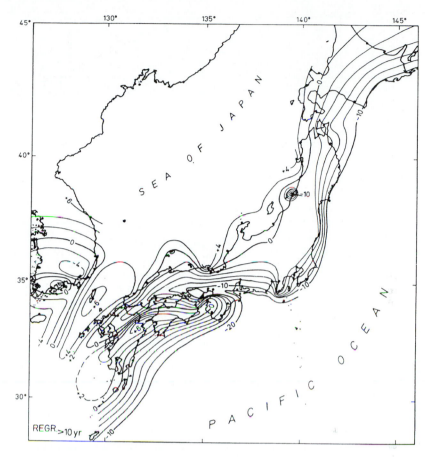

Figure 4-31 The average annual movement of the land relative to the sea, based on analyses of 86 tide-gauge records (stations shown as solid dots) in Japan and South Korea. Contours are at 2 mm/year intervals. [From *Geological Society of America Bulletin* 97, D. G. Aubrey and K. O. Emery. Reproduced with permission of the publisher, the Geological Society of America, Boulder, Colorado USA. Copyright © 1986.]

TABLE 4-2 GLOBAL EUSTATIC RISE IN SEA LEVEL DETERMINED FROM TIDE GAUGES BY VARIOUS STUDIES

Study	Rate (mm/year)
Gutenberg (1941)	1.1 ± 0.8
Hicks (1978)	1.5 ± 0.3
Emery (1980)	3.0
Gornitz et al. (1982)	1.2
Barnett (1984)	2.3 ± 0.2
Gornitz and Lebedeff (1987)	1.2 ± 0.3
Braatz and Aubrey (1987)	1.0 ± 0.1
Peltier and Tushingham (1989)	2.4 ± 0.9
Douglas (1991)	1.8 ± 0.1

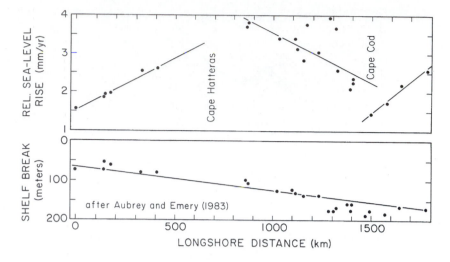

Figure 4-32 Results from the eigenfunction analysis of tide-gauge sea-level data for the east coast of the United States. Regression lines denote three main segments having different sea-level trends, patterns that are not reflected in depths at the shelf break. [From *Geological Society of America Bulletin 97*, D. G. Aubrey and K. O. Emery. Reproduced with permission of the publisher, the Geological Society of America, Boulder, Colorado USA. Copyright © 1986.]

changes will average out, leaving only the global rise in sea level. The results of the various studies differ mainly in which tide gauges are included in the analyses—more specifically, which are excluded due to their "anomalous" behavior. For example, Hicks (1978) eliminated data that are obviously affected by localized subsidence or emergence and then combined the data from the remaining 27 tidal stations within the United States in an attempt to obtain an average sea-level rise. The result for the period 1940–1975 indicates that sea level increased at an average rate of 1.5 ± 0.3 mm/year. Gornitz, Lebedeff, and Hansen (1982) analyzed tide-gauge stations from throughout the world, excluding those having records shorter than 20 years and stations in seismically active areas and where subsidence is locally strong. The remaining 193 stations were divided into 14 regions on the basis of geographic proximity, these regions were then analyzed to determine patterns of relative sea-level change, and finally the fourteen regions were combined to determine the average global pattern. The final result was a 1.2 mm/year rise, a 12-cm increase in sea level during the past century. Although the results from the series of studies listed in Table 4-2 have not established a precise value for the global rise in sea level, the cluster of values suggests that sea level has been rising at a rate of 1–2 mm/year, 10–20 cm per century, with a value of 1.5 mm/year representing a reasonable "best guess." Although these magnitudes may not at first seem particularly significant, they can result in shoreline retreat rates that are 50–100 times as great, so the rise is an important factor in the long-term erosion of many coasts.

The cause of this 1–2 mm/year rise in the mean level of the sea is uncertain. The main contributions are likely the melting of glaciers and the thermal expansion of the ocean's water. Meier (1984) has estimated that melting of mountain glaciers has contributed about 28 mm of water to the level of the sea during the period 1900–1960, representing a rate of about 0.5 mm/year. This could be an underestimate, as it is based on mass-balance data for only twenty-five glaciers, and regions that could have contributed most substantially to the sea-level rise are underrepresented in the data. The error limits are wide, but it seems likely that although the partial melting of small mountain glaciers has contributed significantly to the global rise in sea level, they can account for only roughly one-third of the observed rate of rise. This estimate does not include the contributions from melting of glaciers in Greenland and Antarctica. Malzer and Seckel (1976) suggest that the Greenland ice sheet is presently growing in the center and shrinking at its margins. The net change is likely small, placed by van der Veen (1988) at less than 0.01 percent of the total ice mass. Even a 1 percent decrease during the past century would have resulted in only a 6-cm rise in sea level, roughly half of the observed rise. Furthermore, the growth of the East Antarctic ice sheet may have counteracted any contribution from Greenland. According to Hollin (1970) and Allison (1979), the interior regions of East Antarctica are actually growing at a rate that corresponds to a drop in sea level of approximately 1 mm/year or 10 cm if extrapolated over the past century. The mass balance of the West Antarctic ice sheet is rather ambiguous, although it appears that the volume of ice has been increasing (Bentley, 1984). In balance, there appears to have been an increase in polar-ice volume during this century and therefore cannot account for the observed rise in sea level.

The thermal expansion of ocean water due to a small increase in average temperature is potentially an important factor in explaining the observed sea-level rise. Sea-water temperatures in the near surface above the thermocline appear to have increased by about 1 °C since early in the century (Barnett, 1983). However, thermal expansion resulting from a 1 °C temperature increase would account for only roughly 2 cm of the 15 cm observed sea-level rise during that time period.

It is apparent from these rough estimates that we are unable to positively account for the various contributions that have produced the observed 10–20-cm increase in sea level experienced during the last 100 years. In spite of this inability to reasonably account for the contributors of the observed sea-level rise, the tide-gauge records do indicate the importance of a climatic cause underlying the general increase. This is demonstrated by the longest records in existence. Tide records obtained in Amsterdam since 1682 give a much longer view of sea-level change than provided by most tide gauges that have been in operation for less than a century (Fairbridge, 1960). The North Sea basin, of which the Netherlands is a part, has been subsiding at a rate of about 10 cm/century. This means that only half of the 20 cm/century rate recorded by the Amsterdam gauge represents a true rise in sea level. The long record of sea-level change reveals that the fluctuations faithfully reflect climatic history. The level rose between 1725 and 1770 at about the present rate, then fell between 1800 and 1850 during a period of exceptionally cold winters. With the present warming spell, the sea level is again rising.

In summary, analyses of tide-gauge records reveal changes in the mean level of the sea ranging from daily fluctuations in response to meteorological events, to monthly variations due to seasonal cycles in water temperatures and other oceanic and atmospheric conditions. The tide-gauge records also demonstrate the existence of a long-term global rise in the level of the sea that is on the order of 15 cm/century, a rise that must be a response to climatic changes, continuing those that have occurred throughout the ice ages, but possibly also reflecting human-induced factors such as global warming due to the greenhouse effect.

LONG-TERM GLOBAL CHANGES IN SEA LEVEL

Much of the 10–20 cm/century rise in global sea level determined from tide gauges likely results from a continued melting of glaciers that has taken place during the last several thousand years with the waning of the ice ages. Drowned beaches, mastodon teeth found on the continental shelf (Whitmore et al., 1967), and submerged peat beds (Emery et al., 1967) all demonstrate that some 20,000 years ago, at the time of the maximum advance of glaciers, the sea surface stood some 100 m or more below its present level. During much of the Quaternary ice age, approximately the last 3 million years, a substantial portion of Earth's total supply of water was locked up within ice sheets covering large areas of the continents. The loss of this water from the oceans resulted in lowered sea levels, exposing what is now the continental shelf. During the last 20,000 years, the continental glaciers have slowly receded and the resulting rise in sea level has been a major force in shaping our coasts.

A timetable of changing sea levels has been obtained by dating material that has a known narrow relationship to past stands of the sea. The carbon-14 technique has been employed mainly for such dating. Carbon-14 has a half-life of 5,360 years and can provide reliable estimates of ages up to about 20,000 years BP, the reliability decreasing markedly for greater ages. Radioactive uranium and thorium can be used to date more ancient materials. The application of the carbon-14 technique to establish sea-level variations requires carbon material formed near some former level of the sea. The most commonly used indicators are shallow-water mollusks (Curray, 1960), salt-marsh and freshwater peat, coralline algae, hermatypic corals, beach rock, lagoonal oolites (Milliman and Emery, 1968), and fossil intertidal organisms (van Andel and Laborel, 1964). If these materials are found on land above the present sea level or underwater on the continental shelf and have not been disturbed subsequent to their formation, then dating provides an indication of the level of the sea at that time. It is apparent that such materials indicate only a relative sea level at that location, not an absolute (eustatic) global sea-level change, since in the meantime the level of that portion of Earth upon which the materials rest could also have changed. The likelihood of such land-level changes has been one of the major difficulties in determining the long-term global variations in sea levels. Thus, data from continental areas identified as being stable are preferred, while areas known to be tectonically active or undergoing isostatic rebound have, until recently, been avoided. Even then, much uncertainty remains, as no area can be considered truly stable.

Using such data, a number of investigators have developed chronologies of sea-level variations, some extending over the past 50,000 years or more (Shepard, 1963; Curray, 1965; Shepard and Curray, 1967; Milliman and Emery, 1968; Kraft, Biggs, and Halsey, 1973). Figure 4-33(a) shows the sea-level curve from Curray (1965), based on data derived from different areas of the world; Figure 4-33(b) from Shepard and Curray (1967) focuses on the sea-level transgression that has occurred during the last 7,000 years. The long-term curve indicates that sea level stood some 130 m lower than at present about 15,000–20,000 years BP, that there was a rapid rise (8 mm/year) until about 7,000 years BP and then a slowing in the rise to about 1.5 mm/year until the sea reached approximately its present level some 2,000 years BP. The uncertainty in such curves, reflected in the scatter of the original data, increases with age due to land-elevation changes and the increased probability of displacements of the dated material. The study of Dillon and Oldale (1978) in particular demonstrates the potential effects on the sea-level curves of elevation changes due to subsidence over the thousands of years involved in such graphs. They traced submerged shorelines along the continental shelf of the U.S. east coast, and found that north of New Jersey the strand lines are inflected downward to greater water depths, evidence for their greater subsidence since formation between 9,000 and 15,000 years BP. This subsidence was attributed to the relaxation of the crust after removal of the weight of the continental glaciers, the area farther to the north experiencing rebound while

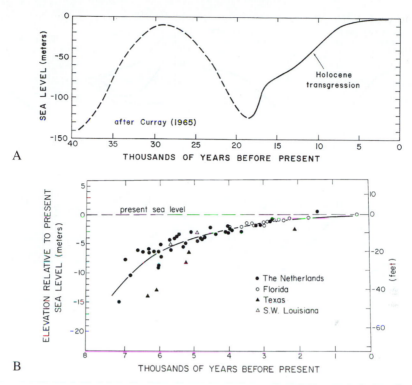

Figure 4-33 (a) Curve for variations in sea level during the past 40,000 years, based on carbon-14 dates compiled by Curray (1965). The dashed curve is estimated from only limited data. [Adapted with permission of Princeton University Press, from J. R. Curray, Late Quaternary History, Continental Shelfs of the United States, *The Quaternary of the United States.* Copyright © 1965 Princeton University Press.] (b) Data and sea-level curve for the past 8,000 years based on a compilation by Shepard and Curray (1967) of data from "stable" areas of the world. [Adapted with permission of Pergamon Press, from F. P. Shepard and J. R. Curray, *Progress in Oceanography.* Copyright © 1967 Pergamon Press.]

the continental shelf area south of the former ice sheets is subject to increased subsidence. The effect of this subsidence was to displace the dated shells, etc., used to establish sea-level curves, to erroneously greater depths. Based on their interpretation, Dillon and Oldale concluded that the low stand of sea level 15,000–20,000 years BP was never lower than 100 m, rather than reaching 130 m as given by the curve in Figure 4-33(a).

Further evidence for the effects of land-elevation changes on the Holocene transgression sea-level curves is provided by Hopley (1978), who compiled fifteen such curves derived from various parts of the world's coastlines, the results showing significant differences suggesting that such curves may be only regionally valid. This diversity of results is also indicated by the modeling results of Clark, Ferrell, and Peltier (1978), analyses that account for land-elevation changes that result from the removal of water from the ocean during the glacial advance and then its return to the ocean with subsequent melting. Based on the model calculations, six regions were identified characterized by different patterns of variation in relative sea-level rise during the last 10,000 years (Fig. 4-34). In Zones II, III, and IV, the sea-level transgression is much like that documented in Figure 4-33 based on data obtained

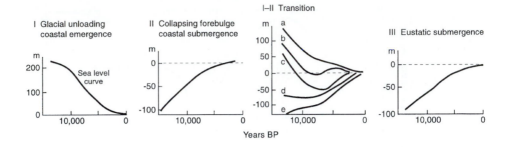

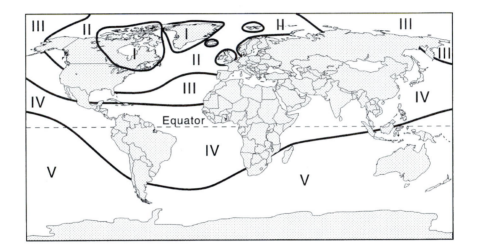

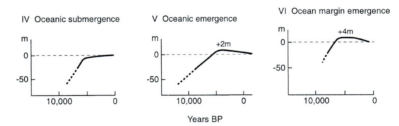

Figure 4-34   Model analyses for the relative sea-level changes expected in six regions of Earth in response to the melting of continental glaciers and isostatic and geoid changes. [Adapted with permission of Academic Press, from J. A. Clark, W. E. Farrell and W. R. Peltier, Global Changes in Post-Glacial Sea Level: A Numerical Calculation, *Quaternary Research 9*. Copyright © 1978 Academic Press.]

on continental shelves of those regions. Zone I is dominated by high rates of rebound, following removal of the ice cover. Most unusual are the results for Zones IV, V, and VI, where the model calculations of Clark, Farrell, and Peltier show either stable sea levels for approximately the last 5,000 years or sea levels that actually have been higher in the past.

Prior to the modeling analyses of Clark, Farrell, and Peltier (1978), there had been much controversy regarding evidence for higher stands of sea levels during the past few thousand years, with evidence offered by some investigators, while being discounted by others. For example, Block (1965) provided historical and archeological evidence for higher stands of sea level within the Mediterranean Sea, van Andel and Laborel (1964) found evidence on the coast of Brazil, and there were a number of accounts of raised beaches on islands in the South Pacific. In their review of the data that indicated higher sea levels, Shepard and Curray (1967) suggested that many of the dates showing high levels were incorrect and actually represented previous high stands in sea levels such as that which occurred some 30,000 years BP during the previous glacial interlude, shown by the dashed portion of the sea-level curve in Figure 4-33(a). They also noted that anomalous dates for terraces above the present level of the sea have resulted from using shells from Indian middens, shells transported by birds, and blocks of material thrown up by storms. This controversy was based on the view that the Holocene sea-level rise would be eustatic, that is, effectively the same everywhere except in tectonically active areas or where there has been appreciable glacial rebound. We see from the model analyses of Clark et al. that this assumption is invalid, and that his results actually predict higher sea levels in those areas where there is documented evidence for their occurrence during approximately the last 5,000 years.

The sea-level curve of Figure 4-33(a), based on dated material from continental shelves, shows a fairly smooth increase in sea level, the average rate having been approximately 8 mm/year until about 7,000 years BP after which it slowed. In contrast, based on documented advances and retreats of continental glaciers, Fairbridge (1961) produced a sea-level curve that showed many fluctuations within the general rise in sea level during the past 20,000 years, fluctuations that could not be documented from the continental shelf data due to its extreme scatter. Studies such as those of Carter, Carter, and Johnson (1986) did argue for an episodic rise in sea level, based on the presence of submerged shorelines and the general stratigraphy of shelf sediments. However, more definite confirmation was obtained from old, submerged coral reefs on Barbados (Fairbanks, 1989; Bard et al., 1990a, 1990b), New Guinea (Edwards et al., 1993), Tahiti (Bard et al., 1993), and throughout the Caribbean (Blanchon and Shaw, 1995). These studies show that at least twice during the period from 15,000 to 7,000 years BP, and possibly three times, the rate of sea-level rise was so great that corals forming the reefs could not grow fast enough to keep pace and became drowned out, to be replaced by new reefs in a more landward position once the rate of sea-level rise decreased. Focusing on corals having a known, narrow depth range where they live, dating of this carbonate material before and after each episode has provided estimates of the timing and magnitude of the sea-level rise. Blanchon and Shaw (1995) termed these episodes "catastrophic rise events," or CREs. Based on their detailed analyses of a number of reef sites, they determined that CRE1 started at about 14,200 years BP and involved a rise of some 13.5 ($\pm$2.5) m within just a few hundred years; CRE2 started at 11,500 years and had a magnitude of 7.5 ($\pm$2.5) m—these values are largely in agreement with those from the earlier studies cited above. The exact rate of sea-level rise of the CREs is unknown but must have exceeded 45 mm/year, the maximum possible upward growth rate of the corals. From this, the duration of CRE1 must have been less than 290 years, with the rate of rise having been greater than about 4.7 cm/year; CRE2 would have taken place in less than 140 years, with a rise of 5.4 cm/year. These rates are some 250–500 times the estimated 1–2 mm/year present rate of sea-level rise.

There is uncertainty as to the causes of these rather abrupt increases in sea level spanning a few centuries. Their origin is generally attributed to the destabilization and collapse of the continental ice sheets over Canada and northern Europe, but the sudden melting of ice in Antarctica may also have been important. There is direct evidence from deep-sea sediments for the release of huge fleets of icebergs into the North Atlantic, and the formation of a fresh-water layer on the ocean's surface that altered its overall circulation. Large glacial lakes were formed in Canada and Siberia, and then released to the ocean with the farther retreat of the ice that had acted as a dam. By 8,400 years BP, the two largest glacial lakes in northern Canada joined to form one lake with a surface area of some 700,000 km^2, and when it broke through into the Hudson Bay lowland, it is estimated that some 100,000 km^3 of lake water rapidly flowed into the sea, instantaneously raising the global sea level by 0.2–0.4 m (Dawson, 1992).

These catastrophic rises in sea level would have had profound effects on the coast, in addition to the drowning of coral reefs. Evidence indicates that during the periods of relatively slow rise in sea level between the CREs, the shoreline was sufficiently stable for a long enough time to build up appreciable beach ridges and other coastal forms, but like the coral reefs, these became drowned during the CREs and largely preserved on what is now the continental shelf. This would account for the presence of submerged beach ridges as investigated by Carter, Carter, and Johnson (1986) who had argued for episodic sea-level changes. Other interesting examples are provided by Fletcher and Sherman (1995), who relate the preservation of submerged wave-eroded notches and other features on the continental shelf of Hawaii to CREs, and the formation and preservation of ridge deposits on the south Florida margin at water depths ranging from 50 to 124 m, investigated by Locker et al. (1996).

Going still further into the past, there have been many cycles of glacial advances and retreats with accompanying changes in sea levels. Some of the interglacial high stands are represented by marine terraces and beach ridges high above the present sea. For example, Oaks and Coch (1963) have identified five depositional terraces or ridges on the Atlantic coastal plain of Virginia and North Carolina at levels ranging from 5 to 15 m above the sea, and correlated them with maximum sea-level elevations that have occurred since Miocene times. Elevated marine terraces can be found throughout the world. They are seen along almost the entire length of the U.S. Pacific coast and have been studied in the Mediterranean and along the English coast by Zeuner (1959) and in Australia by Gill (1961) and Jennings (1961). There is an enormous amount of literature on marine terraces; a list has been compiled by Richards and Fairbridge (1965) and has been reviewed by Guilcher (1969).

There has been some debate as to the relative heights of the successive interglacial sea levels and their relationship to the marine terraces. Along some coasts, "stairways" of terraces exist, with the oldest and most poorly preserved "step" being topmost and the youngest the lowest. Early investigators interpreted this as an indication that there has been a progressive lowering of successive high stands of the sea. Later workers recognized that a staircase of terraces generally results from the tectonic uplift of the land, a gentle uplift that preserves successively older terraces at higher levels. The terrace level would then be attributable to the combined effects of tectonic uplift and actual (eustatic) sea-level cycles. The best stairways of terraces are indeed associated with tectonically active areas such as the U.S. west coast and the Mediterranean.

Detailed analyses of terrace ages within stairways have allowed investigators to separate out the tectonic rise from the cycles of global sea levels. The earliest study of this type was that of Broecker et al. (1968), who utilized uplifted coral reefs on Barbados. That island is rising from the sea at a rate fast enough to separate coral reefs that formed at successive high stands of sea level, even when the stands were below the present level. Three distinct high stands of the sea represented by coral-reef terraces were dated by Broecker et al. at about 122,000, 103,000, and 82,000 years BP. Assuming a uniform rate of uplift, the stand at 122,000 years BP was about 6 m above the present level, in agreement with other investigations, while the next two high levels were each about 13 m below present sea level. This would explain why terraces of these two latter dates are not found in many areas. Further work on the uplifted Barbados coral reefs has been completed by Mesolella et al. (1969), Bender et al. (1979), and Radtke and Grün (1990). These latter studies extended the analysis, the so-called "Barbados model," to the more elevated coral reefs that are as much as 700,000 years old.

There has been a parallel series of studies of uplifted coral reefs in New Guinea (Veeh and Chappell, 1970; Chappell, 1974; Bloom et al., 1974; Chappel and Polach, 1976; Hearty and Aharon, 1988). Veeh and Chappell found dates of high stands that correspond to those found on Barbados. In addition, they found high stands at 180,000–190,000 years BP and at 35,000–50,000 years BP, as well as the present high stand commencing about 6,000 years ago. Again, only the high stand at about 120,000–140,000 years BP was above the present sea level. Chappell (1974) further documented evidence from New Guinea terraces. A well-developed series of coral terraces was investigated, rising to more than 600 m, progressively uplifted by tectonic activity. Terrace correlation and radiometric dating gave sea-level maxima at 30,000; 40,000– 50,000; 60,000; 80,000; 105,000; 120,000; 140,000; 185,000; and 220,000 years BP. Many of these agree with those from Barbados but others have been added. Using a least-squares search, a best-estimate sea-level curve and corresponding tectonic-uplift pattern were derived from eleven series of terrace elevations. According to the results in Figure 4-35, most of the inter-glacial high-water levels and their associated terraces stood below the present sea level, even the 30,000–40,000 years BP level generally considered to have been close to the present sea level. The present elevations of these terraces above the sea are therefore due to tectonic uplift. Figure 4-35 correlates the interglacial sea-level highs with Emiliani's (1970) paleotemperature curve, which is inferred from the oxygen isotopes contained within deep-sea foraminifera shells. The correlation is convincing, showing the expected parallel variations in sea level and temperatures. More recent comparisons of this type have been undertaken by Chappell and Shackleton (1986) and Tooley (1993), again showing strong correlations.

For obvious reasons, not as much information is available on low stands of the sea. The evidence comes from nickpoints and terraces observed in depth soundings and during submersible dives near the outer shelf and upper slope, and from submerged coral reefs. Worldwide occurrences of certain terrace levels indicate a connection with sea level rather than with any local structural control. Terraces are fairly common at depths of 130 m, which may

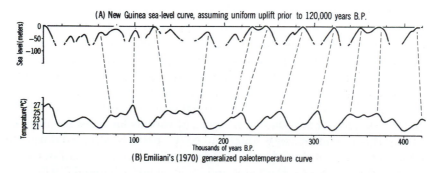

Figure 4-35 Maxima in sea levels based on uplifted coral-reef terraces in New Guinea, compared with the paleotemperature curve of Emiliani (1970). [From *Geological Society of America Bulletin* 85, J. Chappell. Reproduced with permission of the publisher, the Geological Society of America, Boulder, Colorado USA Copyright © 1974.]

correspond to the last low stand of the sea about 15,000–20,000 years BP [see Jongsma (1970), for example]. Extending the Barbados work to submerged coral reefs, Fairbanks (1989) dated corals at various levels from sixteen drilled cores, determining that the sea was 121 ± 5 m below the present level during the last glacial maximum and obtained a curve for the sea-level rise up to the present that is very similar to Figure 4-33(a) established by Curray (1965). In northern Australia, van Andel and Veevers (1967) found terraces at 170–175 m depth and at about 200 m, which were interpreted as having resulted from lowered sea levels. Veeh and Veevers (1970) dated the 175-m-deep terrace in eastern Australia at 13,600–17,000 years BP, suggesting that sea level dropped to below the 130-m depth usually considered. Jongsma (1970) dated the 200-m deep terrace at an age of 170,000 years BP or greater. Schwartz (1972) concluded that seamounts in the North Pacific show evidence for a maximum lowering of the sea to between 220 and 250 m. Pratt and Dill (1974) also provided evidence for very low glacial sea levels derived from the Bering shelf and southern Australia. Echosounding records show a continuous terrace at both locations at a depth of 240 m. In southern Australia there is also an indication of a terrace at the 200 m depth. Pratt and Dill reviewed evidence from other parts of the world for such considerable low stands of the sea.

One effect of these Quaternary sea-level changes has been a series of rapid transgressions and regressions of the shore. Each regression took the shore to approximately the edge of the present continental shelf or possibly deeper. This condition for the area of the North Sea is illustrated in Figure 4-36, showing a wide expanse of land that is now under the sea. Each successive transgression caused the shoreline to migrate across the continental shelf. It is generally considered that the broad, relatively flat coastal plains and continental shelves have resulted from this series of transgressions (Curray, 1969). Both erosion and sediment deposition are concentrated in the nearshore, and the repeated migration of this region across the continental margin acted to bevel it off to a nearly uniform inclined plane.

In these discussions, we have considered that the sea-level variations were produced by the advance and retreat of glaciers, and this is largely the case. Water-temperature changes throughout the Quaternary amounted on average to some 5 °C, a change that would account for only about a 10 m variation in sea level due to the associated thermal expansion and contraction—small (but not negligible) compared with the observed change. Factors other than glacial advances and retreats, therefore, have had only a small influence on the level of the sea, at least over spans of time on the order of hundreds of thousands to millions of years. This is confirmed by the correspondence between the glacial stages on the continents, ice cores obtained in Greenland and Antarctica, and the history of sea-level change. Each time the glaciers advanced, the sea level dropped. The climatic changes that produced the glacial advances and retreats are believed to have been due to variations in the amount of solar radiation striking Earth. This amount changes in a systematic fashion because of the tilting of Earth's axis, precession of the equinoxes, and the variation in the eccentricity of Earth's orbit. These variations have periods of 41,000, 21,000, and 90,000 years, respectively. Milankovitch (1938) calculated the fluctuations in radiation received at various latitudes for the past million years and showed that a sequence of warmer and colder periods occurred during that time. These variations have been correlated with interglacial and glacial phases. For example, the studies by Broecker et al. (1968) on Barbados and by Veeh and Chappell (1970) in New Guinea of interglacial high stands of the sea support the Milankovitch hypothesis of astronomical controls over the ice age periodicity. Although the Milankovitch hypothesis seems to explain the periodicity, it does not offer an explanation for the actual onset of glaciation.

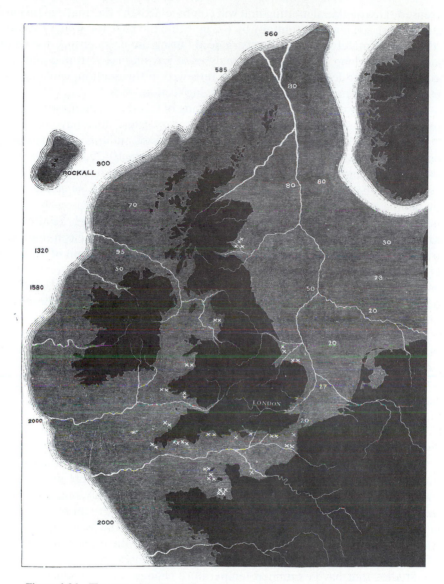

Figure 4-36   The area of the North Sea now covered by water, depicted as land with river drainages that existed during low stands of sea level such as occurred about 20,000 years BP.

GREENHOUSE WARMING AND FUTURE LEVELS OF THE SEA

Of considerable concern to society is the on-going rise in the level of the sea and what elevations it might achieve during the twenty-first century. We have become aware of the potential peril of greenhouse warming, a global increase in average temperatures that may already be underway due to carbon dioxide and other gases pumped into the atmosphere by human activities. In addition to altering the global climate, it is predicted that there will be increased rates of glacial melting that will combine with the thermal expansion of the ocean's water to raise sea levels potentially by 1–2 m within the next 100 years. This rate, far in excess of the approximately 15 cm rise experienced during the past century, would obviously have major impacts along our coasts. The analyses of greenhouse warming and scenarios for the accompanying sea-level increase will be reviewed in this section, while analyses of the potential impacts are considered in the following section.

The burning of fossil fuels and the clearing of forests has increased atmospheric carbon dioxide by about 15 percent during the past century. There must have been a major input of CO_2 after about 1850 from the clearing of hundreds of millions of hectacres of forest to increase agricultural land areas. Forests incorporate some 10–20 times more carbon per unit area than cropland or pastureland. This is one of the concerns regarding the continued deforestation of Earth, in particular of the rain forests. In addition, since the Industrial Revolution, people have been burning increased quantities of fossil fuels such as coal, oil, and gas. One by-product is carbon dioxide, which is reentering the atmosphere after having been tied up in fossil materials for millions of years. The first definitive measurements demonstrating the increase in CO_2 in the atmosphere were obtained on the summit of Mauna Loa in Hawaii and at the South Pole station of the U.S. Antarctic Program, sites chosen because they are free from local contamination (Keeling and Bacastow, 1977; Keeling et al., 1989). The measurements show a progressive increase in the CO_2 content of the atmosphere over the decades and further demonstrate that the rate of rise is increasing with time.

Carbon dioxide and other gases (including nitrous oxide, chlorofluorocarbons, and methane), introduced into the atmosphere by natural processes and human activities, are known as *greenhouse gases* in that, while they allow the warming rays of the sun to reach Earth, they block the excess heat that would normally reradiate back out to space. Carbon dioxide and other greenhouse gases alter the heat balance of Earth by acting as a one-way window, much like the effect of glass windows in a greenhouse. The CO_2 is transparent to radiation at visible wavelengths, so the light from the sun is able to pass through the atmosphere to warm Earth and the oceans. The heated ground and water reradiate longer wavelengths, infrared radiation that is blocked by the CO_2 (and by greenhouse windows), preventing its escape back into space. This is the greenhouse effect. If there were no carbon dioxide in the atmosphere, heat would escape from Earth much more readily, and the oceans would be a solid mass of ice. Venus provides an example of an extreme greenhouse effect where there is too much CO_2, raising the surface temperature to 400 °C. The increase in the amount of CO_2 in the atmosphere due to human activities acts to close the atmospheric infrared "window," and presumably will raise globally averaged temperatures on Earth.

Studies that have examined temperature changes have produced roughly similar results, indicating that there has been a general warming of Earth during the past century (Hansen and Lebedeff, 1988; Jones, 1988). Figure 4-37, from Hansen and Lebedeff, shows a significant increase in temperature from 1880 to 1940, an increase of about 0.5 °C. After 1940 temperatures decreased slightly until 1970, only to make another significant jump so that the total increase this century has been on the order of 0.7 °C. The causes of these changes are uncertain, and some investigators have argued that they may not be real. Warming due to increased greenhouse gases should be more gradual than observed. It is possible that the measured temperature increase is natural, being part of the global warming that has occurred during the past century as Earth emerged from the Little Ice Age, an unusually cold period during the sixteenth and seventeenth centuries when most mountain glaciers advanced. The "urban-island" effect may also be a significant factor, as most temperature records come from

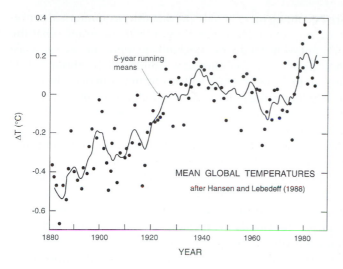

MEAN GLOBAL TEMPERATURES
after Hansen and Lebedeff (1988)

5-year running means

Figure 4-37 Changes in global temperatures between 1880 and 1987. [Used with permission of American Geophysical Union, from J. Hansen and S. Lebedeff, Global Surface Air Temperatures: Update through 1987, *Geophysical Research Letters 15,* p. 323. Copyright © 1988 American Geophysical Union.]

cities that can create their own localized climate. Records range from 0.1 to 0.5 °C temperature increases per century, depending on the size of the city, suggesting that the urban-island effect does have a significant impact on the temperature record. Such effects contribute to the uncertainty in establishing how much of the temperature increase is caused by normal climatic warming versus warming due to increases in the CO_2 content of the atmosphere. Although many studies have used the parallel increases in atmospheric CO_2 and measured temperatures as evidence that greenhouse warming is underway, the magnitude of the warming and its possible causes are still poorly established.

Further evidence comes from increasing water temperatures of the oceans (Barnett, 1983; Jones, Wigley, and Wright 1986), where sea-surface temperatures appear to have increased by about 1 °C since early in the century. However, it is possible that the recorded increase is again artificial and can be attributed to improved instrumentation and sampling; the database for ocean temperatures is handicapped by poor spatial and temporal coverage, and by various types of systematic biases such as changing methods in making the measurements. Furthermore, as noted earlier, the 1 °C temperature increase and associated thermal expansion can account for only roughly 2 cm of the 15 cm observed sea-level increase that has occurred during this century.

As we continue to use fossil fuels and cut down forests, more and more CO_2 will be added to the atmosphere, with the potential for increased greenhouse warming during the next century, and perhaps resulting in accelerated increases in the global level of the sea. By projecting future uses of fossil fuels and continued deforestation, a committee of the U.S. National Research Council concluded in 1983 that carbon dioxide levels will probably double by late in the next century, causing an increase in the average Earth temperature between 1.5 and 4.5 °C. During the past decade, increasingly sophisticated models have been developed to examine climate changes, largely in response to the growing concerns regarding accelerated greenhouse warming (Schlesinger, Gates, and Han, 1985; Hansen et al., 1988). Models can take several forms, from purely empirical to statistical or to dynamical or mixtures thereof. In some analyses, crude ocean models have been coupled with atmospheric models. Most model studies to date involve a calibration for the present-day conditions and then are run for various altered states of the atmosphere that might occur in the future. Certain aspects of the global system of ocean and atmosphere are becoming increasingly predictable, as numerical representations of the driving processes are improved. However, there is little assurance that the predictions are correct due to uncertainties in feedback mechanisms (cloudiness, snow-ice albedo, etc.). The models suggest that the average greenhouse warming will be about 2–4 °C by the second half of the next century but that this increase will not be uniform over the globe. The increases are expected to be much higher in the polar regions, about three times the global average. These models demonstrate the complexity of Earth's atmosphere/ocean system, indicating how difficult it is to make predictions regarding climate changes and the resulting potential increases in sea level. The increase in CO_2 and resulting elevated temperatures would generate more water vapor in the atmosphere and the clouds would move to higher levels, both positive feedbacks that could increase the greenhouse effect and further increase warming. The warming would be expected to melt snow and ice, reducing Earth's albedo, its reflectivity. More solar radiation would then be absorbed, leading to a further increase in temperature. In the opposite direction, the greater cloud cover would be expected to reflect more incoming solar radiation and may limit a further temperature increase.

One of the major uncertainties in the models is the interaction between the atmosphere and oceans. The ocean acts as a giant regulator, not only of CO_2 but also of climate, and thus occupies a central role in the analyses. The absorption of CO_2 into the ocean is a slow process, so that much of the CO_2 introduced into the atmosphere will remain there for a considerable period of time; each year the ocean on average takes up an amount of CO_2 approximately equal to 40 percent of the fossil fuel CO_2 added to the atmosphere. This makes possible the progressive rise in CO_2 in the atmosphere, leading to the potential increase in temperatures resulting from greenhouse warming. It appears that the last glaciation was accompanied by massive CO_2 transfers into and out of the oceans, the cause, consequences, and explanation of which are still poorly understood but demonstrate the importance of the exchange between the atmosphere and oceans.

The ocean can be represented as a two-layer system: a well-mixed surface layer with an average depth of about 100 m, while below is the cold water of the deep ocean. The exchange of heat between these two layers is very slow, so initially only the near-surface mixed layer would be heated as the climate becomes warmer. The increase in sea level (Δh) is related to the change of temperature (ΔT) within the thickness (D) of the upper mixed layer of the ocean according to the relationship:

$$\Delta h = \alpha D \Delta T \tag{4.5}$$

where α is the coefficient of thermal expansion (Wyrtki, 1990). The value of α depends strongly on the water temperature, varying from 0.0008 $(C°)^{-1}$ near freezing temperatures to 0.0034 at a temperature of 30 °C. As a consequence of this variability, a temperature increase in the tropics potentially has more effect on sea-level rise than a comparable temperature increase in polar waters. On the other hand, the mixed-layer thickness D affected by a temperature increase will be thinner in the tropics and thickest at high latitudes, and the atmosphere/ocean models indicate that the temperature increase will be greatest in the polar regions. Table 4-3 (from Wyrtki) assesses these relative factors and the resulting sea-level rise to be expected during the next century. It is seen that the effects of thermal expansion leading to a sea-level rise are greatest in the polar regions, while being minimal in the tropics.

An expansion of the tropics due to greenhouse warming would decrease the impact of thermal expansion but may lead to an increase in the frequency and intensity of tropical

TABLE 4-3 ESTIMATES OF SEA-LEVEL INCREASE, Δh, RESULTING FROM THE THERMAL EXPANSION OF THE MIXED LAYER OF THE OCEAN [Adapted with permission of University Press of Hawaii, from K. Wyrtki, Sea Level Rise: The Facts and the Future, *Pacific Science 44.* Copyright © 1990 University Press of Hawaii.]

Region	Temp (°C)	ΔT (°C)	α	D(m)	Δh (cm)
Tropics	30	1	0.0034	100	3.4
Mid-latitudes	10	3	0.0017	400	20.0
Polar regions	0	4	0.0008	1,000	32.0

storms and hurricanes. One can estimate that a 10 percent increase in the strength of winds over the subtropical gyre would increase the circulation by a corresponding amount (Wyrtki, 1990). This increased circulation would raise sea level in the center of the gyre, while the levels would drop along its periphery, that is along the coastlines of the continents.

Accounting for potential increases of glacial melting due to greenhouse warming is difficult, particularly so in view of our poor understanding of the present-day changes in the mass balances of ice within glaciers. As discussed earlier, Meier (1984) found that the melting of alpine glaciers accounted for a 2.8 cm rise in sea level during the period 1900–1961, while at the same time the air temperature rose by about 0.5 °C. On this basis, he suggested that when air temperatures increase by 1.5–4.5 °C during the next century, sea level will rise 8–25 cm due to the increased melting of mountain glaciers. More of a problem is what the great ice sheets on Greenland and Antarctica will do, with some analyses concluding that these ice sheets will actually grow and therefore extract water from the ocean and lower sea levels, while other studies predict that there will be enhanced melting that contributes to the sea-level rise. The loss of ice due to melting of the Antarctic ice sheet is likely to increase due to greenhouse warming, but this loss could be compensated for by the expected increase of snowfall in the interior. At present, snowfall is restricted because the cold air contains very little water vapor. But when the air becomes warmer, it will be able to carry more moisture to the inland regions so that snow accumulation should increase. Thus, the effect of climatic warming could lead to the growth of the Antarctic ice sheet.

Particularly significant is the possibility of the melting of the unstable West Antarctic ice sheet, which many glaciologists believe is unstable because much of it is below sea level (Clark and Lingle, 1977; Mercer, 1978). Its melting would produce a sea-level rise of 5–6 m within a relatively short period of time, an increase that would inundate much of our coast (for example, half of Florida would be covered by the sea). It is possible that the higher level of the sea 120,000 years BP differed from the present condition in the breakup of the West Antarctic ice sheet. If all of the remaining glaciers were to melt, sea level would rise by some 60 m, covering New York City and most coastal areas (Donn, Farrand, and Ewing, 1962), but this melting would require many thousands of years.

Such analyses serve as the basis for many of the assessments that have been made in predictions of potential sea levels during the next century resulting from greenhouse warming. Several of these predictions are graphed in Figure 4-38 as to evaluated levels at various times during the century. All predictions involve an acceleration of sea-level rise compared with the rate experienced during the last century, indicated by the dashed line in the graph. Analyses by the Environmental Protection Agency in the United States (Hoffman, Keyes, and Titus, 1983) attempted to first project the extent of greenhouse warming into the future, and then based on that prediction, estimated that sea-level rises of 50–340 cm could be expected by the year 2100, equivalent to average rates of 5 to 30 mm/year. These are the

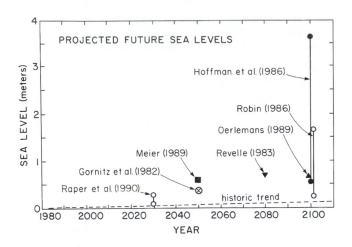

Figure 4-38 Predicted sea levels for the twenty-first century that have been made by various investigators. Many of the predictions span large ranges due to uncertainties in the analyses. The dashed line represents a continuation of the 1.5 mm/year sea-level rise that has been measured during this century.

greatest rates that have been predicted by the various studies and are large since they include the partial melting of the West Antarctic ice sheet. The Committee of the National Research Council on Carbon Dioxide Assessment suggests a rate of 7 mm/year by the year 2100 (Revelle, 1983). The recent estimates by van der Veen (1988) are lower, 2.8–6.6 mm/year by 2085 AD, rates that are still about 2–4 times greater than the 1–2 mm/year rise that has prevailed during the past 100 years. The assessments for the various contributing processes, as evaluated by van der Veen and in the previous analyses of Revelle (1983) and Hoffman, Wells, and Titus (1986) are summarized in Table 4-4; a more thorough review of analyses of this type is provided by Warrick (1993). The large ranges of values are a reflection of the considerable uncertainties in each individual assessment, leading to a very large uncertainty in the final predicted sea-level rise. Hoffman et al. (1986) include the disintegration of the West Antarctica ice sheet, and if that potential contribution is removed, the results obtained by the series of studies are somewhat closer in their total assessments.

Predictions of greenhouse-related sea-level rise remain highly uncertain, though all studies agree that an acceleration beyond the present rate is likely. In general, the projections have decreased in more recent estimates compared with those made a decade ago, resulting from refinements in the analyses. The most recent "official" projection made in December 1995 by the International Project on Climate Change (IPCC) is a best estimate sea-level increase of 0.46 m by the year 2100, with the uncertainty range extending from 0.15 to 0.95 m.

TABLE 4-4   ASSESSMENTS OF CONTRIBUTING FACTORS TO A SEA-LEVEL RISE DURING THE NEXT CENTURY RESULTING FROM GREENHOUSE WARMING

	Revelle (1983)	Hoffman et al. (1986)	van der Veen (1988)
Year in which the atmos. CO_2 doubled	2060	2056	2085
Associated global warming	3 °C	2–4 °C	2–4 °C
Base trend	17		15
Thermal expansion of ocean	30	15–34	8–16
Melting of mountain glaciers	12	7–18	10–25
Greenland ice balance	12	2–7	0–10
Antarctica response		6–43	−5–0
Total (net rates)	71	30–102	28–66

These projections are about 25 percent lower than those made by IPCC in 1990. The uncertainties are large, and the processes are poorly understood, so that it is not possible to reasonably assess what the future holds in the way of increased levels of the sea as a result of potential greenhouse warming.

Several studies have attempted to determine whether it is possible to identify in the tide-gauge records any increase in the rate of sea-level rise (an acceleration) that might be attributed to greenhouse warming. Emery (1980) concluded that the available data do indicate an accelerated rise, finding higher rates of sea-level rise for the decade of the 1970s compared with rates found by earlier studies using pre-1970 data. He recognized the implications regarding global warming but was cautious in stating that it is too soon to conclude whether the rise is related to the greenhouse effect. Analyses by Gornitz and Lebedeff (1987) of global sea level suggest an inflection of the trend in the mid-1930s, with a higher rate since that time. However, Hicks and Hickman (1988) concluded that rates for 1940–1962 versus 1962–1986 depend on regional groupings of tide gauges and show no consistent acceleration. The consensus seems to be that noise in the data precludes confident conclusions with regard to whether global sea level is rising at accelerated rates in recent decades (Wyrtki, 1990). Barnett (1984) concluded that the detection of a response in sea level associated with greenhouse warming will be difficult due to the large, natural variability in the data produced by glacial and tectonic processes, and many more years of measurements will be required before this is possible. Thus, we may be well into the twenty-first century before we can establish whether there is an accelerated rise in sea level due to greenhouse warming, and before we can establish with any degree of certainty what those elevated levels might be.

In summary, we are inadvertently conducting a great geophysical experiment as a result of our introduction of greenhouse gases into the atmosphere. About the only certain facts are the actual measurements of atmospheric carbon dioxide and some fairly reliable data on the rate at which we are using fossil fuels and cutting forests. The problem is otherwise obscured by many unknowns and uncertainties. There appears to have been increases in temperatures of the atmosphere and ocean during this century, but it cannot be affirmed that these are responses to the greenhouse gases. We are unable to quantitatively account for the observed rise in sea level during this century with respect to contributions by the melting of glaciers and the thermal expansion of the ocean's water due to the temperature increase. This is a rather unsatisfactory basis upon which to predict sea-level increases during the twenty-first century and accounts for the great disparity in the estimates made by various studies. The level of the sea is likely to increase during the next 100 years, but we cannot be certain whether it will be another 10–20 cm as experienced during this century or whether it might be 100–200 cm as predicted by some as a response to greenhouse warming.

THE COASTAL RESPONSE

The results of the on-going increase in sea level and the anticipated impacts of accelerated increases due to greenhouse warming are the inundation of low-lying lands and the erosion of our shores. The gradual rise in mean sea level will progressively raise the zone of coastal flooding, so areas formerly considered safe will be affected by extreme tidal elevations and storm surge. The effects of inundation will be most immediate along low-lying coasts. The city of Venice, built on a series of lagoonal islands, is already being adversely affected by the progressively higher elevations of storm surges brought on by increasing mean levels of the sea, enhanced by local subsidence resulting from groundwater extraction (Pirazzoli, 1991). Several times each year, the city becomes submerged by tens of centimeters of lagoonal water (Fig. 4-39). The greatest impacts due to simple inundation can be expected in areas of dense populations living on low deltaic plains, the nation of Bangladesh being a prime example. The low-relief atoll islands could disappear entirely beneath the rising sea; Roy and Connell (1991) have developed models for the expected response of atolls to accelerated rates of sea-level rise. In the Netherlands, more than 7 million people live in areas that are already below sea level, protected only by dikes, so that scenarios for water-level increases due to greenhouse warming are of immediate concern.

In addition to simple inundation by the increased sea level, it is anticipated that there will be enhanced rates of shoreline erosion. This expectation has resulted in renewed focus on developing models that attempt to predict shoreline recession rates as a function of the water-level increase. Best known is the model developed by Bruun (1962, 1988) that simply involves an upward and landward translation of the equilibrium beach profile, while con-

Figure 4-39   San Marco Square in Venice flooded by a storm surge in the north Adriatic Sea during October 5, 1992. At the time, the 23rd International Coastal Engineering Conference was being held in the city. A description of the weather conditions and tide levels can be found in the proceedings (pp. 3496–3503) of the conference. A low-pressure center was located directly over Italy, so that strong winds blew northward along the length of the Adriatic, creating the storm surge. During its peak, San Marco Square was covered by 46 cm of water, and over 50 percent of the city was flooded. This water level is not unusual, as during the years 1980–1989, it was exceeded twelve times and its return period is close to 1 year. The most extreme flooding occurred on November 4, 1966, when the water depth in San Marco Square was 114 cm. Of concern is that increasing levels of the sea, together with the continued subsidence of the city, will greatly exacerbate the problem during the next century.

serving the sand volume. Several investigators have modified the original Bruun model to incorporate other components in the budget of sediments that can affect the total amount of sand within the migrating beach. A similar model, considering the landward migration of an entire barrier island, has been developed by Dean and Maurmeyer (1983). The objective of this section is to examine these models and to consider some of the studies that have attempted to test them under laboratory and field conditions. The review presented here is a summary of the longer, more detailed consideration given this topic by Komar et al. (1991).

The first and best-known of the models that relate shoreline retreat to an increase in sea level is that proposed by Bruun in 1962. Bruun (1988) provides a recent rederivation of the relationships, as well as a discussion of the assumptions involved in the model and its uses and misuses. This model is still the most commonly used, particularly in management analyses of potential shoreline changes resulting from increased sea levels. The basics of the Bruun model is illustrated in Figure 4-40. The model assumes that with a rise in sea level, the equilibrium profile of the beach and shallow offshore moves upward and landward. The analysis is two dimensional and assumes: (1) the upper beach is eroded due to the landward translation of the profile; (2) the material eroded from the upper beach is transported immediately offshore and deposited, such that the volume eroded is equal to the volume deposited; and (3) the rise in the nearshore bottom as a result of this deposition is equal to the rise in sea level, thus maintaining a constant water depth in the offshore. Following these assumptions, Bruun derived the basic relationship for the shoreline retreat rate, R, due to an increase in sea level, S, obtaining

$$R = \frac{L_*}{B + h_*} S \tag{4.6}$$

where L_* is the cross-shore distance to the water depth h_*, taken by Bruun as the depth to which nearshore sediments exist (as opposed to finer-grained continental shelf sediments). Those parameters are illustrated in Figure 4-40, where it is apparent that the depth h_* is that required to ensure sediment continuity, that the two-dimensional volume of sand deposited in the offshore equals the volume eroded from the upper portion of the beach profile. The vertical dimension B in Equation (4.6) represents the berm height or other elevation estimate of the eroded area that may include dunes backing the beach. The relationship can also be expressed as

$$R = \frac{1}{\tan \theta} S \tag{4.7}$$

where $\tan\theta = (B + h_*)/L_*$ is the average slope of the profile along the cross-shore width L_*. In that $\tan \theta \approx 0.01–0.02$ is the range of average slopes for many coastal sites, Equation (4.7) yields $R = 50S–100S$, proportionalities that are commonly used as a "rule of thumb" to

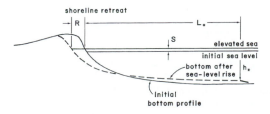

Figure 4-40   The net change in beach profile position due to a rise in sea level, S, according to the Bruun model, resulting in a zone of offshore deposition and erosion of the upper beach, with an overall recession rate, R.

calculate expected shoreline retreat rates or distances (R) from a rise in sea level (S). In this form, the relationship in particular demonstrates that a small increase in mean sea level can be expected to result in a substantial shoreline retreat.

Dean and Maurmeyer (1983) have generalized the Bruun rule to account for the landward and upward migration of an entire barrier island. This is depicted in Figure 4-41, where it is assumed that the barrier island accretes vertically at the same rate as the rise in sea level. Their derived relationship, with minor corrections (Dean, 1991), is

$$R = \frac{L_{*o} + W + L_{*L}}{h_{*o} - h_{*L}} S \qquad (4.8)$$

where L_{*o} and L_{*L} represent the widths of the active nearshore zones on the ocean and lagoon sides, and h_{*o} and h_{*L} are the associated water depths. W is the width of the barrier island (Fig. 4-41) that is considered to remain constant. This equation reduces to the basic Bruun rule [Equation (4.6)] for the case of no deposition on the barrier island or within the lagoon (i.e., when W, L_{*L}, and $h_{*L} = 0$). Equation (4.8) for an entire barrier island always predicts a greater retreat rate R than does the Bruun model, because sand is added to the island to maintain its vertical position relative to sea level and also to the lagoon side to maintain the width of the island. Furthermore, the net vertical dimension $h_{*o} - h_{*L}$ contributing sand during the island retreat is reduced compared with the Bruun rule, which considers h_{*o} alone, also leading to a higher calculated retreat.

All of the above analyses are two-dimensional treatments that conserve the quantity of sand within the cross-shore profile. The investigators were aware of this assumption and provide some discussion of potential longshore movements of sand that might affect the cross-shore balance. Such a consideration involves the development of a budget of sediments for the beach section being analyzed, with various potential sand gains and losses that can alter the total sand volume within the profile. The barrier-island model of Dean and Maurmeyer (1983) already introduces two-dimensional components of the sediment budget in having accounted for island overwash and inlet processes removing sediments from the ocean beach. Considering the third dimension in the longshore, Hands (1980, 1983) and Dean and Maurmeyer (1983) produced a modified Bruun rule that accounts for longshore sediment transport into and out of the beach section being analyzed, and also accounts for some potential losses of sediment offshore in cases where the eroded material is too fine to remain on the beach.

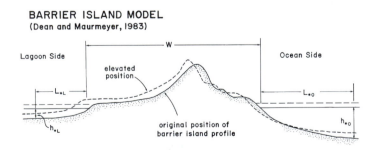

BARRIER ISLAND MODEL
(Dean and Maurmeyer, 1983)

Figure 4-41 Model to account for the response of an entire barrier island to a rise in sea level. [Adapted with permission of CRC Press, from R. G. Dean and E. M. Maurmeyer, Models for Beach Profile Response, *Handbook of Coastal Processes and Erosion*, p. 159. Copyright © 1983 CRC Press.]

A basic assumption of the Bruun-type models is the existence of an equilibrium beach profile and that this profile is maintained or eventually achieved following a change in water level. It is clear from the derivations and accompanying discussions that the focus is a long-term equilibrium involving decades of profile adjustments to rising sea levels. With that long view inherent in applications of the models, the occurrence of temporary profile fluctuations due to storms and other factors generally can be ignored.

Nearly all studies involving data collection to test models relating shoreline erosion to a sea-level increase have focused on the Bruun model. The first tests involved laboratory wave-flume investigations conducted by Schwartz (1965, 1967). The tests consisted of generating waves until an equilibrium profile developed, followed by an increase in water level and a renewal of wave activity until a new equilibrium was achieved. Water-level increases ranged from 1 to 6 cm. It was found that with a rise in water level, there is a shoreward displacement of the entire beach profile, with the upper beach eroding while deposition occurs on the adjacent offshore bottom; these changes conform with the profile modifications hypothesized by Bruun (1962). Although the wave-flume studies of Schwartz demonstrated an overall profile shift due to a rise in the mean-water level, no comparisons were made between observed shoreline recession rates and those predicted by the Bruun model, Equations (4.6) and (4.7).

The most straightforward field test of the equations derived by Bruun (1962) would be to compare predicted versus observed shoreline recession rates at various coastal sites, employing the closest tide gauge to determine the local relative change in sea level. The graph of Figure 4-42 from Dean (1990) shows the results of such a comparison involving statewide averaged erosion rates versus the average sea-level rise determined from tide gauges within the boundaries of the state. The data scatter is considerable, and accretion is found in three states (New York, Delaware, and Georgia) even though relative sea level has been rising. The rule of thumb proportionalities $R = 50S$ and $R = 100S$ from the Bruun relationships are seen to be poor predictors of statewide erosion. In a similar analysis, but confined to the shores of Chesapeake Bay, Rosen (1978) evaluated the Bruun equations as predictors of shoreline recession. Within that 336-km length of shoreline, there is a considerable variation in sea-level rise recorded from tide gauges, ranging from −0.46 to +5.43 mm/year. For the entire bay, the measured erosion differed by only 3 percent from that predicted by Bruun's Equation

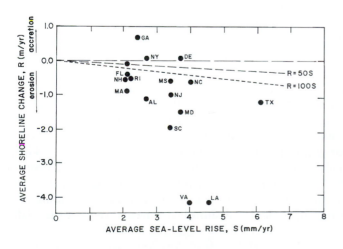

Figure 4-42 State-wide rates of shoreline erosion or accretion versus averaged local relative sea-level changes. [Adapted with permission of John Wiley & Sons Inc., from R. G. Dean, Beach Response to Sea Level Change, *Ocean Engineering Science* v. 9, p. 885. Copyright © 1990 John Wiley & Sons.]

(4.6) for a 100-year time span. However, on the eastern shore the difference was +58 percent and on the western shore it was −7 percent (the positive percentage indicates that the predicted erosion exceeded the measured). Individual counties ranged from +224 percent to −68 percent, demonstrating that the Bruun equations can be seriously in error when used for site-specific estimates.

It is unclear what has caused the large scatter seen in Figure 4-42 for the state-by-state comparison of Dean (1990) and in the analyses of Rosen (1978) in Chesapeake Bay. It likely results from other factors in the overall budget of sediments that could create considerable variability in shoreline changes beyond those predicted as a response to an increase in sea level and would account for those areas having a net accretion in spite of an increase in sea level. Everts (1985) has made the most concerted effort to account for the overall sediment budget in a study of shoreline recession along Smith Island, Virginia, and along the barrier islands of North Carolina. On Smith Island, sediment moves landward from the littoral zone by overwash and ephemeral inlet processes, and significant volumes of sand are removed from the island due to longshore transport. It was found that such sediment-budget factors account for 47 percent of the shore retreat, with the remainder attributed to the sea-level increase. Having undertaken sediment-budget corrections, the calculated shoreline recession due to the rise in relative sea level (−5.5 m/year) was almost exactly the same as the actual change (−5.6 m/year). However, a similar analysis for the North Carolina barrier islands found that the predicted shoreline change (−1.7 m/year) is about 121 percent of the measured rate. Even with that amount of disagreement, it is apparent that inclusion of the overall budget of sediments permits improved comparisons between predicted and measured shoreline recession rates.

The field experiments having the greatest control over the beach response to water-level changes are those that have been undertaken in Lake Michigan (Dubois, 1975; Hands, 1979, 1980, 1983). There have been significant water-level fluctuations in the Great Lakes, involving changes on the order of 1 m over time spans of 5–10 years, the corresponding rates of rise being 10–30 cm/year, more than an order-of-magnitude greater than the 1–2 mm/year rate of rise in ocean levels during this century (Fig. 4-1). These rates and time periods make it possible to document the shoreline responses in studies limited to a decade of measurements.

In the study of Hands (1979, 1980, 1983), twenty-five beach profiles spread over 50 km of shore were monitored for 8 years. Four of the survey series occurred during a period of increasing lake levels, while the last two took place during declining levels. The maximum water-level change between surveys was 0.39 m. Figure 4-43 illustrates the observed profile changes. In keeping with the equilibrium assumption, the bars migrated landward and maintained a nearly constant depth beneath the gradually rising lake levels. Hands did confirm that deposition balanced the erosion, verifying the conservation of sediment volume within the changing profile as required by the Bruun model. The observed shoreline retreat distances are compared in Figure 4-44(a) with those predicted by the Bruun Equation (4.6). The comparisons span three time intervals determined by surveys: 1969–1971, 1969–1975, and 1969–1976. The lake-level hydrograph for the study period is given in Figure 4-44(b), which shows a gradual rise to a high stand during 1973–1974, followed by a rapid drop in lake levels. The water levels in 1971, 1975, and 1976 were all higher than during the 1969 base survey, but the surveys of 1975 and 1976 occurred during the period of subsiding lake levels. It is seen in Figure 4-44(a) that the calculated shoreline retreats were substantially

LAKE MICHIGAN (after Hands, 1983)

A. Shore Retreat, 1967-1975

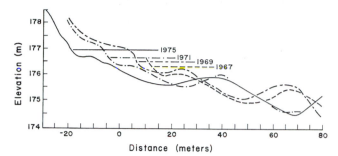

B. Shore Recovery, 1975-1976

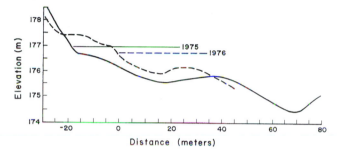

Figure 4-43 Measured profile adjustments on a Lake Michigan beach determined by Hands (1980, 1983), showing that the bars and shoreline shifted landward during a period of rising lake levels. [From E. B. Hands, *Prediction of Shore Retreat and Nearshore Profile Adjustments to Rising Water Levels on the Great Lakes*, U.S. Army Coastal Engineering Research Center Technical Memo No. 80-7.]

higher than those measured for the 1969–1971 and 1969–1975 comparisons, but good agreement was achieved for the 1969-1976 total span represented by the surveys. Hands noted that the shoreline recession rates had dramatically decreased as lake levels declined after 1974, and the beach prograded for the first time at twelve of thirty-four survey sites. Hands attributed the higher calculated than measured shoreline erosion to the profile retreat lagging behind the lake-level rise. By this interpretation, the rising water levels established a potential for erosion, but realization of that potential required cross-shore sediment redistributions that were dependent on storm-wave energy. The eventual convergence of measured and predicted retreat, 3 years after annual lake levels had peaked, suggested that several storm seasons may be required before beach profiles are able to adjust to significant changes in mean water levels. Evidence for this lag is that shoreline recession persisted after the lake level peaked. Hands felt that the crucial proof of the Bruun model and evidence for its usefulness lay in the 1976 final agreement between predicted and measured shoreline recession distances.

A lag time of the beach response, as suggested by Hands (1979, 1980, 1983), likely does account for part of the disagreement between the measured recession and that predicted by the Bruun equations. The profile shifts on Lake Michigan beaches involve the migration of very large bars with appreciable sand movements, occurring under limited wave-energy conditions. However, it is unclear whether a lag on the order of 3 years is reasonable or whether part of the disagreement between the measured and computed recession distances

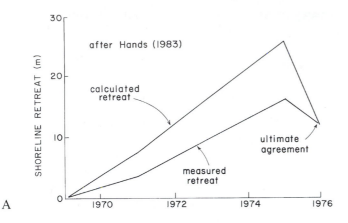

A

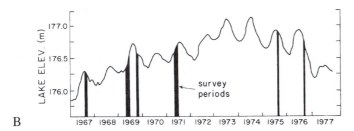

B

Figure 4-44 (a) Calculated shoreline re-treat distances on Lake Michigan beaches derived from the Bruun rule, Equation (4.6), versus measured distances. (b) Hydrograph showing lake-level changes and survey periods. [From E. B. Hands, *Prediction of Shore Retreat and Nearshore Profile Adjustments to Rising Water Levels on the Great Lakes,* U.S. Army Coastal Engineering Research Center Technical Memo No. 80-7.]

represents a partial failure of the Bruun model. The ultimate agreement, after 6 years of pro-file surveys, between the measured and computed distances, is more likely to have resulted from the reduced lake levels in the last 2 years of the study than to any basic validity of the Bruun model. It is probable that had the lake levels continued to rise during the time frame of the study, the disagreement between measured recession and that predicted by the Bruun equations would have persisted and might have continued to diverge as suggested by the trends seen in Figure 4-44(a).

In their review of these models relating shoreline recession to the rise in sea level, Ko-mar et al. (1991) concluded that the Bruun (1962) model is correct insofar as its prediction of an upward and landward translation of an equilibrium profile but that the Bruun Equa-tions (4.6) and (4.7) generally provide a poor prediction of the resulting rates of shoreline erosion. The poor predictions point to the importance of the overall budget of sediments, sediment contributions and losses that tend to overwhelm the small shoreline changes that result from modest increases in sea levels. An additional problem is that the model of Bruun, and the related models such as the barrier-migration analysis of Dean and Maurmeyer (1983), cannot account for any time lag of the beach response to an elevated water level. This perhaps is not a problem in predicting shoreline recessions over time spans of decades to a century, as would occur in response to greenhouse warming, but as we have seen, the lag has been a problem in analyzing beach responses in the Great Lakes that take place within a decade. Where lags in the beach response are important, Komar et al. suggest that beach and dune-erosion models may be applicable, models that are reviewed in Chap-ter 7. Although those models were developed to predict shoreline and dune recession under the elevated water and waves associated with individual storms, they could be employed to assess the longer-term integrated recession and accompanying profile changes in response to elevated water levels like those that have occurred in the Great Lakes.

SUMMARY

There are many processes that affect the level of the sea over time scales ranging from hours to decades to centuries. Foremost are the tidal cycles that produce continuous water-level changes that can amount to several meters per day. The level of the tides can directly affect the profile of the beach (Chapter 7) and thereby can alter the patterns of wave breaking in the nearshore and the distributions of currents and sediment transport. The longer-term variations reflect changes in the mean level of the sea, having averaged out the tidal fluctuations. There are differences from day to day due to meteorological conditions, the most significant being the storm surge that can accompany hurricanes and typhoons. There are monthly cycles in the mean level due to seasonal effects of changing water temperatures and other oceanic and atmospheric processes. Of particular interest and concern are the yearly changes in the mean elevation of the sea that reflect a progressive rise in the global eustatic level, a rise that is progressively inundating low-lying coastal areas and eroding shorelines. The concern is that this rise will accelerate during the twenty-first century in response to global warming resulting from the greenhouse effect, possibly leading to a dramatic increase in coastal flooding and erosion. In the last section of this chapter we reviewed the models designed to predict the resulting erosion and found that more work is needed in order to improve their predictive capabilities. As we approach the twenty-first century, we are faced with many uncertainties as to the expected magnitude of greenhouse warming and the accompanying rise in sea level, and are in a relatively poor position to predict the erosional response along our coasts.

QUESTIONS AND PROBLEMS

1. The ancients believed that an angel moving his foot in and out of the water produced tides. How would you, in a scientific fashion, go about testing this hypothesis? (Do not be superficial, the hypothetical angel can actually explain many of the observed properties of tides.)

2. If the Moon is on the opposite side of Earth from you, does this position act to make you weigh more or less? Why?

3. The length of Lake Superior is approximately 500 km and has an average depth of approximately 80 m. What is the natural period of oscillation of the lake, and would one expect the development of significant tides?

4. If the tidal range is 5 m and the still-water depth is 30 m, according to Equation (4.3), how much faster is the tidal-wave crest (high tide) traveling than the trough (low tide)?

5. Assume that the tidal level at a coastal site varies in a simple sinusoidal fashion, the period being 12 hours and the range 3 meters. With a uniformly sloping beach, would the water level reside longer in the vicinity of the high-tide, mid-tide, or low-tide positions on the beach? Evaluate the distribution of time spent by the water level at various increments of distance along the profile of the beach if its slope is 1 in 20 (that is, 2.9 °).

6. The flow in and out of a bay, separated from the ocean by a sand spit except for a narrow inlet, is largely tide dominated. What will be the controlling factors in determining the cross-sectional area of the inlet channel? Devise a conceptual graph showing the expected dependence of the section area on the controlling parameters. How would you collect data to show this dependence and quantitatively establish the graph? [See O'Brien (1931) for the solution.]

7. From the curve in Figure 4-33(a) develop a graph for the annual mean rate (meters per century) of sea-level rise. How do the rates for the past century compare with estimates derived from tide-gauge records and those predicted for the next century as a response to greenhouse warming?

8. Large variations in sea levels have occurred throughout geologic history, at times when there were no cycles of glacial expansion and retreat as occurred during the last few million years. Could such changes be due to small variations in water temperatures, on the order of 0.001–0.01 °C, but affecting the entire depth of the sea (average depth $D = 4,000$ m)? What other global factors might produce large changes in the level of the sea?

9. For a coastal area having a land relief of $B = 2$ m and an estimated offshore closure at $h_* = 10$ m and $L_* = 125$ m, estimate the expected shoreline retreat that might correspond to the predicted magnitudes of sea-level rise (Fig. 4-38) that potentially could occur during the twenty-first century.

REFERENCES

ALLISON, I. (1979). The Mass Budget of the Lambert Glacier Drainage Basin, Antarctica. *Journal of Glaciology* 22: 223–235.

AUBREY, D. G. and K. O. EMERY (1983). Eigenanalysis of Recent United States Sea Levels. *Continental Shelf Research* 2: 21–22.

AUBREY, D. G. and K. O. EMERY (1986). Relative Sea Levels of Japan from Tide-Gauge Records. *Geological Society of America Bulletin* 97: 194–205.

BARD, E., B. HAMELIN, R. G. FAIRBANKS, and A. ZINDLER (1990a). Calibration of the [14]C Timescale over the Past 30,000 Years Using Mass Spectrometric U-Th Ages from Barbados Corals. *Nature* 345: 405–410.

BARD, E., L. D. LABEYRIE, J.-J. PICHON, M. LABRACHERIE, M. ARNOLD, J. DUPRAT, J. MOYES, and J.-C. DUPLESSY (1990b). The Last Deglaciation in the Southern and Northern Hemispheres: A Comparison Based on Oxygen Isotopes, Sea Surface Temperature Estimates, and Accelerator [14]C Dating from Deep-Sea Sediments. In *Geological History of the Polar Oceans: Arctic Versus Antarctic,* U. Bleil and J. Thiede (editors). Pp. 405–415. Dordrecht: Kluwer Academic Publishers.

BARD, E., M. ARNOLD, R. G. FAIRBANKS, and B. HAMELIN (1993). [230]Th-[234]U and [14]C Ages Obtained by Mass Spectrometry on Corals. *Radiocarbon* 35: 191–199.

BARNETT, T. P. (1983). Recent Changes in Sea Level and Their Possible Causes. *Climate Change* 5: 15–38.

BARNETT, T. P. (1984). The Estimation of "Global" Sea Level Change: A Problem of Uniqueness. *Journal of Geophysical Research* 89(C5): 7980–7988.

BENDER, M. L., R. G. FAIRBANKS, F. W. TAYLOR, R. K. MATTHES, J. G. GODDARD, and W. S. BROECKER (1979). Uranium-Series Dating of the Pleistocene Reef Tracts of Barbados, West Indies. *Geological Society of America Bulletin* 90: 577–594.

BENTLEY, C. R. (1984). Some Aspects of the Cryosphere and Its Role in Climatic Change. *Geophysical Monograph* 29: 207–220.

BLANCHON, P. and J. SHAW (1995). Reef Drowing During the Last Deglaciation. Evidence for Catastophic Sea-Level Rise and Ice-Sheet Collapse: *Geology* 23: 4–8.

BLAHA, J. P. (1984). Fluctuations of Monthly Sea Level as Related to the Intensity of the Gulf Stream from Key West to Norfolk. *Journal of Geophysical Research* 89(C5): 8033–8042.

BLOCK, M. R. (1965). A Hypothesis for the Change of Ocean Levels Depending on the Albedo of the Polar Ice Caps. *Palaeogeography, Palaeoclimatology, Palaeoecology* 1: 127–142.

BLOOM, A. L., W. S. BROECKER, J. M. A. CHAPPELL, R. K. MATTHEW, and K. J. MESOLELLA (1974). Quaternary Sea Level Fluctuations on a Tectonic Coast: New ^{230}Th/^{234}U Dates from the Huon Peninsula, New Guinea. *Quaternary Research* 4: 185–205.

BRAATZ, B. V. and D. G. AUBREY (1987). Recent Relative Sea-Level Change in Eastern North America. In *Sea-Level Fluctuations and Coastal Evolution,* D. Nummedal et al. (editors). Special Publication No. 41, pp. 29–46. Soc. Econ. Paleontologists and Mineralogists, Tulsa, Oklahoma.

BROECKER, W. S., D. L. THURBER, J. GODDARD, T.-L. KU, R. K. MATTHEWS, and K. J. MESOLELLA (1968). Milenkovitch Hypothesis Supported by Precise Dating of Coral Reefs and Deep-Sea Sediments. *Science* 159: 297–300.

BRUUN, P. (1962). Sea Level Rise as a Cause of Shore Erosion. *Journal of Waterways and Harbors Division, Amer. Soc. Civil Engrs.* 88: 117–130.

BRUUN, P. (1988). The Bruun Rule of Erosion by Sea-Level Rise: A Discussion of Large-Scale Two and Three-Dimensional Useages. *Journal of Coastal Research* 4: 627–648.

CARTER, R. M., L. CARTER, and D. P. Johnson (1986). Submergent Shorelines in the SW Pacific: Evidence for an Episodic Post-Glacial Transgression: *Sedimentology* 33: 629–649.

CHAPPELL, J. (1974). Geology of Coral Terraces, Huon Peninsula, New Guinea: A Study of Quaternary Tectonic Movements and Sea-Level Changes. *Geological Society of America Bulletin* 85: 553–570.

CHAPPELL, J. and H. A. POLACH (1976). Holocene Sea-Level Change and Coral-Reef Growth at Huon Peninsula, Papua New Guinea. *Geological Society of America Bulletin* 87: 235–240.

CHAPPELL, J. and N. J. SHACKLETON (1986). Oxygen Isotopes and Sea Level. *Nature* 324: 137–140.

CLANCY, E. P. (1969). *The Tides, Pulse of the Earth.* Garden City, NY: Anchor.

CLARK, J. A. and C. S. LINGLE (1977). Future Sea-Level Changes Due to West Antarctic Ice Sheet Fluctuations. *Nature* 269: 206–209.

CLARK, J. A., W. E. FARRELL, and W. R. PELTIER (1978). Global Changes in Post-Glacial Sea Level: A Numerical Calculation. *Quaternary Research* 9: 265–287.

CURRAY, J. R. (1960). Sediments and History of Holocene Transgression, Continental Shelf, Northwest Gulf of Mexico. In *Recent Sediments Northwest Gulf of Mexico,* F. P. Shepard, F. B. Phleger, and Tj. H. van Andel (editors), pp. 221–266. Amer. Assoc. Petrol. Geologists, Tulsa, Oklahoma.

CURRAY, J. R. (1965). Late Quaternary History, Continental Shelfs of the United States. In *The Quaternary of the United States,* H. E. Wright and D. G. Frey (editors), pp. 723–735. New Jersey: Princeton University Press.

CURRAY, J. R. (1969). History of Continental Shelves. In *The New Concepts of Continental Margin Sedimentation,* D. J. Stanley (editor). AGI Short Course Lecture Notes.

DAVIES, J. L. (1964). A Morphogenic Approach to World Shorelines. *Zeitschrift für Geomorphologie* 8: 127–142.

DAWSON, A. G. (1992). *Ice Age Earth: Late Quaternary Geology and Climate.* New York: Routledge.

DEAN, R. G. (1990). Beach Response to Sea Level Change. "Ocean Engineering Science." In *The Sea,* edited by B. Le Méhauté and D. M. Hanes (editors). vol. 9 pp. 869–887. New York: John Wiley & Sons.

DEAN, R. G. (1991). Equilibrium Beach Profiles: Characteristics and Applications: *Journal of Coastal Research* 7: 53–84.

DEAN, R. G. and E. M. MAURMEYER (1983). Models for Beach Profile Response: In *Handbook of Coastal Processes and Erosion,* P. D. Komar (editor), pp. 151–166. Boca Raton, FL: CRC Press.

DEAN, R. G. et al. (1987). *Responding to Changes in Sea Level.* Washington, D.C.: National Academy Press.

DEFANT, A. (1958). *Ebb and Flow.* Ann Arbor, MI: University of Michigan Press.

DILLON, W. P. and R. N. OLDALE (1978). Late Quaternary Sea-Level Curve: Reinterpretation Based on Glaciotectonic Influence. *Geology* 6: 56–60.

DONN, W. L., W. R. FARRAND, and M. EWING (1962). Pleistocene Ice Volumes and Sea-Level Lowering. *Journal of Geology* 70: 206–214.

DOUGLAS, B. C. (1991). Global Sea Level Rise. *Journal of Geophysical Research* 94(C4): 6981–6992.

DUBOIS, R. N. (1975). Support and Refinement of the Bruun Rule of Beach Erosion. *Journal of Geology* 83: 651–657.

EDWARDS, R. L., J. W. BECK, G. S. BURR, D. J. DONAHUE, J. M. A. CHAPPELL, A. L. BLOOM, E. R. M. DRUFFEL, and F. W. TAYLOR (1993). A Large Drop in Atmospheric $^{14}C/^{12}C$ and Reduced Melting in the Younger Dryas, Documented with ^{230}Th Ages of Corals. *Science* 260: 962–968.

EMERY, K. O. (1980). Relative Sea Levels from Tide-Gauge Records. *Proceedings of the National Academy of Sciences* 77: 6968–6972.

EMERY, K. O., R. L. WIGLEY, A. S. BARTLETT, M. RUBIN, and E. S. BARGHOORN (1967). Freshwater Peat on the Continental Shelf. *Science* 158: 1301–1307.

EMERY, K. O. and D. G. AUBREY (1991). *Sea Levels, Land Levels, and Tide Gauges.* New York: Springer-Verlag.

EMILIANI, C. (1970). Pleistocene Paleotemperatures. *Science* 168: 822–825.

ENFIELD, D. B. and J. S. ALLEN (1980). On the Structure and Dynamics of Monthly Mean Sea Level Anomalies along the Pacific Coast of North and South America. *Journal of Physical Oceanography* 10: 557–578.

ENFIELD, D. B. and J. S. ALLEN (1983). The Generation and Propagation of Sea Level Variability along the Pacific Coast of Mexico. *Journal of Physical Oceanography* 13: 1012–1033.

EVERTS, C. H. (1985). Sea Level Rise Effects. *Journal of Waterway, Port, Coastal and Ocean Engineering, Amer. Soc. Civil Engrs.* 111: 985–999.

FAIRBANKS, R. G. (1989). A 17,000-Year Glacio-Eustatic Sea Level Record: Influence of Glacial Melting Rates on the Younger Dryas Event and Deep-Ocean Circulation. *Nature* 342: 637–642.

FAIRBRIDGE, R. W. (1960). The Changing Level of the Sea. *Scientific American* 202: 70–79.

FAIRBRIDGE, R. W. (1961). Eustatic Changes in Sea Level. *Physics and Chemistry of the Earth* 4: 99–185.

FAIRBRIDGE, R. W. and W. S. NEWMAN (1968). Postglacial Coastal Subsidence of the New York Area. *Zeitschrift für Geomorphologie* 12: 296–317.

FLETCHER, C. H. and C. E. SHERMAN (1995). Submerged Shorelines on O'ahu, Hawai'i: Archive of Episodic Transgression during the Deglaciation?: In "Holocene Cycles: Climate, Sea Levels and Sedimentation," *Journal of Coastal Research,* Special Issue 17, pp. 141–152.

FLICK, R. E. and D. R. CAYAN (1985). Extreme Sea Levels on the Coast of California. *Proceedings of the 19th Coastal Engineering Conference, Amer. Soc. Civil Engrs.* pp. 886–898.

GILL, E. D. (1961). Changes in the Level of the Sea Relative to the Level of the Land in Australia. *Zeitschrift für Geomorphologie*, Supp., 3: 73–79.

GORNITZ, V., S. LEBEDEFF, and J. HANSEN (1982). Global sea level trend in the past century: *Science* 215: 1611–1614.

GORNITZ, V. and S. LEBEDEFF (1987). Global Sea-Level Changes during the Past Century. In *Sea-Level Fluctuations and Coastal Evolution,* D. Nummedal et al. (editors). Special Publication No. 41, pp. 3–16. Soc. Econ. Paleontologists and Mineralogists, Tulsa, Oklahoma.

GUILCHER, A. (1969). Pleistocene and Holocene Sea Level Changes. *Earth Science Review* 5: 69–97.

GUTENBERG, B. (1941). Changes in Sea Level, Postglacial Uplift, and Mobility of the Earth's Interior. *Geological Society of America Bulletin* 52: 721–772.

HANDS, E. B. (1979). *Changes in Rates of Shore Retreat, Lake Michigan,* 1967–1976. Coastal Engineering Res. Center, Tech. Memo. No. 79–4.

HANDS, E. B. (1980). *Prediction of Shore Retreat and Nearshore Profile Adjustments to Rising Water Levels on the Great Lakes.* Coastal Engineering Res. Center, Tech. Memo. No. 80–7.

HANDS, E. B. (1983). The Great Lakes As A Test Model for Profile Responses to Sea Level Changes: In *Handbook of Coastal Processes and Erosion,* P. D. Komar (editor). Pp. 176–189, Boca Raton, FL: CRC Press.

HANSEN, J. and S. LEBEDEFF (1988). Global Surface Air Temperatures: Update through 1987. *Geophysical Research Letters* 15(4): 323–326.

HANSEN, J., I. FUNG, A. LACIS, D. RIND, S. LEBEDEFF, R. RUEDY, G. RUSSELL, and P. STONE (1988). Global Climate Changes As Forecast by Goddard Institute for Space Studies Three-Dimensional Model. *Journal of Geophysical Research* 93(D8): 9341–9364.

HEARTY, P. J. and P. AHARON (1988). Amino Acid Chronostratigraphy of Late Quaternary Coral Reefs: Huon Peninsula, New Guinea, and the Great Barrier Reef, Australia. *Geology* 16: 579–583.

HICKS, S. D. (1972). On the Classification and Trends of Long Period Sea Level Series. *Shore & Beach* 40: 20–23.

HICKS, S. D. (1978). An Average Geopotential Sea Level Series for the United States. *Journal of Geophysical Research* 83(C3): 1377–1379.

HICKS, S. D. and L. E. HICKMAN (1988). United States Sea Level Variation through 1986. *Shore & Beach* 56: 3–7.

HOFFMAN, J. S., D. KEYES, and J. G. TITUS (1983). *Projecting Future Sea Level Rise: Methodology, Estimates to the Year 2100, and Research Needs.* Washington, D.C.: U.S. Environmental Protection Agency.

HOFFMAN, J. S., J. B. WELLS, and J. G. TITUS (1986). Future Global Warming and Sea Level Rise. In *Iceland Coastal and River Symposium,* G. Sigbjarnason (editor). Pp. 245–266.

HOLLIN, J. T. (1970). Is the Antarctic Ice Sheet Growing Thicker? *IASH Publication 96,* pp. 363–374.

HOPLEY, D. (1978). Sea Level Change on the Great Barrier Reef: An Introduction. *Philosophical Transactions Royal Society of London,* Series A, 291: 159–166.

HUYER, A., W. E. GILBERT, and H. L. PITTOCK (1983). Anomalous Sea Levels at Newport, Oregon During the 1982–83 El Niño. *Coastal Oceanography and Climatology News* 5: 37–39.

JENNINGS, J. N. (1961). Sea Level Changes in King Island, Bass Strait. *Zeitschrift für Geomorphologie* 3: 80–84.

JONES, P. D. (1988). Hemispheric Surface Air Temperature Variations—Recent Trends and an Update to 1987. *Journal of Climate* 1: 654–660.

JONES, P. D., T. M. L. WIGLEY, and P. B. WRIGHT (1986). Global Temperature Variations Between 1861 and 1984. *Nature* 322: 430–434.

JONGSMA, D. (1970). Eustatic Sea Level Changes in the Arafura Sea. *Nature* 228: 150–151.

KEELING, C. D. and R. O. BACASTOW (1977). *Energy and Climate.* Washington, D.C.: National Academy of Sciences, pp. 72–95.

KEELING, C. D. et al. (1989). A Three-Dimensional Model of Atmospheric CO_2 Transport Based on Observed Winds: 1. Analysis of Observational Data. In *Aspects of Climate Variability in the Pacific and Western Americas,* D. H. Peterson (editor). *Geophysical Monographs of the American Geophysical Union* 55: 165–236.

KOMAR, P. D. (1986). The 1982-83 El Niño and Erosion on the Coast of Oregon. *Shore & Beach* 54: 3–12.

KOMAR, P. D. and D. B. ENFIELD (1987). Short-Term Sea-Level Changes and Coastal Erosion. In *Sea-Level Fluctuations and Coastal Evolution,* D. Nummedal et al. (editors). Special Publication No. 41, pp. 17–27. Soc. Econ. Paleontologists and Mineralogists, Tulsa, Oklahoma.

KOMAR, P. D. et al. (1991). The Response of Beaches to Sea-Level Changes: A Review of Predictive Models. *Journal of Coastal Research* 7: 895–921 (report of the Scientific Committee on Ocean Research Working Group 89).

KOMAR, P. D. and S. M. SHIH (1993). Cliff Erosion along the Oregon coast: A Tectonic-Sea Level Imprint Plus Local Controls by Beach Processes. *Journal of Coastal Research* 9: 747–765.

KRAFT, J. C., R. B. BIGGS, and S. D. HALSEY (1973). Morphology and Vertical Sedimentary Sequence Models in Holocene Transgressive Barrier Systems. In *Coastal Geomorphology,* D. R. Coates (editor), pp. 321–354, Binghamton, NY: State University of New York.

LEATHERMAN, S. P. and J. J. MØLLER (1990). Hurricane Hugo's Impact on the South Carolina Beaches. *Symposium on Coastal Geomorphology, Littoral Drift, Climate Development, Greenhouse Effects, Sea Level Rise and Its Consequences, Erosion and Protection* Denmark: Skagen, pp. 332–355.

LOCKER, S. D., A. C. HINE, L. P. TEDESCO, and E. A. SHINN (1996). Magnitude and Timing of Episodic Sea-Level Rise During the Last Deglaciation. *Geology* 24: 827–830.

MALZER, H. and H. SECKEL (1976). Das Geometrische Nivellement Über das Inlandeis—Hohenänderungen Zwischen 1959 und 1968 im Ost-West-Profil der EGIG. *Zeitschrift für Gletsch. Glacialgeolie* 11: 245–252.

MAUL, G. A., F. CHES, M. BUSHNELL, and D. A. MAYER (1985). Sea Level Variation as an Indicator of Florida Current Volume Transport: Comparisons with Direct Measurements. *Science* 227: 304–307.

MEIER, M. (1984). Contribution of Small Glaciers to Global Sea Level, 1984. *Science* 226: 1418–1421.

MERCER, J. H. (1978). West Antarctic Ice Sheet and CO_2 Greenhouse Effect—A Threat of Disaster. *Nature* 271: 321–325.

MESOLELLA, K. J., R. K. MATTHEWS, W. S. BROECKER, and D. L. THURBER (1969). The Astronomical Theory of Climatic Change: Barbados Data. *Journal of Geology* 77: 250–274.

MILANKOVITCH, M. (1938). Astronomische Mittel zur Erforschung der erdgeschichtlichen Klimate. In *Handbuch der Geophysik,* B. Gutenberg (editor). Vol. 9, pp. 593–698. Berlin.

MILLIMAN, J. D. and K. O. EMERY (1968). Sea Levels During the Past 35,000 years. *Science* 162: 1121–1123.

MITCHELL, C. E., P. VINCENT, R. J. WELDON, and M. A. RICHARDS (1994). Present-Day Vertical Deformation of the Cascadia Margin, Pacific Northwest, United States. *Journal of Geophysical Research* 99(B6): 12,257–12,277.

NAMIAS, J. and J. C. K. HUANG (1972). Sea Level at Southern California: A Decadal Fluctuation. *Science* 177: 351–353.

NUMMEDAL, D. (1983). Barrier Islands. In *Handbook of Coastal Processes and Erosion,* P. D. Komar (editor). Pp. 77–121. Boca Raton, FL: CRC Press.

OAKS, R. A. and N. K. COCH (1963). Pleistocene Sea Levels, Southeastern Virginia. *Science* 140: 979–984.

O'BRIEN, M. P. (1931). Estuary Tidal Prisms Related to Entrance Areas. *Journal of Civil Engineering* 1: 738–793.

PELTIER, W. R. and A. M. TUSHINGHAM (1989). Global Sea Level Rise and the Greenhouse Effect: Might They Be Connected? *Science* 244: 806–810.

PIRAZZOLI, P. A. (1991). Possible Defenses Against a Sea-Level Rise in the Venice Area, Italy. *Journal of Coastal Research* 7: 231–248.

PRATT, R. M. and R. F. DILL (1974). Deep Eustatic Terrace Levels: Further Speculations. *Geology* 2: 155–159.

RADTKE, U. and R. GRÜN (1990). Revised Reconstruction of Middle and Late Pleistocene Sea-Level Changes Based on New Chronologic and Morphologic Investigations in Barbados, West Indies. *Journal of Coastal Research* 6: 699–708.

REVELLE, R. R. (1983). Probable Future Changes in Sea Level Resulting from Increased Atmospheric Carbon Dioxide. In *Changing Climate,* pp. 433–448. Washington, D.C.: National Academy Press.

RICHARDS, H. G. and R. W. FAIRBRIDGE (1965). *Annotated Bibliography of Quaternary Shorelines, 1945–64.* Academy of Natural Science, Philadelphia, Publication No. 6.

ROSEN, P. S. (1978). A Regional Test of the Bruun Rule on Shoreline Erosion. *Marine Geology* 26: M7–M16.

ROY, P. and J. CONNELL (1991). Climatic Change and the Future of Atoll States. *Journal of Coastal Research* 7: 1057–1075.

SAURAMO, M. (1958). Die Geschichte der Ostsee. *Ann. Acad. Sci. Fenn. Series A.*

SCHLESINGER, M. E., W. L. GATES, and Y.-J. HAN (1985). The Role of the Ocean in CO$_2$-Induced Climatic Warming: Preliminary Results from the OSU Coupled Atmosphere-Ocean GCM. In "Coupled Atmosphere-Ocean Models." *Proceedings of the 16th International Liege Colloquium,* New York: Elsevier.

SCHWARTZ, M. (1965). Laboratory Study of Sea-Level as a Cause of Shore Erosion. *Journal of Geology* 73: 528–534.

SCHWARTZ, M. (1967). The Bruun Theory of Sea-Level Rise as a Cause of Shore Erosion. *Journal of Geology* 75: 76–92.

SCHWARTZ, M. (1972). Seamounts as Sea-Level Indicators. *Geological Society of America Bulletin* 83: 2975–2980.

SHEPARD, F. P. (1963). Thirty-Five Thousand Years of Sea Level. In *Essays in Marine Geology, in Honor of K.O. Emery,* pp. 1–10. Los Angeles: University Southern California Press.

SHEPARD, F. P. and J. R. CURRAY (1967). Carbon-14 Determination of Sea Level Changes in Stable Areas. In *Progress in Oceanography, The Quaternary History of the Ocean Basins,* Oxford: Pergamon Press. Vol. 4, pp. 283–291.

STOMMEL, H. M. (1953). Examples of the Possible Role of Inertia and Stratification in the Dynamics of the Gulf Stream System. *Journal of Marine Research* 12: 184–195.

TOMLINSON, L. (1858). *Recreations in Astronomy.* London, England: Parker.

TOOLEY, M. J. (1993). Long Term Changes in Eustatic Sea Level. In *Climate and Sea Level Change: Observation, Projections and Implications,* R. A. Warwick (editor), pp. 81–107. New York: Cambridge University Press.

VAN ANDEL, TJ. H. and J. LABOREL (1964). Recent High Relative Sea Level Strand near Recife, Brazil. *Science* 145: 580–581.

VAN ANDEL, TJ. H. and J. J. VEEVERS (1967). *Morphology and Sediments of the Timor Sea. Bull. Bureau Miner. Resour. Geol. Geophys. Australia* Vol. 83.

VAN DER VEEN, C. J. (1988). Projecting Future Sea Level. *Surveys in Geophysics* 9: 389–418.

VEEH, H. H. and J. CHAPPELL (1970). Astronomical Theory of Climatic Change: Support from New Guinea. *Science* 169: 862–865.

VEEH, H. H. and J. J. VEEVERS (1970). Sea Level at -175 m Off the Great Barrier Reef 13,600 to 17,000 Years Ago. *Nature* 226: 536–537.

VINCENT, P. (1989). *Geodetic Deformation of the Oregon Cascadia Margin.* M.S. dissertation, Eugene, OR: University of Oregon.

WARRICK, R. A. (1993). Climate and sea level change: a synthesis. In *Climate and Sea Level Change: Observations, Projections and Implications,* R. A. Warrick et al. (editors), pp 3–21. New York: Cambridge University Press.

WEMELSFELDER, P. J. (1953). The Disaster in the Netherlands Caused by the Storm Flood of February 1, 1953. *Proceeding of the 4th Coastal Engineering Conference, Amer. Soc. Civil Engrs.,* pp. 256–271.

WHITMORE, F. C., K. O. EMERY, H. B. S. COOKE, and D. J. SWIFT (1967). Elephant Teeth from the Atlantic Continental Shelf. *Science* 156: 1477–1481.

WOOD, F. J. (1977). *The Strategic Role of Perigean Spring Tides in Nautical History and North American Coastal Flooding, 1635–1976.* Washington, D.C.: NOAA, U.S. Department of Commerce.

WUNSCH, C., D. V. HANSEN, and B. D. ZETLER (1969). Fluctuations of the Florida Current Inferred from Sea Level Records. *Deep Sea Research* 16: 447–470.

WYRTKI, K. (1973). Teleconnections in the Equatorial Pacific Ocean. *Science* 180: 66–68.

WYRTKI, K. (1975). Fluctuations of the Dynamic Topography in the Pacific Ocean. *Journal of Physical Oceanography* 5: 450–459.

WYRTKI, K. (1977). Sea Level During the 1972 El Niño. *Journal of Physical Oceanography* 7: 779–787.

WYRTKI, K. (1983). Sea Level in the Equatorial Pacific in 1982. *Tropical Ocean-Atmosphere Newsletter*, No. 16, pp. 6–7.

WYRTKI, K. (1984). The Slope of Sea Level Along the Equator During the 1982/1983 El Niño. *Journal of Geophysical Research* 89(C6): 10,419–10,424.

WYRTKI, K. (1990). Sea Level Rise: The Facts and the Future. *Pacific Science* 44: 1–16.

ZEUNER, F. E. (1959). *The Pleistocene Period.* London, England: Hutchinson.

5

The Generation of Waves and Their Movement Across the Sea

Dumont d'Urville even asserts that he has seen waves above 108 feet high, to the depths of which the ship descended as into a valley, and M. Fleuriot de Langle attests to the truth of this assertion.

Elisee Reclus
The Ocean, Atmosphere, and Life (1873)

When a breeze sweeps over the smooth surface of a pond, it quickly disturbs the water and forms small ripples. If the wind continues, genuine waves develop. The longer and harder the wind blows, the larger the waves that are generated. Winds are always blowing over some portion of the vast expanse of the sea, so waves are ever present. At a particular coastal site, waves may arrive simultaneously that were generated by several storms, some of which were located thousands of kilometers away.

Wind-generated waves are important as energy-transfer agents—they first obtain their energy from the wind, then transfer it across the expanse of the ocean, and finally deliver it to the coastal zone where the energy can be the primary cause of erosion or may generate a variety of nearshore currents and sediment-transport patterns. This transfer of energy is depicted schematically in Figure 5-1. The generation of waves depends on the speed of the wind and on the duration and aerial extent of the storm. Waves in the storm area itself are termed *sea* and are highly complex. There are many different heights all at once, and waves seem to suddenly appear and then as abruptly disappear. This complex pattern results because a storm does not simply generate one set of waves but instead a whole spectrum of waves having a range of periods and heights. However, as the waves leave the storm area, they become more regular and develop into *swell,* having more uniform heights and distances between crests. With this regularity, one can follow individual waves for considerable distances as they travel across the sea. It is the swell that transfers energy across the ocean and delivers it to the coastal zone where the waves finally break and expend their energy in the surf.

The objectives of this chapter are to examine the generation of waves, including the development of relationships between storm factors and the resulting wave heights and energies, to present formulae that describe the motions of regular swell waves as they travel across the ocean, and finally to examine changes in the waves as they shoal while approaching the shore. In the following chapter we will consider their energy expenditure within the nearshore, first by breaking, then as bores crossing the surf zone, and finally as run-up on the beach.

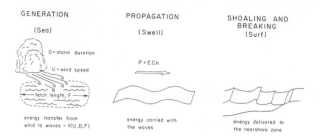

GENERATION
(Sea)

PROPAGATION
(Swell)

SHOALING AND
BREAKING
(Surf)

D = storm duration

U = wind speed

fetch length, F

P = ECn

energy transfer from
wind to waves = f(U,D,F)

energy carried with
the waves

energy delivered to
the nearshore zone

Figure 5-1   Schematic diagram showing the generation of irregular waves (sea) by a storm, the propagation of regular swell waves across the ocean, and the delivery of the energy to the nearshore when the waves break and form surf.

PERIODIC WAVES

The motion of regular waves, as in the case of swell, is periodic, that is, the motion is repetitive through fixed periods of time. At some stationary position—say, the pile in Figure 5-2—a succession of wave crests (or troughs) pass at intervals of time T, the *wave period*. Alternately, the repetition of waves can be expressed in terms of the wave *frequency, $f = 1/T$,* in a sense the number of waves per unit time. If L is the *wave length* [the horizontal distance between successive crests (or troughs)], then the wave must move a distance L in the time T so its speed of propagation, C, is simply

$$C = \frac{L}{T} \tag{5.1}$$

This relationship holds for all periodic waves since it is based solely on their geometry. C is variously termed the *phase velocity* or *celerity* of the wave. It will be seen later that there is a direct relationship between the wave length and the period of the waves when they are in deep water and thus between the celerity and wave period—the greater the wave period, the longer the wave length and the faster their movement in crossing the deep ocean. When the waves are in shallow water, the water depth becomes the controlling parameter in their movement—the shallower the water, the slower the wave celerity.

The other parameter required to describe this simple periodic wave is the *wave height* H, the vertical distance from trough to crest (Fig. 5–2). The wave height is independent of other wave parameters and to a first approximation the reverse is true as well. To higher approximations the wave height causes a slight increase in the wave celerity C.

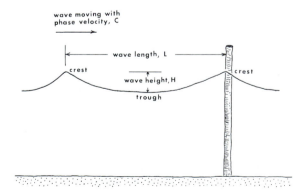

wave moving with
phase velocity, C

wave length, L

crest

wave height, H

crest

trough

Figure 5-2   The parameters that, along with the period T, describe simple oscillatory wave motions.

In the deep ocean, if one follows the motion of a cork floating on the water surface as waves pass, it will be observed that the cork rises and falls and at the same time moves back and forth, describing a circular motion whose diameter is the wave height H and whose period is T. The cork makes no net advance in the direction of wave motion. Waves therefore transfer energy across the water surface, often for thousands of kilometers, but with a negligible net drift of the water itself—the water does not move along with the wave form.

The irregularity of the sea in the storm area is due to the simultaneous generation of waves having a range of periods and heights rather than the simple pattern depicted in Figure 5-2. The complexity of the sea comes from the superposition of many waves. An observed wave crest within the sea is generally the result of several individual wave crests temporarily coming together, adding to yield the observed high wave. However, that addition of waves is fleeting so the crest quickly disappears as the individual waves continue on their separate journeys. Because of the dependence of the wave celerity on the period, with the speed of movement increasing with the period, the longer-period waves travel the fastest as they leave the area of generation. They accordingly sort themselves out by period, a process termed *wave dispersion,* and this is the primary factor in converting irregular sea into a regular swell where the waves have a specific period T, as in Figure 5-2, or only a very narrow range of periods. By this sorting process, the first swell arriving from a distant storm will have the longest periods since they have traveled the fastest, and the periods will progressively decrease over a day or two with the arrival of subsequent waves.

Wind-generated waves are generally limited to periods less than 20 sec. They are often referred to as *gravity waves* in that the force of gravity is important to their movement. Another component of wind-generated waves is the *capillary waves,* having periods of a fraction of a second and wave lengths of but a few centimeters; they are distinguished from the longer-period waves because water surface tension is important to their motion, as well as gravity. In the direction of long-period motions, greater than 20 sec but less than about 5 min, are the *infragravity waves.* They result from wave/wave interactions and turn out to be extremely important in the nearshore, being a factor that in part controls the run-up of water at the shoreline (Chapter 6), may have a role in the generation of nearshore currents (Chapter 8), and have been offered as an explanation for the formation of certain types of sand bars and beach cusps (Chapter 11).

WAVE MEASUREMENTS, ANALYSIS, AND DATA SOURCES

As noted above, a storm generates waves having a range of heights and periods, and more than one storm may be present so that waves generally arrive at a beach simultaneously from different directions. The measurement and analysis of the waves is therefore more complex than simply determining the period with a stopwatch and the height with a meter stick. Instead, the measurement of waves requires sophisticated instrumentation that can survive the rigors of the marine environment and demands complex analyses to unravel the mixtures of waves arriving from multiple storms.

The Measurement of Waves

The variety of wave-measurement systems is illustrated in Figure 5-3, taken from Earle and Bishop (1984) who provide a useful guide to the measurement of ocean waves and their

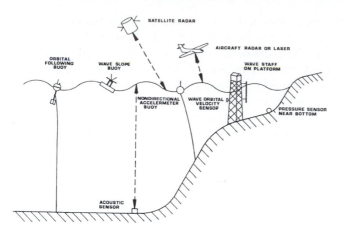

Figure 5-3 Diverse methods that have been used to obtain measurements of waves. [Used with permission of Endeco Inc., from M. D. Earle and J. M. Bishop, *A Practical Guide to Ocean Wave Measurement and Analysis,* p. 28. Copyright © 1984 Endeco Inc.]

analysis. The measurement techniques vary from *in situ* sensors such as buoys and staffs to the remote sensing of the waves by satellites and aircraft. Wave-measuring equipment can be categorized into those that determine the heights and periods of the waves, but not their directions, versus those that obtain fully directional wave measurements. The three general categories of *in situ* non-directional wave measurement devices are surface piercing, pressure sensing, and surface following. The surface piercing type generally consists of a staff fixed in position, attached to a pier or offshore platform, or in studies of waves within the nearshore, held in position by guy wires. The operating principle of a wave staff is that it is a component of an electrical circuit that produces voltage changes proportional to the length of the staff that is submerged. In its simplest form, the staff may have a series of bare contacts at equal intervals along its length. The staff plus the water makes a complete electrical circuit, the resistance and therefore the current of the circuit varying with the water level and the number of staff contacts under water. Resistors in the circuit are selected so that the current flow is proportional to the submerged length of the staff. The fluctuating water level is therefore translated into a fluctuating electrical current that may be recorded. Other staffs operate by electrically scanning in sequence the equally spaced metal contacts. As each contact is scanned, an electrical pulse is transmitted if the contact is under water—no pulse occurs if the contact is above water. The most common resistance wire staff, which can be constructed fairly easily, consists of a wire that is spirally wound about an insulating support such as a rigid plastic pipe; in this form, the staff is a variable resistor in a voltage divided circuit to which an alternating current is applied. Still another approach is to use a pair of vertical wires, the capacitance between them depending on the lengths of the wires underwater.

Pressure-type gauges convert the pressure fluctuations associated with the varying water-surface elevations into an electrical current or voltage signal. The conversion from water pressure to an electric current is accomplished through the use of strain gauges, thermocouples, or coils in a magnetic field. Pressure sensors are generally placed on or near the sea floor, this being one of their advantages in that they are less vulnerable to damage. The depth of placement depends on the waves that are to be measured, as the pressure fluctuations associated with the surface waves decrease with depth, the rate of decrease being greater for the shorter-period waves. Because of this, the shorter-period waves may be preferentially filtered out by the water column, but this may be an advantage in that the small chop from local winds is eliminated, smoothing the record of the dominant swell. This pressure decrease with depth limits the depth range in applications of pressure sensors to generally less than 20 m. In all cases the record obtained by the sensor must be corrected for the depth to obtain the proper wave heights and energy at the surface.

A variety of buoys have been used to follow the surface movements of the water associated with the passage of waves (Steele and Johnson, 1979). Buoys are particularly useful in deep water where pressure sensors cannot be used and where structures are not available for attachment of other types of wave sensors. One type of buoy is spherical, held by a slack mooring line, sometimes with an elastic rubber cord, so the buoy is free to move and follow the wave motions. A well-known example of a spherical buoy is the WAVERIDER buoy manufactured in the Netherlands. Inside the buoy is an accelerometer that measures the vertical accelerations, which when integrated once give the vertical velocity, while a second integration provides a record of wave elevations. Other buoys have the shape of a disc, which tilts from the horizontal and thereby yields measurements of wave slopes that in turn provide information on heights and periods of the waves. Such buoys can also be used to obtain directional measurements of waves, generally doing this by measuring the pitch, roll and heave and utilizing the theory developed by Longuet-Higgins, Cartwright, and Smith (1963) to analyze the wave directions, periods, and heights from the motion. Large disc-type buoys with diameters up to 12 m are used operationally at many U.S. offshore locations by the National Data Buoy Center (NDBC) of the National Oceanic and Atmospheric Administration (NOAA). One significant advantage in the use of buoys is that they may be designed to report to satellites and for this reason have been utilized in most long-term wave data collection programs.

In principle, the direction of wave approach can be determined from pairs of staffs, pressure gauges, or some other devise, separated by some tens of meters. This is illustrated in Figure 5-4(a) for a pair of staffs within the nearshore, oriented parallel with the shoreline. When waves arrive at some angle to the shore, their crests will on average pass one staff at a slightly different time than at the second staff. This difference produces a phase shift in the records obtained by the two staffs, and this phase shift together with a knowledge of the celerity of the waves permits calculations of the wave approach angles with respect to the orientation of the staffs and shoreline. An extended version is provided by arrays of sensors, as illustrated in Figure 5-4(b), making possible the measurement of waves from any direction. The example illustrated here, which employs four pressure sensors, is the system used by the Coastal Data Information Program initiated in 1975 at the Scripps Institution of Oceanography (SIO) to monitor wave conditions along U.S. coasts (Seymour, Sessions, and Castel, 1985). This system, used in water depths less than 15 m, has a cable leading ashore to provide power and to deliver the measured data to a land-based recorder. In the standard mode of operation, each instrument array along the coast is interrogated once every 6 hours, when the central station at SIO initiates a telephone call to the shore station using an autodialer and normal telephone lines. The shore station responds by answering the call and transmitting the collected data.

Certain types of wave-measurement arrays combine pressure sensors to determine wave heights and periods, with an electromagnetic current meter to determine wave directions through measurements of the orbital velocities generated by the waves (Helmsley, McGehee, and Kucharski, 1991). More generally, electromagnetic current meters like those in Figure 5-5 have been used extensively in nearshore studies to measure all forms of water motions in order to understand their origins. Electromagnetic current meters are based on Faraday's law of electrodynamic induction—a conductor moving within a magnetic field generates an electrical potential at right angles to both the direction of motion and the magnetic-field axis. The electric potential is proportional in magnitude to the velocity of the conductor,

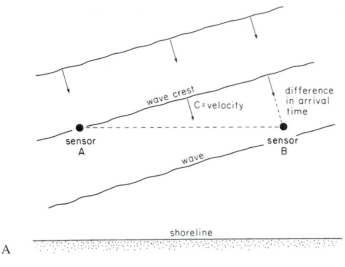

A

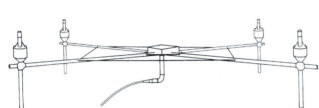

B

Figure 5-4 (a) A pair of wave staffs placed parallel to the shoreline so that the directions of wave approach toward the beach can be determined from differences in arrival times at the two sensors. (b) An offshore array consisting of four pressure sensors, used to determine the directions of waves as well as their heights and periods. [From Automated Remote Recording and Analysis of Coastal Data, R. J. Seymour, M. H. Sessions and D. Castel, *Journal of Waterway, Port, Coastal and Ocean Engineering*, 1985. Reproduced with permission from the American Society of Civil Engineers.]

and this is the key to its use in measuring water currents. In this application the current meter develops its own magnetic field using a solenoid, while the water is the conductor whose movement induces an electric potential that is proportional to the flow velocity. The response is not dependent on the conductivity of the water and available instruments work in fresh water as well as in the sea. The flowmeter is equipped with electrodes to measure the induced potential, the electrodes being placed symmetrically around the solenoid in the plane normal to the magnetic field. With this arrangement, the electrodes act in pairs to measure orthogonal components of flow in that plane. A single instrument can measure only two velocity components, so a determination of all three components requires two meters.

Electromagnetic current meters have a variety of shapes and sizes, but generally are either spherical, cylindrical, or disc shaped; the spherical design is shown in Figure 5-5. Being small, their presence does not seriously disturb the flow. Commercially available meters are able to measure velocities in the range 0 to 3 m/sec and have a nearly instantaneous time response and so can be used where there are rapid changes in magnitudes as required for measuring orbital motions under waves. Huntley (1979) has reviewed their use in studying nearshore waves and currents and indicates that they are able to measure the entire range of current motions from steady flow to oscillatory motions with periods as low as 0.1 sec.

A variety of remote sensing techniques for measuring waves is illustrated in Figure 5-3; Earle and Bishop (1984) provides a discussion of each. Those involving the use of radar from land, aircraft, and satellites represent ongoing research rather than being established tools. The use of sophisticated narrow-beam contouring radar essentially provides an instantaneous topographic map of the sea surface, and it has been demonstrated that this technique has the potential for providing high-resolution directional measurements of waves.

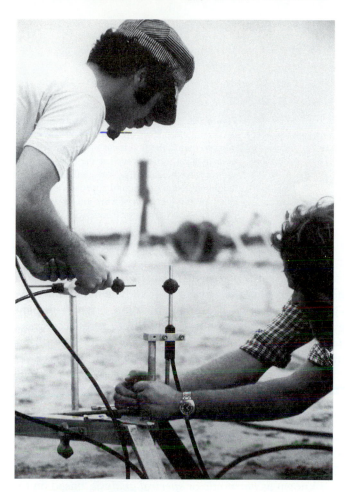

Figure 5-5 A Marsh-McBirney electromagnetic current meter that can measure two components of the water flow, including that due to wave orbital motions and any superimposed current. [Courtesy of D. Huntley]

The Analysis of Wave Data

The analysis of wave records obtained from any of the above devices usually takes one of two possible paths (Fig. 5-6). A statistical analysis of wave heights can be performed, noting the maximum wave height in the record, the average height, or a root-mean-square wave height. A commonly used statistical wave height is the *significant wave height,* defined as the average of the highest one third of the waves measured over a stated interval of time, usually 17 or 20 minutes (Thompson and Vincent, 1985). It is designated by $H_{1/3}$ or H_s. The number of waves to be averaged is determined by dividing one-third of the time duration of wave observations by the *significant wave period,* in turn defined as the average period of the highest one-third of the waves determined from large, well-defined groups of waves. The use of significant-wave parameters was based on the impression that in many applications the larger waves are more "significant" than the small waves, and H_s thereby provides a more representative measure of wave heights than, for example, the average wave height. It also has been shown that H_s roughly corresponds to a visual estimate of a representative wave height in that the observer naturally tends to weight his observation toward the larger waves. The significant wave period has less physical meaning than the height, and can lead to

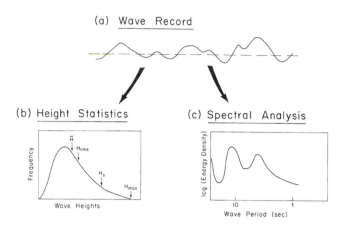

Figure 5-6 Schematic diagram illustrating the two main approaches used to analyze wave measurements, the first focusing on determinations of wave statistics such as the root-mean-square wave height (H_{rms}) or significant wave height (H_s), and the second approach used to calculate the spectrum of wave energy over the range of wave periods or frequencies.

appreciable errors if an attempt is made to employ it in calculations using the theoretical wave equations.

Other possible wave statistics are the *average wave height,* $\overline{H}$; the *root-mean-square wave height,* H_{rms}; the average of the highest 10 percent of the waves, $H_{1/10}$; and the maximum wave height H_{max}, which occurs during a given time interval. Assuming that there is a narrow band of wave periods and that the energy comes from a large number of different sources whose phase is random, Longuet-Higgins (1952) developed theoretical relationships for the ratios of the various wave statistics. A comparison of his theoretical values with measurements is given in Table 5-1, where it is seen that there is relatively good agreement. Cartwright and Longuet-Higgins (1956) extend the consideration to waves having any range of periods.

The above approach to derive statistics for the wave heights and periods can be useful in many applications, but more generally an analysis technique is required that unravels the waves generated by different storms and better describes the complete distributions of wave energies and periods. Basically, the problem is to work backward from the complexity of measured waves to determine the simple components whose summation yields that complexity. The procedure by which this is done is known as *harmonic* or *spectral analysis,* based on the mathematics of Fourier, who, in 1807, showed that any curve can theoretically be broken down into a series of sine waves having different lengths and amplitudes. The computational techniques were first developed in communications, but have been found to be applicable to the analysis of water waves (Munk, Snodgrass, and Tucker, 1959). The computations are long and tedious, and therefore must be performed on a computer. An example of the results is shown in Figure 5-7, obtained from a pair of wave staffs (K and L) placed parallel to the shore just beyond the breaker zone as diagrammed in Figure 5-4(a). The analysis presents the

TABLE 5-1 RELATIONSHIP BETWEEN SIGNIFICANT WAVE HEIGHT (H_s) AND OTHER WAVE HEIGHT STATISTICS

	$\overline{H}/H_s$	H_{rms}/H_s	$H_{1/10}/Hs$	H_{max}/H_s
Theoretical prediction by Longuet-Higgins (1952) for a narrow spectrum	0.64	0.71	1.27	1.53–1.85*
Analysis of 25 wave pressure records by Putz (1952)	0.62	—	1.29	1.87
Observations of Goodknight and Russell (1963) of broad-spectrum hurricane waves	0.60	0.69	1.25	1.57

*The value depends on the wave period and on the length of the record.

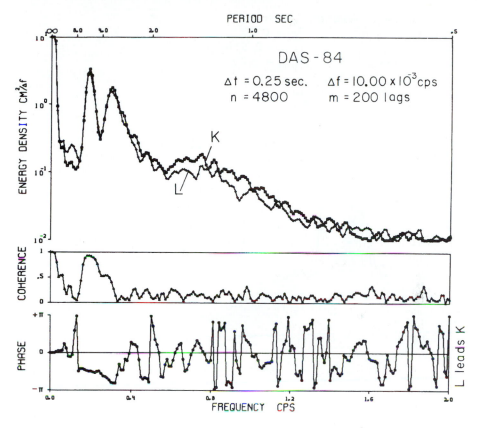

Figure 5-7 Wave spectra obtained from a pair of wave staffs, *L* and *K*, placed seaward of the breaker zone and parallel to the shoreline. Pronounced energy peaks appear at the periods 3.6 and 6.6 sec, corresponding to two wave trains approaching the beach.

energy density per unit frequency interval for each frequency or period; the "energy" units are cm^2 per frequency interval Δf, which can be converted into true energy through multiplication by $\rho g/8$, where ρ is the density of water, and g is the acceleration of gravity. In the example of Figure 5-7, there are two pronounced energy peaks centered on the periods 3.6 and 6.6 sec, demonstrating the presence of two dominant wave trains (the peak with a near-infinite period is the tide, recorded as a very long wave). The 3.6 sec peak represents locally generated waves and thus is broader since a range of wave periods is still being formed. The 6.6 sec waves have traveled beyond their area of generation, and this has narrowed their spectrum, so the waves are swell characterized by a specific period. The total energy in each individual wave train can be obtained by summing the energy densities under its peak. The phase relationship shown below the spectra in Figure 5-7 is obtained by cross-correlating the pair of staffs that are parallel to shore, and as described earlier, can be used to determine the direction of approach of the wave trains. The coherence is a measure of the reliability of the spectra and the phase measure—the higher the coherence the better the results.

The most complete description of waves is provided by a *directional wave spectrum*, which includes the direction of wave approach as well as the wave energy at a specific period or frequency (Borgman, 1979). This is illustrated in Figure 5-8 by a directional spectrum

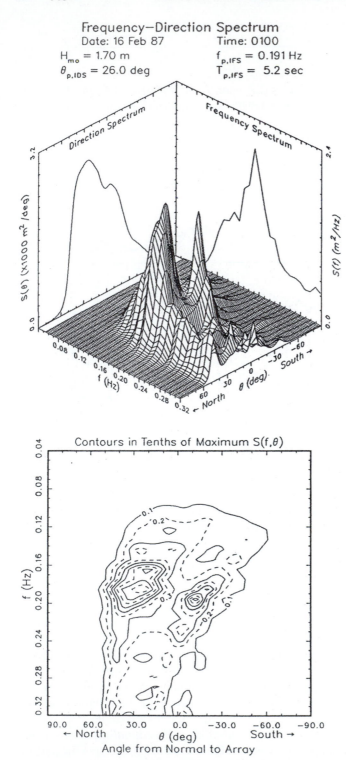

Figure 5-8   An example of a directional wave spectrum obtained at the Field Research Facility, Duck, North Carolina. [From C. E. Long, *Use of Theoretical Wave Height Distributions in Directional Seas,* U.S. Army Coastal Engineering Research Center Technical Report No. CERC-91-6.]

obtained at the Field Research Facility, Duck, North Carolina, utilizing a linear array of pressure sensors extending parallel to the shore (Long, 1991). The three-axis plot includes the wave energy density and frequency, as before, but now also includes the wave direction (north and south quadrants, with the $0°$ angle representing waves arriving from directly offshore). This particular example shows two major energy peaks, differing slightly in their central frequencies and periods, and approaching the coast with distinctly different angles. A standard two-dimensional energy versus frequency spectrum would not have distinguished these two peaks, demonstrated by the projected summation into the $S(f)$ energy versus frequency two-dimensional plane shown in the graph. This example illustrates the importance of analyses that fully account for the directions of wave advance as well as their energies and periods, since the approach angles are extremely important to the generation of longshore currents and sediment transport (Chapters 8 and 9).

Sources of Wave Data

Several countries have established programs to systematically collect wave data along their coasts: Japan, England (Draper, 1979), Canada (Wilson and Baird, 1981), Egypt (Nafaa, Fanos, and Elwany, 1991), and the United States. The sources of wave data for the United States include NOAA buoys of the NDBC, and the Coastal Data Information Program operated by the Scripps Institution of Oceanography. Florida has established the Coastal Data Network of the University of Florida to obtain wave measurements and other coastal data. In addition to these programs that obtain direct measurements of waves on a daily basis, the Wave Information Study (WIS) of the U.S. Army Corps of Engineers has documented the wave climate along the U.S. ocean coast and within the Great Lakes by hindcasting the wave conditions, that is, by analyzing the energies of the waves that would have been generated by measured storm winds and storm durations.

The NDBC buoy program acquires wave measurements from a large number of stations in the oceans and Great Lakes (Steele and Johnson, 1979; NDBC, 1992). The stations include deep-ocean locations, continental shelf stations, and experimental stations. All stations have the capacity to yield spectra of the measured waves including spectral-peak periods, the corresponding significant wave height, the average zero up-crossing wave period (Goda, 1974), as well as meteorological data (wind speeds and directions). The measurements are obtained hourly and transmitted via satellite to the land base. In addition to the moored buoys, NDBC also has a Coastal Marine Automated Network and Coastal System (C-MAN) that collects data on waves and meteorological conditions at numerous coastal sites, utilizing offshore platforms, lighthouses, and various beach locations.

The Coastal Data Information Program (CDIP) operated by the Scripps Institution of Oceanography is presently collecting wave data at 16 locations along the west coast of the United States, mostly from California with one station in Oregon and one in Washington (Seymour, Sessions, and Castel, 1985). As discussed above, measurements are obtained from directional arrays of pressure sensors [Fig. 5-4(b)] located in depths less than 15 m, as well as from deep-water moored buoys. In general, measurements are made four times each day, and analysis products include wave spectra and significant wave parameters. The data are summarized in monthly reports, and the total data set for each site is available on computer disc.

Prior to the general availability of wave measurements from U.S. coasts, the U.S. Army Corps of Engineers established the Littoral Environment Observations (LEO) program to obtain nearshore data at minimal cost (Sherlock and Szuwalski, 1987). Volunteer observers made daily estimates that included the breaker height, wave period, direction of wave approach, wind speed, wind direction, current speed, and current direction. The wave heights and directions were estimated visually, while the other measurements were made with simple equipment. The quality of the data depends on the skill of the individual observer, and therefore differs from site to site. The data generally can be used to establish a rough wave climate for a site, but should not be used for quantitative assessments of nearshore processes such as calculations of currents and sediment transport rates.

The WIS was initiated by the Corps of Engineers to yield a long-term wave climate for the U.S. coast (Helmsley and Brooks, 1989). Rather than involving direct measurements of the waves, which generally are unavailable for a sufficiently long time to adequately establish the wave climate, the WIS data instead are based on hindcast procedures, that is, the assessment of the wave conditions for documented weather patterns; the techniques are discussed below as to how wave heights and periods are predicted from wind speeds, storm durations, and aerial extents (fetches) of storms. The WIS analyses have been divided into three main phases. In Phase I the deep water wave data were hindcast for a spatial grid on the order of 2° along the coast. Phase II utilized the same meteorological information but at a finer scale (0.5°) to better resolve the sheltering effects of the continental geometry and at a time step of 3–6 hours. Phase II wave estimates are available for 71 stations along the Atlantic coast, 53 stations along the Pacific coast of California, Oregon and Washington, and 50 stations along the Gulf coast. Comparable data are available for the Great Lakes. Phase III involved the transformation of the Phase II wave data into shallow water; such transformations are available for the Atlantic coast and the Pacific coast north of Point Conception, California. Figure 5-9 is an example of WIS data sites for analyses of the three Phases of hindcast data, progressively moving the estimates closer to the shore. The data are listed in a number of reports available from the Coastal Engineering Research Center of the Corps of Engineers, and generally include directional wave spectra as well as significant-wave parameters hindcast at 3–6 hour intervals for some 20–30 years. The reports usually also contain summary statistics such as average monthly wave heights and periods and probabilities of extreme-wave statistics such as the significant wave height and period of the projected 100-year storm.

Hubertz, Driver, and Reinhard (1991) provide an example of the use of WIS data to analyze the wave climate of a specific location, in that case of the Great Lakes. In comparisons with buoy measurements of waves on the Great Lakes, systematic differences were found with the WIS hindcasts, requiring a recalibration of the analysis and a recalculation of the WIS data. Similar differences were found between WIS hindcasts and buoy measurements for the east coast (Miller and Jensen, 1990; Hubertz et al., 1994), again resulting in a recalculation of the WIS data. Because of the recalibrations, the WIS hindcasts now provide satisfactory descriptions of the wave climate for the Great Lakes and east coast of the United States. This is not the case for the west coast, where comparisons between WIS hindcasts and buoy data again established that systematic differences exist, with the significant wave heights of the WIS data being 30 percent too high (Hubertz, et al., 1992; Tillotson and Komar, 1997); unfortunately, the west coast WIS data have not been recalculated, so this systematic error remains.

By whatever method the waves are measured or hindcast, the resulting data on wave heights and periods can be analyzed in terms of their statistics and probabilities of occurrence. One valuable analysis is to determine the probability of severe wave conditions that might produce extreme erosion along the coast and damage structures such as jetties or offshore oil platforms. For such costly structures, it is required that they be able to survive an extreme event such as the combination of wave heights and periods that occur only once in 100

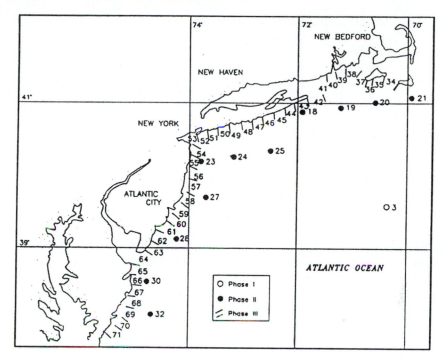

Figure 5-9 Example chart for the northeast coast of the United States showing the locations of available wave-data hindcasts derived from the WIS program of the U.S. Army Corps of Engineers. [Courtesy of U.S. Army Corps of Engineers]

years during an exceptionally severe storm. Inexpensive structures may be designed to withstand waves having only a 10–20 year recurrence interval. Unfortunately, one rarely has 100 years of actual wave measurements or even hindcast data upon which to base such an analysis. Instead, the analysis must be based on a shorter-term record of the wave climate, and then extrapolation to the expected 100-year recurrence event. Borgman (1963) provides a useful description of the fundamental concepts in working with extreme but rare events, Wang and LeMéhauté (1983) specifically address the problem of the duration of measurements needed to project long-term wave statistics, while Earle and Bishop (1984) provide a summary of the techniques and applications. Figure 5-10 gives an example product from Earle and Bishop, based on 30 years of hindcasts of all severe storms and hurricanes off Cape Hatteras, North Carolina. The procedure to generate such a graph is to rank the storms in order from lowest to highest significant wave heights, and then to assign probabilities to each by means of a plotting position equation such as $P_n = n/(N + 1)$, where P_n is the cumulative probability for a significant wave height less than or equal to the nth ranked height, and N is the total number of ranked values. For an annual series which has one maximum height per year, the return period is given by $R = 1/[1 - P(H)]$, where $P(H)$ is the cumulative probability for a significant wave height less than or equal to H. From this relationship, expected return periods of 10, 25, 50, and 100 years, respectively, correspond to the cumulative probabilities 0.90, 0.96, 0.98, and 0.99. This is the basis for the graph in Figure 5-10 for the Cape Hatteras wave data. Extreme-value data like that dealt with here often follow a theoretical probability distribution such as the Weibell distribution, and the graph paper of Figure 5-10 has axes designed so that any data

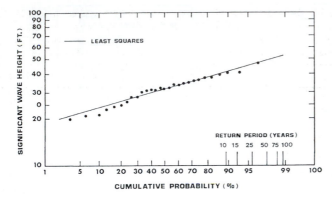

Figure 5-10 The analysis of extreme annual wave heights, extrapolated to estimate the height that might be expected to have a recurrence interval of 100 years, the height used in the design of many ocean structures. [Used with permission of Endeco Inc., from M. D. Earle and J. M. Bishop, *A Practical Guide to Ocean Wave Measurement and Analysis*, p. 55. Copyright © 1984 Endeco Inc.]

which follow the Weibell distribution form a straight line. With the data having established a straight line, it is easier to extrapolate the trend to probabilities corresponding to return periods of interest. In the example of Figure 5-10, the measurements were limited to 30 years but extrapolation yields a significant wave height of approximately 52 feet (15.8 m) for a 100-year return interval. Since this is the significant wave height, the average height of the one-third highest waves, still higher waves would occur during that extreme event; therefore, the probable maximum wave height for a given number of waves or the highest one-tenth wave height generally is selected for design purposes as in the construction of an offshore oil platform, utilizing the ratios given in Table 5-1 to determine these larger wave heights from the extrapolated significant wave height.

Storm waves on the open sea can be an awesome sight. From the bridge of the liner *Majestic,* Cornish (1934) observed waves in the North Atlantic and estimated that their heights reached 23 m. Rudnick and Hasse (1971) reported on waves observed from *Flip,* a research vessel that can be "flipped" into a vertical position by flooding its stern compartments to make it an ideal wave-observing platform; during an exceptional storm north of Hawaii, Rudnick and Hasse observed a maximum wave height of 24 m, with the significant wave height being 15 m. Offshore oil rigs equipped with automatic wave-recording instruments have reported the measurement of remarkable wave conditions during intense storms. Off the coast of Oregon, waves up to 18 m, and one 29-m wave, were reported by Watts and Faulkner (1968), and Rogers (1966) documented seas with waves of 15 m occurring under winds gusting up to 70 m/sec (150 mph). The highest wave reliably measured on the open ocean was seen from the *U.S.S. Ramapo,* a 146-m long naval tanker, while traveling from Manila to San Diego in February 1933 (Bascom, 1980). A wind of about 30 m/sec (65 mph) blew steadily from astern, generating waves in line with the direction of the ship's progress. The officer on watch had the presence of mind to line up the wave crests with the crow's nest and the horizon, and simple trigonometry enabled him to calculate the wave heights. He noted a sequence of increasing heights of 24, 27, 30, and 33 m and finally the maximum height of 34 m (112 feet).

A number of studies have investigated specific storms and wave conditions that were particularly destructive to coasts. The extreme 1962 Ash Wednesday storm that caused considerable destruction along the east coast of the United States was documented by Bretschneider (1964), while Dolan, Inman, and Hayden (1990) analyzed the extreme northeaster that struck the east coast during March 1989, and FitzGerald, van Heteren, and Montello (1994) documented the destruction resulting from the Halloween Eve storm of 1991.

Dolan, Lins and Hayden (1988) have analyzed 1,349 northeasters that occurred between 1942 and 1984 in order to better understand the characteristics of these storms that have resulted in a great deal of erosion along the eastern U.S. shore. On the west coast, major storms occurred during the 1982–1983 El Niño, causing much destruction; they have been analyzed by Earle, Bush, and Hamilton (1984); Seymour et al. (1984); and Komar (1986). Strong winds and extremely large waves occurred during a January 1988 southeaster, affecting the Pacific coast from Baja California to San Francisco; the articles by Strange, Graham, and Cayan (1989) and Seymour (1989), respectively, analyzed the meteorological and wave conditions, while other papers in that special issue of *Shore & Beach* examined the destructive impacts of the storm on the California coast. There are many papers that document the properties of hurricanes and the resulting destruction and coastal erosion. The most thoroughly investigated storm was Hurricane Hugo that during September 1989 produced considerable shoreline damage on Caribbean islands and then struck the southeast coast of the United States; it is documented by a series of papers in a special volume of the *Journal of Coastal Research* [see Finkl and Pilkey (1991)].

WAVE GENERATION AND PREDICTION

Although the observation is common, the generation of waves by the wind blowing over an initially still surface of water is a perplexing problem, the physics of which remains poorly understood. When the wind first begins to blow, small capillary waves or ripples are formed having periods less than 1 sec and heights of only a couple of centimeters. As time passes, waves with longer periods are formed, but small ripples continue to be present so that waves having a range of periods exist. The longer-period waves have longer wave lengths, and this also permits them to achieve greater heights without breaking. Concomitant with the progressive increase in wave periods present in the area of generation is an increase in wave heights. The longer and harder the wind blows, the larger the resulting waves.

The Processes of Wave Generation

In view of the significance of storms and the waves they generate, it is important to be able to predict the resulting wave characteristics—their heights, energies, and range of periods. At present, the generally accepted theory to account for the growth of waves is the combined Miles-Phillips mechanism, which incorporates two processes of energy transfer from wind to waves. In the initial growth stage, the portion developed by O. M. Phillips, there is a linear increase in wave energy with time due to a resonance between wind-induced atmospheric pressure fluctuations and the developing waves. Waves can be initiated from a calm sea by this mechanism. A second, more efficient growth mechanism, proposed by J. Miles, operates on waves already present, and involves an interactive coupling between the wind and waves. Starting from calm conditions, the initial generation of waves by the Phillips mechanism involves a linear growth of the wave energy, but there is then a shift to an exponential growth as predicted by the Miles theory.

Suppose that a turbulent flow of air begins to blow over an initially quiet water surface. Within the wind will be turbulent velocity eddies that are felt as gusts. Associated with these eddies are fluctuations in air pressure that act on the water surface in such a way as to produce small waves. It has been observed that as a gust moves over the water surface, a group of five to ten waves of uniform height and period may be generated. Phillips (1957, 1958a, 1958b, 1958c) extensively developed the role played by these pressure fluctuations associated with gusts. The waves develop by means of a resonance mechanism that occurs when a component of the surface pressure distribution moves at the same speed as the free-surface waves, thereby permitting the transfer of more energy from the wind to the waves.

Miles (1957, 1959) analyzed a simple logarithmic velocity profile of the wind over a water surface on which there already is present a small sinusoidal wave. His analysis examined the pressure variations on the water surface that result from the perturbation of the airflow by the sinusoidal surface. These variations produce an air pressure distribution that is greatest over the troughs and least over the crests. This pressure distribution causes the airflow in the lee of the crest to turn back, since it is flowing toward the higher pressure in the next trough, leading to a flow separation and crest-lee eddy as diagrammed in Figure 5-11. Jeffreys (1925, 1926) had earlier attempted to explain wave generation by a *sheltering theory,* assuming a wind pattern over the waves as seen in Figure 5-11. With this pattern, the normal air pressure at the water surface differs between the windward and leeward faces of any wave crest, with the wind pressing more strongly on the windward slope and aiding in the natural downward motion of the water surface, transmitting an extra force that results in the growth of the wave. Similarly, the weakened air pressure in the lee of the crest acts to enhance the upward motion of the water surface immediately in front of the moving crest, again resulting in the transfer of energy from wind to wave. The principal advance of the analyses of Miles over this earlier work of Jeffreys is that Miles did not *a priori* assume a certain degree of sheltering as did Jeffreys but instead directly calculated the flow separation and sheltering from the physical model.

Miles (1960) combined the theories of wave generation by turbulent pressure fluctuations (Phillips, 1957) and by surface sheltering (Miles, 1957) into the modern theory for the energy transfer from wind to waves. However, another important aspect is the transfer of energy between waves of different periods. We have seen that as wave generation continues, more and more energy appears at the longer periods within the entire wave spectrum. One mechanism of this transfer, suggested by Stewart (1967) and developed by Longuet-Higgins (1969), results when the short-period waves steepen and break on the crests of the longer period waves. Since the longer-period waves are traveling faster in deep water, they in effect sweep up the energy and momentum contained within the shorter-period waves that are being overtaken. Accordingly, the short-period waves (i.e., ripples) are continuously regenerated by the winds through the Miles-Phillips mechanisms, with the energy then transferred to the longer-period waves of the spectrum by the sweeping mechanism of Longuet-Higgins.

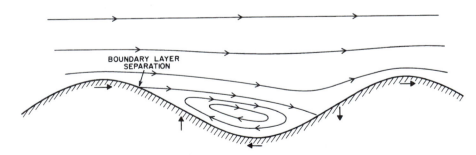

Figure 5-11   The sheltering effect of wave crests, which produces a pressure distribution of the air over the water surface that results in the transfer of energy from wind to waves.

The growth of waves is eventually limited by breaking, a process that dissipates their energy. An equilibrium can eventually be achieved wherein this energy loss by breaking balances the addition of new energy being transferred from the wind to the waves. In the prediction of waves, it is important to account for both the progressive increase in energy with time, as well as this equilibrium condition.

Wave Predictions—the Significant Wave Approach

In the absence of a full understanding of the physics of wave generation, semiempirical approaches have been developed for wave prediction. The analyses are semitheoretical, in that, theory is involved in their formulation, semiempirical, in that, the actual predictive relationships require data for the evaluation of various coefficients. The development of wave-prediction techniques, therefore, are based on the simultaneous collection of data on winds and the waves they generate. As new data accumulate, the relationships are modified and so continue to evolve. The resulting approaches can be classified on the basis of their results as (1) significant-wave methods and (2) wave-spectrum methods; this division follows the schematic diagram of Figure 5-6, discussed earlier.

The heights and periods of the waves generated are governed by the wind velocity U and the duration or time that the wind blows, D. The third important factor is the *fetch*, the distance F over which the wind blows (Fig. 5-12). The fetch distance restricts the time during which individual waves are moving under the action of the wind and therefore governs the time during which energy can be transferred from wind to waves. Both the significant-wave and spectrum methods relate the predicted wave parameters to these three storm conditions.

The significant-wave approach was introduced by Sverdrup and Munk (1946, 1947), and later revised with more data by Bretschneider (1952, 1957). The forecasting relationships have acquired the abbreviated name *S-M-B methods* after Sverdrup, Munk, and Bretschneider. The approach yields predictions of the significant wave height H_s and significant wave period T_s from the storm conditions: wind velocity U, fetch distance F, and storm duration D (Fig. 5-12). Bretschneider (1959) and Wiegel (1964) presented empirical graphs in terms of the dimensionless ratios gF/U^2, gD/U, gH_s/U^2, and gT_s/U; recent versions of such empirical

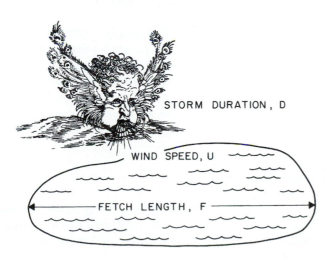

STORM DURATION, D

WIND SPEED, U

FETCH LENGTH, F

Figure 5-12 The principal storm factors important in the generation of waves.

curves are shown in Figure 5-13, as presented in the *Shore Protection Manual* of the Corps of Engineers (CERC, 1984). Here the wind "velocity" is given by

$$U_A = 0.71 U^{1.23} \tag{5.2}$$

where U is the measured or estimated wind speed (m/sec), and U_A is a "wind-stress factor" that is related to the stress (force/unit area) exerted by the wind on the water surface but has units of velocity (m/sec). The velocity of the wind increases systematically with distance above the water, so as used in the empirical relationships, it is necessary that U specifically be the value at 10 m elevation. CERC (1984) provides detailed information on how to estimate U from weather charts, how to modify the value so it does correspond to the 10-m elevation, and also provides correction procedures that account for temperature differences between the air and water. Once these procedures have been followed to determine the wind speed U, Equation (5.2) is used to convert the speed into the wind-stress factor U_A, which appears in the empirical relationships for wave prediction. The curves in Figure 5-13 relate gH_s/U_A^2 to gF/U_A^2, thereby accounting for the dependence of the significant wave height on the wind speed and its growth with increasing storm fetch distance, F. To the right in the diagram, the curves level off at values that depend on gd/U_A^2, where d is the water depth, thereby accounting for depth limitations to wave growth. The "deep-water wave" line is for wave generation where there are no depth limitations, yielding the empirical relationship

$$\frac{gH_s}{U_A^2} = 1.6 \times 10^{-3} \left(\frac{gF}{U_A^2} \right)^{1/2} \tag{5.3}$$

which accounts for the increase in wave height H_s with fetch distance, F. The upward limit of this deep-water relationship is the line given by

$$\frac{gH_s}{U_A^2} = 2.433 \times 10^{-1} \tag{5.4}$$

for "fully developed seas" (Fig. 5-13), the equilibrium condition of wave generation where the distance of storm fetch is no longer limiting the growth of the waves. This equilibrium condition can be affected by shallow water depths, d, and this accounts for the series of horizontal lines that depend on gd/U_A^2. The complex equation that yields gH_s/U_A^2 for shallow-water wave conditions is inserted into the diagram of Figure 5-13.

These relationships can be used to develop nomograms that simply relate the significant wave height to the wind-stress factor U_A and fetch length F; Figure 5-14 is an example from the *Shore Protection Manual* that gives both the significant wave height and period, and also includes a set of lines for the minimum duration of the storm required to achieve the wave conditions where the fetch is otherwise the limiting factor. Although such wave-prediction analyses have focused on the significant wave height H_s, the ratios presented in Table 5-1 can be used to determine the other wave statistics ($H_{1/10}$, H_{rms}, and H_{max}) as well as the complete statistical distribution of wave heights.

Wave Predictions—the Wave Spectrum Approach

The S-M-B approach to predict wave heights and periods came into being during World War II when simple wave predictions were needed. This approach is still used when a simple representation of the wave climate is adequate. However, more sophisticated applications re-

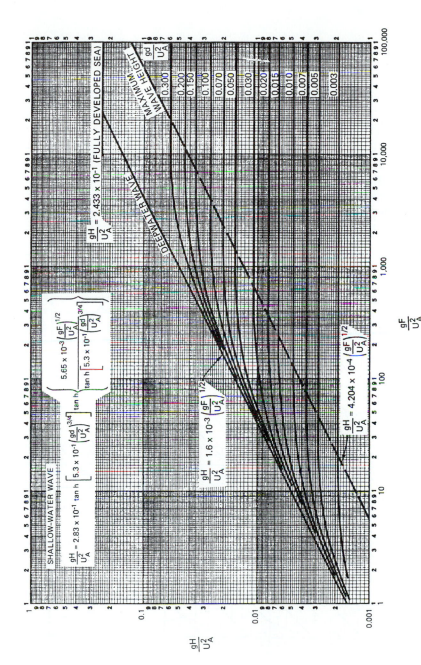

Figure 5-13 The empirical curves relating the dimensionless ratios gH/U_A^2 versus gF/U_A^2, where H is the significant wave height generated by a storm having a fetch F and wind-stress factor U_A. [From CERC, Shore Protection Manual, 1984, U.S. Army Corps of Engineers]

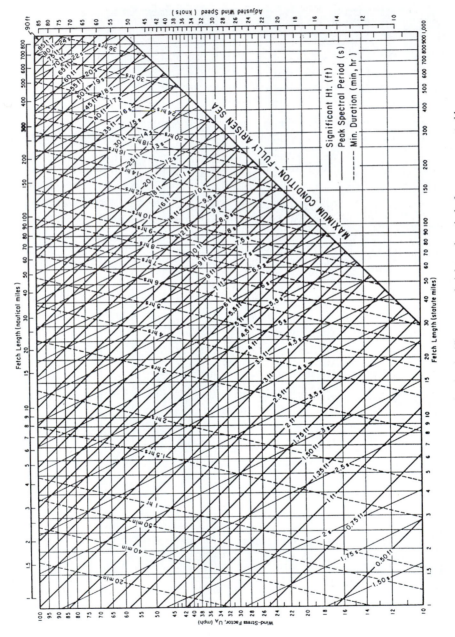

Figure 5-14 A nomogram for the significant wave height and period of waves generated by a wind-stress factor U_A and fetch. [From CERC, Shore Protection Manual, 1984, U.S. Army Corps of Engineers]

quire the characterization of the waves by their spectra, providing a measure of the wave energy at each period or frequency. Therefore, the more modern approaches for wave prediction attempt to produce the wave-energy spectrum for a given wind speed, duration, and fetch.

As discussed above, the generation of waves involves the transfer of energy from wind to waves, with the initial growth of short-period, high-frequency ripples and the subsequent transfer of that energy to longer-period, lower-frequency waves. This growth pattern is supported by clusters of spectra of waves being generated, as seen in Figure 5-15 from the work of Phillips (1958b). Each wave spectrum in a developing sea can be thought of as being divided into two portions, an equilibrium range of higher frequencies (lower periods), where the energy content is at a maximum, and a growth range of lower frequencies (longer periods), where energy is still being added. In the equilibrium range it is found that the decrease in energy with wave frequency follows a f^{-5} frequency dependence (shown in Figure 5-15 as the σ^{-5} convergence of the spectra, where $\sigma = 2\pi/T = 2\pi f$ is the radian frequency). In this range the waves are saturated with energy and can grow no higher, so that as more energy is added, it is immediately transferred to lower frequencies on the other side of the peak within the growth range of the spectrum or is lost through wave-breaking turbulence. Because of the shift of the peak period during wave growth, the growth range also progressively shifts to lower frequencies (longer periods), the equilibrium range expanding to occupy periods that previously were in the growth range. The cluster of lines on the left of Figure 5-15 in the growth range are representative of wave spectra for which equilibrium has not been achieved.

Wave prediction schemes that yield spectra must account for the progressive shift in the peak period separating the equilibrium and growth ranges and must also account for the total energy within the spectrum and how it depends on wind speed, fetch, and duration. The earliest approaches to predicting spectra are provided by the studies of Pierson, Neumann and James (1955), yielding the *P-N-J method,* and by Darbyshire (1963), Bretschneider (1963), and Pierson and Moskowitz (1964). The P-N-J method predicts that the peak period is related to the wind velocity by $T_{peak} = 0.78U$, where U is the wind speed (m/sec), while the total energy within a fully developed sea is proportional to the fifth power of the wind speed. A comparatively simple wave-spectrum formulation is that provided by Liu (1971), based on combined data from Lake Michigan, laboratory studies, and from the ocean. His equation for the wave spectrum is

$$E'(\sigma) = \frac{0.4g^2}{F_o^{1/4}\sigma^5} \exp\left[-5.5 \times 10^3 \left(\frac{g}{U_* F_o^{1/3}\sigma}\right)^4\right] \tag{5.5}$$

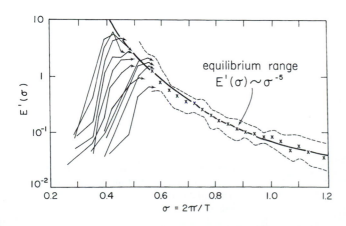

Figure axis labels: $E'(\sigma)$ (vertical), $\sigma = 2\pi/T$ (horizontal), with values 0.2, 0.4, 0.6, 0.8, 1.0, 1.2 and equilibrium range $E'(\sigma) \sim \sigma^{-5}$

Figure 5-15 A cluster of spectra of wind-generated waves obtained at different times. On the right the curves merge in the range of frequencies known as the equilibrium range, where wave conditions are already fully developed. The heavy line depicts a f^{-5} decrease in wave energy with increasing frequency *f*. On the left is a cluster of lines representing the spectra at low frequencies (high periods) in the growth range where equilibrium has not been achieved. [Adapted with permission of Cambridge University Press, from O. M. Phillips, The Equilibrium Range in the Spectrum of Wind-Generated Waves, *Journal of Fluid Mechanics, Vol. 4*, p. 432. Copyright © 1958 Cambridge University Press.]

for fetch-limited, deep-water waves. Here U_* is the friction wind velocity, similar to U_A in being a measure of the stress exerted by the wind; $F_o = gF/U_*^2$ is the dimensionless fetch parameter, and $\sigma = 2\pi/T$ is the radian frequency. Figure 5-16 shows a series of spectra calculated with Equation (5.5) for a wind speed $U_{10} = 20$ m/sec at the standard elevation of 10 m above the water surface and for various fetch distances ranging from 50 to 2000 km. The spectra are wide at early stages of development without a pronounced peak but become narrower with the front face steeper as the fetch distance increases. The spectral energy density $E'(s)$ approaches an equilibrium state at high frequencies as the fetch increases, just as observed in the measured spectra of Figure 5-15, but the low-frequency waves in the growth range continue to increase in energy, not having reached a fully developed sea state. All of these behaviors of the Liu spectrum given by Equation (5.5) agree with observed changes in wind-wave spectra as waves are generated.

One of the most commonly used spectral formulations is the JONSWAP *spectrum*, derived from the Joint North Sea Wave Project conducted during 1968–1969 off the west coast of Denmark (Hasselmann et al., 1973, 1976). As shown in Figure 5-17, waves were recorded at a series of offshore distances in a line normal to the shore of the island of Sylt, while the lower graph contains a series of spectra measured when the winds were blowing in the offshore direction, showing how the wave spectrum develops downwind. Formulation of the JONSWAP spectrum incorporates an energy transfer between various frequency components and limitations due to breaking, leading to an energy-saturated equilibrium spectrum. Its general form is

$$E(f) = \frac{\alpha g^2}{(2\pi)^4 f^5} \exp\left[-1.25\left(\frac{f_m}{f}\right)^4\right]\gamma \exp\left[\frac{-(f/f_m - 1)^2}{2\omega^2}\right] \tag{5.6a}$$

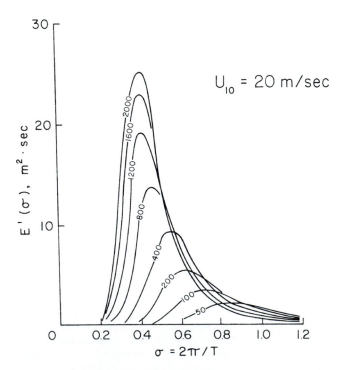

$U_{10} = 20$ m/sec

Figure 5-16   The fetch-limited spectra of Liu (1971) calculated with Equation (5.5) for a wind speed of 20 m/sec and for a series of fetch lengths in kilometers.

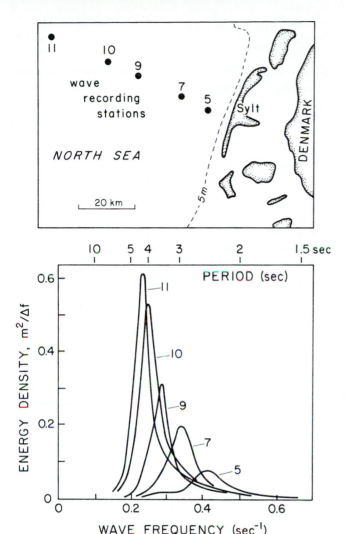

Figure 5-17 A series of spectra of waves measured at recording stations offshore from the island of Sylt on the North Sea coast of Denmark during the JONSWAP experiments. At the time of measurement, the winds were directed from land toward the offshore, resulting in the growth of the spectra from stations 5 through 11. [From K. Hasselmann, et al., Measurements of Wind-wave Growth and Swell Delay During the Joint North Sea Wave Project, *Deutsche Hydrographische Zeitschrift,* Supplement A8, 1973.]

for the energy density $E(f)$ as a function of the frequency $f = 1/T$, where f_m is the frequency of the spectral peak, α is the Phillips "constant," and γ and ω are coefficients that govern the peakedness and asymmetry of the spectrum. Based on comparisons with numerous measured spectra, it was established that although the coefficients are widely scattered, they can be taken approximately as constants having the values

$$\gamma = 3.3$$

$$\omega = 0.07 \text{ when } f < f_m$$

$$= 0.09 \text{ when } f > f_m$$

However, the proportionality coefficient α has a weak dependence on the fetch distance F and wind velocity U according to

$$\alpha = 0.076\left(\frac{gF}{U^2}\right)^{-0.22} \tag{5.6b}$$

while the nondimensional peak frequency (Uf_m/g) and total energy within the spectrum $(E_T g^2/U^4)$ have stronger dependencies on the fetch distance:

$$\frac{Uf_m}{g} = 0.076\left(\frac{gF}{U^2}\right)^{-0.33} \tag{5.6c}$$

$$\frac{E_T g^2}{U^4} = 0.076\left(\frac{gF}{U^2}\right) \tag{5.6d}$$

Reduced from their nondimensional forms, these relationships indicate that $f_m \propto (FU)^{-0.33}$ or the corresponding peak period $T_m \propto (FU)^{0.33}$, and the total energy is $E_T \propto FU^2$; the period and energy therefore increase with fetch and wind speed as expected. The JONSWAP spectrum of Equation (5.6) is similar to those proposed by earlier investigations, including the Liu (1971) spectrum of Equation (5.5), and has similar dependencies on the storm fetch and wind speed; its main improvement is its being supported by more data but in particular due to its development having been based on more refined models of energy transfer from wind to waves and between waves of different frequencies due to nonlinear interactions. The full analyses of these aspects of the development can be found in Hasselmann et al. (1973, 1976).

The JONSWAP spectrum is limited in its application to waves generated in deep water. Bouws et al. (1985) have modified the form to a finite-depth wind-wave spectrum, which they term the *TMA spectrum* (named after the three data sets used to establish the spectrum—the Texel, Marsen, and Arsloe data). The resulting spectral equation is the JONSWAP formulation multiplied by a factor given as a function of depth and wave frequency. Modified forms and values for the $\alpha, \gamma,$ and ω coefficients of the JONSWAP spectrum were derived by fitting the TMA spectra form to more than 2,800 measured spectra covering a wide range of conditions. Figure 5-18 illustrates the form of the TMA spectrum for a fixed peak frequency $f_m = 0.1 \text{ sec}^{-1}$ and generated by the same wind speed and other storm factors. Only the water depth of wave generation is varied in the series of spectra, and it is seen that this dependence is very strong, with waves generated in shallow water depths having much less total energy as indicated by the areas under the respective curves. The TMA spectrum represents the most broadly based and applicable wave spectrum available today.

WAVE THEORIES

To understand the movement of waves as they cross the deep sea after they leave the area of generation, and then follow their transformations as they enter shallow water near the coast, it is necessary to have mathematical relationships that predict the velocity of wave movement, the change in wave height with depth, and equations for the wave energy and power. In developing mathematical theories for these wave properties, equations also are derived that describe the orbital motions and velocities followed by water particles beneath the waves, the movement of which is important to the entrainment and transport of sediment on the sea floor.

There are several wave theories that commonly are used in applications, ranging from the comparatively simple linear Airy-wave theory to the complex cnoidal theory, with soli-

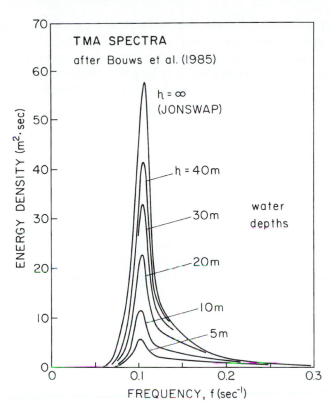

Figure 5-18 The equilibrium TMA spectrum calculated for a fixed peak frequency but for waves generated in various water depths for the same storm conditions, showing that the depth represents a strong constraint on the total energy of the generated waves. [Adapted with permission of American Geophysical Union, from E. Bouws, H. Günther, W. Rosenthal and C. L. Vincent, Similarity of the Wind Wave Spectrum Infinite Depth Water, 1. Spectral form, *Journal of Geophysical Research 90*, p. 979. Copyright © 1985 American Geophysical Union.]

tary waves being a special case involving a single, isolated wave as the name implies. The emphasis here will be on the Airy- and Stokes-wave theories, which are, respectively, applicable in deep and shallow water and have reasonably simple mathematical relationships describing their motions. Because of its complexity, little consideration will be given to cnoidal waves even though they potentially have the widest range of applications, while there will be a brief examination of solitary wave theory as it has been used in some applications. More extensive reviews of the wave theories with complete derivations can be found in Kinsman (1965), LeMéhauté (1976), and Dean and Dalrymple (1984).

Linear Airy Wave Theory

Derivations of relationships describing the internal motions of fluids all begin with the differential forms of the equations of motion, the Navier-Stokes equations, mathematical relationships that balance the changes in momentum of the flowing water against the forces causing those changes. However, all of the derivations that focus on the movement of waves neglect friction, so the full momentum equations reduce to the Euler equation that contains only terms for the water-particle velocities and local pressures. A second relationship that must be included in analyses of wave motions is a continuity equation, one that ensures the conservation of water mass. In order to simultaneously solve these equations, additional simplifying assumptions are required; the greater the number of assumptions, the simpler the resulting equations but the greater the potential departure from the actual wave motions one is trying to describe. Yet the simplest of the wave theories, Airy-wave theory, can still be used in many applications, particularly in describing swell waves of approximately uniform heights in deep water or the linear summation of many sinusoidal waves to generate a complex sea. In the derivation of Airy-wave theory, it is assumed that the wave height is much smaller than the wave length and water depth. With this approximation, the water-surface elevation $\eta(x,t)$ obtained in the solution is found to be

$$\eta(x,t) = \frac{H}{2} \cos(kx - \sigma t) \tag{5.7}$$

where H is the wave height, x is the coordinate axis in the direction of wave advance, t is time, $k = 2\pi/L$ is the *wave number,* where L is the wave length, and $\sigma = 2\pi/T$ is the *radian fre-quency* where T is the wave period. These parameters and others involved in the Airy solu-tion are illustrated in Figure 5-19, while Table 5-2 summarizes the principal equations of interest derived in the Airy-wave solution. The profile of the Airy wave is sinusoidal either with distance x at a fixed time or with time t at a fixed position ($x = $ constant).

A particularly fundamental and important relationship to emerge from the Airy-wave derivation is the *dispersion equation,*

$$\sigma^2 = gk \tanh(kh) \tag{5.8}$$

or substituting the identities $\sigma = 2\pi/T$ and $k = 2\pi/L,$

$$L = \frac{g}{2\pi} T^2 \tanh\left(\frac{2\pi h}{L}\right) \tag{5.9}$$

where h is the water depth. This relationship is difficult to solve directly in that it contains the wave length L on both sides of the equation, and for good measure L is contained within the hyperbolic tangent on the right side. Figure 5-20 contains a graph of the hyperbolic functions

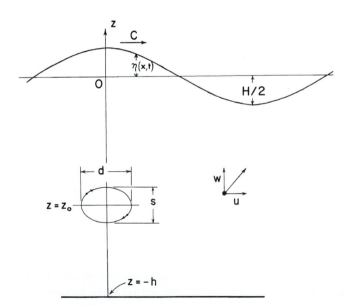

Figure 5-19   The notation associated with Airy-wave theory. The origin of the coordinate system is at the still-water surface, with z positive up-ward and the sea floor at $z = -h$. The water sur-face is sinusoidal, and each water particle moves with a horizontal velocity u and vertical velocity w in an elliptical orbit of major diameter d and minor diameter s.

TABLE 5-2 EQUATIONS DERIVED FROM LINEAR AIRY-WAVE THEORY

Parameter	General Expression	Deep Water	Shallow Water
Surface elevation	$\eta(x,t) = \dfrac{H}{2}\cos(kx - \sigma t)$		
Phase velocity	$C = \dfrac{gT}{2\pi}\tanh\left(\dfrac{2\pi h}{L}\right)$	$C_\infty = \dfrac{gT}{2\pi}$	$C_s = \sqrt{gh}$
Wave length	$L = \dfrac{gT^2}{2\pi}\tanh\left(\dfrac{2\pi h}{L}\right)$	$L_\infty = \dfrac{gT^2}{2\pi}$	$L_s = T\sqrt{gh}$
Horizontal orbital diameter	$d = H\dfrac{\cosh[k(z_o + h)]}{\sinh(kh)}$	$d = He^{kz_o}$	$d = \dfrac{HL_s}{2\pi h} = \dfrac{HT}{2\pi}\sqrt{\dfrac{g}{h}}$
Vertical orbital diameter	$s = H\dfrac{\sinh[k(z_o + h)]}{\sinh(kh)}$	$s = He^{kz_o}$	$s = 0$
Horizontal orbital velocity	$u = \dfrac{\pi H}{T}\dfrac{\cosh[k(z_o + h)]}{\sinh(kh)}\cos(kx - \sigma t)$	$u = \dfrac{\pi H}{T}e^{kz}\cos(kx - \sigma t)$	$u = \dfrac{H}{2}\sqrt{\dfrac{g}{h}}\cos(kx - \sigma t)$
Vertical orbital velocity	$w = \dfrac{\pi H}{T}\dfrac{\sinh[k(z_o + h)]}{\sinh(kh)}\sin(kx - \sigma t)$	$w = \dfrac{\pi H}{T}e^{kz}\sin(kx - \sigma t)$	$w = 0$

for the r range of interest. From this graph it is apparent that as $r = kh = 2\pi h/L$ becomes large, $\tanh(2\pi h/L) \approx 1$, so that Equation (5.9) reduces to

$$L_\infty = \frac{g}{2\pi}T^2 \tag{5.10}$$

The subscript ∞ is used here to denote this as the *deep-water approximation,* so-called because of the assumption that h is large compared with the wave length (one commonly sees the sub-

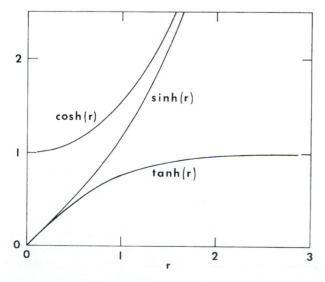

Figure 5-20 Values of the hyperbolic functions.

script 0 used to denote the deep-water condition, as in L_0, but this subscript is used here to denote the wave-orbital motions at the sea floor). This deep-water approximation is generally limited to $h > L_\infty/2$, which limits the error in its use at less than 5 percent compared with utilizing the general Equation (5.9). Employing the basic relationship of Equation (5.1), $C = L/T$, the corresponding deep-water phase velocity is

$$C_\infty = \frac{g}{2\pi} T \tag{5.11}$$

There is also a *shallow-water approximation* to Equation (5.9) when $h < L_\infty/20$ (5 percent error), in which case $\tanh(2\pi h/L) \approx 2\pi h/L$ (Fig. 5-20), so the general relationship of Equation (5.9) now reduces to

$$L_s = T\sqrt{gh} \tag{5.12}$$

and the phase velocity is

$$C_s = \sqrt{gh} \tag{5.13}$$

The subscript s denotes this shallow-water approximation. In contrast to the deep-water condition where the wave length and phase velocity depend only on the wave period, in shallow water the primary dependence is on the water depth. As will be discussed in subsequent sections, these dependencies have important ramifications for the dispersal of waves in deep water by their respective periods, while the depth control in shallow water can affect the direction of wave movement by wave refraction.

Summarizing the regions of application of the approximations:

Deep water	$h/L_\infty > 1/2$
Intermediate water (general equations)	$1/4 > h/L_\infty > 1/20$
Shallow water	$h/L_\infty < 1/20$

$$\tag{5.14}$$

In the intermediate depth range, where we cannot use the deep-water or shallow-water approximations if the errors are not to exceed 5 percent, it is necessary to employ the general Equation (5.9). Tables have been developed, derived from the solution of this equation, which can be found in Wiegel (1954, 1964) and in the *Shore Protection Manual* (CERC, 1984). The variations of L/L_∞ and C/C_∞ with h/L_∞ obtained from these tables are graphed in Figure 5-21; the other curves presented in this graph will be discussed below. If an equation is desired rather than a set of tables or a graph, an approximate alternative to the general relationship of Equation (5.9) is given by

$$L = L_\infty \left[\tanh\left(\frac{2\pi h}{L_\infty}\right) \right]^{1/2} \tag{5.15}$$

derived by Eckart (1952). This equation may be used in intermediate water depths, and can be directly solved for the wave length L at the depth h, having calculated the deep-water wave length L_∞ with Equation (5.10). The Eckart relationship of Equation (5.15) more clearly shows that as the water depth decreases while waves approach the coast, the wave length L and corresponding phase velocity $C = L/T$ will be reduced from their deep-water values.

The Airy-wave solution also provides a description of the water-particle movement at depth beneath the surface, the horizontal and vertical components of the velocity denoted,

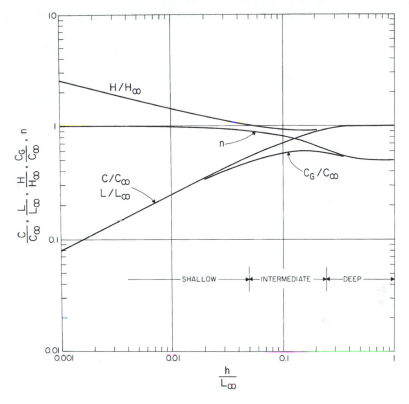

Figure 5-21 Shoaling transformations for Airy waves as functions of the ratio of the water depth, h, to the deep-water wave length L_∞ that depends only on the wave period.

respectively, by u and w, as depicted in Figure 5-19, and also of the trajectories of the particle paths. The equations obtained in the solution are listed in Table 5-2, and, as above, there are deep-water and shallow-water approximations to the general expressions. According to the general solution, the water particles follow elliptical paths, as shown in Figures 5-19 and 5-22, having horizontal and vertical axial diameters denoted by d and s. The ellipse is centered at a water depth $z = z_0$, which is a negative value as specified by the coordinate system employed where $z = 0$ at the still-water surface and $z = -h$ represents the sea floor (Fig. 5-19). The deep-water approximation is

$$d = s = He^{kz_o} = He^{\frac{2\pi z_o}{L_\infty}} \tag{5.16}$$

which represents a circular motion since $d = s$, with the orbital motion at the water surface ($z_0 = 0$) being $d = s = H$, the height of the wave (Fig. 5-22)—this corresponds to the motion of a floating cork as described at the beginning of this chapter. The size of the orbital diameter decreases exponentially with depth since z_0 has negative values, and the rate of decrease with depth depends on the wave length $L_\infty = (g/2\pi)T^2$ and hence on the wave period. The results establish that the motions of short-period waves decrease with depth much more rapidly than long-period waves. The orbital velocities show similar dependencies, Table 5-2, but like the profile expression of Equation (5.8), they are a function of either the sine or

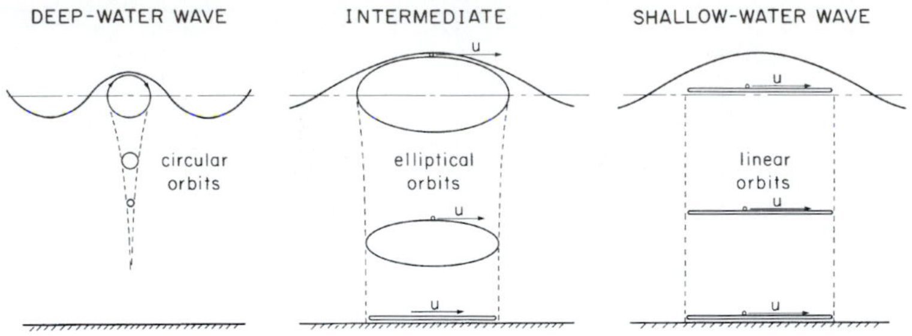

Figure 5-22 A diagram of the orbital motions of Airy waves in deep water where the orbits are circular, and in intermediate water depths where the orbits are elliptical but become flatter and smaller as the bottom is approached. In shallow water Airy theory predicts that all water motions consist of to-and-fro horizontal movements, uniform with depth.

cosine of $(kx - \sigma t)$, since they vary spatially at a fixed time t or with time at a fixed position defined by x and z.

In intermediate water depths, the water particles follow elliptical paths (Fig. 5-22), with the sizes of the ellipses decreasing downward through the water column and becoming flatter as the bottom is approached since there can be no water movement through the bottom itself. Accordingly, at the bottom $(z_0 = -h)$ the vertical orbital diameter is $s = 0$ and velocity $w = 0$, while the horizontal components are

$$d_0 = \frac{H}{\sinh(kh)} \tag{5.17a}$$

$$u_0 = \frac{\pi H}{T \sinh(kh)} \cos(kx - \sigma t) \tag{5.17b}$$

Since $\sinh(kh)$ increases with increasing $kh = 2\pi h/L$ (Fig. 5-20), both d_0 and u_0 at the sea floor will decrease with increasing water depth h. The greater the wave height H, the larger the bottom orbital diameter and velocity. The existence of a horizontal velocity at the bottom results from the assumption of zero viscosity and friction in deriving Airy-wave theory. This cannot strictly be the case, in that the velocity right at the sediment interface must be zero due to the no-slip condition between solids and fluids. In reality there is a thin boundary layer with a velocity gradient at the bottom, with the velocity decreasing to zero at the bed itself. Above the boundary layer, which is generally only some 1–3 cm thick, the viscous effects of the fluid are negligible so the derived equations for the orbital motions are found to provide a satisfactory description of the actual water motions under waves (so long as the wave height is small compared with the water depth and wave length so that Airy-wave theory is applicable).

In shallow water the orbital diameter reduces to $s = 0$ and

$$d = \frac{HT}{2\pi} \sqrt{\frac{g}{h}} = \frac{H}{kh} \tag{5.18}$$

indicating that the ellipses have flattened to horizontal motions throughout the water column (Fig. 5-22), not just at the bottom. In shallow water kh is always less than 0.6, so the orbital diameter is greater than the wave height. There is no dependence on z, so the orbital diameter is constant from the surface to the bottom, and the horizontal orbital velocity u is likewise independent of z. This is a rather unsatisfactory description of the water-particle motions actually observed beneath waves in shallow water and results from the simplifying assumptions made in deriving Airy-wave theory, in particular, the assumption that the wave height is negligibly small compared with the water depth. More advanced analyses such as the Stokes-wave theory do not make this assumption, and accordingly provide a better description of particle motions and other attributes of waves in shallow water.

Thus far we have reviewed the Airy-wave equations for the surface expression of the motion and the internal orbital motions. Airy theory also provides equations for the energy contained within the waves and for the propagation of that energy as the waves cross the ocean. Although there is no net movement of water during wave propagation, since the water particles always return to their starting positions according to Airy-wave theory, the motion of the wave itself constitutes a transfer of energy over the sea surface. The displacement of the water surface away from the flat still-water condition gives the wave form a potential energy. At the same time, the orbital motions of the water under the wave constitute a kinetic energy. Integrating both potential and kinetic energies along the full length of the wave yields the total *energy density E* of the Airy wave as

$$E = \frac{1}{8}\rho g H^2 \tag{5.19}$$

and is thus related directly to the wave height H; ρ is the density of water. This relationship gives the "density" of the energy in the sense that it has been summed (integrated) over one wave length, so it becomes the total energy per unit length of wave crest. In the metric system its units are ergs/cm^2 or Newtons/m^2 (energy/length^2). As waves cross the continental shelf and pass into progressively shallower water, the wave height H changes, generally increasing until the waves reach the nearshore and break. This means that E will also vary during wave shoaling, not being conserved as is usually thought of for energy (since it is not the energy alone but an energy density). Instead, it is the flux of energy that is approximately constant, the rate at which the energy density is carried along by the moving waves. This *energy flux* is given by

$$P = ECn = EC_g \tag{5.20}$$

where C is the speed of individual waves (the phase velocity or celerity), and

$$n = \frac{1}{2}\left[1 + \frac{2kh}{\sinh(2kh)}\right] \tag{5.21}$$

The velocity $C_g = Cn$ is the speed at which the energy is carried along, the so-called *group velocity* since it is the rate of movement of groups of waves as distinguished from individual waves that move with the phase velocity C. The graph of n versus h/L_∞ is presented in Figure 5-21. In deep water $n = 1/2$ but increases in value as the waves travel into water of intermediate depth, becoming $n = 1$ in shallow water. This means that, in deep water, individual waves advancing with their phase velocity C are traveling at twice the speed of the group as a whole, that is, faster than the wave energy is being propagated. The effect of this difference can be seen in laboratory wave channels where a finite number of waves are generated in deep water. It has been observed that the leading wave progressively declines in height and eventually disappears as it outruns the energy of the group, while a new wave is formed at the end of the group from the energy that is being "left behind" by the waves moving with the celerity C. The important ramifications to wave transformations and propagation in the deep ocean will be discussed later.

The wave energy flux of Equation (5.20) is denoted by P because it also represents the *wave power* or, more correctly, the power per unit wave-crest length. Snyder, Wiegel, and Bermel (1957) have shown that Equation (5.20) is a correct representation of the power or energy flux of the waves, demonstrated by generating waves at one end of a laboratory channel and measuring the transmitted power at the other end.

Also associated with the wave advance is a flux or transmission of momentum. Longuet-Higgins and Stewart (1960, 1964) have defined the *radiation stress* as "the excess flow of momentum due to the presence of the waves." The radiation stress is the "excess" momentum in that the dynamic pressure is used, the hydrostatic pressure being subtracted from the absolute pressure. This ensures that the assessed momentum is due solely to the presence of the waves. If the x-axis is again placed in the direction of wave advance, while the y-axis is parallel to the wave crests, then there are two non-zero components to the radiation stress: the x- and y-fluxes of x-momentum and y-momentum. The radiation stress or momentum flux across the plane $x = $ constant (parallel to shore) in the direction of wave advance (the x-direction) is found to be given by

$$S_{xx} = E\left[\frac{2kh}{\sinh(2kh)} + \frac{1}{2}\right] = E\left(2n - \frac{1}{2}\right) \tag{5.22}$$

Despite the orbital velocity component parallel to the wave crest being zero, there is still a flux of momentum in the y-direction because the pressure departs from the hydrostatic when the waves are present. This flux of y-momentum across the plane $y = $ constant is

$$S_{yy} = E\left[\frac{kh}{\sinh(2kh)}\right] = E\left(n - \frac{1}{2}\right) \tag{5.23}$$

In deep water $n = 1/2$ so that $S_{xx} = E/2$ and $S_{yy} = 0$; in shallow water with $n = 1$, $S_{xx} = 3E/2$, and $S_{yy} = E/2$.

Radiation stress has proved to be a very powerful tool in the analysis of a variety of wave-related phenomena. In Chapter 6, radiation stress will be used to predict changes in the mean water level (set-down and set-up) associated with the arrival of waves in the nearshore, and in Chapter 8 it will be seen that the longshore-directed component of the radiation stress is responsible for the generation of longshore currents. Other applications have been to the generation of surf beat, the interactions of waves with steady currents, and the steepening of short gravity waves on the crests of longer waves (Longuet-Higgins and Stewart, 1964).

Stokes-Wave Theory

The limitations of the simple Airy-wave theory result mainly from the assumption in its derivation that the wave height is small compared with the water depth and wave length. Such an assumption must fail as the waves enter intermediate to shallow-water depths where wave heights tend to increase as the depth decreases. A wave theory such as the Stokes derivation should then be used, one that has not made an assumption that wave heights are negligibly small. This solution was accomplished by Sir George Stokes (1847, 1880), who provided second- and third-order solutions, respectively, representing more accurate but more complex analyses; the main equations of interest here are listed in Table 5-3. We immediately see the difference between the Airy and Stokes waves in their respective profiles, both in the equations and in the graphical depiction given in Figure 5-23. The theoretical Stokes-wave profile is characterized by narrow crests while the troughs are broad and flat, a profile that is comparable to the shapes of actual ocean waves as they enter intermediate to shallow water and approach the shore. In deep water the second-order general solution for the profile given in Table 5-3 reduces to

$$\eta_\infty = \frac{H_\infty}{2}\cos 2\pi\left(\frac{x}{L_\infty} - \frac{t}{T}\right) + \frac{\pi H_\infty^2}{4L_\infty}\cos 4\pi\left(\frac{x}{L_\infty} - \frac{t}{T}\right) \tag{5.24}$$

where the first term is identical to Equation (5.7) obtained in the Airy-wave solution, while the second term has the effect of enhancing the crest amplitude and subtracting from the trough amplitude so as to yield the profile shape depicted in Figure 5-23. If the wave steepness H/L is small, the Stokes profile approaches that given by Airy theory. As the water depth decreases and the waves enter intermediate to shallow water, the peaking of the wave crests become still more pronounced and the troughs are flatter.

The general equations for the water particle velocities (Table 5-3) similarly include additional terms beyond that given by the Airy solution. The additional term is positive under

TABLE 5-3   EQUATIONS DERIVED FROM STOKES WAVE THEORY

Wave Profile (second order)

General
$$\eta = \frac{H}{2}\cos(kx - \sigma t) + \frac{\pi}{2}\frac{H^2}{L}\frac{\cosh(kh)[2 + \cosh(2kh)]}{[\sinh(kh)]^3}\cos[2(kx - \sigma t)]$$

Deep water
$$\eta_\infty = \frac{H_\infty}{2}\cos\left[2\pi\left(\frac{x}{L_\infty} - \frac{t}{T}\right)\right] + \frac{\pi H_\infty^2}{4L_\infty}\cos\left[4\pi\left(\frac{x}{L_\infty} - \frac{t}{T}\right)\right]$$

Wave Celerity (third order)

General
$$C = \frac{gT}{2\pi}\tanh\left(\frac{2\pi h}{L}\right)\left[1 + \left(\frac{\pi H}{L}\right)^2\frac{5 + 2\cosh(4\pi h/L) + 2\cosh^2(4\pi h/L)}{8\sinh^4(2\pi h/L)}\right]$$

Deep water
$$C_\infty = \frac{gT}{2\pi}\left[1 + \left(\frac{\pi H_\infty}{2L_\infty}\right)^2\right]$$

Orbital Velocities (second order)

Horizontal
$$u = \frac{\pi H}{T}\frac{\cosh[k(z + h)]}{\sinh(kh)}\cos(kx + \sigma t) + \frac{3}{4}\left(\frac{\pi H}{L}\right)^2 C\frac{\cosh[2k(z + h)]}{[\sinh(kh)]^4}\cos[2(kx - \sigma t)]$$

Vertical
$$w = \frac{\pi H}{T}\frac{\sinh[k(z + h)]}{\sinh(kh)}\sin(kx - \sigma t) + \frac{3}{4}\left(\frac{\pi H}{L}\right)^2 C\frac{\sinh[2k(z + h)]}{[\sinh(kh)]^4}\sin[2(kx - \sigma t)]$$

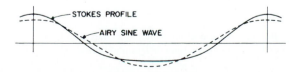

STOKES PROFILE

AIRY SINE WAVE

Figure 5-23   The theoretical profile of the Stokes wave compared with a sinusoidal Airy wave of the same height and length.

the wave crest and trough, and negative 1/4 and 3/4 wave lengths from the crest. As graphed in Figure 5-24, the effect is to increase the magnitude but shorten the duration of the velocity under the crest but decrease the magnitude and lengthen the duration of the velocity under the trough. This asymmetry of the onshore versus offshore velocities has been observed under waves in shallow water and is believed to be important to the cross-shore transport of sediments (Chapter 7).

The second-order Stokes equation for the phase velocity is the same as derived by Airy theory, but the third-order equation given in Table 5-3 again includes an additional term that accounts for the steepness H/L of the waves. In deep water the equation is approximately

$$C_\infty = \frac{g}{2\pi}T\left[1 + \left(\frac{\pi H_\infty}{2L_\infty}\right)^2\right] \tag{5.25}$$

which indicates that the finite steepness of the waves causes a slight increase in the phase velocity over that calculated with Airy theory.

A fundamental and interesting difference between Stokes and Airy waves is that the Stokes orbital motions of water particles do not close but instead lead to a nonperiodic current or mass transport of water in the direction of wave advance. The associated *Stokes drift* current is given by

$$U = \frac{1}{2}\left(\frac{\pi H}{L}\right)^2 C\frac{\cosh[2k(z + h)]}{[\sinh(kh)]^2} \tag{5.26}$$

This relationship was derived with no consideration of viscosity and predicts a current in the direction of wave advance at all water depths z, not accounting for any return flow. Longuet-Higgins (1953) has formulated the same problem for a channel of finite length and constant depth and accounted for the fluid's viscosity and the presence of a thin boundary layer at the bed. Because of the finite length of the channel, continuity must be considered and the mass

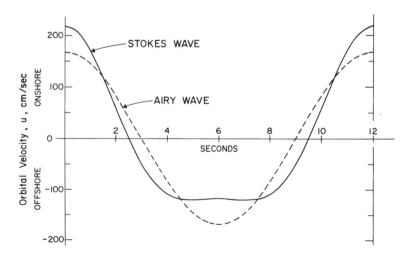

Figure 5-24 A comparison of the theoretical bottom orbital velocities under the Stokes wave with that of the Airy wave having the same height and period ($H = 4$ m, $h = 15$ m, $T = 12$ sec).

transport of water in the direction of wave advance is then balanced by a return flow. Longuet-Higgin's solutions yield velocity distributions such as those diagrammed in Figure 5-25, there being a net flow in the direction of wave advance near the water surface and at the bottom, balanced by a return flow in the opposite direction at mid-depths. Measurements by Russell and Osorio (1957) of the vertical distribution of mass transport velocities in a laboratory wave tank found reasonable agreement with the theoretical relationships. The net shoreward velocity near the bottom derived by Longuet-Higgins (1953) is given by

$$\overline{U}_o = \frac{5}{4}\left(\frac{\pi H}{L}\right)^2 C \frac{1}{[\sinh(kh)]^2} \tag{5.27}$$

providing predictions that compared favorably with the wave-tank measurements of Russell and Osorio. One might expect that this wave-induced mass transport could be important to the onshore transport of sediment, but this has never been conclusively demonstrated. In general, the magnitude of $\overline{U}_o$ is small and therefore is easily masked by other currents occurring in shallow water.

For any given water depth and wave period or length, there is an upper limit to the height of the Stokes wave beyond which it becomes unstable and breaks. The Stokes criterion for wave breaking is that the water-particle velocity at the crest is just equal to the propagation velocity C of the wave form. It is apparent that if this particle velocity at the crest exceeds C, then the wave would topple forward and break. Stokes (1880) determined that this breaking condition corresponds to a crest angle of 120°, and Mitchell (1893) established that in deep water this condition can also be expressed as a limiting wave steepness

$$\left(\frac{H_\infty}{L_\infty}\right)_{max} = 0.142 \approx \frac{1}{7} \tag{5.28}$$

This limit in steepness for deep-water waves is agreed upon by most investigators. For waves in intermediate to shallow water depths, Miche (1944) gives the limiting steepness as

$$\left(\frac{H}{L}\right)_{max} = 0.142\tanh(kh) \tag{5.29}$$

a relationship that has been shown to be satisfactory in applications.

It is apparent that the Stokes theory provides a considerable advance over the Airy-wave theory in accounting for the heights and steepness (H/L) of waves. In deep water the differences are relatively small, but as the waves enter intermediate to shallow water where the growing wave heights become important, the Stokes theory provides a better

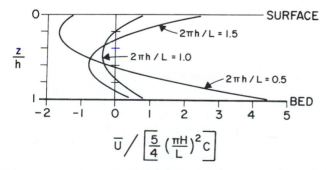

Figure 5-25 Theoretical wave-drift velocities in a laboratory channel of finite length. Negative values indicate flow up channel, opposite in direction to the wave advance. [Adapted with permission of The Royal Society, from M. S. Longuet-Higgins, Mass Transport in Water Waves, *Philosophical Transactions Royal Society of London*, p. 572. Copyright © 1953 The Royal Society.]

description of the wave motions, including the profile shape and asymmetry of orbital velocities beneath the crest and trough. Finally, having accounted for the heights of waves, the
Stokes theory has been able to consider the instability of overly steep waves that can lead to
their breaking. The focus here has been mainly on the second-order Stokes theory, and although it provides a marked improvement over Airy theory in describing wave motions in
intermediate to shallow water, its use in some instances can still lead to unacceptably large
errors. In this case, one must turn to the more complex higher-order Stokes solutions or to
the cnoidal wave theory.

Waves in Shallow Water—Cnoidal and Solitary Wave Theories

Stokes theory begins to fail when waves are in very shallow water where there is a combination of high wave steepness and ratios of wave height to water depth that are close to unity.
At this stage, one should utilize cnoidal wave theory but the analyses employing that theory
are sufficiently complex that it is seldom used. Instead, some investigators have resorted to
the use of solitary wave theory, rationalizing that in the nearshore the waves have distinct
crests separated by wide, flat troughs and are in effect a series of solitary waves.

    The theory of cnoidal waves was first developed by Korteweg and de Vries (1895), with
later contributions to the theory having been made by Keulegan and Patterson (1940) and
Keller (1948). The cnoidal wave is periodic, with the profile given by

$$\eta = H cn^2\left[2K(\kappa)\left(\frac{x}{L} - \frac{t}{T}\right), \kappa\right] \tag{5.30}$$

in which $K(\kappa)$ is the complete elliptic integral of the first kind of modulous κ, and η is the vertical coordinate of the water surface above the trough level at the horizontal coordinate x.
The term $cn(r)$ is the Jacobian elliptic function of r and accounts for the name cnoidal (analogous to sinusoidal). Figure 5-26 illustrates one profile compared with the laboratory wave-
flume measurements of Taylor (1955). Masch and Wiegel (1961) have developed graphs of
the principal properties of the cnoidal wave for given values of h, H, and L; these graphs generally are used in lieu of equations whenever the cnoidal theory is applied.

    It is unfortunate that the mathematics of the cnoidal wave theory is so difficult, as it po-
tentially has the widest range of applications. When κ becomes zero, $cn(r,\kappa) = \cos(r)$ and

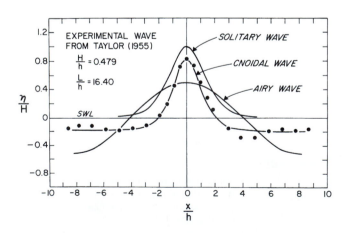

Figure 5-26 A comparison of an experimental
wave measured by Taylor (1955) with the
theoretical forms of the cnoidal wave, solitary
wave, and Airy wave. [Adapted with permission
of Prentice-Hall, from R. L. Wiegel, *Oceanographical Engineering*. Copyright © 1964
Prentice-Hall.]

$K(\kappa) = \pi/2$, so the cnoidal wave reduces to the linear Airy wave. When $\kappa = 1$, its other extreme value, the period and wave length become infinite such that the cnoidal wave becomes equivalent to the solitary wave. At intermediate values of the modulous k between 0 and 1, the wave profile of Equation (5.30) has crests separated by flattened troughs, comparable to the Stokes wave solution. However, due to the complexity of the cnoidal wave, its range of application is limited and preference is given to one of the simpler wave theories.

The solitary wave, as its name suggests, is a progressive wave consisting of a single crest; it is not oscillatory like the other waves examined. There is no wave period or wave length associated with the solitary wave. It would appear, then, that the solitary wave would not be particularly useful in describing the periodic waves found in the ocean. However, when ocean waves enter shallow water, their crests peak up and are separated by wide flat troughs and so appear much like a series of individual solitary waves. It is this similarity that first suggested such an application of the solitary wave (Bagnold, 1947; Munk, 1949). We already have seen in the other wave theories that in shallow water the period is not particularly significant but rather the water depth becomes important. Therefore, we are not particularly oriented toward the periodicity of the waves in shallow water, so consideration of a solitary wave seems reasonable. Because of its similarity to real waves and because of its simplicity, the solitary wave has seen some application to nearshore studies.

The character of the solitary wave was first described by Russell (1844) who produced such waves in the laboratory by suddenly releasing a mass of water at one end of a wave tank. The first theoretical consideration was that of Boussinesq (1872), while Rayleigh (1876) and McCowan (1891) developed the theory further and obtained higher approximations. The profile and notation of the solitary wave is shown in Figure 5-27, and its equation relative to the moving crest is

$$\eta = H \mathrm{sech}^2\left(\sqrt{\frac{3}{4}\frac{H}{h}}\frac{x}{h}\right) \tag{5.31}$$

where η is the vertical coordinate above the still-water line at a horizontal distance x from the crest; the height H and water depth h are also in reference to the still-water level (Fig. 5-27). The wave velocity C is given by

$$C = \sqrt{gh}\left[1 + \frac{1}{2}\frac{H}{h} - \frac{3}{20}\left(\frac{H}{h}\right)^2 + \ldots\right] \tag{5.32a}$$

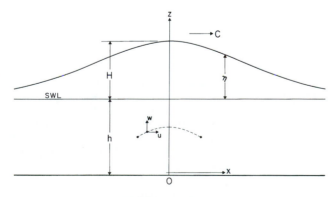

Figure 5-27 The notation associated with a solitary wave. SWL represents the still-water level.

which is greater than the shallow-water phase velocity of the Airy wave ($C = \sqrt{gh}$) due to the inclusion of terms that depend on $\gamma = H/h$. This relationship is nearly equal to

$$C = \sqrt{gh\left(1 + \frac{H}{h}\right)} = \sqrt{g(h + H)} \tag{5.32b}$$

determined empirically by Russell (1844) and obtained theoretically as first-order approximations to the solution of the equations of wave motion. As seen in Figure 5-28, the two relationships for C depart somewhat at higher values of H/h, with the laboratory measurements of Daily and Stephen (1952, 1953) showing better agreement with Equation (5.32a); however, the differences are small, especially for $H/h < 0.4$, so that for many practical purposes the simpler Equation (5.32b) is satisfactory.

As the solitary wave advances into shallow water, the wave height progressively increases until a condition is reached where the wave becomes unstable and breaks. This instability is again reached when the particle velocity at the crest equals the phase velocity C; the angle at the crest is again 120°, as found in the Stokes wave analysis. Using these criteria, various investigators have theoretically derived critical breaking ratios in the range

$$\gamma_b = \left(\frac{H}{h}\right)_{max} = \frac{H_b}{h_b} = 0.73 - 1.03 \tag{5.33}$$

although the 0.78 value determined by McCowan (1894) is most commonly cited. Field measurements by Sverdrup and Munk (1946) on ocean beaches with very low slopes agreed with the $\gamma_b = 0.78$ value. Ippen and Kulin (1954) conducted laboratory experiments with solitary waves that examined the effects of a sloping bottom on the γ_b value. With a slope of only 0.023, the critical ratio was about $\gamma_b = 1.2$, while on the steepest slope studied (0.065), values of γ_b reached as high as 2.8, the value depending on the initial wave height. Therefore, a sloping bottom can produce significant departures from the theoretical values for γ_b. Based on data extrapolations, Kishi and Saeki (1966) indicated that γ_b would not be reduced to the 0.78 theoretical value of McCowan until bottom slopes became gentler than about 0.007. The critical ratio $\gamma_b = H_b/h_b$ in general has been adopted as a breaking criterion for waves in shallow water, irrespective of any connection with solitary wave theory. Additional experimental studies with periodic waves to examine how this breaking criterion varies with the bottom slope and initial wave steepness will be presented in Chapter 6.

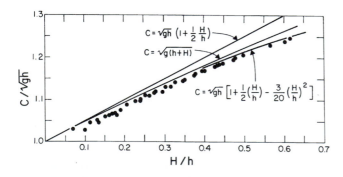

Figure 5-28 The velocity of the solitary wave compared with various theoretical equations. [Adapted from The Solitary Wave, J. W. Daily and S. C. Stephen, *Proceedings 3rd of the Coastal Engineering Conference*, 1952. Reproduced with permission from the American Society of Civil Engineers.]

Limits of Applications

Several wave theories have been reviewed here, with some indication as to their limits of applications, which depend on how successful they are in describing the actual motions of waves. This will be firmed up at this point, yielding Figure 5-29, which gives the fields of applications in terms of the ratios H/h and h/L. In the construction of this graph, the widest possible regions are given to the simpler wave theories. For example, the difficult cnoidal wave theory is given only a restricted region of application within its potentially much greater field, while top preference is given to the comparatively simple Airy-wave theory.

The limits to wave existence are indicated in Figure 5-29 by the $\gamma_b = H_b/h_b = 0.78$ ratio from solitary wave theory and the more general wave-breaking relationship of Equation (5.29) from Miche (1944). The limit of application of Airy-wave theory is taken to be

$$\frac{1}{2}(u^2 + w^2) < 0.05gH \tag{5.34}$$

based on the simplifying assumption in the Euler equation, which indicates that the orbital velocities u and w evaluated at the surface must be small, here placed at less than 5 percent of gH. Muir Wood (1969) has shown that this is equivalent to

$$\frac{H}{L} < \frac{1}{16}\tanh\left(\frac{2\pi h}{L}\right) \tag{5.35}$$

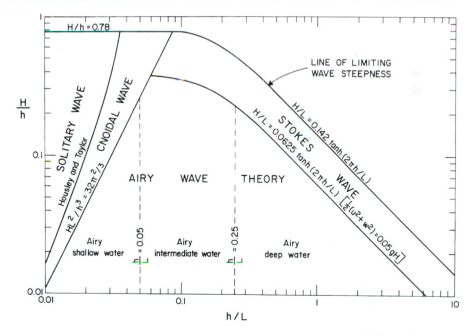

Figure 5-29 The areas of applications of the various wave theories as a function of the ratios H/h and h/L, where preference is given to the simpler theories such as Airy waves.

which is graphed in Figure 5-29 as the line separating the regions of application of Airy-wave theory and Stokes-wave theory. The use of the limiting line

$$\frac{HL^2}{h^3} = \frac{32\pi^2}{3}$$

(5.36)

is based on cnoidal wave theory as developed by Keller (1948) and Littman (1957), with this specific criterion having been suggested by Longuet-Higgins (1956) as another limit to the application of linear Airy-wave theory. Finally, Housley and Taylor (1957) defined the regions of application of solitary waves versus oscillatory wave theories in a graph of H/h versus $T\sqrt{g/h}$, based on comparisons of theoretical and measured phase velocities. The equation of the line separating the regions of application in Figure 5-29 is

$$\frac{H}{h} = \frac{1600}{(T\sqrt{g/h})^{2.5}}$$

(5.37)

Here the curve is shown separating solitary and cnoidal wave theories, while Housley and Taylor had eliminated cnoidal theory altogether, the line instead separating the solitary and Airy-wave theories. A cnoidal theory region is included to illustrate how that theory becomes solitary theory in one direction and Airy theory in the other. In applications, one could carry the Airy and Stokes regions over into the cnoidal area.

The approach in Figure 5-29 is but one scheme to define the limits of applications of the several wave theories, an approach that is only a slight modification of that presented by Muir Wood (1969). Dean (1970) and LeMéhauté (1976) provide alternatives based on H/gT^2 (essentially the wave steepness H/L) versus h/gT^2. They also include in their considerations the *Stream Function* wave theory developed by Dean (1965) and summarized by Dean (1990) with applications, an approach that obtains numerical solutions of the equations that govern wave motions, without the necessity of making further simplifying assumptions. Such numerical results provide a better description of wave motions than do the analytical wave theories, but the Stream Function analyses exist only as listings within tables rather than having the convenience of equations as provided by the mathematical solutions of Airy-wave theory. However, in many respects, the Stream Function analysis is easier to apply than complex mathematical solutions such as the cnoidal wave theory.

WAVE PROPAGATION AND TRANSFORMATION

Now that we have equations to describe the energy and motions of waves, let us return to an examination of their movement across the deep ocean after their generation by a storm, and their subsequent transformation as they enter shallow water and approach the coast. These topics are important, recalling from the depiction in Figure 5-1 that the travel of waves across the sea delivers the energy derived from the storm to the nearshore zone where it plays a role in a range of processes, including the generation of nearshore currents and the movement of sediments along beaches.

Deep-Water Wave Propagation

Once waves leave the area of generation and are no longer under the influence of the wind, they begin to sort themselves out by wave dispersion and thereby become more regular. It has been established that the rate at which the wave energy and wave group as a whole travels is given by the group velocity

$$C_g = Cn = \frac{C}{2} = \frac{1}{2}\frac{gT}{2\pi}$$

(5.38)

The rate of movement across the sea therefore depends on the wave period, with the longer-period waves traveling faster than the short-period waves. The important factor is that although the entire spectrum of waves in the area of generation is complex, the individual groups of sine waves that make up the spectrum travel according to this relationship. The result is that, starting together in the generation area, the longer-period waves outrun and leave behind the shorter-period waves. This sorting by wave period, termed *wave dispersion,* is the main factor in producing the more uniform swell from the initially complex sea in the area of generation. Dispersion therefore produces a narrowing of the energy spectrum, as shown schematically in Figure 5-30, with a strong energy peak corresponding to the wave period of the swell. The greater the distance of travel from the area of generation, the narrower the spectrum becomes.

Each wave period in the spectrum has a group of waves associated with it, moving outward from the generation area and, to a first approximation, moving independently of waves having different periods. The different groups with separate periods have nonidentical travel speeds; therefore, although they start out together in the generation area, they soon become separated as travel proceeds. Within each group, waves continuously arise in the rear of the group, travel forward through it, and then die away as they approach the front. The wave group has a permanence that is not shared by individual waves. Individual waves can be followed for only a short time, while the progress of the wave groups can be traced across entire oceans.

Let us follow the progress of one separate wave group associated with the period T as it moves from the generation area to the coastline shown in Figure 5-31. Idealized, the group of

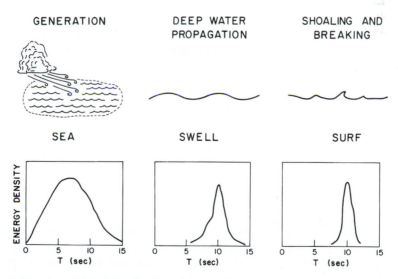

Figure 5-30 The change in wave spectra from the storm area of wave generation where the spectrum is wide, to a spectrum of swell waves narrowed by processes of wave dispersion and dissipation, to the surf that also consists of a narrow spectrum.

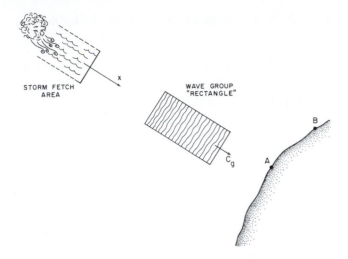

STORM FETCH
AREA

WAVE GROUP
"RECTANGLE"

C_g

A

B

Figure 5-31   The wave group "rectangle" associated with one wave period in the complete spectrum as it moves from the generation area to the measurement point A on the coast.

waves will occupy a rectangular area whose width is the width of the storm front and whose length is governed by the storm duration and total fetch length. Let the edge of the conceptual rectangle be at $x = 0$ at the time $t = 0$ of first wave formation. The group outlined by the rectangle will move downwind in the positive x direction at a speed given by Equation (5.38). If the storm duration is 30 hours, then the windward edge of the rectangle would pass the line $x = 0$ approximately 30 hours later than the front of the rectangle, and its total length would be $C_g \times 30$ hours. The edges of the rectangle would not actually remain sharp, as some of the wave disturbance would leak into the surrounding still water. A few low waves would run ahead of the leading edge, and wave energy would likewise diffuse sideways, making the conceptual edges of the rectangle rounded rather than sharp. These effects would be small, however, and waves would not reach point B in Figure 5-31 due to this process. We will see later, however, that waves might actually reach point B from the storm area, but because of the angular spreading of the waves, not all of the generated waves are initially moving in the mean fetch direction. Even with this angular spread the waves are directed down a rather narrow corridor and much of the ocean will not be affected. Waves generated by a storm are not analogous to a pebble thrown into a pond, where the waves are circular and spread in every direction, and the energy per unit crest length of a wave quickly decreases. Because the spread of ocean waves is limited, the wave group will still retain much of its energy when it reaches a distant shore.

Returning to the single conceptual wave group in Figure 5-31 having a period T, if R is the distance from the leading edge of the storm fetch to the point A on the coast, then the time t_{ob} of first observation of arrival of the waves would be

$$t_{ob} = \frac{R}{C_g} = \frac{4\pi R}{gT} \tag{5.39}$$

It can be seen that there is an inverse linear relationship between t_{ob} and T. Therefore, if 10-sec period waves arrive in 2 days from the storm area, 5-sec waves will arrive in 4 days. If t_{ob} is plotted against T for a given storm, there is a straight-line relationship (see Fig. 5-33, discussed below). For a line source of waves, the final arrival time of waves of period T will be determined by the storm duration D, being $t_{ob} + D$. There would be some error in this simple estimate if the fetch length were large, such that one could not consider a simple line source.

Since there is a range of wave periods produced in the generation area—an entire spectrum—there would likewise be a series of conceptual rectangles such as that discussed above, one for each period. There is of course a continuum, not a discrete number having individual periods as conceptualized here. However, we can picture the wave dispersive process as being many different rectangles moving at different velocities, arriving at the coastal point A in Figure 5-31. The first rectangle to arrive would have the longest wave period, followed by rectangles of progressively shorter wave periods.

This process of wave dispersion by period was first clearly demonstrated in the study of Barber and Ursell (1948). Figure 5-32 contains a series of wave spectra they measured at 2-hour intervals for 2.5 days from a wave recorder at Pendeen, Cornwall, England. Swell from a tropical storm off the east coast of the United States was first detected in the spectra as an energy peak at 19 sec at 1900 hours on June 30, 1945. In subsequent spectra it is seen that the mean peak period progressively decreased on account of dispersion, reaching approximately 13 sec at the end of July 2. Figure 5-33 graphs this progressive shift of the maximum and minimum periods that limit the width of the spectrum peak associated with the storm. The irregularities in the points about the straight lines were demonstrated to result from the effects of the passage of the waves through tidal currents near the wave recorder. Similar results were obtained for other storms, the most distant being off Cape Horn, some 11,000 km from Cornwall.

Subsequent observations have confirmed the global nature of swell propagation. Wiegel and Kimberly (1950) demonstrated that southerly swell observed throughout the summer at Oceanside, California, arrives from Southern Hemisphere storms between 40 and 65°S and between 120 and 160°W. Wave heights reached 3–4 m, and periods were from 12–18 sec. Munk and Snodgrass (1957) obtained the energy spectra of the very low amplitude, long-period ($T = 20$ sec) forerunners of swell arriving at Guadelupe Island west of Baja California, Mexico, from storms in the Indian Ocean, entering the Pacific along the great circle route between Antarctica and Australia, a travel distance of some 15,000 km. Snodgrass et al. (1966) traced waves generated by storms near Antarctica, south of New Zealand, by a series of stations in a great circle route across the entire Pacific to Yakutat, Alaska, a distance of 10,000 km. All of these studies demonstrate that deep-ocean swell does in fact propagate with the group velocity appropriate to its period, and that the waves follow great circle paths as they travel across the deep ocean.

As the waves travel these immense distances, how much of their energy is lost? It appears that relatively little is lost, judging from the sizes of swell waves that sometime reach the west coast of the United States from storms in the far South Pacific. Snodgrass et al. (1966) particularly examined wave attenuation during travel. Figure 5-34 from their study gives typical spectra at distances of 0 km (the inferred spectra at the end of the storm fetch), at 1,000 km from the storm area, and at 10,000 km distance. It is seen that most of the attenuation occurs close to the area of generation, and is most pronounced in the short-period (high-frequency) portion of the spectrum. With farther travel beyond the first 1,000 km, little additional attenuation occurs. There are several possible mechanisms by which waves traveling in the mean fetch direction may lose energy as they travel across the deep ocean:

1. Internal viscous damping,
2. Angular spreading of the waves as they leave the localized storm area,
3. Contrary winds blowing against the moving waves, and
4. Wave-wave interactions as waves of the same storm or of separate storms combine and interact.

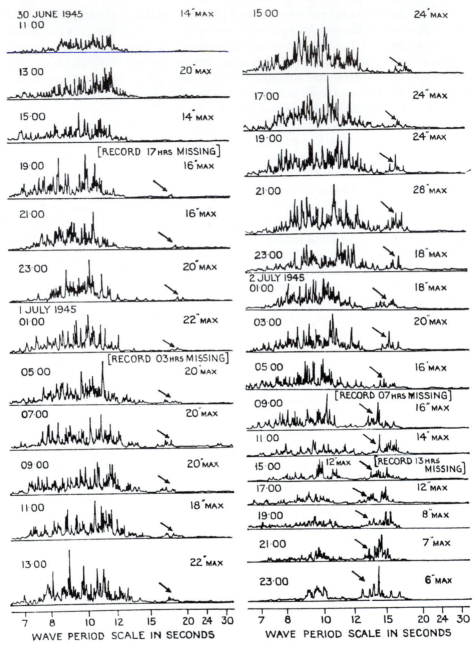

Figure 5-32 Wave spectra obtained at 2–hour intervals from a wave recorder at Pendeen, Cornwall, England. The arrows point to the spectral peaks resulting from swell generated by a tropical storm off the east coast of the United States. With time, there is a progressive shift in periods and changes in the sizes (energies) of the peaks associated with that storm. [Used with permission of The Royal Society, from N. F. Barber and F. Ursell, The Generation of Ocean Waves and Swell, 1: Waves Periods and Velocity, *Philosophical Transactions of the Royal Society,* p. 544. Copyright © 1948 The Royal Society.]

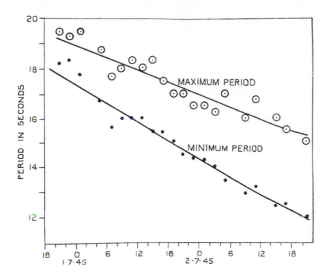

Figure 5-33 The progressive shift of the maximum and minimum periods that limit the width of the spectral peak associated with the storm documented by the spectra in Figure 5-32. [Used with permission of The Royal Society from N. F. Barber and F. Ursell, The Generation of Ocean Waves and Swell, 1: Waves Periods and Velocity, *Philosophical Transactions of the Royal Society,* p. 545. Copyright © 1948 The Royal Society.]

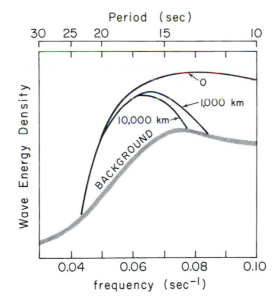

Figure 5-34 Wave spectra obtained from a storm in the South Pacific at distances 0, 1,000, and 10,000 km from the storm front. [Adapted with permission of The Royal Society from D. Snodgrass, G. W. Groves, K. F. Hasselmann, G. R. Miller, W. H. Munk and W. H. Powers, Propagation of Ocean Swell Across the Pacific, *Philosophical Transactions of the Royal Society,* p. 492. Copyright © 1966 The Royal Society.]

The effect of the internal viscous damping of water waves of small amplitude has been studied mathematically by Lamb (1932), for waves in deep water, and by Hough (1896) in shallow water. Keulegan (1950, p. 730) presents a particularly good discussion of internal viscous effects. The expression for the decay in the wave height due to internal friction is

$$H = H_i \exp\left(\frac{-8\pi^2 \nu t}{L_\infty^2}\right)$$

(5.40a)

or substituting $L_\infty = gT^2/2\pi$ for the deep-water wave length,

$$H = H_i \exp\left(\frac{-32\pi^4 \nu t}{g^2 T^4}\right) \tag{5.40b}$$

where H_i is the initial wave height at time $t = 0$, and H is the height at time t; ν is the kinematic viscosity of water ($\nu = 0.01$ cm^2/sec for typical water temperatures). The wave height falls off exponentially with time, the rate of decay depending strongly on the wave period since it enters the equation with a fourth-power exponent. Viscous damping, therefore, preferentially removes the short-period waves, and leaves the longer-period waves largely unaffected. Viscous damping as well as wave dispersion could be effective in enhancing the regularity of the waves once they leave the area of generation. The time required for the wave height to decrease to one-half its original value ($H = H_i/2$) is given by

$$t_{1/2} = 0.0088 \frac{L_\infty^2}{\nu} \tag{5.41a}$$

$$t_{1/2} = \frac{0.0088}{4\pi^2} \frac{g^2 T^4}{\nu} \tag{5.41b}$$

Figure 5-35 contains a plot of $t_{1/2}$ versus T, based on Equation (5.41b). It can be seen from this graph that a 5-sec wave will travel for some 2,660 hours (a deep-water distance of about 37,400 km) before its height is reduced to half the original. A wave with a period of 1 sec

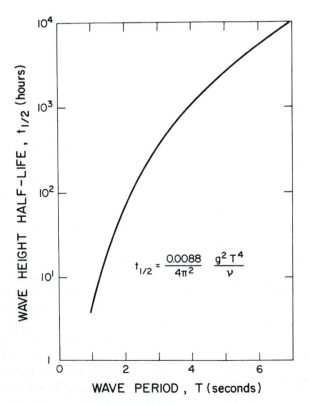

Figure 5-35   The wave half-life according to Equation (5.41), the time required for the wave height to decrease by one-half due to internal viscous damping.

would travel for only 4.3 hours (a distance of 12 km) for the same result. It is apparent that viscous damping is important in removing the very short-period waves from the spectrum but has little effect on the long-period waves. This process is most important close to the area of wave generation and plays a role in converting the initially irregular sea of the storm area into regular swell. However, Snodgrass et al. (1966) concluded that viscous damping accounted for only a small portion of the attenuation of the spectra given in Figure 5-34.

Not only do waves disperse by period in the predominant direction of advance, they also spread sideways over a broader area because they start out with slightly different directions. Because of the variability of the wind about some mean direction, waves are generated having a corresponding variability in direction. Waves in the generation area not only move in the direction of the mean wind but at various other angles. One important effect of angular spreading is that a position such as point B in Figure 5-31 may receive swell from the generation area, even though it is to one side of the direction of the mean fetch. The other important effect is that like dispersion by wave period, angular spreading will diminish the wave energy arriving at some forecast point. The wave energy arriving at either points A or B in Figure 5-31 will be only a fraction of what was present in the storm area.

Contrary winds blowing against the moving waves may produce some loss in energy. However, the profile of swell in deep water is low and gentle, so the waves do not offer much form drag to an opposing wind. In most instances of swell in deep water, the loss of energy due to opposing winds is probably negligible.

It was seen earlier that within the area of wave generation, their growth is limited by breaking, mainly of the short-period waves. Furthermore, waves may interact with one another to cause the energy of a given wave to scatter into waves of almost the same period but traveling in slightly different directions from the original waves (Phillips, 1960; Hasselmann, 1963). The overall effect is similar to angular spreading due to variable wind directions but is much more effective within and near the area of wave generation. After dispersion separates the waves by period and narrows the spectral peak, scattering by wave-wave interactions will no longer be as effective. Snodgrass et al. (1966) attributed the changes seen in Figure 5-34 primarily to a combination of continued wave breaking and wave-wave scatter, even after the waves had left the generation area.

In summary, as waves leave their immediate area of generation, they become sorted by wave dispersion with the longest-period waves traveling the fastest. As a result, at some distant observation point, the longest-period waves arrive first, followed by progressively shorter-period waves. Acting together, wave dispersion and viscous damping narrow the wave spectrum and change the complex sea in the generation area into regular swell. Nonlinear effects such as wave breaking and wave-wave scattering continue for a short distance beyond the generation area and are important to wave-energy dissipation. Once the waves have developed into a regular swell, however, they may travel for thousands of kilometers across the entire ocean basin with relatively little additional loss of energy. Waves generated by storms off Antarctica may accordingly be measured on the beaches of Alaska.

Shoaling Transformations of Waves

In deep water the profile of the swell is nearly sinusoidal, with low rounded crests. As the waves enter intermediate to shallow water, they undergo a systematic transform, as depicted in Figure 5-36. The wave velocity and length progressively decrease and the height increases; only the wave period remains constant. In water depths not far beyond the breaker zone, the

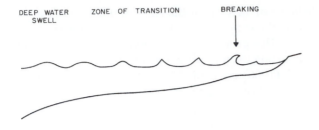

DEEP WATER   ZONE OF TRANSITION   BREAKING
   SWELL

Figure 5-36   Illustrative transformation of waves as they enter progressively shallower water and approach the coast.

wave train consists of a series of peaked crests separated by relatively flat troughs. Finally, the crests become oversteepened, unstable, and the waves break. The height of a breaking wave may be substantially larger than its height in deep water. The transformation is most readily apparent for long-period swell since they "feel bottom" first and generally are the most nearly sinusoidal to begin with. Locally generated short-period waves are initially steep, even in deep water, so the transformation is not as apparent.

Although solutions are available for waves traveling over a sloping bottom (see Stoker, 1957), they are complicated and seldom used. A simpler approach is to use the wave solutions discussed earlier, assuming they apply even though they were developed for a flat horizontal bottom (Rayleigh, 1911). Using Airy-wave theory, we have

$$\frac{L}{L_\infty} = \frac{C}{C_\infty} = \tanh\left(\frac{2\pi h}{L}\right) \approx \left[\tanh\left(\frac{2\pi h}{L_\infty}\right)\right]^{1/2} \tag{5.42}$$

which yields the variation in the wave length L and phase velocity C with changing water depth h. These variations have been graphed in Figure 5-21, where it can be seen that the wave length and phase velocity systematically decrease with decreasing water depth, that is, as the beach is approached. Similarly, we have variations in n [Equation (5.21)] and hence changes in the group velocity $C_g = Cn$ as the depth decreases. The graphical presentation in Figure 5-21 shows n smoothly increasing from its 1/2 value in deep water to $n = 1$ in shallow water. The curve for C_g/C_∞ is the product of the n and C/C_∞ curves and has the unusual feature of there being a slight increase in the group velocity as the waves first enter intermediate water depths (Fig. 5-21), followed by the expected decrease as the waves enter shallow water.

The variations in the heights of the shoaling waves can be calculated from a consideration of the energy flux. If the losses of energy due to bottom drag and reflection are negligible, generally a reasonable assumption for waves outside the nearshore region, then the energy flux is constant and we have

$$P = ECn = (ECn)_\infty = \text{constant} \tag{5.43}$$

where $(ECn)_\infty$ is the energy flux evaluated in deep water. Without energy losses, the energy flux in the shoaling waves remains equal to its value in deep water. Using the energy relationship for Airy waves,

$$E = \frac{1}{8}\rho g H^2$$

the ratio of the wave height H in water of depth h to the deep-water wave height H'_∞ is found to be

$$\frac{H}{H'_\infty} = \left(\frac{1}{2n}\frac{C_\infty}{C}\right)^{1/2} \tag{5.44}$$

Both n and the ratio C/C_∞ are established functions of the water depth [Equations (5.21) and (5.42)], so H/H'_∞ will similarly depend on the depth. The resulting calculated curve for H/H'_∞ is given in Figure 5-21 and for the most part shows the expected increase in heights as the waves pass into shallow water. Examining the curve more closely, it can be seen that the curve predicts that there should be a small initial decrease in the heights of the waves as they enter intermediate water depths, the heights becoming slightly less than the deep-water wave height ($H/H'_\infty < 1$). This temporary height reduction is brought about by the increase in $C_g = Cn$, noted above. Since ECn must remain constant, the increase in the rate at which the energy is carried forward (C_g) must result in a temporary decrease in the energy density E and thus in the wave height H. This decrease, followed by a rapid increase in wave heights, has been observed in laboratory wave studies (Fig. 5-37) and sometimes can be seen on natural ocean beaches when there is a very regular swell.

The wave steepness H/L also varies in the shoaling waves. The steepness temporarily drops slightly below its deep-water value as the waves pass through intermediate water depths and then rapidly increases as H increases while L is reduced. The sudden increase in steepness along with the increase in height are the most striking features of shoaling waves. The steepness increases rapidly until a point is reached where the waves become unstable and break.

The above analyses of wave transformations as they enter intermediate to shallow water depths have been based entirely on Airy-wave theory, which is not particularly applicable in shallow water. With its simple equations, Airy theory made the analysis straightforward and does predict the correct general trends as to changes in phase velocities, heights, etc., as the waves shoal. However, it can be expected that the actual magnitudes of the wave changes during shoaling would differ from those predicted by Airy theory, increasing in error as the waves approach the breaker zone. This departure is particularly shown in laboratory studies of wave transformations, illustrated by Figure 5-37 from the early study of Iversen (1952a, and 1952b). The results shown are typical of his measurements and reveal an initial height decrease below the deep-water wave height, a decrease that is actually greater than predicted by Airy-wave theory, with an even greater departure in the subsequent increase in wave heights as the breaker zone is approached. Part of the disagreement results from frictional drag on the waves, produced by the side walls of the laboratory wave channel as well as bottom friction. A modified curve attempting to account for such effects is included in Figure 5-37, showing some improvement in the agreement with the measurements, but it is still apparent that most of the departure results from the failure of Airy-wave theory in shallow water.

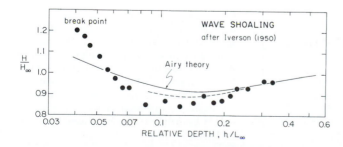

Figure 5-37 Measured wave-height variations relative to the deep-water height H_∞ during shoaling in a laboratory channel, demonstrating a marked departure from the predicted transformation by Airy-wave theory. [Adapted from Waves and Breakers in Shoaling Water, H. W. Iversen, *Proceedings of the 3rd Coastal Engineering Conference*, 1952. Reproduced with permission from the American Society of Civil Engineers.]

Flick, Guza, and Inman (1981) have undertaken a comparable wave-channel study that included measurements of heights and orbital velocities of shoaling and breaking waves, and also included direct comparisons with the Stokes- and cnoidal-wave theories. Figure 5-38 shows a representative comparison that demonstrates excellent agreement between measurements and theory for variations in wave heights during shoaling up to the point of wave breaking. The Stokes theory employed here is the third-order solution, a more advanced analysis than the second-order equations presented in Table 5-3 and discussed earlier. Flick and co-workers concluded that the third-order Stokes solution predicts the wave height changes and orbital velocities in the offshore at depths where the Ursell number $U_r = (H/h)/(kh)^2$ is less than about 1. Shoreward of the Stokes region and up to the break point, cnoidal theory provides a better prediction of the harmonics of the wave form. These harmonics are found in spectra of the waves and relate to the overall shapes of the waves, including the peakedness of the crests and their asymmetry as the waves approach the breaking condition and either plunge or spill forward (Chapter 6). Figure 5-38 includes comparisons between the harmonics of the measured waves and curves derived from the Stokes and cnoidal wave theories, providing an illustration of the conclusions of Flick, Guza, and Inman, regarding the regions of application of these two wave theories. The theories account for the progressive development of the crest peakedness as the waves shoal, but do not predict the asymmetry when the waves begin to break.

The most sophisticated analyses of wave transformations during shoaling account for the full spectrum of wave energies as well as the development of harmonics due to the changing wave forms. Figure 5-39, from Guza and Thornton (1980), illustrates the changing patterns of spectra for waves measured successively in a series of water depths ranging from 10.2 to 1.78 m as they approached Torrey Pines Beach, San Diego, California. This example shows the dramatic shift in the energy of ocean swell into the series of harmonics as the waves move toward the shore. The total energy density is not constant since there is an overall increase in wave heights during the shoaling process, but it can be seen in Figure 5-39 that essentially all of the energy increase is contained within the growth of the harmonics, not within the original peak frequency of the swell's spectra. The series of studies by Freilich and Guza (1984) and Elgar and Guza (1985a, 1985b, 1986) examine such changes in wave spectra during shoaling, and utilize field measurements like those in Figure 5-39 to test the validity of various theoretical models that have been developed to account for these transformations.

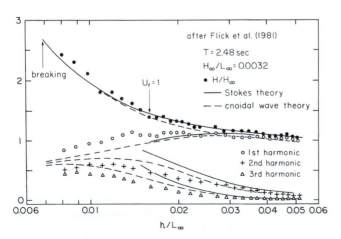

Figure 5-38 Measured wave-height variations, expressed as the ratio H/H_∞, compared with third-order Stokes-wave theory and the cnoidal theory. Comparisons are also made with the harmonics of the wave form determined from spectra, related to the degree of peakedness of the wave crests. [Adapted with permission of American Geophysical Union, from R. E. Flick, R. T. Guza and D. L. Inman, Elevation and Velocity Measurements of Laboratory Shoaling Waves, *Journal of Geophysical Research 86*, p. 4153. Copyright © 1981 American Geophysical Union.]

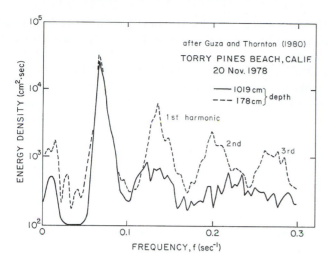

Figure 5-39 Spectra of shoaling waves measured at 1019–cm and 178–cm water depths, showing the development of higher harmonics in shallow water due to the waves having more peaked crests and flatter troughs. [Adapted with permission of American Geophysical Union, from R. T. Guza and E. D. Thornton, Local and Shoaled Comparisons of Sea Surface Elevations, Pressures and Velocities, *Journal of Geophysical Research 85*, p. 1526. Copyright © 1980 American Geophysical Union.]

In summary, based on the above review, when swell travels from deep to shallow water as the waves approach the shore, they undergo systematic transformations that generally increase their heights and reduce their wave lengths and celerities, resulting in increases in wave steepness up to the point where the waves become unstable and break. When employing one of the wave theories to describe the motions, linear Airy theory with its sinusoidal profile of low steepness is generally thought to be limited in application to the deep-water conditions, while the Stokes- or cnoidal-wave theories are required in intermediate to shallow water depths where the steepness of the waves increases. The progressive transformations in the spectra reflect these changes in wave forms during shoaling. Established analysis models to evaluate wave transformations during shoaling accordingly shift from the use of Airy- to Stokes- or cnoidal-wave theories as the waves approach the coast. Somewhat surprising then are the results of measurement programs like those of Kamphuis (1991, 1994), which have found that simple linear Airy-wave theory can be used for the entire wave transformation up to the point of breaking. Kamphuis' measurements were made in a laboratory wave basin, where irregular waves were generated that have a JONSWAP spectrum in deep water. However, the analyses focused on variations in the significant wave heights during shoaling, not on the details of the wave characteristics or their spectra. In terms of changing wave heights during shoaling, the variations measured in the laboratory agreed closely with linear Airy-wave theory, suggesting that this simple wave analysis could be used rather than turning to more complex theories when the waves enter shallow water. Kamphuis then reanalyzed the available field data on wave transformations during shoaling and obtained mixed results, where in some instances linear Airy theory alone was adequate, while in other instances the nonlinear wave theories were needed. He was unable to establish the circumstances under which one approach was preferable over the other.

Kamphuis (1991, 1994) attributed his measurement results to the energy dissipation of the waves as they shoal, an energy loss that is not included in the wave theories. This potential energy loss comes from: (1) bottom friction acting on the orbital motions of the waves; (2) percolation of the water into and out of the sediment; and (3) spectral saturation of the irregular waves, leading to some wave breaking even in the offshore. He established in his wave-basin experiments, where the bottom was covered with sand, that bottom friction and percolation were small, leading him to conclude that the energy loss resulted from a small amount of wave breaking in the offshore.

In some instances the frictional drag at the bottom exerted on the orbital motions of the waves can affect changes in their heights and energies as the waves shoal (Putnam and Johnson, 1949). The wave theories of Airy, Stokes and others were derived with an assumption of zero viscosity, that is, they neglected any frictional drag at the bottom. This yielded solutions that predict finite horizontal orbital velocities at the bottom, for example Equation (5.17b) for Airy-wave theory. However, the existence of frictional drag produces a thin boundary layer with a velocity gradient, the velocity at the bottom itself being $u = 0$ due to the no-slip condition of water adjacent to any solid. Within the boundary layer, the horizontal velocity increases within a span of 1–3 cm from the bottom, until the horizontal orbital velocity achieves the value predicted by the wave equations. A review of the properties of the flow boundary over smooth and rough beds can be found in Sleath (1984). In the present context, the drag on the bottom is important because it represents a loss of wave energy and power. As simplified from the relationship given by Madsen, Poon, and Graber (1988), the onshore (x-coordinate) gradient of the wave power P and thus of the wave heights is balanced by the bottom friction according to

$$\frac{dP}{dx} = \frac{1}{4}\rho g H \frac{dH}{dx} C_g = -\frac{1}{16}\rho f_w u_0^3 \tag{5.45}$$

in which f_w is a dimensionless frictional-drag coefficient and u_0 is the bottom orbital velocity generated by the waves. Graphs for f_w have been developed by Jonsson (1966) and Kamphuis (1975), based on experiments where the bottom consists of fixed grains of uniform size that determine the boundary roughness. Madsen and co-workers (1988, 1990) focused on the more natural condition of there being a mobile bed of sediments that can be transported and molded into bedforms, such as ripple marks, that determine the bottom roughness (Madsen Poon, and Graber, 1988; Madsen, Mathisen, and Rosengaus, 1990). They demonstrated that the sizes and overall geometry of the sediment ripples, as created by the wave-orbital motions and any superimposed currents, govern the frictional drag. For example, simple periodic waves like those generated in a laboratory wave channel tend to form sharp-crested sand ripples on the bottom, and this produces roughly twice the drag compared with ripple marks formed under random waves where the ripples have more rounded crests. In the latter case, Madsen and co-workers (1990) found $f_w \approx 0.1$, which implies that frictional drag on waves can still be significant and should be included in analyses of wave shoaling. In applications, however, frictional drag is commonly neglected unless waves of long period cross a wide, shallow continental shelf where it is clear that energy losses will be significant.

The analyses by Resio (1987, 1988) indicated that much of the energy loss that has been attributed to bottom friction results from spectral transformations in shallow water where irregular wave breaking is the major energy-loss mechanism. His comparisons with various data sets from the field under storm conditions show good agreement with the theory. Thus, his mechanism of energy losses in shoaling waves is likely to be important for storm waves generated along the coast or where the waves are otherwise irregular with a broad spectrum, as found in the wave-basin experiments of Kamphuis (1991, 1994). However, this process of energy loss would be relatively insignificant for nearly monochromatic, low-steepness swell generated by distant storms, in which case bottom friction and percolation likely account for the energy losses.

The presence of kelp in the immediate offshore appears to dissipate some of the energy of waves as they approach the coast. This observation has led to experiments to examine the potential use of artificial seaweed to reduce the wave energy along eroding beaches (Price, Tomlinson, and Hunt, 1968). Kobayashi, Raichle, and Asano (1993) have undertaken a theoretical analysis of wave dissipation by underwater vegetation such as kelp, deducing that there will be an exponential decay in the wave height with distance as the waves pass through the kelp. Good agreement was found between their theoretical relationship and wave-channel data for artificial kelp, although the calibrated drag coefficients varied widely and appeared to be affected by the nature of the kelp motions. Elwany et al. (1995) have obtained field measurements of wave dissipation in a *Macrocystis* kelp forest on the coast of southern California. They compared wave data in the offshore with waves measured in the lee of the kelp and also obtained comparable wave measurements in a control area down the coast where no kelp is found. The particular kelp bed involved in their study was 350-m wide and had an average plant density of about 10 plants per 100 m^2, typical of kelp beds in southern California. They found that the measured spectra, significant wave heights, mean-wave directions, and total radiation stress differed only slightly between the offshore kelp and control stations, and were similar at the onshore sites with and without kelp. Therefore, the effect of the 350-m wide kelp bed on surface waves could not be detected in their measurements. They concluded that kelp plants have evolved a hydrodynamically streamlined form that enables them to bend with the flow, both at the small scale of individual fronds and at the larger scale of the plant stalk than spans the complete water column.

WAVE REFRACTION AND DIFFRACTION

Upon entering intermediate to shallow water, waves are subject to *refraction* in which the direction of their travel changes with decreasing water depth in such a way that the crests tend to become more nearly parallel with the depth contours. For straight coasts having parallel contours, the wave crests become more closely parallel to the shoreline (Fig. 5-40). This

Figure 5-40 The refraction of swell as the waves approach the nearshore, bending so they become more closely parallel to the shore as they enter shallow water. [Courtesy of R. Wiegel]

refraction of waves results from their phase-velocity dependence on the water depth once they enter intermediate to shallow water, the lower the depth the slower the rate of advance. As illustrated in Figure 5-41, this results in the rotation of the wave crests with respect to the depth contours. In that diagram, the portion of wave crest at B is in deeper water than at A, and accordingly moves faster and therefore reaches B′, which represents a greater shift in position than the movement from A to A′, where the velocity is slower. This trend continues such that the wave crest progressively rotates and becomes more nearly parallel with the shore.

The refraction of water waves is analogous to the bending of light rays, and the change in direction is related to the change in the wave phase velocity through the same *Snell's law*:

$$\frac{\sin \alpha_1}{C_1} = \frac{\sin \alpha_2}{C_2} = \text{constant} \tag{5.46}$$

where α_1 and α_2 are angles between adjacent wave crests and the respective bottom contours, while C_1 and C_2 are the phase velocities at the two depths. For the simple case of a straight shoreline with parallel contours, the angle at any depth can be related to the deep-water angle of approach α_∞ with

$$\sin \alpha = \frac{C}{C_\infty} \sin \alpha_\infty \tag{5.47}$$

As the phase velocity C decreases relative to its value in deep water as the shore is approached, the angle α will correspondingly decrease from its deep-water value.

With an irregular bottom, wave refraction may cause either a spreading or a convergence of the wave energy. This effect can be examined best by concentrating on the *wave rays*, the lines drawn normal to the wave crests and therefore in the direction of wave advance and energy propagation (Fig. 5-42). On a straight coast with parallel contours, refraction will cause the rays to spread out, producing a similar spreading of the energy so the wave heights are reduced at the shoreline in comparison to what they would have been without refraction. The reason for this energy reduction is that the energy flux or power between adjacent rays is constant (neglecting bottom friction) and spreading due to refraction requires that the same amount of power be extended over a greater length of wave crest. The opposite is true for ray

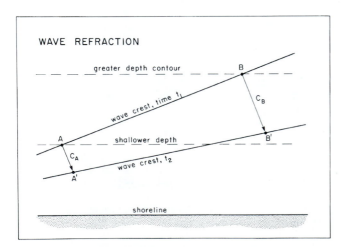

Figure 5-41 Wave refraction is brought about by the faster movement of waves in deep water than in shallow water. During a certain time interval, the wave crest at point B in deeper water moves farther than the crest at point A in shallower water, causing the wave crest to swing around and become more nearly parallel to the shore.

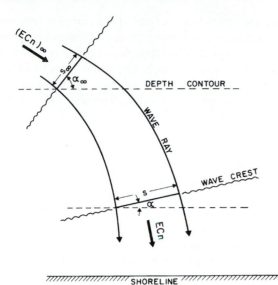

DEPTH CONTOUR

WAVE RAY

WAVE CREST

SHORELINE

Figure 5-42 The conservation of the wave energy flux ECn between two rays orthogonal to the wave crests.

convergence. If s_∞ is the spacing of the rays in deep water and s is the spacing later in the shoaling waves (Fig. 5-42), then

$$P = ECns = (ECns)_\infty = \text{constant} \tag{5.48}$$

Going through the same development that led to Equation (5.44) for shoaling in the absence of refraction, we now obtain

$$\frac{H}{H_\infty} = \left[\frac{1}{2n}\frac{C_\infty}{C}\right]^{1/2}\left[\frac{s_\infty}{s}\right]^{1/2} \tag{5.49}$$

Inclusion of wave refraction introduces the ratio s_∞/s, which was not present in Equation (5.44). In practice, tables for wave transformations give the ratio H/H'_∞, which does not account for refraction; this is also the ratio graphed in Figure 5-21. That ratio must then be multiplied by the correction coefficient

$$K_r = \left(\frac{s_\infty}{s}\right)^{1/2} \tag{5.50}$$

to account for refraction. For straight coasts with parallel contours, simple geometry considerations give

$$\frac{s_\infty}{s} = \frac{\cos \alpha_\infty}{\cos \alpha} \tag{5.51}$$

so the refraction coefficient K_r is simple to compute. In applications with irregular bathymetry and refraction, it is necessary to construct a wave-refraction diagram and to directly assess the variations in s along the shoreline compared with their offshore spacings. These measured s values are then used to evaluate K_r along the shoreline, and Equation (5.49) is then used to determine the wave-height variations.

Irregular bottom topography can cause waves to be refracted in a complex way and produce significant variations in wave heights and energies along the coast. Waves refract and diverge over the deep water of a submarine canyon or other depression in the sea floor, so the waves on the beach shoreward of the canyon are reduced in height while those to either side, where the rays converge, are somewhat higher (Fig. 5-43). An example of this divergence in the lee of submarine canyons is provided by the wave refraction diagram of Figure 5-44 for the area offshore from La Jolla, California (Munk and Traylor, 1947). Both the patterns of wave rays and crests are included in the diagram, but the rays are not all equally spaced in deep water. The rays approaching the canyons are initially placed closer together to better show their extreme divergence and the spreading of the energy, leading to small wave heights along the shore in the lee of each canyon. In contrast, there is some convergence of the rays to either sides of the canyons and in the shallower water between the two canyons. This would lead to enhanced wave heights and energies along those stretches of shore. As will be seen in Chapter 8, the resulting longshore variations in wave-breaker heights on this beach in La Jolla control the patterns of nearshore currents, with offshore-directed rip currents preferentially being positioned in the lees of the canyons where the wave heights are lowest. Waves also refract and bend toward headlands because of the offshore shoal area associated with the headland (Fig. 5-43). The wave energy is therefore concentrated on the headland, and the wave

A. REFRACTION OVER A CANYON

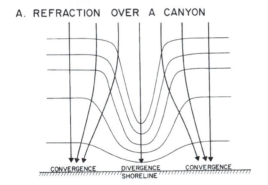

B. REFRACTION AT A HEADLAND

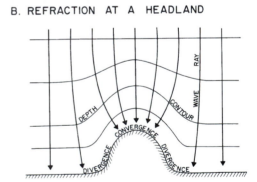

Figure 5-43 The convergence and divergence of wave rays over a submarine canyon and at a headland, resulting from wave refraction.

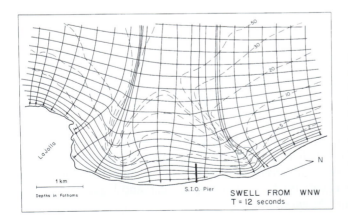

Figure 5-44 Wave refraction over submarine canyons and along the headland at La Jolla, California. The wave crests advance more rapidly over the deep water of the canyons, resulting in a spreading of the wave rays and a decrease in the wave energy in the lee of each canyon. [Adapted with permission of The University of Chicago Press, W. H. Munk and M. A. Traylor, Refraction of Ocean Waves: a Process Linking Underwater Topography to Beach Erosion, *Journal of Geology 55*, p. 10–11. Copyright © 1947 The University of Chicago Press.]

heights there may be substantially larger than in the adjacent embayments where ray spreading has reduced the wave energy.

Wave convergence or divergence due to wave refraction may be important in deciding where to construct a pier or other structure along a coast. The Standard Oil Company's oil-loading wharf at El Segundo, California, was built before wave refraction techniques came into general practice (Dunham, 1950). Unfortunately, the wharf was unwittingly located in an area of strong wave convergence (Fig. 5-45), produced by a shallower portion of the continental shelf to the south of a submarine canyon. Under certain wave directions and periods, large breaking waves developed at the wharf site, closing it for extended periods of time. The construction of wave-refraction diagrams may also be useful in analyses of long-term shoreline erosion patterns. This is illustrated by the investigation of Goldsmith (1976), who examined the potential for erosion along the coast of Delaware south to North Carolina by calculating refraction patterns of waves from various directions, noting the shoreline areas of greatest wave convergence and potential for erosion. The role of refraction over shoals that concentrate the energy of the waves on the shore, sometimes resulting in erosion, has been particularly significant along the coasts surrounding the North Sea where shoals are common. In Chapter 2 it was noted that on the East Anglia coast of England, Robinson (1980) demonstrated the significance of an offshore shoal (Sizewell Bank) to the long-term erosion of the coast, including the ancient town of Dunwich (Fig. 2-3). Of special interest, the slow northward migration of the shoal has caused periods of erosion of beaches and cliffs to alternate with periods of beach accretion, so that while Dunwich has undergone massive erosion over the centuries, cliff retreat during the past 50 years has been minor. MacDonald and O'Connor (1994) have further examined the effects of shoals in the North Sea on wave refraction and coastal erosion, demonstrating that wave dissipation over the shallow-water shoals as well as energy convergence due to refraction can be important. Figure 5-46 from their study contrasts the alongcoast variations in wave energy in the lee of a shoal, with the distribution that would exist if the shoal was not present. The shoal causes some dissipation of the wave energy, so if it were not present, the total amount of energy reaching the shore would be greater, and this is apparent in the slightly higher energy of the dashed curve in Figure 5-46 compared with the average level of the solid curve which shows the distribution of energy

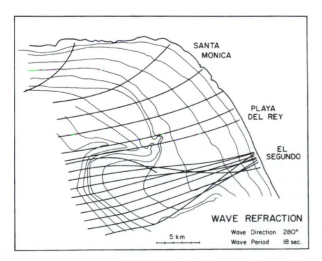

SANTA
MONICA

PLAYA
DEL REY

EL
SEGUNDO

WAVE REFRACTION

Wave Direction   280°
5 km
Wave Period   18 sec.

Figure 5-45 Wave refraction near El Segundo, California, which produces huge waves at the oil-loading wharf constructed there. [Adapted from Refraction and Diffraction Diagrams, J. W. Dunham, *Proceedings of the 1st Coastal Engineering Conference,* 1950. Reproduced with permission from the American Society of Civil Engineers.]

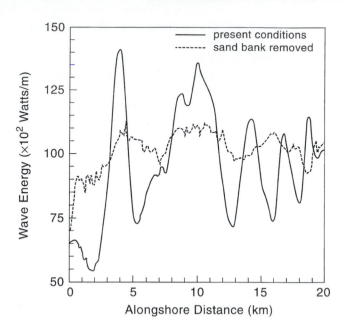

Figure 5-46   The effects of an offshore shoal on the alongcoast distribution of wave energy reaching the shoreline due to the effects of the shoal on refraction and in dissipating some of the wave energy. With the shoal (present conditions), there is a strong variation in the wave energy reaching the shore, with the peaks occurring where the wave rays converge, while the troughs correspond to zones of ray spreading. The analyses included calculations of the energy distribution if the shoals were not present (sand bank removed), which evens out the alongcoast energy since refraction becomes less important. However, the total energy reaching the coast is greater in the absence of the shoals, since their presence causes some wave-energy dissipation. [Courtesy of N. J. MacDonald]

with the shoal present. Although the shoal does act to dissipate some of the wave energy, of greater importance is the focusing of the energy by refraction, resulting in the energy peaks graphed in Figure 5-46 where the energy is roughly twice that reaching the coast in the troughs (areas of ray divergence). Much of the study of MacDonald and O'Connor focuses on how such patterns of wave refraction over shoals and the resulting shoreline erosion might change during the next century due to potential elevated sea levels, the changes in water depths over the shoals altering both the wave dissipation and refraction patterns.

It can be seen in Figure 5-45 that some of the wave rays in that example cross one another, a common problem in diagrams for wave refraction over a complex bottom bathymetry. In this situation the distance s between the adjacent rays decreases to zero, and according to Equation (5.49), the wave height H should increase to an infinite value at the crossing point. This obviously does not happen. Instead, as the wave height increases and the water-surface slope steepens near the crossing point, energy will travel laterally along the wave crest and secondary waves may develop. A review of the complex wave behavior near wave-ray crossings can be found in Pierson (1972).

Related to refraction is the *diffraction* of water waves, a phenomenon where the energy is transferred laterally along a wave crest and is most noticeable where an otherwise regular train of waves is interrupted by a barrier such as a small island or breakwater (Fig. 5-47). Such structures cut off the wave energy, creating a shadow zone, usually desired for sheltering boats. The diffraction process causes wave energy to leak into the shadow zone, at times creating undesirable wave conditions within the harbor. Wave diffraction can also be an important control on the shapes of shorelines, in large part accounting for the arcuate shapes of pocket beaches where there is only a narrow opening to the sea, and also for the curvature of shorelines partially sheltered by breakwaters and jetties.

In recent years there have been considerable advances in analyses of the combined wave refraction and diffraction, in part, due to the increased access to powerful computers that permit the development of detailed numerical models. Rather than yielding wave-ray

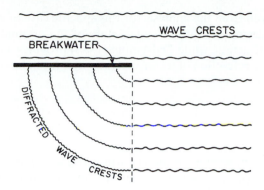

Figure 5-47   Wave diffraction into the shadow zone of an offshore breakwater.

diagrams, as presented above, the most recent numerical analyses obtain solutions for a finite number of grid cells that cover the area of interest. This is illustrated in Figure 5-48 by Ebersole (1985) for the refraction of waves in the vicinity of Oregon Inlet on the coast of North Carolina. Figure 5-48(a) gives the wave directions at each grid cell, while 5-48(b) contours the

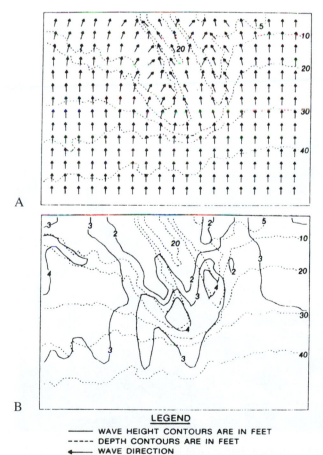

Figure 5-48   Analysis of the wave refraction and diffraction in the area of Oregon Inlet, North Carolina, where the results are given for grid cells in terms of wave directions (a) or as contours of wave heights (b). Different results are obtained for each combination of offshore wave direction, period and wave energy. [From Refraction-Diffraction Model for Linear Waves, B. A. Ebersole, *Journal of Waterway, Port, Coastal and Ocean Engineering,* 1985. Reproduced with permission from the American Society of Civil Engineers.]

LEGEND
——— WAVE HEIGHT CONTOURS ARE IN FEET
- - - - DEPTH CONTOURS ARE IN FEET
◀——— WAVE DIRECTION

wave heights over the area as derived from the analysis for this specific case of waves arriving from deep water. Kirby and Dalrymple (1986) provide a state-of-the-art analysis that includes both wave refraction and diffraction, while Dalrymple (1988) provides a comparable model, but one that includes only the refraction of water waves so that it can be run on a personal computer.

TSUNAMI

Thus far we have considered only waves generated by winds, those that are the most important energy source in the nearshore and generally are responsible for coastal erosion problems. However, before concluding this chapter we need to recognize the existence of the destructive *tsunami* (Fig. 5-49) that can be produced by a displacement of the sea floor at the time of an earthquake or explosive volcanic eruption. The sudden upward or downward movement of a portion of the seabed momentarily raises or lowers the overlying water surface. That disruption of the water then travels outward from its site of origin as a series of waves. In the deep ocean, the heights of these waves are small, typically less than a meter. However, tsunamis have long periods, the time between successive waves in a series being some 10–20 minutes. One significance of this long period is that they are in effect shallow-water waves even in the deepest part of the ocean. When the tsunami waves reach the continental shelf, their rate of movement slows considerably, which in turn causes their heights to increase. As the tsunami waves shoal across the wide continental shelf, they progressively increase in height, until at the shore itself they have been reported to achieve heights of 10–20 m. Their breaking and run-up on a sloping beach can carry water well beyond the normal reach of the ocean.

Figure 5-49   The tsunami and earthquake destruction of Lisbon, Portugal, in 1755.

The most common sources of large tsunamis are earthquakes in the north Pacific around Alaska and Japan and the west coast of South America, particularly Chile. Two struck the U.S. west coast during the 1960s—on March 28, 1964 and on May 16, 1968 (Wilson and Tørum, 1968). The 1964 tsunami caused extensive damage along the coasts of Oregon and California, with the greatest destruction having occurred in Crescent City in northern California, where one wave washed about 500 m inland, destroying twenty-nine blocks of the business district and causing some $29 million in damage. Hawaii is even more prone to tsunamis. The most recent tsunami of significance to reach the shores of Hawaii were generated in the sea floor of the Aleutian Trench (1946 and 1957), from Kamchatka (1952), and from Chile (1960). Waves during the 1946 event exceeded 10 m in Hilo, flooding much of the city and killing 150 people (Wiegel, 1964). The 1960 earthquake in Chile, magnitude 9.5, is the largest quake ever recorded. It generated a tsunami that killed more than 1,000 people in Chile, 61 in Hawaii, and 199 in Japan.

Because of their very long periods, tsunami are in effect shallow-water waves even as they cross the deep ocean. Accordingly, their phase velocities should be on the order of

$$C = \sqrt{gh} \approx \sqrt{9.81 \text{ m/sec} \times 4{,}000 \text{ m}}$$

$$\approx 200 \text{ m/sec}$$

$$\approx 700 \text{ km/hr}$$

$$\approx 17{,}000 \text{ km/day}$$

This phase velocity is roughly 400 miles per hour and demonstrates the rapidity with which tsunami spread across the sea. Such high rates have been verified by wave arrival times at coastal sites distant from the earthquake source. Such an analysis is illustrated by Figure 5-50 for the 1964 tsunami that was generated by an earthquake offshore from Alaska (Wilson and Tørum, 1968). This diagram shows the wave front at 1–hour intervals (Greenwich Mean

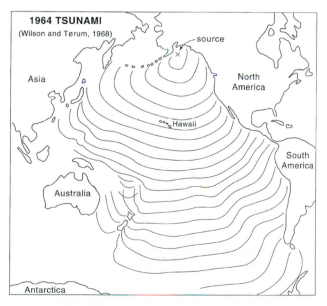

Figure 5-50   The calculated advance of tsunami waves generated by the 1964 earthquake in the offshore south of Alaska. Each successive crest position represents 1 hour of movement. [From B. W. Wilson and A. Tørum, *The Tsunami of the Alaskan Earthquake, 1964,* U.S. Army Coastal Engineering Research Center Technical Memo N. 25, 1968.]

Time), based on a wave-refraction analysis for its movement in the deep ocean and measured arrival times at the many tide stations along the coasts of the Pacific Ocean. Some bending of the wave crests due to refraction is seen to have occurred in the deepest ocean basin but in particular as the waves cross the continental slope and shelf.

The records of this tsunami in the tide gauges of Santa Monica and Los Angeles in southern California are shown in Figure 5-51. These records demonstrate that a tsunami is not simply a single wave or even a series of 5–10 waves but instead that waves from the disturbance can continue to arrive for many hours. Although the wave heights tend to decrease with time, it is not necessarily the case that the first wave will be the largest; in the Santa Monica record of Figure 5-51, it appears that the fourth wave had the greatest height. These respective records from Santa Monica and Los Angeles, obtained only some 75 km apart, also reveal that the sizes of waves and even their general patterns can be spatially variable along the coast. Munk, Snodgrass, and Tucker (1959) found that the spectra are similar at one tide gauge for different tsunami but dissimilar at different tide gauges for a particular tsunami, as seen here in Figure 5-51. This observation led them to conclude that local resonance determined by the geometry of the offshore continental shelf and coastline configuration, are extremely important to the response at a specific site. Because of the control by the offshore submarine topography and the orientation and elevation of the coast, with possible resonance

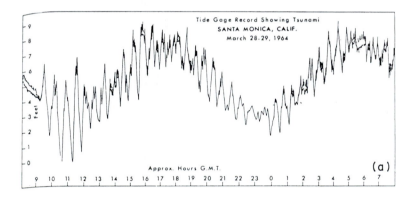

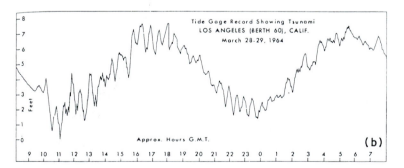

Figure 5-51 Tide-gauge records at Santa Monica and Los Angeles, California, showing the presence of a series of tsunami waves lasting for some 24 hours, superimposed on the broad tidal cycles. [From B. W. Wilson and A. Tørum, *The Tsunami of the Alaskan Earthquake, 1964,* U.S. Army Coastal Engineering Research Center Technical Memo N. 25, 1968.]

effects, the patterns of destruction by tsunami can be highly variable along a coast and are often difficult to understand. Maximum destruction often takes place along the shores of bays and estuaries rather than on the open ocean, due to the funneling and constriction of the tsunami waves and their run-up.

SUMMARY

In this chapter we have concentrated on waves in the deep ocean, first on their generation by storms, then on their propagation across the sea as regular swell, and finally as they approach the coast where the waves deliver their acquired energy as breakers and surf. In reviewing wave generation, it was found that the wave heights, energies, and periods achieved are dependent on the storm conditions of wind speeds, duration, and fetch distance. Wave-prediction analyses are available that yield either the significant wave height and period or the complete spectrum of generated waves. Following generation by the storm, the waves are sorted by a dispersion process wherein their rates of propagation across the sea depend on the wave periods, with the higher period waves traveling faster. This dispersion, together with some viscous dissipation of the short-period waves and spreading of their energy into other directions, results in the conversion of the broad-spectrum sea in the area of generation to regular swell having a comparatively narrow spectrum that can be characterized by a single dominant period. It is this regular swell that is most amenable to mathematical description, that is, by one of the theories of wave motion such as the Airy- or Stokes-wave solutions. The equations derived in these wave theories are particularly important in describing the transformations of the waves as they cross the continental shelf and enter the range of intermediate to shallow-water depths. The waves generally slow down as they shoal, with their heights and thus the wave steepness increasing. This trend eventually results in their instability, causing the waves to break as they approach the shore. This line of review and analyses will be continued in the next chapter, where we consider the expenditure of the wave energy within the nearshore, first by breaking, then as bores crossing the surf zone, and finally as run-up on the beach.

PROBLEMS

1. From the surface expression for the Airy wave, Equation (5.7), show that the maximum value of the slope $\partial \eta / \partial x$ of the water surface is $\pi H/L$.

2. There is evidence that oceans existed on the planet Mars in the distant past. The acceleration of gravity on Mars is 3.72 m/sec^2, about one-third that of Earth. How would this difference affect the generation and motions of waves?

3. Deduce the general expressions for d_0 and u_0, the orbital diameter and velocity at the seafloor ($z = -h$) for Airy waves. Confirm that s and w are zero at the bottom.

4. Waves generated by a storm range in periods from 5 to 20 sec. Assuming deep-water conditions for the entire travel distance, apply Airy theory to construct a rough graph of travel time versus T for beaches 1,000 km and 3,000 km from the storm area. How much later do the 5-sec waves arrive at the 3,000-km beach than the 20-sec waves? Compare this with their arrival times at the 1,000-km distant beach.

5. Waves arriving at a beach from a distant storm progressively decrease in wave period, from 10 to 5 sec, with the 5-sec waves arriving 10 hours later than the 10-sec waves. Assuming deep-water wave conditions for the entire travel distance, how far away was the storm?

6. Calculate and graph the deep-water Airy-wave orbital diameters at water depths $z = -2, -5, -10,$ and -20 m for the following wave trains:

Wave #1	Wave #2	Wave #3
$T = 5$ sec	$T = 10$ sec	$T = 10$ sec
$H = 1$ m	$H = 1$ m	$H = 3$ m

7. Using Airy-wave theory, determine the wave length L and phase velocity C for waves whose period is 10 sec, traveling in a water depth $h = 25$ m.

8. Compare the maximum wave heights that can be achieved by 1-sec, 5-sec, and 10-sec waves in deep water, instability with breaking limiting any further growth.

9. Proportionally, how much more power must be derived from the wind to increase the height of a $T = 10$-sec wave from $H = 0.5$ to 1 m, than to increase a 4-sec wave through the same height range?

10. Winds blow across a circular deep-water lake of diameter 2 km, generating waves. What is the maximum time that the wind could transfer energy into waves of periods 2 sec, 4 sec, and 10 sec? What does this imply regarding the growth of waves within that range of periods?

11. A wave group of period 10 sec has an overall length of 1,000 km. What is the approximate life span of a single wave within the group, assuming deep-water conditions? How far will the wave group as a whole have traveled during that time?

12. It has been suggested that the detection of long-period but low-amplitude waves, those having periods of 20 sec, could serve as "forerunners" to predict the imminent arrival of larger storm waves in the period range 10–15 sec. Undertake calculations to test this possibility. For example, how much warning would the forerunners provide if the storm distance is 1,000 km versus 5,000 km?

13. A storm generates waves with periods ranging up to 15 sec. What would be the widths of the spectra at the distances 1,000 km, 2,000 km, and 5,000 km after 125 hours of storm activity, assuming the storm continues for the full time?

14. Which of the following quantities increase, decrease, or remain essentially constant during the shoaling of regular waves: $C, C_g = Cn, T, E, H, ECns$?

15. Uniform swell in deep water has a period of 12 sec and approaches the coast with an angle of 25° with respect to the shoreline trend. When the waves finally break on the beach, the water depth at breaking is 2.5 m and the breaker height is approximately the same as the depth. Calculate the wave-breaker angle and also the K_r refraction coefficient.

16. Tsunamis in the deep ocean have heights on the order of 1 m, but because of their long periods they are effectively in "shallow water." Assuming one can correctly apply shallow-water Airy theory to describe the tsunami propagation across the ocean, calculate the energy flux (power) of the waves in the deep ocean, where the water depth is 4,000 m, and estimate the height of the wave as it reaches the coast and breaks, assuming that breaking occurs where the wave height is approximately equal to the water depth.

17. Write a computer program to calculate the JONSWAP spectrum of Equation (5.6), using the various relationships provided to evaluate the spectrum from the storm fetch and wind velocity. Run the program for the storm factors $F = 1,000$ km and $U = 5$ m/sec and for the larger storm with $F = 5,000$ km and $U = 10$ m/sec.

18. Develop a computer program to compute the characteristics of Airy waves, including the profile, celerity, energy density, power, and maximum horizontal orbital velocity and diameter d_0 at the sea floor. Use Equation (5.15) as an acceptable approximation for the wave length. Set up the program

to accept input for the wave period, height, and water depth. Undertake test runs using various input-parameter combinations to assure that the program is yielding correct results.

19. Write a computer program to compute the surface profile and horizontal orbital velocity at the seafloor ($z = -h$) for a second-order Stokes wave, with the wave height, period and water depth being given. Use Equation (5.15) as an acceptable approximation for calculation of the wave length and then use $C = L/T$ to determine the wave celerity (these approximations are acceptable for the 2nd order solution). Undertake test runs using the same input combinations employed in Problem 18 above.

REFERENCES

Bagnold, R. A. (1947). Sand Movement by Waves: Some Small-Scale Experiments with Sand at Very Low Density. *Journal Institute of Civil Engineers* 27: 447–469.

Barber, N. F. and F. Ursell (1948). The Generation of Ocean Waves and Swell, I: Wave Periods and Velocity. *Philosophical Transactions of the Royal Society of London,* Series A, 240: 527–560.

Bascom, W. (1980). *Waves and Beaches.* Garden City, NY: Anchor Books.

Borgman, L. E. (1979). Directional Wave Spectra from Wave Sensors. In *Ocean Wave Climate,* M. D. Earle and A. Malahoff (editors), pp. 269–300. New York: Plenum Press.

Borgman, L. E. (1963). Risk Criteria. *Journal of the Waterways and Harbors Division, Amer. Soc. Civil Engrs.* 89: 1–35.

Boussinesq, J. (1872). Théorie des Ondes et de Remous Qui Se Propagent le Long d'un Canal Rectangulaire Horizontal, en Communiquant Au Liquide Contenu dans Ce Danal des Vitesses Sensiblement Paralleles de la Surface au Fond. *Jour. Math. Pures et Appliquées* (Lionvilles, France), 17: 55–108.

Bouws, E., H. Günther, W. Rosenthal, and C. L. Vincent (1985). Similarity of the Wind Wave Spectrum in Finite Depth Water, 1: Spectral Form. *Journal of Geophysical Research* 90(C1): 975–986.

Bretschneider, C. L. (1952). The Generation and Decay of Wind Waves in Deep Water. *Transactions American Geophysical Union* 33: 381–389.

Bretschneider, C. L. (1957). Revisions in Wave Forecasting—Deep and Shallow Water. *Proceedings of the 6th Coastal Engineering Conference, Amer. Soc. Civil Engrs.,* pp. 30–67.

Bretschneider, C. L. (1959). *Wave Variability and Wave Spectra for Wind Generated Gravity Waves.* Beach Erosion Board Technical Memo No. 118, pp. 192. U.S. Army Corps of Engineers.

Bretschneider, C. L. (1963). A One-Dimensional Gravity Wave Spectrum. In *Ocean Wave Spectra.* Pp. 41–56. Englewood Cliffs, NJ: Prentice-Hall.

Bretschneider, C. L. (1964). The Ash Wednesday East Coast Storm, March 5–8, 1962. *Proceedings of the 9th Conference on Coastal Engineering, Amer. Soc. Civil Engr.,* pp. 617–659.

Cartwright, D. E. and M. S. Longuet-Higgins (1956). The Statistical Distribution of the Maxima of a Random Function. *Proceedings of the Royal Society of London,* Series A, 237: 212–232.

CERC (1984). *Shore Protection Manual.* (2 volumes) Coastal Engineering Research Center, Waterway Experiment Station, Corps of Engineers.

Cornish, V. (1934). *Ocean Waves and Kindred Physical Phenomena.* London, England: Cambridge University Press.

Daily, J. W. and S. C. Stephen (1952). The Solitary Wave. *Proceedings of the 3rd Coastal Engineering Conference, Amer. Soc. Civil Engrs.,* pp. 13–30.

Daily, J. W. and S. C. Stephen (1953). Characteristics of the Solitary Wave. *Transactions American Society Civil Engineers* 118: 575–587.

Dalrymple, R. A. (1988). A Model for the Refraction of Water Waves. *Journal of Waterway, Port, Coastal, and Ocean Engineering, Amer. Soc. Civil Engrs.* 114(4): 423–435.

DARBYSHIRE, J. (1963). The One-Dimensional Wave Spectrum in the Atlantic Ocean and Coastal Waters. In *Ocean Wave Spectra*, pp. 27–31. Englewood Cliffs, NJ: Prentice-Hall.

DEAN, R. G. (1965). Stream Function Representation of Nonlinear Ocean Waves. *Journal of Geophysical Research* 70: 4561–72.

DEAN, R. G. (1970). Relative Validities of Water Wave Theories. *Journal Waterways and Coastal Engineering, Amer. Soc. Civil Engrs.* 96(WW1): 105–119.

DEAN, R. G. (1990). Stream Function Wave Theory and Applications. In *Handbook of Coastal and Ocean Engineering,* J. B. Herbich (editor), Vol. 1, pp. 63–94. Houston, TX: Gulf Publishing Company.

DEAN, R. G. and R. A. DALRYMPLE (1984). *Water Wave Mechanics for Engineers and Scientists.* Englewood Cliffs, NJ: Prentice-Hall.

DOLAN, R., H. LINS, and B. HAYDEN (1988). Mid-Atlantic Coastal Storms. *Journal of Coastal Research* 4: 417–433.

DOLAN, R., D. L. INMAN, and B. HAYDEN (1990). The Atlantic Coast Storm of March 1989. *Journal of Coastal Research* 6: 721–725.

DRAPER, L. (1979). A Note on the Wave Climatology of UK Waters. In *Ocean Wave Climate,* M. D. Earle and A. Malahoff (editors), pp. 327–331. New York: Plenum Press.

DUNHAM, J. W. (1950). Refraction and Diffraction Diagrams. *Proceedings of the 1st Coastal Engineering Conference, Amer. Soc. Civil Engrs.,* pp. 33–49.

EARLE, M. D. and J. M. BISHOP (1984). *A Practical Guide to Ocean Wave Measurement and Analysis.* Marion, MA: Endeco Inc.

EARLE, M. D., K. A. BUSH, and G. D. HAMILTON (1984). High-Height Long-Period Ocean Waves Generated by a Severe Storm in the Northeast Pacific Ocean During February 1983. *Journal of Physical Oceanography* 14: 1286–1299.

EBERSOLE, B. A. (1985). Refraction-Diffraction Model for Linear Waves. *Journal of Waterway, Port, Coastal, and Ocean Engineering, Amer. Soc. Civil Engrs.* 111(6): 939–953.

ECKART, C. (1952). The Propagation of Waves from Deep to Shallow Water. In *Gravity Waves.* National Bureau of Standards, Circular No. 521, pp. 165–173.

ELGAR, S. and R. T. GUZA (1985a). Shoaling Gravity Waves: Comparison Between Field Observations, Linear Theory, and a Nonlinear Model. *Journal of Fluid Mechanics* 158: 47–70.

ELGAR, S. and R. T. GUZA (1985b). Observations of Bispectra of Shoaling Surface Gravity Waves. *Journal of Fluid Mechanics* 161: 425–448.

ELGAR, S. and R. T. GUZA (1986). Nonlinear Model Predictions of Bispectra of Shoaling Surface Gravity Waves. *Journal of Fluid Mechanics* 167: 1–18.

ELWANY, M. H. S., W. C. O'REILLY, R. T. GUZA, and R. E. FLICK (1995). Effects of Southern California Kelp Beds on Waves. *Journal of Waterway, Port, Coastal, and Ocean Engineering, Amer. Soc. Civil Engrs.* 121: 143–150.

FINKL, C. W. and O. H. PILKEY (1991). Impacts of Hurricane Hugo. September 10–22, 1989. *Journal of Coastal Research* Special Issue No. 8.

FITZGERALD, D. M., S. VAN HETEREN, and T. M. MONTELLO (1994). Shoreline Processes and Damage Resulting from the Halloween Eve Storm of 1991 Along the North and South Shores of Massachusetts Bay, USA. *Journal of Coastal Research* 10: 113–132.

FLICK, R. E., R. T. GUZA, and D. L. INMAN (1981). Elevation and Velocity Measurements of Laboratory Shoaling Waves. *Journal of Geophysical Research* 86(C5): 4149–4160.

FREILICH, M. H. and R. T. GUZA (1984). Nonlinear Effects on Shoaling Surface Gravity Waves. *Philosophical Transactions of the Royal Society of London,* Series A, 311: 1–41.

GODA, Y. (1974). Estimation of Wave Statistics from Spectral Information. *International Symposium on Ocean Wave Measurements and Analysis, Amer. Soc. Civil Engrs.,* pp. 320–337.

GOLDSMITH, V. (1976). Wave Climate Models for the Continental Shelf: Critical Links Between Shelf Hydraulics and Shoreline Processes. *Beach and Nearshore Sedimentation,* Special Publication No. 24, pp. 24–47. Soc. Econ. Paleontologists and Mineralogists, Tulsa, Oklahoma.

GOODKNIGHT, R. C. and T. L. RUSSELL (1963). Investigation of the Statistics of Wave Heights. *Journal of Waterways and Harbors Division, Amer. Soc. Civil Engrs.,* 89: 29–54.

GUZA, R. T. and E. D. THORNTON (1980). Local and Shoaled Comparisons of Sea Surface Elevations, Pressures and Velocities: *Journal of Geophysical Research* 85(C3): 1524–1530.

HASSELMANN, K. (1963). On the Non-Linear Energy Transfer in a Gravity Wave Spectrum. *Journal of Fluid Mechanics* 15: 273–281 (Part 2), 385–398 (Part 3).

HASSELMANN, K. et al, (1973). Measurements of Wind-wave Growth and Swell Decay During the Joint North Sea Wave Project (JONSWAP). *Deutsche Hydrographische Zeitschrift,* Supplement A8, 12: 95.

HASSELMANN, K, D. B. ROSS, P. MULLER, and W. SELL (1976). A Parametric Wave Prediction Model. *Journal of Physical Oceanography* 6: 200–228.

HELMSLEY, J. M. and R. M. BROOKS (1989). Waves for Coastal Design in the United States. *Journal of Coastal Research* 5: 639–663.

HELMSLEY, J. M., D. D. McGEHEE, and W. M. KUCHARSKI (1991). Nearshore Oceanographic Measurements: Hints on How to Make Them. *Journal of Coastal Research* 7: 301–315.

HOUGH, S. S. (1896). On the Influence of Viscosity on Waves and Currents. *Proceedings of the London Mathematical Society* 28: 264–288.

HOUSLEY, J. G. and D. C. TAYLOR (1957). Application of the Solitary Wave Theory to Shoaling Oscillatory Waves. *Transactions American Geophysical Union* 38: 56–61.

HUBERTZ, J. M., D. B. DRIVER, and R. D. REINHARD (1991). Wind Waves on the Great Lakes: A 32 Year Hindcast. *Journal of Coastal Research* 7: 945–967.

HUBERTZ, J. M., B. A. TRACY, J. B. PAYNE, and A. CIALONE (1992). *Verification of Pacific Ocean Deepwater Hindcast Information.* WIS Report 29, Coastal Engineering Research Center, U.S. Army Corps of Engineers, Vicksburg, Mississippi.

HUBERTZ, J. M., R. M. BROOKS, W. A. BRANDON, and B. A. TRACY (1994). Hindcast Wave Information for the U.S. Atlantic coast. *Journal of Coastal Research* 10: 79–100.

HUNTLEY, D. A. (1979). Electromagnetic Flowmeters in Nearshore Field Studies. In *Workshop on Instrumentation for Currents and Sediments in the Nearshore Zone,* Nat. Res. Council Canada, pp. 47–60.

IPPEN, A. T. and G. KULIN (1954). The Shoaling and Breaking of the Solitary Wave. *Proceedings of the 5th Coastal Engineering Conference, Amer. Soc. Civil Engrs.,* pp. 27–49.

IVERSEN, H. W. (1952a). Waves and Breakers in Shoaling Water. *Proceedings of the 3rd Conference on Coastal Engineering, Amer. Soc. Civil Engrs.,* pp. 1–12.

IVERSEN, H. W. (1952b). Studies of Wave Transformation in Shoaling Water, Including Breaking. In *Gravity Waves,* National Bureau of Standards Circular No. 521, pp. 9–32.

JEFFREYS, H. (1925). On the Formation of Water Waves by Wind. *Proceeding of the Royal Society of London,* Series A, 107: 189–206.

JEFFREYS, H. (1926). On the Formation of Water Waves by Winds. *Proceedings of the Royal Society of London,* Series A, 110: 241–247.

JONSSON, I. G. (1966). Wave Boundary Layers and Friction Factors. *Proceedings of the 10th Coastal Engineering Conference, Amer. Soc. Civil Engrs.,* pp. 127–148.

KAMPHUIS, J. W. (1975). Friction Factors Under Oscillatory Waves. *Journal of Waterways, Harbors, and Coastal Engineering Division, Amer. Soc. Civil Engrs.* 101: 135–144.

KAMPHUIS, J. W. (1991). Wave Transformation. *Coastal Engineering* 15: 173–184.

KAMPHUIS, J. W. (1994). Wave Height from Deep Water Through Breaking Zone. *Journal of Waterway, Port, Coastal, and Ocean Engineering, Amer. Soc. Civil Engrs.* 120: 347–367.

KELLER, J. B. (1948). The Solitary Wave and Periodic Waves in Shallow Water. *Comm. Appl. Math.* 1(4): 323–329.

KEULEGAN, G. H. (1950). Wave Motion. In *Engineering Hydraulics,* H. Rouse (editor). Chapter 11, New York: John Wiley & Sons.

KEULEGAN, G. H. and G. W. PATTERSON (1940). Mathematical Theory of Irrotational Translation Waves. *Journal of Research of the National Bureau of Standards* 24, RP1272.

KINSMAN, B. (1965). *Wind Waves*: Englewood Cliffs, NJ: Prentice-Hall.

KIRBY, J. T. and R. A. DALRYMPLE (1986). An Approximate Model for Nonlinear Disperison in Monochromatic Wave Propagation Models. *Coastal Engineering* 9: 545–561.

KISHI, T. and H. SAEKI (1966). The Shoaling, Breaking, and Runup of the Solitary Wave on Impermeable Rough Slopes. *Proceedings of the 10th Coastal Engineering Conference, Amer. Soc. Civil Engrs.*, pp. 322–348.

KOBAYASHI, N., A. W. RAICHLE, and T. ASANO (1993). Wave Attenuation by Vegetation. *Journal of Waterway, Port, Coastal, and Ocean Engineering, Amer. Soc. Civil Engrs.* 119: 30–48.

KOMAR, P. D. (1986). The 1982–83 El Niño and Erosion on the Coast of Oregon. *Shore & Beach* 54: 3–12.

KORTEWEG, D. J. and G. de VRIES (1895). On the Change of Form of Long Waves Advancing in a Rectangular Canal, and on a New Type of Long Stationary Waves. *Philosophical Magazine,* Series 5, 39: 422–443.

LAMB, H. (1932). *Hydrodynamics,* 6th edition. Cambridge, MA: Cambridge University Press.

LEMÉHAUTÉ, B. (1976). *An Introduction to Hydrodynamics and Water Waves.* Dusseldorf: Springer-Verlag.

LITTMAN, W. (1957). On the Existence of Periodic Waves Near Critical Speed. *Comm. Pure App. Math.* 10: 241–269.

LIU, P. C. (1971). Normalized and Equilibrium Spectra of Wind Waves in Lake Michigan. *Journal of Physical Oceanography* 1: 249–257.

LONG, C. E. (1991). *Use of Theoretical Wave Height Distributions in Directional Seas.* U.S. Army Corps of Engineers, Coastal Engineering Research Center Technical Report No. CERC-91–6.

LONGUET-HIGGINS, M. S. (1952). On the Statistical Distribution of the Height of Sea Waves. *Journal of Marine Research* 11: 245–266.

LONGUET-HIGGINS, M. S. (1953). Mass Transport in Water Waves. *Philosophical Transactions of the Royal Society of London,* Series A, 245: 535–581.

LONGUET-HIGGINS, M. S. (1956). The Refraction of Sea Waves in Shallow Water. *Journal of Fluid Mechanics* 1: 163–176.

LONGUET-HIGGINS, M. S. (1969). A Nonlinear Mechanism for Generation of Sea Waves. *Proceedings of the Royal Society of London,* Series A, 311: 371–389.

LONGUET-HIGGINS, M. S., D. E. CARTWRIGHT, and N. D. SMITH (1963). Observations of the Directional Spectrum of Sea Waves Using the Motions of a Floating buoy. In *Ocean Wave Spectra,* pp. 111–132. Englewood Cliffs, NJ: Prentice-Hall.

LONGUET-HIGGINS, M. S. and R. W. STEWART (1960). Changes in the Form of Short Gravity Waves on Long Waves and Tidal Currents. *Journal of Fluid Mechanics* 8: 565–583.

LONGUET-HIGGINS, M. S. and R. W. STEWART (1964). Radiation Stresses in Water Waves: A Physical Discussion, with Applications. *Deep-Sea Research* 11: 529–562.

MACDONALD, N. J. and B. A. O'CONNOR (1994). Influence of Offshore Banks on the Adjacent Coast. *Proceedings of the 24th Coastal Engineering Conference, Amer. Soc. Civil Engrs.*, pp. 2311–2324.

MADSEN, O. S., Y.-K. POON, and H. C. GRABER (1988). Spectral Wave Attenuation by Bottom Friction: Theory. *Proceedings of the 21st Coastal Engineering Conference, Amer. Soc. Civil Engrs.*, pp. 492–504.

MADSEN, O. S., P. P. MATHISEN, and M. M. ROSENGAUS (1990). Movable Bed Friction Factors for Spectral Waves. *Proceedings of the 22nd Coastal Engineering Conference, Amer. Soc. Civil Engrs.*, pp. 420–429.

MASCH, F. D. and R. L. WIEGEL (1961). *Cnoidal Waves: Tables of Functions.* Engineering Foundation Council on Wave Research, Berkeley, California.

McCOWAN, J. (1891). On the Solitary Wave. *Philosophical Magazine,* Series 5, 32: 45–58.

McCOWAN, J. (1894). On the Highest Wave of Permanent Type. *Philosophical Magazine,* Series 5, 38: 351–357.

MICHE, R. (1944). Undulatory Movements of the Sea in Constant and Decreasing Depth. *Ann. de Ponts et Chaussees* (May–June, July–August), pp. 25–78, 131–164, 270–292, 369–406.

MILES, J. (1957). On the Generation of Surface Waves by Shear Flows. *Journal of Fluid Mechanics* 3: 185–204.

MILES, J. (1959). On the Generation of Surface Waves by Shear Flows, Part 2. *Journal of Fluid Mechanics* 6: 568–582.

MILES, J. (1960). On the Generation of Surface Waves by Turbulent Shear Flows. *Journal of Fluid Mechanics* 7: 469–478.

MILLER, H. C. and R. E. JENSEN (1990). *Comparison of Atlantic Coast Wave Information Hindcasts with Field Research Facility Gauge Measurements.* TR CERC-90–17, U.S. Army Engineer Waterways Experiment Station, Vicksburg, Mississippi.

MITCHELL, J. H. (1893). On the Highest Waves in Water. *Philosophical Magazine,* Series 5, 36: 430–437.

MUIR WOOD, A. M. (1969). *Coastal Hydraulics.* London, England: MacMillan.

MUNK, W. H. (1949). The Solitary Wave Theory and Its Application to Surf Problems. *New York Academy of Science Ann.* 51: 376–424.

MUNK, W. H. and F. E. SNODGRASS (1957). Measurements of Southern Swell at Guadalupe Island. *Deep Sea Research* 4: 272–286.

MUNK, W. H., F. E. SNODGRASS, and M. J. TUCKER (1959). Spectra of Low Frequency Ocean Waves. *Bulletin Scripps Institution of Oceanography* 7(4): 283–362.

MUNK, W. H. and M. A. TRAYLOR (1947). Refraction of Ocean Waves: A Process Linking Underwater Topography to Beach Erosion. *Journal of Geology* 55: 1–26.

NAFAA, M. G., A. M. FANOS, and M. A. ELWANEY (1991). Characteristics of Waves Off the Mediterranean Coast of Egypt. *Journal of Coastal Research* 7: 665–676.

NDBC (1992). *National Data Buoy Center 1992 Annual Report.* National Data Buoy Center, Stennis Space Center (Mississippi), U.S. Department of Commerce, National Oceanic and Atmospheric Administration.

PHILLIPS, O. M. (1957). On the Generation of Waves by Turbulent Wind. *Journal of Fluid Mechanics* 2: 417–445.

PHILLIPS, O. M. (1958a). Wave Generation by Turbulent Wind over a Finite Fetch. *Proceedings of the 3rd U.S. National Congress Applied Mechanics,* pp. 785–789.

PHILLIPS, O. M. (1958b). The Equilibrium Range in the Spectrum of Wind-Generated Waves. *Journal of Fluid Mechanics* 4: 426–434.

PHILLIPS, O. M. (1958c). On Some Properties of the Spectrum of Wind-Generated Ocean Waves. *Journal of Marine Research* 16: 231–240.

PHILLIPS, O. M. (1960). On the Dynamics of Unsteady Gravity Waves of Finite Amplitude. *Journal of Fluid Mechanics* 9: 193–217.

PIERSON, W. J. (1972). Wave Behavior near Caustics in Models and Nature. In *Waves on Beaches,* R. E. Meyer (editors), pp. 163–180. New York: Academic Press.

PIERSON, W. J., G. NEUMANN, and R. W. JAMES (1955). *Observing and Forecasting Ocean Waves by Means of Wave Spectra and Statistics.* U.S. Dept. of the Navy Hydrog. Office Publ. No. 603.

PIERSON, W. J. and L. MOSKOWITZ (1964). A Proposed Spectral Form for Fully Developed Wind Seas Based on the Similarity Theory of S. A. Kitaigorodskii. *Journal of Geophysical Research* 69 (24): 5181–5190.

PRICE, W. A., K. W. TOMLINSON, and J. N. HUNT (1968). The Effect of Artificial Seaweed in Promoting the Build-Up of Beaches. *Proceedings of the 11th Coastal Engineering Conference, Amer. Soc. Civil Engrs.,* pp. 570–578.

PUTNAM, J. A. and J. W. JOHNSON (1949). The Dissipation of Wave Energy by Bottom Friction. *Transactions American Geophysical Union* 30: 67–74.

PUTZ, R. R. (1952). Statistical Distributions for Ocean Waves. *Transactions American Geophysical Union* 33: 685–692.

RAYLEIGH, L. (1876). On Waves. *Philosophical Magazine,* Series 5, 1: 257–279.

RAYLEIGH, L. (1911). Hydrodynamical Notes. *Philosophical Magazine,* Series 6, 21: 177–187.

RESIO, D. T. (1987). Shallow-Water Waves. II: Data Comparisons. *Journal of Waterway, Port, Coastal, and Ocean Engineering, Amer. Soc. Civil Engrs.* 113(3): 264–281.

RESIO, D. T. (1988). Shallow-Water Waves. I: Theory. *Journal of Waterway, Port, Coastal, and Ocean Engineering, Amer. Soc. Civil Engrs.* 114(1): 50–65.

ROBINSON, A. H. W. (1980). Erosion and Accretion Along Part of the Suffolk Coast of East Anglia, England. *Marine Geology* 37: 133–146.

ROGERS, L. C. (1966). Blue Water 2 Lives Up to Promises. *Oil and Gas Journal* (August), pp. 73–75.

RUDNICK, P. and R. W. HASSE (1971). Extreme Pacific Waves, December 1969. *Journal of Geophysical Research* 76(3): 742–744.

RUSSELL, R. C. H. and J. D. C. OSORIO (1957). An Experimental Investigation of Drift Profiles in a Closed Channel. *Proceedings of the 6th Coastal Engineering Conference, Amer. Soc. Civil Engrs.*, pp. 171–183.

RUSSELL, J. S. (1844). Report on Waves. *14th Meeting of the British Association for the Advancement of Science.* pp. 311–390.

SEYMOUR, R. J. (1989). Wave Observations in the Storm of 17–18 January 1988. *Shore & Beach* 57: 10–13.

SEYMOUR, R. J., R. R. STRANGE, D. R. CAYAN, and R. A. NATHAN (1984). Influence of El Niños on California's Wave Climate. *Proceedings of the 19th Coastal Engineering Conference, Amer. Soc. Civil Engrs.*, pp. 577–592.

SEYMOUR, R. J., M. H. SESSIONS, and D. CASTEL (1985). Automated Remote Recording and Analysis of Coastal Data. *Journal of Waterway, Port, Coastal, and Ocean Engineering* 111: 388–400.

SHERLOCK, A. R. and A. SZUWALSKI, A. (1987). *A User's Guide to the Littoral Environmental Observation Retrieval System.* Instruction Report CERC-87-3, U.S. Army Corps of Engineers, Coastal Engineering Research Center, Vicksburg, Mississippi.

SLEATH, J. F. A. (1984). *Sea Bed Mechanics.* New York: John Wiley & Sons.

SNODGRASS, D., G. W. GROVES, K. F. HASSELMANN, G. R. MILLER, W. H. MUNK, and W. H. POWERS (1966). Propagation of Ocean Swell Across the Pacific. *Philosophical Transactions of the Royal Society of London,* Series A, 259: 431–497.

SNYDER, C. M., R. L. WIEGEL, and C. J. BERMEL (1957). Laboratory Facilities for Studying Water Gravity Wave Phenomena. *Proceedings of the 6th Coastal Engeering Conference, Amer. Soc. Civil Engrs.*, pp. 231–251.

STEELE, K. and A. JOHNSON (1979). Data Buoy Wave Measurements. In *Ocean Wave Climate,* M. D. Earle and A. Malahoff (editors), pp. 301–316. New York: Plenum Press.

STEWART, R. W. (1967). Mechanics of the Air-Sea Interface. *Physics of Fluids,* Supplement, 10: 547–555.

STOKER, J. J. (1957). *Water Waves.* New York: Interscience.

STOKES, G. G. (1847). On the Theory of Oscillatory Waves. *Transactions Cambridge Philosophical Society* 8: 441 [also in *Mathematical and Physical Papers,* London, England: Cambridge University Press (1880). Vol. 1, pp. 197–229].

STOKES, G. G. (1880). On the Theory of Oscillatory Waves. In *Mathematical and Physical Papers,* Vol. 1, pp. 314–326. London, England: Cambridge University Press.

STRANGE, R. R., N. E. GRAHAM, and D. R. CAYAN (1989). Meteorological Development of the Unusually Severe Coastal Storm During January 16–18, 1988. *Shore & Beach* 57: 3–9.

SVERDRUP, H. U. and W. H. MUNK (1946). Theoretical and Empirical Relations in Forecasting Breakers and Surf. *Transactions American Geophysical Union.* 27: 828–836.

SVERDRUP, H. U. and W. H. MUNK (1947). *Wind, Sea, and Swell Theory of Relationships in Forecasting.* U.S. Dept. of the Navy Hydrogr. Office Publ. No. 601, Washington, D.C.

TAYLOR, D. C. (1955). *An Experimental Study of the Transition Between Oscillatory and Solitary Waves.* Masters thesis, Cambridge, MA: MIT.

THOMPSON, E. F. and C. L. VINCENT (1985). Significant Wave Height for Shallow Water Design. *Journal of Waterway, Port, Coastal, and Ocean Engineering* 111: 828–842.

TILLOTSON, K. and P. D. KOMAR (1997). The Wave Climate of the Pacific Northwest (Oregon & Washington): A Comparison of Data Sources. *Journal of Coastal Research* 13: 440–452.

WANG, S. and B. LeMEHAUTE (1983). Duration of Measurements and Long-Term Wave Statistics. *Journal of Waterway, Port, Coastal,* and *Ocean Engineering, Amer. Soc. Civil Engrs.* 109: 236–249.

WATTS, J. S. and R. E. FAULKNER (1968). Designing a Drilling Rig for Severe Seas. *Ocean Industry* 3(11): 28–37.

WIEGEL, R. L. (1954). *Gravity Wave Tables of Functions.* Council on Wave Research, Engineering Foundation, University of California, Berkeley.

WIEGEL, R. L. (1964). *Oceanographical Engineering.* Englewood Cliffs, NJ: Prentice-Hall.

WIEGEL, R. L. and H. L. KIMBERLY (1950). Southern Swell Observed at Oceanside, California. *Transactions American Geophysical Union* 31(5): 717–722.

WILSON, B. W. and A. TØRUM (1968). *The Tsunami of the Alaskan Earthquake, 1964.* Engineering evaluation. U.S. Army Coastal Engineering Research Center Technical Memo No. 25.

WILSON, J. R. and W. F. BAIRD (1981). *Canadian Wave Climate Study Organization and Operation.* Manuscript Report Series No. 59, Marine Sciences and Information Directorate, Department of Fisheries and Oceans, Ottawa, Canada.

6

Wave Breaking and Surf-Zone Processes

Wave after wave, each mightier than the last
Til last, a ninth one, gathering half the deep
And full of voices, slowly rose and plunged
Roaring, and all the wave was in a flame.

Alfred, Lord Tenneyson, 1809–1892
The Coming of Arthur

Waves acquire energy from the wind blowing over the vast expanse of the sea. Storms can impart a tremendous amount of power to the waves, which then travel for thousands of kilometers to the coastal zone. Wave energy, accumulated over a large area of the sea, is then dissipated in the relatively narrow surf zone at the coast. Much of the energy is expended by wave breaking in the nearshore, at times providing an inspiring display of natural forces as the waves crash against rocky coasts or surge across sandy beaches. This is by far the most important energy input into the coastal zone and is responsible for the generation of nearshore currents and the transport of sediments, processes that control the morphology of the beach.

The focus of this chapter is on the processes of wave-energy dissipation in the nearshore. We will examine wave breaking, the movement of bores across the surf zone, and the properties of wave run-up at the shoreline. It will be seen that the presence of waves alters the mean level of the water in the surf zone, elevating it as wave set-up shoreward of the breaker zone. Of particular significance is the generation of infragravity water motions in the nearshore, motions that have periods in excess of 20 sec. In many instances the infragravity energy is in the form of edge waves that are trapped to the sloping beach. Edge waves can directly affect the patterns of run-up on the beach face, and as will be explored in later chapters, they may also play a role in the generation of nearshore currents and have profound effects on the beach morphology, including the formation of beach cusps and crescentic bars.

WAVE BREAKING IN THE NEARSHORE

When waves reach a beach and enter water that is approximately as deep as the waves are high, they become unstable and break with the crest thrown forward as the wave disintegrates into bubbles and foam (Fig. 6-1). A common notion is that waves break because they drag on the bottom until they trip forward and the crest topples downward. This is not the case. Experiments have established that the friction is small, and computer models of wave transformations show the same pattern leading to breaking, while having neglected friction

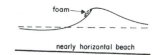

I. SPILLING BREAKERS

foam

nearly horizontal beach

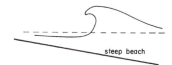

II. PLUNGING BREAKERS

steep beach

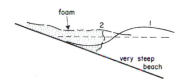

III. SURGING BREAKERS

foam

2   1

very steep
beach

Figure 6-1 The plunging breaker, which, in spectacular fashion, curls over and throws its mass forward toward the beach.

Figure 6-2 Three types of breaking waves that occur in the nearshore, depending on the beach slope and wave steepness.

in the analysis. Instead, a wave breaks when it becomes overly steep, especially near the peak of its crest, because the velocities of water particles in the wave crest exceed the velocity of the wave form so that the crest surges ahead.

Three common types of breakers are recognized (Fig. 6-2): spilling, plunging, and surging. With *spilling breakers* each wave gradually peaks until the crest becomes unstable and cascades down as "white water"—bubbles and foam. With *plunging breakers,* the shoreward face of the wave becomes vertical, curls over, and plunges forward and downward as an intact mass of water (Figs. 6-1 and 6-2). *Surging breakers* peak up as if to plunge, but then the base of the wave surges up the beach face so that the crest collapses and disappears. Galvin (1968) identified a fourth type of breaking wave (Fig. 6-3), a *collapsing breaker,* which is intermediate between the plunging and surging types. There is actually a continuum of breaker types grading from one to another, at times making it difficult to apply such classifications. Furthermore, on a given day at a beach, it is common to see some waves break by plunging while others are spilling, depending on their individual heights and interactions with other waves and the sea floor.

In general, spilling breakers tend to occur on beaches of very low slope with waves of high steepness values; plunging waves are associated with steeper beaches and waves of intermediate steepness; surging occurs on high-gradient beaches with waves of low steepness. Based on observations of waves breaking in laboratory channels, Galvin (1968) found fairly good breaker-type predictions using either $H_\infty/L_\infty S^2$ or H_b/gT^2S, where the ∞ subscript refers to the deep-water wave parameters, H_b is the breaker height, and S is the beach slope. As these dimensionless ratios increase (Fig. 6-4); the breaker type changes from surging to plunging to

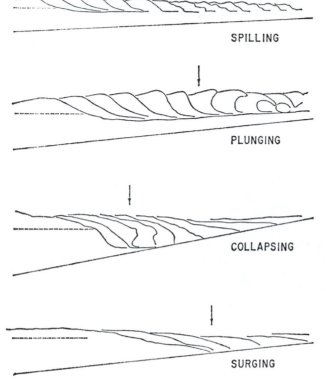

SPILLING

PLUNGING

COLLAPSING

SURGING

Figure 6-3 Breaking waves on a beach, their profiles having been traced from high-speed moving pictures obtained in a glass-walled laboratory channel. The arrows point to the initial positions of breaking. [Used with permission of American Geophysical Union, from C. J. Galvin, Breaker Type Classification on Three Laboratory Beaches, *Journal of Geophysical Research 73*, p. 3652. Copyright © 1968 American Geophysical Union.]

spilling. These ratios employed by Galvin combine the beach slope S with the wave steepness, expressed alternately as H_∞/L_∞ or H_b/gT^2 (since $L_\infty \propto gT^2$). These ratios have been rearranged by Battjes (1974) into deep-water and inshore forms of the Iribarren number:

$$\xi_\infty = \frac{S}{(H_\infty/L_\infty)^{1/2}} \tag{6.1a}$$

$$\xi_b = \frac{S}{(H_b/L_\infty)^{1/2}} \tag{6.1b}$$

In reference to the Iribarren number, the breaker-type classification data of Galvin (1968) form the following limits according to the reanalysis of Battjes (1974):

spilling	$\xi_\infty < 0.5$	$\xi_b < 0.4$
plunging	$0.5 < \xi_\infty < 3.3$	$0.4 < \xi_b < 2.0$
surging	$\xi_\infty > 3.3$	$\xi_b > 2.0$

These divisions have been further supported by Okazaki and Sunamura (1991), based on an extensive set of laboratory measurements on uniformly sloping beaches. Smith and Kraus (1991) experimented with waves breaking over bars and found slightly different ξ_∞ ranges for

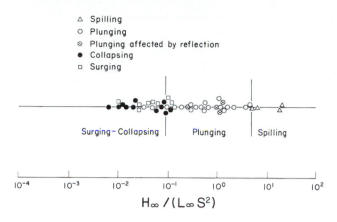

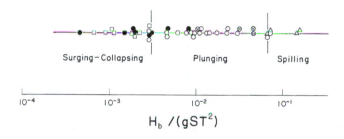

Figure 6-4   The wave breaker types identified in Figure 6-3, interpreted as functions of dimensionless ratios containing wave heights, lengths, periods, and the beach slope S. [Adapted with permission of American Geophysical Union, from C. J. Galvin, Breaker Type Classification on Three Laboratory Beaches, *Journal of Geophysical Research 73*, p. 3658. Copyright © 1968 American Geophysical Union.]

the wave-breaker types. The lower transition values found by Smith and Kraus indicate that some waves that would break by spilling on a plane beach will plunge if a bar is present, and some waves that would plunge on a plane slope will collapse on a barred profile.

There has been little attempt to test these breaker-type classification schemes on natural beaches. Weishar and Byrne (1978) obtained films of breaking waves at Virginia Beach, Virginia, and analyzed the breaking characteristics of 116 waves. They concluded that neither the dimensionless ratios of Galvin (1968) nor the Iribarren number ranges proposed by Battjes (1974) adequately discriminate between plunging and spilling breakers. The problem is that on natural beaches with a spectrum of wave periods and heights, one generally sees a mixture of breaker types at any given time, and this is what Weishar and Byrne documented. The limits developed by Galvin and Battjes are based on uniform waves generated in the laboratory, and do not adequately account for the variability of waves on ocean beaches. Undoubtedly the trends are right, with there being a tendency to shift from spilling to plunging to surging as ξ_∞ and ξ_b increase, but the limits based on laboratory results need to be revised for field conditions and can never be expected to provide a precise division for breaker types.

In Chapter 5 we have seen that there are many theoretical criteria that purport to define the condition at which waves become unstable and break. The most basic of the criteria is that the velocity of the water in the very peak of the wave crest is equal to or slightly exceeds the phase velocity of the wave form, the rate at which the wave itself is moving forward. It is clear that if the particle velocity at the crest exceeds the velocity of the wave, the crest will topple forward and break, just as observed. Measurements by Iversen (1952) of breaking waves in a laboratory wave channel confirmed that the particle velocities near the crest are approximately equal to the wave celerity. An example is shown in Figure 6-5 for Iversen's

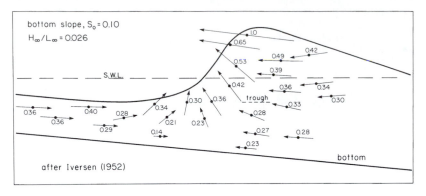

Figure 6-5 Measurements of the internal velocity field of an incipient breaking wave, determined from photographs of neutrally buoyant particles. The measured velocities u have been normalized as $u_p/\sqrt{gY_b}$, where Y_b is the total water depth from the wave crest to the bottom. [Adapted from H. W. Iversen, *Gravity Waves*, National Bureau of Standards Circular No. 521, 1952.]

measurements of the velocities of particles internal to a wave that is at its incipient point of breaking, with its landward face nearly vertical and ready to curl over into a plunging breaker. The particle velocities are expressed in the dimensionless form $u_p/\sqrt{gY_b}$, where u_p is the measured particle velocity and, Y_b is the vertical distance from the wave crest to the bottom. According to Equation (5.33), $\sqrt{gY_b}$ is proportional to the celerity of the wave since $Y_b \approx (H_b + h_b)$. It can be seen that the dimensionless ratio at the wave crest is 1.0, indicating that the actual magnitude of the water-particle velocity is on the order of the wave celerity and that it exceeds the velocities in the remaining body of the breaker, so that the crest will surge forward and plunge. Also seen in Figure 6-5 is the complexity of the wave-breaking process on steep beaches where the backwash down the beach face from a previously broken wave collides with the incoming wave. Other studies that have measured velocities internal to breaking waves include Miller and Zeigler (1964); Adeyemo (1970); Stive (1980); Hedges and Kirkgoz (1981); Mizuguchi (1986); Okayasu, Shibayama, and Mimura (1986); Matsunaga, Takehara, and Awaya (1988). Greated, Skyner, and Bruce (1992) have developed a particle image velocimetry (PIV) technique, which they used to obtain detailed measurements of the velocity field within breaking waves.

The mathematical analysis of the details of wave breaking is difficult because the equations of motion and associated boundary conditions are nonlinear and must be applied to a free surface with an unknown initial position that needs to be determined as part of the solution. There have been a number of attempts to numerically model the developing form of the breaking wave, most notably by Biesel (1951) and Longuet-Higgins and Cokelet (1976). Figure 6-6 from the latter study shows the computed results for the development of a plunging breaker, where the numerical model shows the same curving over of the wave crest as is actually observed in plunging waves. Figure 6-7 shows a series of breaker profiles calculated by Gaughan and Komar (1975), based on the numerical approach of Biesel and neglecting bottom friction. The series of computed breaking waves can be classified as ranging from spilling to plunging, depending on the bottom slope assumed in the model. Gaughan and Komar undertook a number of computations for breaking waves, including a range of beach slopes and initial wave steepnesses, and showed that classifying the results as spilling versus plunging breakers for the most part corresponded to the criteria established by Galvin (1968), as given in Figure 6-4.

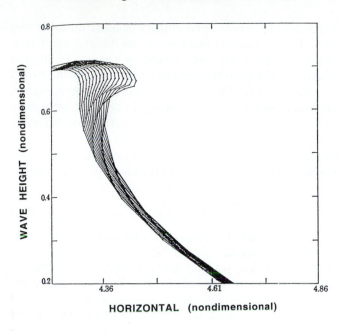

Figure 6-6   Numerically computed profiles of the crest of a breaking wave at small time increments. [Adapted with permission of The Royal Society, from M. S. Longuet-Higgins and E. D. Cokelet, The Deformation of Steep Surface Waves in Water: 1. A Numerical Method of Computation, *Proceedings of The Royal Society of London*, p. 19. Copyright © 1976 The Royal Society.]

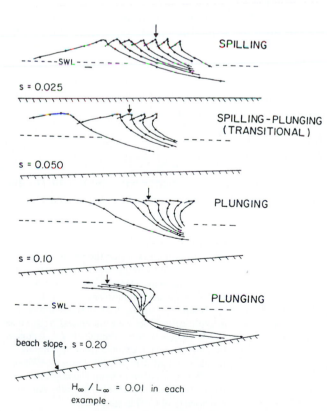

Figure 6-7   Computed sequences of wave profiles during breaking on sloping beaches, forming patterns that can distinguish between spilling and plunging breakers that correspond to observed transformations like those in Figure 6-3. [Used with permission of American Geophysical Union, M. K. Gaughan and P. D. Komar, The Theory of Wave Propagation in Water of Gradually Varying Depth and the Prediction of Breaker Type and Height, *Journal of Geophysical Research 80*, p. 2993. Copyright © 1975 American Geophysical Union.]

It has been seen in Chapter 5 that solitary waves theoretically become unstable and break at a critical value of the ratio $\gamma_b = H_b/h_b$, although there was some disagreement as to its exact value (0.73–1.03). Measured values of the critical γ_b for solitary waves generated in laboratory wave channels vary considerably, the ratio depending principally on the beach slope (Ippen and Kulin, 1954; Kishi and Saeki, 1966). This is also the case for periodic waves, as shown by the curves in Figure 6-8, based on the review by Weggel (1972) of laboratory measurements, principally those of Iversen (1952). The critical value of γ_b for wave breaking depends on both the beach slope and the deep-water wave steepness. For a given wave steepness, the greater the beach slope the higher the value of $\gamma_b = H_b/h_b$ at breaking. More recently, in a review of seventeen data sets obtained by various investigators in laboratory experiments, Kaminsky and Kraus (1993) derived the empirical formula

$$\gamma_b = 1.20\xi_\infty^{0.27} \tag{6.2}$$

a dependence on the deep-water form of the Iribarren number of Equation (6.1a), which includes the beach slope and deep-water wave steepness.

The critical $\gamma_b = H_b/h_b$ ratio for wave breaking is used in many applications. One example is the analysis of wave transformations as they progress from deep water toward the shore. The analysis follows the wave heights as they progressively increase until they reach a height versus depth ratio that corresponds to γ_b, at which point it is assumed that the waves become unstable and break. Another application comes in the design of structures such as seawalls or jetties. In that case, the maximum possible wave height that can impact the structure, the *design wave*, is governed by the water depth in front of the structure, waves having greater heights than governed by the ratio $\gamma_b = H_b/h_b$ having broken farther offshore.

In some applications it is desirable to calculate wave-breaker heights from their deep-water parameters without going through the entire shoaling transformations as developed in Chapter 5. Such a formulation was first derived by Munk (1949), who equated the energy flux at the breaker zone to the energy flux in deep water. Munk applied solitary wave theory to the breaker zone with $\gamma_b = H_b/h_b = 0.78$ as the breaking criterion and used Airy-wave theory for the deep-water energy flux. This led to the relationship

$$\frac{H_b}{H_\infty} = \frac{1}{3.3(H_\infty/L_\infty)^{1/3}} \tag{6.3}$$

for the breaker height H_b that corresponds to waves in deep water with the height H_∞ and steepness H_∞/L_∞ (wave refraction has been neglected but can be accounted for as discussed in

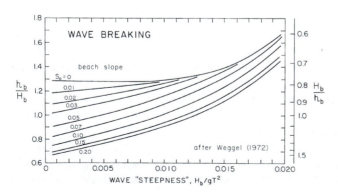

Figure 6-8 The critical condition for wave breaking in shallow water, the height of the wave compared with the water depth, which depends on the wave steepness and beach slope. Scales for the critical ratio are given as $1/\gamma_b = h_b/H_b$ and $\gamma_b = H_b/h_b$, both of which are useful in applications. The wave "steepness" is given as H_b/gT^2, which differs from the H_b/L_∞ steepness by a factor of 2π. [Adapted from Maximum Breaker Height, J. R. Weggel, *Journal of Waterway, Port, Coastal and Ocean Engineering*, 1972. Reproduced with permission from the American Society of Civil Engineers.]

Chapter 5). Equation (6.3) is plotted in Figure 6-9 along with the field and laboratory wave-breaking data reported in Munk (1949). It can be seen that the relationship agrees with the measurements in the range of low H_∞/L_∞ values but does poorly at higher wave steepnesses. Komar and Gaughan (1972) examined the use of Airy-wave theory for the prediction of wave-breaker heights, in spite of the fact that the theory should not be valid under those conditions. A critical ratio $\gamma_b = H_b/h_b$ again was accepted as a wave-breaking criterion but more as an observational fact not connected with solitary wave theory. Using Airy-wave theory to evaluate the energy flux in both deep water and at the breaker zone, Komar and Gaughan obtained

$$\frac{H_b}{H_\infty} = \frac{0.563}{(H_\infty/L_\infty)^{1/5}} \tag{6.4}$$

The dependence on the steepness is now to the one-fifth power, rather than to the one-third power obtained by Munk in Equation (6.3). The Komar and Gaughan relationship was fitted to laboratory and field data, so it is semiempirical with the 0.563 coefficient derived from the fit to that combined data. Equation (6.4) is shown in Figure 6-9 as the solid curve, where it is seen to fit the data over the entire range of H_∞/L_∞ steepness values, being more successful than Equation (6.3) derived by Munk.

When such analyses are limited to data gathered in the controlled conditions of laboratory wave-channel experiments, as in the studies of LeMéhauté and Koh (1967) and Kaminsky and Kraus (1993), the H_∞/L_∞ exponent is empirically found to be on the order of one-fourth, intermediate between the one-third and one-fifth values given, respectively, by Equations (6.3) and (6.4). Based on a reanalysis of the limited data of Iversen (1952), LeMéhauté and Koh also found a dependence on the beach slope, but this has not been substantiated by subsequent analyses. In their reanalysis of seventeen laboratory data sets, Kaminsky and Kraus empirically obtained the relationship

$$\frac{H_b}{H_\infty} = \frac{0.46}{(H_\infty/L_\infty)^{0.28}} \tag{6.5}$$

This equation does provide the best agreement with existing laboratory data, but as can be seen in Figure 6-9 where it is compared with the field data of Munk (1949), this relationship yields H_b/H_∞ ratios that are about 25 percent too high, over predicting the sizes of the breaker heights. Equation (6.4) of Komar and Gaughan (1972) appears to provide better agreement with the field data.

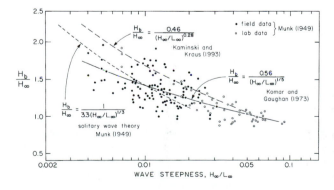

Figure 6-9 The wave breaker height H_b related to the deep-water wave steepness. The curve based on solitary wave theory is from Munk (1949), the "modified Airy-wave theory" curve is from Komar and Gaughan (1972), while the empirical curve for laboratory data is that of Kaminsky and Kraus (1993).

If taken out of its dimensionless form, Equation (6.4) of Komar and Gaughan (1972) becomes

$$H_b = 0.39g^{1/5}(TH_\infty^2)^{2/5} \tag{6.6}$$

which displays better the relationship between the breaking height of a wave in the nearshore and its deep-water height H_∞ and period T. The 0.39 coefficient is again empirical, based on the fit to laboratory data and to the field data of Munk (1949). This form of the relationship is graphed in Figure 6-10 in comparison with laboratory data and the field data of Munk, and also with the measurements of Weishar and Byrne (1978) that were collected subsequent to the formulation of Equations (6.4) and (6.6). The combined data span nearly three orders of magnitude for the breaker heights and confirm the predictive capability of these relationships.

During wave breaking, particularly for plunging waves, each wave may travel a significant horizontal distance from the point of inception of breaking, where the front face of the wave is nearly vertical, to the position where the plunging crest finally touches down. This is shown schematically in Figure 6-11, with X_p representing the plunge distance. Galvin (1969) developed a predictive model for the plunge distance, based on the expected trajectory of the wave crest, which is initially moving horizontally with the celerity C_b, but then moves downward due to the pull of gravity. Based on laboratory data (Fig. 6-11), the resulting predictive relationship is

$$\frac{X_p}{H_b} = 4.0 - 9.25S \tag{6.7}$$

where S is the beach slope. Such a dependence on the beach slope would be expected because the breaker position is nearer to shore on steeper slopes, and the downward plunging waves

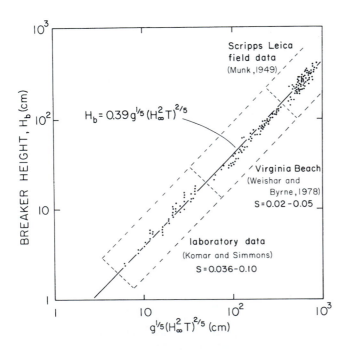

Figure 6-10 A test of Equation (6.6) of Komar and Gaughan (1972) for the prediction of wave-breaker heights from deep-water parameters, compared with laboratory and field data.

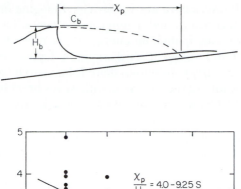

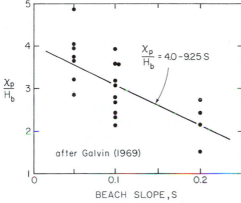

$$\frac{X_p}{H_b} = 4.0 - 9.25\,S$$

after Galvin (1969)

Figure 6-11 The plunge distance X_p of a breaking wave, analyzed as the trajectory of the crest resulting from its horizontal and downward movements. Laboratory wave-channel data show a dependence of X_p/H_b on the beach slope. [Adapted from Breaker Travel and Choice of Design Wave Height, C. J. Galvin, *Journal of Waterway, Port, Coastal and Ocean Engineering*, 1969. Reproduced with permission from the American Society of Civil Engineers.]

will impact the surface sooner if the beach is sloping upward. The measured plunge distances are on average greater than the predictions of the trajectory model developed by Galvin, which predict that $X_p/H_b \approx 2$; the discrepancy is likely due to the simplicity of the free-fall predictions of the model and the observation that the heights of the waves actually increase slightly, following the assumed breaking criterion of a vertical front. In their field study of breaking waves at Virginia Beach, Virginia, Weishar and Byrne (1978) found an average ratio $X_p/H_b = 5.9$, with the range extending from 1 to 10. The measured wave celerities were close to those predicted in the trajectory model of Galvin, but the observed plunge times were systematically greater than predicted. In their laboratory study of wave breaking over bars, Smith and Kraus (1991) found smaller plunge distances for profiles with bars than on uniformly sloping beaches. They developed empirical relationships for the X_p/H_b ratio as functions of the deep-water Iribarren number, ξ_∞, different relationships being obtained for uniform profiles versus profiles with bars. Some of the bar forms included in their measurements are of sufficient relief to represent reefs or offshore submerged breakwaters that would cause waves to break as they approach the shore.

WAVE DECAY IN THE SURF ZONE

On steeply sloping reflective beaches, once waves break they immediately surge up the beach face as run-up, with little intervening surf. In contrast, on lower-sloping dissipative to intermediate beaches, there can be an extensive surf zone, at times being crossed by several

broken wave bores while a few small waves have not yet broken. Plunging breakers characteristically dissipate their energy in a concentrated manner in the region just shoreward of
the breaker line, while spilling breakers dissipate their energy at a slower rate over a greater
surf-zone width. The patterns of surf and wave-energy dissipation depend on the morphology of the beach profile. If the slope is fairly continuous, there is likely to be a uniform dissipation of the waves as they cross the width of the surf zone, with waves breaking at all depths.
However, if the profile is characterized by a series of bars and troughs, there will be a concentration of wave breaking over the shallow depths of the bars, with little or no breaking
over the deeper troughs. Having once broken over the outer bar, waves may reform over an
intervening trough and then break for a second time over another bar or at the base of the
more steeply sloping beach face.

   It is important to be able to analyze such patterns of wave decay within the surf zone.
In addition to understanding how the wave energy is dissipated before it reaches the shore
where it might be destructive to property, the expenditure of this energy plays a direct role in
the generation of nearshore currents, in the cross-shore and longshore transport of sediments,
and in modifications of the beach morphology. Accordingly, models that have been developed
to analyze wave-height and energy variations across the surf zone, those reviewed here, are
employed in analyses of nearshore currents and sediment transport patterns that will be examined in subsequent chapters.

   In analyses of wave-height variations across the surf zone, one has to distinguish between conditions where the waves arriving from deep water are of essentially constant height
and period, like those generated in a laboratory channel, versus the more natural condition
where there is likely a wide spectrum of wave heights and periods. In the former case, the uniform waves will break at essentially a fixed water depth according to the γ_b breaking criterion discussed above, that is, at a fixed distance from the still-water shoreline. The problem
then is one of simply following the decay of their heights by tracking the waves while they
continue to break as they cross the surf zone. Such conditions are best documented in the controlled condition of a laboratory wave channel but are approximately met on natural beaches
when there is a system of nearly uniform swell waves having a narrow spectrum. More generally, however, the natural beach is characterized by the arrival of waves having a large range
of heights, with the largest waves breaking in deeper water while the smaller waves approach
more closely to shore before they break in shallow water. Thus, at any position within the surf,
one observes some waves that have already broken and display the white-water foam of
bores, while other waves are still undergoing their transformations leading to initial breaking. As one proceeds from deep to shallow water on a uniformly sloping beach, there is a progressive increase in the proportion of waves that are breaking versus those that still have not
reached instability and broke.

   Such patterns of broken and unbroken waves within the surf zone have been documented by Thornton and Guza (1983) at Torrey Pines Beach, California, a fine-sand beach
that is gently sloping with minimal development of bars and troughs. The measurements were
taken from a series of wave staffs and electromagnetic current meters placed along a cross-
shore profile from a 10–m water depth in the offshore to the inner surf zone. Figure 6-12 shows
histograms of wave-height distributions at four depths within the surf zone, also distinguishing between waves that have not yet broken versus those that have broken (hatched areas).
It is apparent that more and more waves have broken toward the inner surf zone and that
breaking has progressively affected smaller and smaller waves. The histograms of all waves
present form skewed distributions, characterized by a large number of small waves and a few

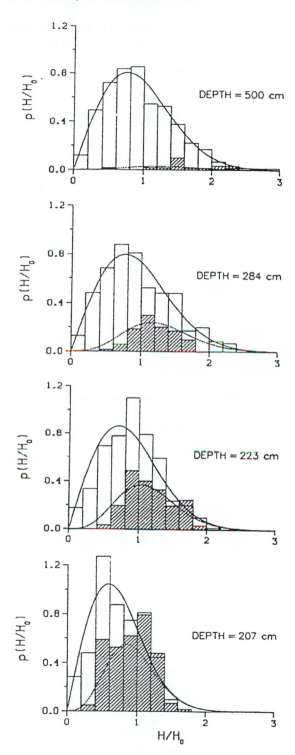

Figure 6-12 Histograms of waves within the surf zone measured at Torrey Pines Beach, California. The hatched portions represent waves that have broken, while the unhatched histograms are for all wave heights at that water depth. The latter are compared with the Rayleigh distribution of Equation (6.8), while the broken waves follow the distribution of Equation (6.9). [Used with permission of American Geophysical Union, from E. B. Thornton and R. T. Guza, Transformation of Wave Height Distribution, *Journal of Geophysical Research 88*, p. 5933. Copyright © 1983 American Geophysical Union.]

large waves that are substantially higher than the mode of each distribution, the dominant size present. Shown for comparison is the theoretical *Rayleigh distribution*, here expressed as

$$p(H) = \frac{2H}{(K_s H_\infty)^2} \exp\left[-\left(\frac{H}{K_s H_\infty}\right)^2\right] \tag{6.8}$$

for the probability of a wave height H, where K_s is the shoaling coefficient and H_∞ is the deep-water root-mean-square wave height. Longuet-Higgins (1952) has shown that the Rayleigh distribution applies to the distribution of wave heights in deep water, while the results in Figure 6-12, from Thornton and Guza, show that the measured wave heights retain this form as they enter shallow water. On the other hand, the distributions of the breaking waves within the surf, the hatched portions of Figure 6-12, are weighted toward the largest waves and tend to be more symmetrical. Thornton and Guza derived the modified distribution for the broken waves

$$p(H) = \left(\frac{H_{rms}}{\gamma h}\right)^n \left\{1 - \exp\left[-\left(\frac{H}{\gamma h}\right)^2\right]\right\} \tag{6.9}$$

where H_{rms} is the root-mean-square of the wave heights measured at the depth h, γ is derived from the depth-limiting condition for wave breaking ($H = \gamma h$), and n is a coefficient to be determined from the observations. The curves given in Figure 6-12 fitted to the broken-wave histograms are for $\gamma = 0.42$ and $n = 2$.

On dissipative beaches of low slope where there is a continuous surf zone of breaking waves and bores, it commonly has been assumed that there is a constant value of γ across the entire surf zone, having values on the order of 0.78–1.3 as established by the initial wave breaking. However, as noted above, direct measurements have found ratios of wave heights to water depths that are substantially less than expected, specifically $\gamma = 0.42$ in the field studies of Thornton and Guza (1982, 1983) at Torrey Pines Beach. Figure 6-13 from their study graphs the root-mean-square wave heights measured at various depths on four different days. Starting from the offshore and following a particular curve for a certain day, the wave height initially increases and then curves over until all waves in shallow water tend to decay in approximately the same manner, following the $H_{rms} \approx 0.42h$ line, indicating that the wave heights are limited by the local water depth. The results show that wave heights within the inner surf zone are essentially independent of wave heights in deep water outside the surf zone. Thornton and Guza describe this inner portion of the surf zone as "saturated" with respect to the energy content of the waves, being controlled by the local water depth rather than by the offshore wave conditions. In the outer surf zone, many waves have not yet broken and the $\gamma = H_{rms}/h$ ratios are less than 0.42 (Fig. 6-13), but as more and more waves break as they progress toward the inner surf zone, the ratio approaches the 0.42 saturation value.

Figure 6-14 from the laboratory wave-channel measurements of Horikawa and Kuo (1966) indicates that the water depth of initial breaking depends on the wave height and corresponds approximately to the $\gamma_b = 0.78$ criterion, but following breaking there is an asymptotic decrease in H toward a lower $\gamma = H/h$ ratio within the surf zone. It is probable that individual waves on natural beaches undergo a similar transformation and rapidly decay in height and energy after they initially break, and then settle down to stable saturated $\gamma = H/h$ ratios on the order of 0.42 as they continue to cross the surf zone.

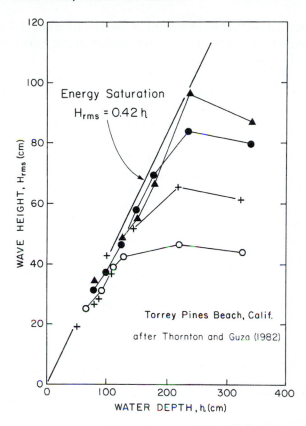

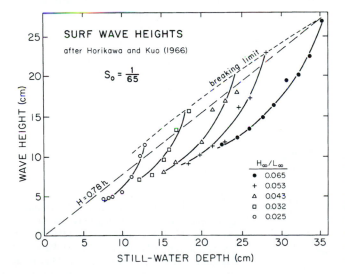

Figure 6-13 Measurements of root-mean-square wave heights across the beach profile at Torrey Pines Beach on 4 days, having different offshore wave conditions. Within the inner surf zone, the measurements converge on the relationship $H_{rms} \approx 0.42h$, implying that the wave heights are "saturated" in their energy content in being controlled by the local water depth rather than by the offshore wave conditions. [Adapted with permission of American Geophysical Union, from E. B. Thornton and R. T. Guza, Energy Saturation and Phase Speeds Measured on Natural Beaches, *Journal of Geophysical Research 87*, p. 9500. Copyright © 1982 American Geophysical Union.]

Figure 6-14 The progressive reduction of wave heights within the surf zone following their initial breaking. [Adapted from A Study of Wave Transformation Inside the Surf Zone, K. Horikawa and C. T. Kuo, *Proceedings of the 10th Coastal Engineering Conference,* 1966. Reproduced with permission from the American Society of Civil Engineers.]

Several investigations have focused on the development of models to account for the decay of wave heights, as seen in Figure 6-14, and the parallel dissipation of the wave energy. Most of these analyses have centered on the energy-flux relationship

$$\frac{\partial(ECn)}{\partial x} = -\varepsilon(x) \tag{6.10}$$

where $\varepsilon(x)$ is the loss in wave energy per unit area per unit time. The investigations have concluded that although there is some loss in energy due to bottom friction as the waves cross the surf zone, most of the loss is associated with wave breaking and the generation of turbulence, either the initial breaking or as the continued breaking of surf bores. The various models that have been developed to analyze cross-shore variations in wave heights differ in the way in which they evaluate $\varepsilon(x)$ in Equation (6.10). For example, Mizuguchi (1980) employed a theoretical equation for energy losses due to laminar-flow dissipation that depends on the molecular fluid viscosity but substituted the kinematic eddy viscosity in order to account for the more important losses due to turbulence. This yielded an equation for $\varepsilon(x)$ that depends on the local wave height and water depth, as well as on the eddy viscosity. Its use in Equation (6.10) yielded a solution for the wave-height variations across a beach profile of arbitrary shape. An example analysis is given in Figure 6-15, showing good agreement with measurements in a laboratory wave channel where the broken waves first progress up a slope and then cross a horizontal section of the profile.

Dally, Dean, and Dalrymple (1985) developed a model where $\varepsilon(x)$ is proportional to the difference between the local energy flux and the "stable" energy flux, such that Equation (6.10) becomes

$$\frac{\partial(ECn)}{\partial x} = -\frac{\kappa}{h}[(ECn) - (ECn)_{stable}] \tag{6.11}$$

where κ is a dimensionless decay coefficient, and h is the local water depth. The model was calibrated using the laboratory measurements of Horikawa and Kuo (1966), where waves first broke over a slope but then passed over a horizontal bottom. As seen in Figure 6-16, the broken waves over the horizontal bottom asymptotically approach a stable height, H_{stable}, defin-

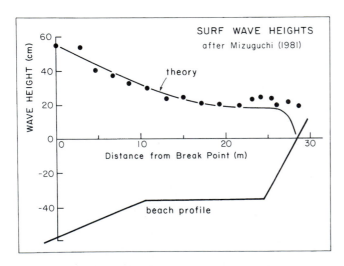

Figure 6-15 Cross-shore variations in wave heights measured in a laboratory wave channel, compared with the theoretical curve based on the model devised by Mizuguchi (1980). [Adapted from A Heuristic Model of Wave Height Distribution in Surf Zone, M. Mizuguchi, *Proceedings of the 17th Coastal Engineering Conference,* 1980. Reproduced with permission from American Society of Civil Engineers.]

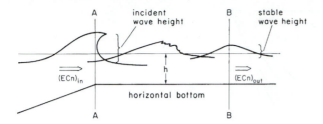

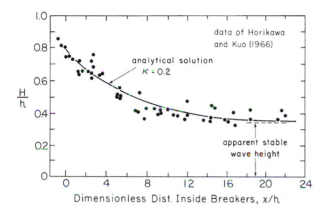

Figure 6-16 Experiments on wave-height variations where the waves initially break over a sloping beach but then achieve a stable height over a horizontal bottom. The data of Horikawa and Kuo (1966) are compared with the theoretical model of Dally and co-workers based on the relationship of Equation (6.11). [Used with permission of American Geophysical Union, from W. R. Dally, R. G. Dean and R. A. Dalrymple, Wave Height Variation Across Beaches of Arbitrary Profile, *Journal of Geophysical Research* 90. Copyright © 1985 American Geophysical Union.]

ing a ratio H_{stable}/h, which is on the order of 0.35–0.40. The curve in Figure 6-16 is based on the analytical solution of Equation (6.11) by Dally and co-workers (1985), with $\kappa = 0.2$ and $H_{stable}/h = 0.35$. Dally and co-workers then extended the analysis to the dissipation of broken waves crossing a uniformly sloping beach. Here the assumption in using Equation (6.11) is that the local energy dissipation on a sloping beach is proportional to the difference between the actual energy flux versus the stable value for that specific water depth. Figure 6-17(a) gives curves obtained in the analytical solutions for a series of $\gamma_b = H_b/h_b$ values for the breaker position, while Figure 6-17(b) provides a comparison with the measurements of Horikawa and Kuo obtained on a beach slope of 1:65. In such a case of wave-height variations over a sloping beach, the wave height is determined by the sum of two opposing factors: the loss of wave height due to energy dissipation and the increase in height due to shoaling. Increasing the beach slope increases the effect of shoaling, resulting in an increase in wave heights. An example analysis by Dally and co-workers over a highly irregular bottom is illustrated in Figure 6-18, based on the prototype-scale wave tank experiments of Saville (1957). With such a complex profile, numerical solutions are required. The results show two zones of rapid decrease in wave heights, corresponding to areas of wave breaking over offshore bars, with reformation of the waves over the deeper water of the intervening longshore trough.

The Dally and co-workers (1985) analysis of wave-height decay in the surf zone has been incorporated into the RCPWAVE wave-propagation model of the U.S. Army Corps of Engineers, a numerical analysis that includes offshore wave transformations, an assessment of the wave breaking condition, and the wave decay within the surf zone (Ebersole, Cialone, and Prater, 1986). The entire package has been submitted to extensive laboratory testing, which

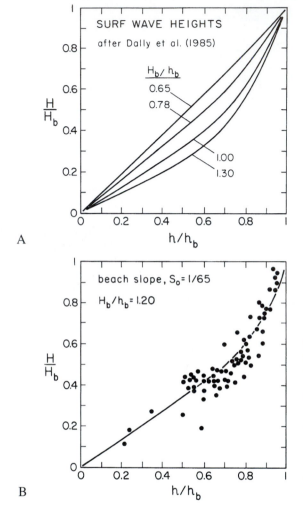

Figure 6-17   (a) The analytical solution of Dally and co-workers (1985) for wave-height variations over a uniformly sloping bottom, (b) compared with the measurements of Horikawa and Kuo (1966) with a bottom slope S = 1:65. [Adapted with permission of American Geophysical Union, from W. R. Dally, R. G. Dean and R. A. Dalrymple, Wave Height Variation Across Beaches of Arbitrary Profile, *Journal of Geophysical Research 90,* p. 11,920. Copyright © 1985 American Geophysical Union.]

generally confirms that there is good agreement between the RCPWAVE predictions and measured wave-height variations in a laboratory wave channel over a non-uniform bottom.

The more recent analyses by Dally have been to extend the application of the model to random waves (Dally and Dean, 1986; Dally, 1990, 1992). The assumption is that within the surf zone, random waves behave as a collection of individual regular waves. Starting from a Rayleigh distribution of wave heights seaward of the surf zone, the transformation is accomplished using linear-wave theory for shoaling and the model of Dally and co-workers (1985) for the decay of the individual waves. The model then predicts distributions of wave heights at various positions across the surf zone and wave statistics at each position. An example result is shown in Figure 6-19 where the numerical solution by Dally and Dean (1986) is compared with the field data of Hotta and Mizuguchi (1980) for transformation in the root-mean-square (H_{rms}), significant ($H_{1/3}$), and highest one-tenth ($H_{1/10}$) wave heights within the distributions; the agreement between theory and measurements is good, except in

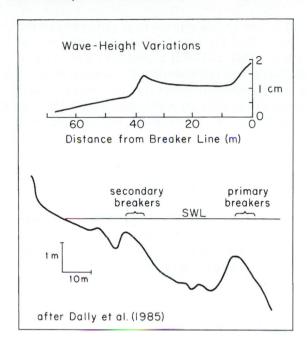

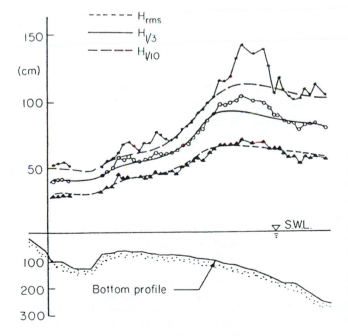

Figure 6-18 Wave variations calculated by the analysis approach of Dally and co-workers (1985) for an irregular beach profile obtained by Saville (1957) in prototype-scale wave tank experiments. [Adapted with permission of American Geophysical Union, from W. R. Dally, R. G. Dean and R. A. Dalrymple, Wave Height Variation Across Beaches of Arbitrary Profile, *Journal of Geophysical Research* *90*, p. 11,926. Copyright © 1985 American Geophysical Union.]

Figure 6-19 The numerical solution of Dally and Dean (1986) of the cross-shore variations in wave statistics, compared with the field data of Hotta and Mizuguchi (1980) obtained on a complex beach profile. Given are the transformations in the root-mean-square (H_{rms}), significant ($H_{1/3}$), and highest one-tenth ($H_{1/10}$) wave heights within the distributions of heights. [From Transformation of Random Breaking Waves on Surf Beat, W. R. Dally and R. G. Dean, *Proceedings of the 20th Coastal Engineering Conference*, 1986. Reproduced with permission from the American Society of Civil Engineers.]

the outer surf zone (mainly in $H_{1/10}$), where the disagreement is attributed to the use of linear-wave theory. In general it is found that the transformations depend on the beach slope and mean wave steepness (Dally, 1990), and in a comparison with data from the Field Research Facility (FRF), Duck, North Carolina, it was established that the presence of offshore winds and wave reflection can significantly affect the patterns of incident wave dissipation (Dally, 1992). A shortcoming of the wave-by-wave analysis developed by Dally is that it is computationally intensive, and is therefore difficult to employ in time-dependent nearshore models. Larson (1995) has modified the analysis so that it requires the transformation of only one representative wave height such as the root-mean-square height, without making any assumptions about the overall distribution of heights. The model was validated against laboratory data from a large-scale wave channel (SUPERTANK) and by field data from the FRF.

Battjes and Janssen (1978), Thornton and Guza (1983), and Battjes and Stive (1985) have developed models for wave dissipation and cross-shore variations of wave heights, models that focus on random waves breaking in different water depths. Following the approach of Battjes and Janssen, Thornton and Guza assessed the $\epsilon(x)$ energy dissipation term in Equation (6.10) with the relationship

$$\varepsilon = \frac{1}{4}\rho g \frac{f(BH)^3}{h} \tag{6.12}$$

for a bore of height H and frequency $f = 1/T$ in a water depth h, where $B \approx 1$ is a breaker coefficient. The average rate of energy dissipation is obtained by integrating Equation (6.12) over the range of wave heights present in the surf and dividing by the total number of waves (broken and unbroken). Figure 6-20 shows the model-evaluated curve compared with measurements obtained at Torrey Pines Beach, California. Here the analysis yields the root-mean-square wave height at each cross-shore position, in recognition of the presence of a range of wave heights at each position, with some waves having broken while others have not. Battjes and Stive (1985) also used a form of Equation (6.12) to assess the energy dissipation of bores, but incorporated into the analysis a Miche-type expression [see Equation (5.29) and discussion in Chapter 5] to define which wave heights have broken and therefore govern the energy losses. The details of the analysis, therefore, differ from that developed by

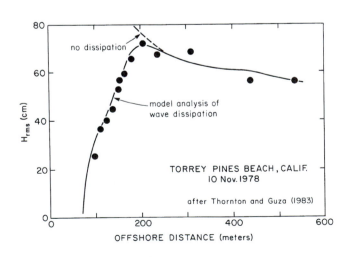

TORREY PINES BEACH, CALIF.
10 Nov. 1978

after Thornton and Guza (1983)

Figure 6-20 Measurements of root-mean-square wave heights across the surf zone at Torrey Pines Beach, California, compared with the curve derived in the analysis of Thornton and Guza (1983) utilizing Equation (6.10) with an integrated form of Equation (6.12) to evaluate the energy dissipation by wave bores. [Used with permission of American Geophysical Union, from E. B. Thornton and R. T. Guza, Transformation of Wave Height Distribution, *Journal of Geophysical Research 88*, p. 5936. Copyright © 1983 American Geophysical Union.]

Thornton and Guza (1983), but the result again is a prediction of the cross-shore variation of the root-mean-square wave heights. An example is shown in Figure 6-21, comparing the theoretical curve derived from a numerical solution with wave measurements in a laboratory wave channel. In this case, random waves are generated in order to better simulate natural wave spectra and the Rayleigh distribution of heights, while the concrete beach profile is in the form of a bar and trough system. The theoretical curve for H_{rms} shows excellent agreement with the measurements, which Battjes and Stive also found in comparisons with field measurements collected on beaches in the Netherlands. In addition to predicting the variation in wave heights across the surf zone, the theoretical analysis of Battjes and Stive predicts the set-up, the increase in the elevation of the mean water level due to the presence of the waves. It can be seen in Figure 6-21 that agreement between the measured and theoretically calculated wave set-up is also very good. Such a prediction of the set-up is an integral part of most of the analyses discussed above, including those by Dally et al. (1985) and Thornton and Guza (1983), but a discussion of wave set-up and its cause has been reserved for later in this chapter.

A number of studies have advanced theoretical models for the propagation of bores across the surf zone (Stoker, 1948; Benjamin and Lighthill, 1954; Keller, Levine, and Whitam 1960; Chester, 1966). A bore represents an abrupt change in water depth, with the front of the bore characterized by varying degrees of white-water breaking as it propagates across the surf zone. Surf bores have been classified by Peregrine (1966) in terms of the ratio of the wave height to water depth, based on considerations of energy losses at its leading edge. For $H/h < 0.28$, the bore is undular and energy is radiated away from the landward advancing bore front in a series of secondary waves. For $0.28 < H/h < 0.75$ the leading edge of the bore may break, although some secondary waves may be present, while for $H/h > 0.75$, the bore

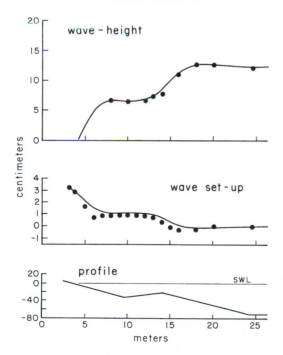

Figure 6-21 Laboratory measurements over a concrete beach profile molded in the form of a bar and trough. The data include the root-mean-square wave heights and the wave set-up, the increase in the mean water elevation in the surf zone, and are compared with curves derived from the theoretical model of Battjes and Stive (1985). [Adapted with permission of American Geophysical Union, from J. A. Battjes and M. J. F. Stive, Calibration and Verification of a Dissipation Model for Random Breaking Waves, *Journal of Geophysical Research 90,* p. 9162. Copyright © 1985 American Geophysical Union.]

is termed "fully developed" and the crest is involved in large-scale turbulent breaking. This classification of surf bores parallels the classification of hydraulic jumps in unidirectional flows such as rivers, a very similar process.

Svendsen (1984a) analyzed the fluxes of momentum (radiation stresses) and wave energy associated with the movement of bores across the surf zone by simple approximations, which include the effects of the surface roller contained within the front of the bore. He employed his "roller model" to predict cross-shore variations in wave heights and set-up, analyses that show good agreement with laboratory measurements. In providing an improved physical basis for the analysis of the momentum and energy of surf bores, the roller model of Svendsen shows considerable promise for use in analyzing a number of surf-zone processes.

In addition to measuring wave-height variations, the laboratory study of Horikawa and Kuo (1966) also examined the celerity of bores, their velocities as they cross the surf zone. They found considerable scatter in the measurements but reported that on average the celerity of bores is greater than predicted by solitary wave theory but is in better agreement with bore theory. This was confirmed by the additional wave-channel studies of Svendsen, Madsen, and Hansen (1978) and Hedges and Kirkgoz (1981). Suhayda and Pettigrew (1977) obtained field measurements of surf bore velocities on a smooth, moderately sloped (0.025) beach in Barbados. A total of twenty-two poles were deployed along the profile, and the movement of bores and their changes in heights were recorded with a camera operating at eighteen frames per second. At the time of measurements, the waves had periods of about 7 sec and heights of 50 cm, forming plunging breakers. Figure 6-22 shows the beach profile (exaggerated vertically) together with the direct measurements of crest and trough elevations. The difference between the crest and trough elevations defines the wave height, which is seen to increase just prior to breaking, but otherwise decreases across the surf zone toward the shore as the bores lose energy. Suhayda and Pettigrew show that the wave-height variations are similar to those found by Horikawa and Kuo (1966) and Sawaragi and Iwata (1974) in their laboratory experiments. Of interest here are the measurements of the celerities of the surf bores, which generally decrease as the bores cross the surf zone, but show some relative increase as the waves approach breaking, especially when the waves reach the base of the beach face and begin to surge as run-up. The upper-most diagram in Figure 6-22 normalizes the measured bore velocities to that predicted by solitary wave theory, given by $C = \sqrt{gY_b}$, where Y_b is the total water depth at the wave-crest position. This ratio is seen to be highly variable across the surf zone, with factors on the order of 1.2 in the zone of wave breaking and where the bores run-up the beach face, indicating that the measured celerities are some 20 percent greater than predicted by solitary wave theory. Similar comparisons were made with the theoretical equation developed by Keller and co-workers (1960) for the speed of a bore, but the disagreement was substantial, leading Suhayda and Pettigrew to conclude that in this instance the waves within the surf zone were not behaving as bores.

Thornton and Guza (1982) also analyzed their data from Torrey Pines Beach, California, to examine wave celerities within the surf zone. Figure 6-23 shows the dependence on the finite wave height, expressed as the difference between the measured celerity C_M compared with that predicted by linear-wave theory, $C_L = \sqrt{gh}$, as a function of the ratio H_{rms}/h. The results indicate that, within the saturated inner surf zone where $\gamma = H_{rms}/h = 0.42$, the phase velocity is on the order of $C \approx 1.15\sqrt{gh}$. This is comparable to the results obtained with solitary wave theory:

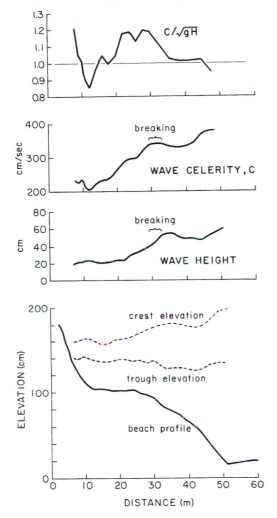

Figure 6-22 Measurements on a beach in Bermuda of cross-shore variations in wave heights and bore velocities C, with the uppermost diagram giving the ratio of the measured velocities compared with those predicted by solitary wave theory. [Adapted with permission of American Geophysical Union, from J. N. Suhayda and N. R. Pettigrew, Observations of Wave Height and Wave Celerity in the Surf Zone, *Journal of Geophysical Research 82,* p. 1420. Copyright © 1977 American Geophysical Union.]

$$C = \sqrt{g(h + H)} = \sqrt{gh(1 + H/h)}$$
$$C = \sqrt{gh(1 + 0.42)} = 1.19\sqrt{gh}$$

Lippmann and Holman (1991) have developed video techniques for measuring phase speeds of breaking waves and bores, and also for determining their approach angles relative of the shoreline trend. The technique is based on the premise that the foam and bubbles at the crests of breakers are bright and contrast with the darker unbroken surrounding water, thereby providing a means for distinguishing between breaking and non-breaking waves. The passage of a breaking wave crest is identified by a local maximum in the image intensity that is recorded by the video camera; a computerized image-processing system is utilized to analyze the measurements. With respect to wave celerity, Lippmann and Holman found in measurements from the FRF in Duck, North Carolina, that the measured surf velocities show reasonably close agreement with velocities predicted by solitary wave theory.

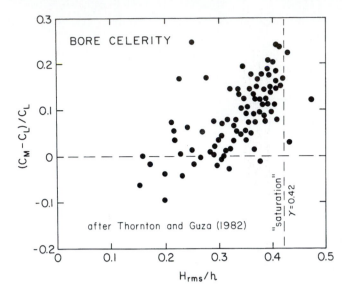

Figure 6-23 Measured velocities (C_M) of surf bores on Torrey Pines Beach, California, which depart from the velocities predicted by linear-wave theory (C_L) depending on the ratio of the local root-mean-square wave height to the water depth within the surf zone. [Adapted with permission of American Geophysical Union, from E. B. Thornton and R. T. Guza, Energy Saturation and Phase Speeds Measured on Natural Beaches, *Journal of Geophysical Research 87*, p. 9507. Copyright © 1982 American Geophysical Union.]

The rate of advance of surf waves toward the shore generally exceeds the velocities of the internal water motions. Figure 6-24 is an example of a measured velocity field beneath gently breaking waves over a bottom of low slope, determined by Divoky, LeMéhauté, and Lin (1970) from photographs of the motions of neutrally buoyant particles suspended in the water, lighted at intervals by a stroboscope. All measurement sets show a fairly uniform horizontal velocity at depth, increasing somewhat toward the surface near the wave crest. The horizontal component of the particle velocity at the crest appears to approach $0.5C$, where C

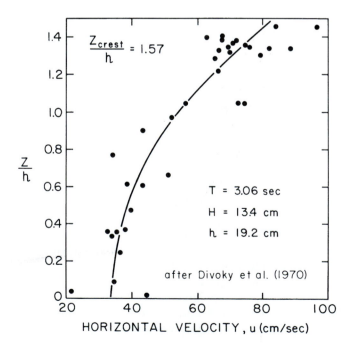

Figure 6-24 Measurements of the horizontal components of water-particle velocities beneath gently breaking waves on a sloping beach. [Adapted with permission of American Geophysical Union, from D. Divoky, B. Le Méhauté and A. Lin, Breaking Waves on Gentle Slopes, *Journal of Geophysical Research 75*, p. 1690. Copyright © 1970 American Geophysical Union.]

is the phase velocity of the surf wave. Cnoidal wave theory was concluded to provide the best fit to the measured velocity distributions like that given in Figure 6-24.

The mass of water carried shoreward by breaking waves in the surf zone may be compensated for by a seaward return flow, the *undertow*. Undertow was observed first by Bagnold (1940) in wave-channel experiments, and analyses of the phenomenon have been undertaken by Dyhr-Nielsen and Sørensen (1970), Svendsen (1984b), Hansen and Svendsen (1984), Stive and Wind (1986), Stive and de Vriend (1987), and Svendsen and Hansen (1988). Undertow consists of a pronounced bottom current flowing in the seaward direction [Figure 6-25(a)] and clearly is fed by the water volume carried toward the shore by the breakers and bores. Undertow develops best in a two-dimensional situation, as can be found within a laboratory wave channel, whereas on a three-dimensional beach the seaward return flow could in part be within seaward-flowing rip currents (Chapter 8). The explanation for the occurrence of undertow involves the elevation of the mean water level within the surf zone, the wave set-up. The set-up produces a seaward-directed pressure gradient of water, which on average is balanced by the momentum of the waves directed toward the shore. However, this balance varies with depth, with the water pressure over balancing the wave momentum close to the bottom, so that the water velocities associated with the seaward-flowing undertow are greatest close to the bottom. This can be seen in Figure 6-25(b) from the laboratory measurements

A   UNDERTOW

B   MEASURED VELOCITIES

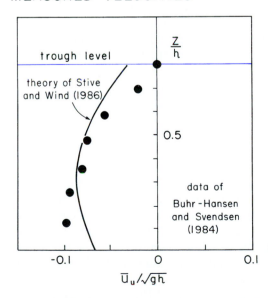

Figure 6-25   (a) The pattern of undertow at depth within the surf zone, representing a seaward return flow to balance the landward transport of water within the crests of breakers and bore. (b) Measurements of the undertow on a laboratory beach by Hansen and Svendsen (1984), where the measured velocity $\bar{u}_u$ is normalized to $\sqrt{gh}$, and h is the local water depth. [Adapted from *Coastal Engineering* 8, I. A. Svendsen, Mass Flux and Undertow in a Surf Zone, 1984, and reprinted from *Coastal Engineering* 10, M. J. F. Stive and H. G. Wind, Cross-shore Mean Flow in the Surf Zone, p. 337, 1986 with kind permission of Elsevier Science-NL, Sara Burgerhartstratt 25, 1055 KV Amsterdam, The Netherlands.]

of Hansen and Svendsen (1984), compared with the theoretical curve of Stive and Wind (1986). Field measurements of undertow have been obtained by Greenwood and Osborne (1990) in the Great Lakes, measurements that confirmed the relationship to the gradient of the cross-shore wave set-up. Smith, Svendsen, and Putrevu (1992) obtained data on cross-shore flows, including undertow, at the FRF using five electromagnetic current meters mounted on a mobile sled that was stationed at various cross-shore positions. The measurements showed strong undertow velocities over the offshore bar, with magnitudes up to 0.5 m/sec but with weaker currents within the shoreward trough and offshore from the bar. Reasonable agreement was found between the measurements and theoretical predictions of cross-shore and vertical variations in flow velocities of the undertow. The laboratory and field measurements, therefore, tend to confirm the theoretical analyses that predict a horizontal circulation involving a net landward flow of water in the wave crests above the level of the wave trough, with the return flow being an undertow confined mainly to the region below the level of the wave troughs. It has been proposed by Dyhr-Nielsen and Sørensen (1970) that undertow is important to the seaward transport of beach sand during storms, and may in some instances be responsible for the formation of offshore bars when it deposits that seaward transported sediment (Chapter 7).

WAVE SET-UP AND SET-DOWN

When waves break on a beach, they produce a *set-up,* a rise in the mean water level above the still-water elevation of the sea. As shown in Figure 6-26 from the laboratory measurements of Bowen, Inman, and Simmons (1968), the set-up is confined to the surf zone shoreward of the point of initial wave breaking and consists of an upward slope of the water in the landward direction. The slope of the set-up water surface is less than the slope of the beach face, so the water intersects the beach at an effective shoreline elevation that is above the still-water shoreline. There is also a zone of *set-down* (Fig. 6-26), a depression in the mean water surface below the still-water level that occurs in the offshore prior to wave breaking, the zone where waves are undergoing rapid transformations in their heights and energy as they shoal.

Interest in wave set-up was inspired by observations during a hurricane in 1938 that struck the east coast of the United States. The maximum mean elevation of the shoreline was found to be 1 m higher in an exposed area of the coast where the wave energy was dissipated as surf than in the calmer water of a sheltered region. This difference could not be explained by variations in storm-surge heights, so it was proposed that the breaking waves and surf caused the elevation in the mean water level. This speculation has been verified by theoretical analyses (Longuet-Higgins and Stewart, 1963, 1964), by detailed laboratory measurements such as those in Figure 6-26 (Saville, 1961; Bowen, Inman, and Simmons 1968; Van Dorn, 1976), and also has been confirmed by field observations (Guza and Thornton, 1981; Holman and Sallenger, 1985; Nielsen, 1988; Hanslow and Nielsen, 1993).

The theoretical analyses of Longuet-Higgins and Stewart (1963, 1964) determined that the momentum flux of the waves, the radiation stress, is responsible for both the wave set-down and set-up on a sloping beach. In simple terms, the set-up is a seaward slope in the water surface that provides a pressure gradient or force that balances the onshore component of the radiation stress. It is therefore the S_{xx} component of radiation stress that is important, the x-directed (shoreward) flux of x-directed momentum due to the waves (Chapter 5). This component is given by

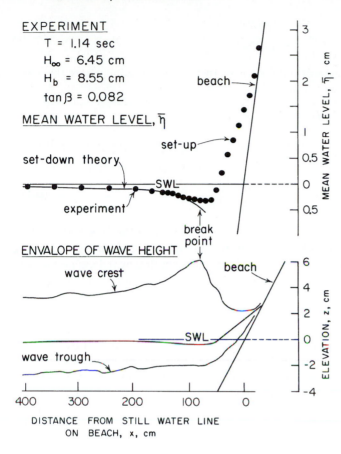

EXPERIMENT
    T = 1.14 sec
    H_∞ = 6.45 cm
    H_b = 8.55 cm
    $\tan \beta$ = 0.082

MEAN WATER LEVEL, $\bar{\eta}$

beach

set-up

set-down theory

experiment

SWL

break point

ENVALOPE OF WAVE HEIGHT

wave crest

beach

SWL

wave trough

400   300   200   100   0

DISTANCE   FROM   STILL   WATER   LINE
ON   BEACH,   x, cm

MEAN WATER LEVEL, $\bar{\eta}$, cm

ELEVATION, z, cm

Figure 6-26   Wave set-down and set-up in the nearshore, departures of the mean water level from the SWL produced by the radiation stress (momentum flux) of the incoming waves. The theoretical curve for the set-down refers to Equation (6.16). [Adapted with permission of American Geophysical Union, from A. J. Bowen, D. L. Inman and V. P. Simmons, Wave "Set-down" and Wave Set-up, *Journal of Geophysical Research 73*, p. 2573. Copyright © 1968 American Geophysical Union.]

$$S_{xx} = \frac{1}{8} \rho g H^2 \left[\frac{2kh}{\sinh(2kh)} + \frac{1}{2} \right] \tag{6.13}$$

where $k = 2\pi/L$ is the wave number, and H and h are, respectively, the local wave height and water depth. In shallow water this becomes

$$S_{xx} = \frac{3}{2} E = \frac{3}{16} \rho g H^2 \tag{6.14}$$

having used $E = \rho g H^2/8$ for the wave energy density. As the waves transform during shoaling, finally break, and then cross the surf zone as bores, changes in this onshore-directed momentum are balanced successively by wave set-down and set-up, displacements in the mean water level. The momentum balance in the cross-shore direction is given by

$$\frac{\partial S_{xx}}{\partial x} + \rho g (\bar{\eta} + h) \frac{\partial h}{\partial x} = 0 \tag{6.15}$$

where $\bar{\eta}$ is the difference between the still-water level and the level in the presence of waves, the set-down or set-up. Outside the breaker zone, where the waves are shoaling and the wave height increases to conserve the energy flux, there is a depression in the sea level to balance the positive gradient $\partial S_{xx}/\partial x$ of the radiation stress, resulting in a set-down given by

$$\overline{\eta} = -\frac{kH^2}{8\sinh(2kh)} \tag{6.16}$$

as derived by Longuet-Higgins and Stewart (1963, 1964). Inside the surf zone where the wave height is depth limited according to

$$H = \gamma(\overline{\eta} + h) \tag{6.17}$$

with γ being approximately constant, the gradient $\partial S_{xx}/\partial x$ is negative, and the solution of Equation (6.15) yields a wave set-up given by

$$\frac{\partial \overline{\eta}}{\partial x} = -\left[\frac{1}{1 + 8/3\gamma^2}\right]\frac{\partial h}{\partial x} \tag{6.18}$$

Here there is a direct proportionality of the set-up slope $\partial\overline{\eta}/\partial x$ with the beach slope $S = dh/dx$, the gradient of the still-water depth. Thus, both set-down and set-up are the products of the shoaling and decay of incident waves, conserving momentum flux by balancing the gradient in the onshore component of the radiation stress, S_{xx}, with the pressure field of the sloping mean sea surface.

The laboratory measurements by Saville (1961) of wave set-up were shown by Longuet-Higgins and Stewart (1963, 1964) to be in reasonable agreement with the above-mentioned theoretical development. However, the more detailed laboratory experiments of Bowen and co-workers (1968), including those in Figure 6-26, established that the theory shows remarkable agreement with measurements of set-down outside the breaker zone and of set-up inside the surf zone. The measurements in Figure 6-26 show the set-down increasing from the offshore to its greatest depression in the water level at the position of wave breaking, followed by a rapid rise in the mean water level of the set-up shoreward of the breaker zone. The curve through the set-down measurements corresponds to the theoretical Equation (6.16). The pattern of the set-up confirms that there is a uniform slope ($\partial\overline{\eta}/\partial x$) that is slightly less than the slope of the beach face, just as predicted by Equation (6.18), derived by Longuet-Higgins and Stewart (1963, 1964).

Additional measurements of wave set-down and set-up have been obtained by Van Dorn (1976) in an extensive series of wave-channel experiments. The results show the same pattern as found by Bowen and co-workers (1968), with the maximum set-down achieved where the uniform waves break and that the set-up slope begins just landward from the break point. Although the slope of the set-up is governed by Equation (6.18), it is apparent from the test series that the elevation of the set-up water surface is greater when the waves are larger and therefore break farther offshore, since the upward slope is initiated in deeper water and can thereby achieve a greater elevation at some cross-shore position. The result is that the effective shoreline formed by the intersection of the set-up surface with the solid beach is at a higher elevation when the wave heights are greater. This trend was also noted by Bowen and co-workers (1968) and is an important factor in the generation of nearshore currents according to the models of Bowen (1969) that will be reviewed in Chapter 8. Van Dorn integrated Equation (6.18) for the set-up slope in the landward direction from the point of intersection between the set-down and set-up ($\overline{\eta} = \overline{\eta}_b$) to obtain a predictive equation for the maximum set-up elevation at the effective shoreline. However, the derived equation was only partially successful in comparisons with direct measurements of maximum set-up elevations, in some instances being off by as much as a factor of 2. In field measurements on various beaches along the coast of Australia, Hanslow and Nielsen (1993) have shown that the slope of the set-up surface increases within the swash zone, compared with the slope in the surf zone; there is some hint of this in Figure 6-26 from the laboratory measurements of Bowen and co-workers (1968). This would in part account for Van Dorn's difficulty in deriving a relationship for the maximum set-up position at the shoreline based on the slope within the surf zone.

It can be seen in Figure 6-26 from the laboratory experiments of Bowen and co-workers (1968) that the effective shoreline position, defined by the intersection of the wave set-up surface and the beach face, can be significantly higher than the still-water level (SWL). In these small-scale laboratory wave-channel experiments, the effective shoreline due to the set-up is between 1 and 4 cm above the still-water shoreline and is shifted landward by roughly 50–100 cm. When scaled up to conditions of an ocean beach, this shift in the elevation and horizontal position of the shoreline due to the existence of wave set-up can be substantial; field measurements that have determined the position of the effective shoreline have demonstrated that the elevation increase can be more than 1 m while the landward shift can be tens of meters (Holman and Sallenger, 1985; Hanslow and Nielsen, 1993). This shift in the position of the effective shoreline is greatest during a storm, so it is apparent that the rise in the mean water level due to wave set-up can be an important factor in shifting the water toward coastal properties, resulting in a greater probability of there being erosional losses during a storm.

Guza and Thornton (1981) measured wave set-up on beaches in southern California (their measurement techniques will be described below when we examine wave run-up). They found that the maximum set-up at the shoreline above the still-water level is

$$\overline{\eta}_{\mathrm{max}} = 0.17 H_{\infty} \tag{6.19}$$

where H_{∞} is the significant height of the incident waves in deep water. Holman and Sallenger (1985) obtained additional measurements from the steeper beach (1:10 average slope) at the FRF in Duck, North Carolina. With offshore wave heights ranging 0.4–4 m, the measured maximum set-up at the shoreline ranged up to 1.6 m, again demonstrating its significance. They found that a direct correlation between the set-up and wave height was highly scattered, but, as is shown in Figure 6-27, the scatter was greatly reduced if the maximum set-up $\overline{\eta}_{\mathrm{max}}$, divided by the deep-water wave height H_{∞}, is related to the

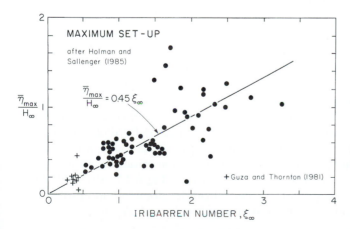

Figure 6-27 Measurements of the maximum set-up, $\overline{\eta}_{\mathrm{max}}$, at the effective shoreline on the beach of the FRF, Duck, North Carolina. The measured set-up is normalized to the deep-water, significant wave height and is compared with the Iribarren number given by Equation (6.1a). [Adapted with permission of American Geophysical Union, from R. A. Holman and A. H. Sallenger, Set-up and Swash on a Natural Beach, *Journal of Geophysical Research 90*, p. 949. Copyright © 1985 American Geophysical Union.]

deep-water form of the Iribarren number of Equation (6.1a). The line fitted to the data in Figure 6-27 yields the relationship

$$\frac{\overline{\eta}_{max}}{H_\infty} = 0.45\xi_\infty \qquad (6.20)$$

having forced the line to pass through the origin. A statistical regression for the data by Holman and Sallenger yielded a non-negligible intercept, but Equation (6.20) is more convenient in analyses. Furthermore, the data included in Figure 6-27 are limited to nearly high-tide conditions. The beach at the FRF has large bars and troughs, and they significantly affect the surf-zone processes during mid to low tides. As a result, the measurements of set-up during mid and low tides were much more scattered than those at high tides, although the gross trends are approximately the same, so that predictions with Equation (6.20) might still be satisfactory. Hanslow and Nielsen (1993) further discuss the effects of the state of the tide on measurements of the maximum wave set-up at the shoreline, but in terms of the water from the run-up entering the porous beach or flowing back out, a factor that will influence the maximum set-up depending on the permeability of the beach sediment. In addition, although the set-up measurements of Hanslow and Nielsen in Australia supported the relationship of Equation (6.20) for steeper beaches, on dissipative beaches of slow slope, they could find no dependence of the maximum set-up on the beach slope as implied by the relationship to the Iribarren number.

The results in Figure 6-27 indicate that depending on ξ_∞, the proportionality coefficient between the maximum set-up and the offshore significant wave height could be on the order of 0.5 or greater, much higher than the 0.17 coefficient found by Guza and Thornton (1981) in Equation (6.19). This suggests the possibility for much higher set-up levels during storm conditions than predicted by the Guza and Thornton relationship. However, during storms, the wave steepness tends to increase and the beach is usually cut back, reducing its slope (Chapter 7), so the Iribarren number is shifted toward lower values. The effect of increasing wave heights on the set-up during a storm is better seen if Equation (6.20) is transformed into its dimensional form. Substituting Equation (6.1a) for the Iribarren number into Equation (6.20) and using $L_\infty = (g/2\pi)T^2$ for the deep-water wave length, the relationship for the maximum set-up elevation at the shoreline becomes

$$\overline{\eta}_{max} = 0.18g^{1/2}SH_\infty^{1/2}T \qquad (6.21)$$

a form that more directly reveals the dependence on the beach slope, the significant deep-water wave height, and the wave period. This relationship is useful in applications to evaluate the maximum set-up at the shoreline and to define where the effective shoreline is relative to the still-water shoreline when large storm waves are present.

It is now recognized that there are uncertainties regarding some of the assumptions made by Longuet-Higgins and Stewart (1963, 1964) in their derivations of Equations (6.16) and (6.18), respectively, for the wave set-down and set-up. They used the linear-wave form of the radiation stress equations, even though the application in most cases is to high waves in shallow water where linear-wave theory should not be used (Chapter 5). Furthermore, we saw in the previous section that Equation (6.17), assumed as a depth limitation to wave heights within the nearshore in the derivation of Equation (6.18), is in some instances not substantiated by direct measurements of wave heights in the surf zone (e.g., Fig. 6-13). Stive and Wind (1982) have focused on these shortcomings in the theoretical development and provide direct assessments of cross-shore variations in the radiation stress (i.e., of $\partial S_{xx}/\partial x$) determined by measuring orbital velocities and pressures within the wave field, while also making detailed measurements of the wave set-down and set-up. An example of the resulting data for S_{xx} variations across the surf zone, and of the set-down and set-up, is given in Figure 6-28; the graph includes comparisons with the theoretical curves based on linear-wave theory, as developed by Longuet-Higgins and Stewart (1963, 1964), and curves based on the nonlinear solutions by James (1974) and Cokelet (1977) that should be more applicable to steep waves within the nearshore. The nonlinear theories clearly are superior to the linear theories, especially in the assessment of the S_{xx} radiation stress. Some discrepancy remains between the theories and measurements of the wave set-down and set-up. The measurements show a concave water surface for the set-up profile, in agreement with the nonlinear theories, but there is an elevation displacement between the measurements and the theoretical curves. This has been a common problem in analyses of wave set-down and set-up because of the uncertainty of where one ends and the other begins. In view of that uncertainty, in a number of such analyses, the theoretical curves have been shifted horizontally in the cross-shore direction in order to maximize agreement between measurements and theory.

Recent analyses of wave set-down and set-up have been part of investigations of wave-height variations across the nearshore, at times over irregular beach profiles. An example of this can be seen in Figure 6-21 from Battjes and Stive (1985). As discussed earlier, those analyses include considerations of the complete probabilities of wave heights and their breaking within the surf zone. Those patterns determine the local gradient of the radiation stress $(\partial S_{xx}/\partial x)$, which governs the local slope of the water surface according to Equation (6.15). This is apparent in the example of Figure 6-21, where the wave dissipation over the offshore

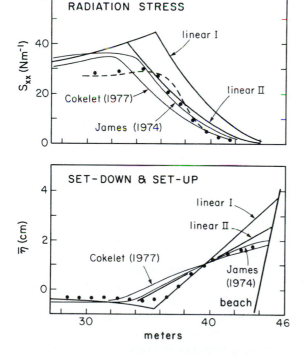

Figure 6-28 Measurements of the S_{xx} component of the radiation stress of shoaling and dissipating waves and of the accompanying wave set-down and set-up. Comparisons are made with theoretical analyses, including two linear models derived by Longuet-Higgins and Stewart (1963, 1964) and the advanced, nonlinear solutions of James (1974) and Cokelet (1977). [Adapted from *Coastal Engineering* 6, M. J. F. Stive and H. G. Wind, A Study of Radiation Stress and Set-up in the Nearshore Region, p. 22, 1982 with kind permission of Elsevier Science-NL, Sara Burgerhartstratt 25, 1055 KV Amsterdam, The Netherlands.]

bar and on the beach face yields greater local slopes in the set-up, while the zone of wave re-formation over the longshore trough produces a horizontal water surface. It also is apparent from this example that the maximum wave set-up at the effective shoreline can be the product of the complex pattern of wave set-up over an irregular bottom topography, so there will not always be a simple relationship solely with the wave conditions and mean beach slope. This factor likely accounts for the scatter of the field measurements of set-up obtained by Holman and Sallenger (1985), discussed above, where the scatter of the data is greatest during mid to low tides when the effects of the profile irregularities are greatest. It might also account for Hanslow and Nielsen (1993) not having found a dependence of the maximum set-up at the shoreline on the slopes of dissipative beaches, suggesting that the irregular topography is more important than the average slope of the beach.

The fundamental cause of wave set-up in the surf zone has been well established in theory (Longuet-Higgins and Stewart, 1963, 1964), which has been verified by both laboratory and field studies. However, uncertainties remain in predictions of the patterns and magnitudes of set-up along the beach profile and in predictions of the maximum set-up at the shoreline. More research is needed as such predictions are important to evaluations of water levels within the surf zone, a factor of significance to the occurrence of property erosion backing the beach, and also in analyses of nearshore currents (including undertow), which are in part driven by spatial variations in mean water elevations as governed by the wave set-up (Chapter 8).

WAVE RUN-UP ON THE BEACH FACE

The swash of waves at the shore represents the cutting edge of the sea, the zone of intense erosion during storms. Because of its importance, a number of investigators have concentrated on making measurements of wave run-up and analyzing the data as functions of the offshore wave conditions and beach morphology. There have been parallel investigations of wave run-up on engineering structures such as rubble-mound jetties, studies that are needed to establish design criteria, including those governing the elevation of the structure so as to limit its overtopping by the run-up of extreme waves.

Field measurements of wave run-up have been obtained by unusual techniques. Guza and Thornton (1981, 1982) utilized an 80-m long dual-resistance wire stretched across the beach profile and held about 3 cm above the sand by non-conducting supports. In principal this technique is the same as that employed by using vertical staffs to measure waves (Chapter 5), but here the staff is laid down across the sloping beach. Holman and Sallenger (1985) obtained run-up measurements by taking videos of the swash over an extended period. The position of the edge of the run-up is then digitized frame-by-frame to follow the time-varying positions of the waterline, an analysis that has been computerized. This "remote-sensing" technique has the advantage that it can simultaneously document the run-up along an extended stretch of shore and is not affected by people on the beach or by flotsam carried by the swash. Holman and Guza (1984) have undertaken direct comparisons of results using the two techniques to measure the run-up, establishing that although there are systematic differences, the results are comparable. Figure 6-29 illustrates another video technique for analyzing wave run-up, termed the "timestack" method, devised by Aagaard and Holm (1989) with the example here from Holland and Holman (1993). The timestack graph documents the time variations in run-up for a fixed profile line, showing the cross-shore location of the run-up edge (dashed line) with time. The intensity patterns seaward of the run-up edge reflect

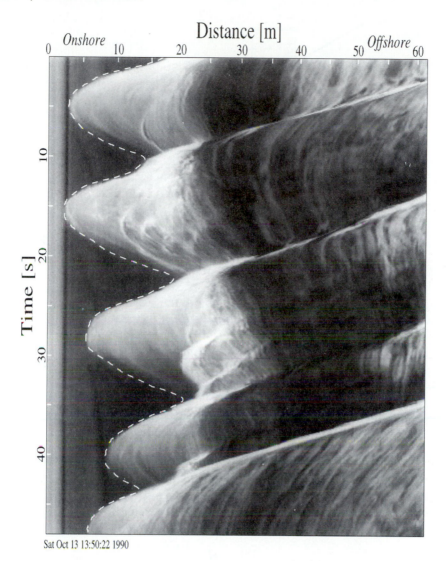

Distance [m]

Onshore   0   10   20   30   40   50   *Offshore*   60

Time [s]

Sat Oct 13 13:50:22 1990

Figure 6-29   Time stacking video documentation of the time variations in the wave run-up along a fixed cross-shore profile. The slanted traces downward and to the left represent bores approaching the shore, while the dashed line is the maximum run-up position. The shades of white versus dark reveal the movement of foam on clear water. [Used with permission of American Geophysical Union, from K. T. Holland and R. A. Holman, The Statistical Distribution of Swash Maxima on Natural Beaches, *Journal of Geophysical Research 98*, p. 10,273. Copyright © 1993 American Geophysical Union.]

changes in the flow field farther offshore but still along this fixed profile line. The slopes of the approximately linear features represent the speeds of incoming run-up bores.

It is the maximum shoreward extent of the wave run-up that has particular significance to the erosion of properties backing the beach and in the overtopping of jetties. It has been found that the total run-up consists of three primary components (Fig. 6-30): (1) the set-up, which determines the mean shoreline position above which the swash of individual waves

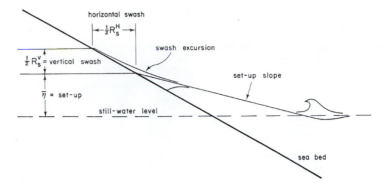

Figure 6-30   The factors that affect the total run-up of swash on a sloping beach, including the set-up that determines the effective shoreline in the presence of waves and the run-up of individual swash beyond that mean shoreline position. The run-up can be analyzed in terms of its vertical or horizontal components.

occurs; (2) fluctuations about that mean, due to the swash of incident waves producing run-up and run-down; and (3) a component in the swash oscillations having periods in excess of 20 sec, infragravity periods beyond the usual range of incident-wave periods. The maximum run-up height achieved by the water is the summation of these components (Fig. 6-30).

Many of the studies of wave run-up have involved engineering structures such as seawalls, jetties, and breakwaters (Chapter 12). Early publications include the paper by Wassing (1957) that reports on model studies of wave run-up on dikes in the Netherlands and that of Hunt (1959) examining run-up on rock-rubble structures. Hunt proposed the simple formula

$$R_{2\%} = 8H_s S \tag{6.22}$$

for the vertical level exceeded by 2 percent of the run-up heights, as a function of the significant wave height H_s at the toe of the structure and the slope of the structure, S. Battjes (1971) has shown that the run-up can be better related to an Iribarren number according to the dimensionless equation

$$\frac{R_{2\%}}{H_s} = C\xi_p \tag{6.23a}$$

where

$$\xi_p = \frac{S}{\sqrt{H_s/L_\infty}} \tag{6.23b}$$

C is a dimensionless coefficient that has to be established experimentally. Ahrens (1981) undertook an extensive series of laboratory experiments, where the cotangent of the structure slope ranged from 1 to 4, and found that $C = 1.61$ when $\xi_p < 2$. Based on field measurements, Grüne (1982) found values of C between 1.33 and 2.86. The measurements of run-up on structures have been reviewed by van der Meer and Stam (1992), who arrived at the curves shown in Figure 6-31, comparing $R_{2\%}/H_s$ to the form of the Iribarren number given by Equation (6.23b), where the wave height is measured at the base of the structure while the wave length is for the deep-water condition. Separate curves are established for smooth versus

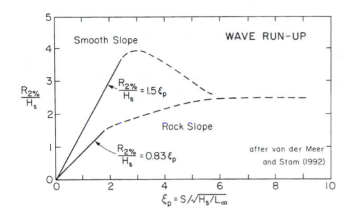

Figure 6-31 Curves based on compilations of run-up measurements on structures in terms of $R_{2\%}/H_s$ versus the Iribarren number, where $R_{2\%}$ is the 2 percent exceedence of the vertical components of the run-up, and H_s is the significant wave height at the base of the structure. [From Wave Runup on Smooth and Rock Slopes of Coastal Structures, J. W. Van der Meer and C. J. M. Stam, *Journal of Waterway, Port, Coastal and Ocean Engineering*, 1992. Reproduced with permission from the American Society of Civil Engineers.]

rough structures, the latter consisting of rock rubble. Simple linear trends between $R_{2\%}/H_s$ and ξ_p are found only at low values of the Iribarren number, with an apparent convergence of the data when $\xi_p > 6$. This graph in particular reveals the effects of the surface roughness, with the run-up being approximately twice as great on the smooth surface compared to a surface covered with rocks.

The use of Equation (6.23) requires an assessment of the slope S, generally taken to be the average slope of the structure or beach face subjected to swash run-up. Where beaches are concave in shape, there can be uncertainties as to how the slope should be defined. This problem has been examined by Mayer and Kriebel (1994) in an extensive series of laboratory wave-channel experiments, the results of which demonstrate the usefulness of Saville's (1958) definition of a composite slope in computations of run-up with Equation (6.23).

There have been comparatively few studies of wave run-up on natural beaches, but the results are interesting in that they were unexpected. In their study of run-up on California beaches, Guza and Thornton (1982) did find the expected result that the excursion distance of the swash fluctuations about the mean set-up level was directly dependent on the incident wave height. When expressed as a "significant" run-up height, R_s, the average of the highest one-third of the run-up levels, it was found that the swash was related to the deep-water significant wave height, H_∞, by the relationship

$$R_s = 0.7H_\infty \tag{6.24}$$

The more surprising result in the study of Guza and Thornton (1982) is that most of the swash excursion occurred at periods that were much greater than the range of incident waves, that is, at periods greater than 20 sec in the infragravity range of motions. In their analysis, Guza and Thornton were able to separate these two swash components through spectral analyses of the run-up records, with one component being the direct swash of the incident wave bores and the other being the infragravity component. The results are shown in Figure 6-32 and demonstrate that the component due to incident wave bores has no dependence on the heights of the waves in deep water, whereas the infragravity motions grow with the offshore wave heights. It is apparent that most of the energy of the swash oscillations is in the infragravity range rather than in the period range of the incident waves, and this is especially so during storms with higher incident waves.

The study of Guza and Thornton (1982) was undertaken on a beach that has a low slope where the waves break offshore and continue to break as bores while crossing the wide surf

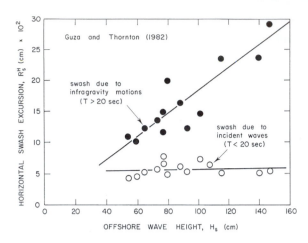

Figure 6-32 The horizontal component of run-up measurements on Torrey Pines Beach, California. The measurements have been separated by spectral analyses into portions due to the incident waves (periods less than 20 sec) and infragravity motions (periods greater than 20 sec). The latter respond to the offshore wave conditions, whereas the swash due directly to the incident bores does not respond to the offshore waves on this dissipative beach. [Adapted with permission of American Geophysical Union, from R. T. Guza and E. B. Thornton, Swash Oscillations on a Natural Beach, *Journal of Geophysical Research 87,* p. 489. Copyright © 1982 American Geophysical Union.]

zone, that is, on a dissipative beach (Chapter 3). On such a beach, the wide surf zone acts as a filter between the offshore waves and the swash at the shoreline. A doubling of the incident wave heights, for example, causes the waves to break at approximately twice the water depth, doubling the distance from the shoreline to the breaker zone. The result is that there is little effect on the swash produced by the individual wave bores, which must now travel twice as far. This is a fundamental attribute of a dissipative beach and also corresponds to the finding of a saturated zone of wave energy within the inner surf zone (Fig. 6-3), where the heights of the bores are depth limited and do not respond to changes in the offshore wave conditions (Thornton and Guza, 1982).

As would be expected, measurements of run-up on steeper reflective beaches are dominated more by the incident waves, and the run-up associated with the direct arrival of individual waves responds to the offshore wave conditions, specifically to the occurrence of a storm. This has been confirmed by the video measurements of wave run-up obtained by Holman and Sallenger (1985) at the FRF in North Carolina. When the total run-up was analyzed, separate trends were obtained for different stages of the tides, largely due to the changes in the effective beach slope of the concave profile found at the FRF, such that the beach is more dissipative at low tide than during high tide. When the tide is low and the FRF beach is dissipative, the run-up associated with the incident waves has little or no response to an increase in the offshore significant wave height, just as was found by Guza and Thornton (1982). In contrast, during the more reflective conditions at times of higher tides, there clearly is a response of the incident-wave run-up to the offshore wave conditions. In both instances, the infragravity component of the run-up responds to the offshore significant wave height, just as was established in Figure 6-32 by Guza and Thornton.

This shift of the relative importance within the run-up of incident waves versus the infragravity component as the beach changes from dissipative to more reflective is further illustrated by changes in the average periodicity of the swash. Figure 6-33 is based on the run-up measurements of Holman and Sallenger (1985) and compares the average period of the measured swash (T_s) to the period of the incident waves (T_i) measured in the offshore. At high ξ_∞ the ratio T_s/T_i is close to 1.0, indicating the dominance of incident waves under these more reflective conditions. As ξ_∞ decreases and the beach become more dissipative, the ratio increases up to values on the order of $T_s/T_i = 2.5$, in response to the greater importance of the infragravity motions. Based on such analyses, Holman and Sallenger concluded that

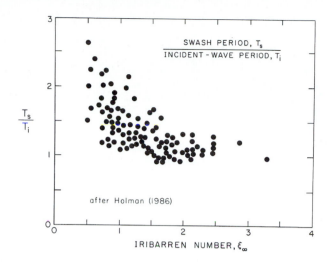

SWASH PERIOD, T_s

INCIDENT - WAVE PERIOD, T_i

after Holman (1986)

IRIBARREN NUMBER, ξ_∞

Figure 6-33 The ratio of the periodicity of the swash as recorded with videos at the FRF, to the period of the incident waves as measured in deep water. As the Iribarren number is reduced and the beach becomes more dissipative, the ratio progressively increases due to the increase in the infragravity motions relative to the incident waves. [Adapted from *Coastal Engineering 9*, R. A. Holman, Extreme Value Statistics for Wave Run-up on a Natural Beach, p. 537, 1986 with kind permission of Elsevier Science-NL, Sara Burgerhartstratt 25, 1055 KV Amsterdam, The Netherlands.]

the infragravity motions start to dominate over the incident waves in the swash when ξ_∞ is less than about 1.75.

Similar to their analysis of set-up, the study of Holman and Sallenger (1985) at the FRF established that the ratio R_s/H_∞ is not a constant as found by Guza and Thornton (1982), Equation (6.24), but again depends on the deep-water Iribarren number ξ_∞. Holman (1986) reanalyzed the data with the objective of deriving predictive relationships for the range of statistics of extreme run-up. Figure 6-34 shows the results for the vertical swash elevations of the run-up above the maximum set-up elevation, that is, above the effective shoreline (not the still-water level). The results are given as the maximum run-up elevation (R_{max}) achieved during 35 min of run-up measurements, and also as the 2 percent exceedence of all run-up elevations ($R_{2\%}$) in the measurements. The latter correlation yields

$$\frac{R_{2\%}}{H_\infty} = 0.45\xi_\infty \qquad (6.25)$$

in terms of the deep-water significant wave height. The lines fitted to the data in Figure 6-34 again are forced to have zero intercepts and therefore differ slightly from the regressions provided by Holman. Similar to the reduction leading to Equation (6.21) for the set-up, Equation (6.25) can be reduced to the simple proportionality

$$R_{2\%} = 0.18g^{1/2}SH_\infty^{1/2}T \qquad (6.26)$$

As noted above, the elevation of $R_{2\%}$ is vertically above the effective shoreline level as determined by the maximum set-up elevation (Fig. 6-30). The sum of the set-up from Equations (6.21) and (6.26) for the run-up yields the total swash elevation

$$R_{2\%}^T = \eta_{max} + R_{2\%} = 0.36g^{1/2}SH_\infty^{1/2}T \qquad (6.27)$$

This would represent the total elevation achieved by the swash of waves above the water level determined by the tide and factors affecting the mean level of the sea at that particular time (Chapter 4). This total run-up above the still-water level is graphed in Figure 6-35 in comparison with the video data of Holman and Sallenger (1985). The results are more scattered than

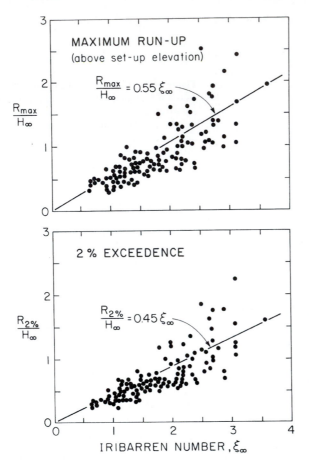

Figure 6-34 The vertical component of the run-up above the effective shoreline determined by the set-up elevation. The run-up elevation is represented as the maximum level R_{max} achieved during the 35-min video record and as the 2 percent exceedence elevation, $R_{2\%}$. Both depend on the deep-water Iribarren number of Equation (6.1a). [Adapted from *Coastal Engineering* 9, R. A. Holman, Extreme Value Statistics for Wave Run-up on a Natural Beach, p. 538, 1986 with kind permission of Elsevier Science-NL, Sara Burgerhartstratt 25, 1055 KV Amsterdam, The Netherlands.]

in Figure 6-34 for the run-up above the maximum set-up elevation, since inclusion of the set-up introduces more variability, especially for data collected during mid to low tides. The relationship for the 2 percent exceedence of the vertical component of the total run-up is

$$\frac{R_{2\%}^{T}}{H_\infty} = 0.92\xi_\infty \tag{6.28}$$

which is equivalent to Equation (6.27), derived by summing the separate components of set-up and run-up beyond the effective shoreline elevation. In general terms, doubling the wave height during a storm produces a $2^{1/2} = 1.41$ factor increase in the total run-up elevation; however, since wave periods also tend to increase during major storms, the combined effects of increased wave parameters could easily double the elevation of the total run-up.

Although only limited data are available, measurements by Nielsen and Hanslow (1991) and Ruggiero et al. (1996) from highly dissipative beaches have been found to disagree with the results of Holman and Sallenger (1985) and Holman (1986) from the more reflective beach at the FRF. Rather than agreeing with relationships such as Equations (6.26) and (6.27), it is found that the run-up elevation depends solely on the incident wave height, not on the incident wave period or beach slope. The resulting correlation is like that of Equa-

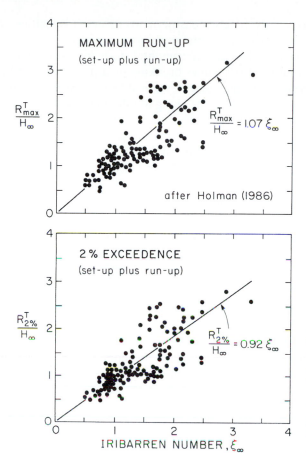

$\dfrac{R^T_{max}}{H_\infty}$

MAXIMUM RUN-UP
(set-up plus run-up)

$\dfrac{R^T_{max}}{H_\infty} = 1.07\,\xi_\infty$

after Holman (1986)

$\dfrac{R^T_{2\%}}{H_\infty}$

2 % EXCEEDENCE
(set-up plus run-up)

$\dfrac{R^T_{2\%}}{H_\infty} = 0.92\,\xi_\infty$

IRIBARREN NUMBER, ξ_∞

Figure 6-35 The vertical component of the total run-up above the still-water elevation, due to the combined effects of wave set-up and the swash run-up beyond the effective shoreline (i.e., $R^T = \eta_{max} + R$). [Adapted from *Coastal Engineering* 9, R. A. Holman, Extreme Value Statistics for Wave Run-up on a Natural Beach, p. 540, 1986 with kind permission of Elsevier Science-NL, Sara Burgerhartstratt 25, 1055 KV Amsterdam, The Netherlands.]

tion (6.24), also obtained from a dissipative beach (Guza and Thornton, 1982), though the proportionality coefficients differ. It should not be surprising that the run-up on a dissipative beach does not depend on the incident wave period, since it has been established that the run-up is controlled primarily by the long-period infragravity energy, with that energy content dependent on the wave heights and energy level of the incident waves (Fig. 6-32).

The run-up statistics employed in the above analyses, R_{max} and $R_{2\%}$, are but two possibilities, those that best reflect the more extreme run-up elevations. Other commonly used statistics include the mean elevation achieved by the run-up, the root-mean-square elevation, and the significant elevation (the average of the highest one-third). These statistical parameters have been related to the Rayleigh distribution, which is skewed such that there are only a few extreme events (Battjes, 1971; Nielsen and Hanslow, 1991; Kobayashi, Cox, and Wurjanto, 1991). Measurements by those studies of all swash run-up elevations during a period of time tend to confirm that they do follow a Rayleigh distribution. However, the measurements by Holland and Holman (1993) of run-up on ocean beaches showed poor agreement with the Rayleigh distribution, so further investigation is needed.

Considering the significance of wave swash as a process important to beach erosion, it is surprising how little study has been undertaken that focuses on the details of the run-up and run-down oscillations. There are relatively few measurements of swash velocities, either of the leading edge of the swash or of velocities internal to the swash wedge.

Roos and Battjes (1976) have undertaken an experimental study in a wave channel to obtain information on the run-up resulting from waves breaking on a plane, smooth beach. They employed a series of wire-resistance gauges placed vertical to the beach face to measure the thickness of the swash lens at several positions along the sloping beach. This allowed them to document the position and velocity of the leading edge of the swash and also to calculate mean velocities at various cross-shore positions internal to the swash; the latter could be inferred from the depth measurements by evaluating the discharge of water required to account for the time variations of water volumes within the swash wedge. Figure 6-36 shows the results for the mean velocity of the swash front, $\overline{C}_{run-up}$, during the entire run-up interval. The velocity data support the empirical relationship

$$\frac{\overline{C}_{run-up}}{\sqrt{gH_s}} = 0.6\sqrt{\xi_p} \tag{6.29a}$$

where

$$\xi_p = \frac{S}{\sqrt{H_s/L_\infty}} \tag{6.29b}$$

and the reference wave height H_s is measured at the base of the slope where the run-up begins. The corresponding time involved in the run-up of the swash to its maximum elevation was found to be given by

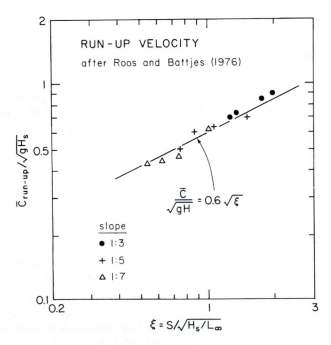

RUN-UP VELOCITY

after Roos and Battjes (1976)

$\frac{\overline{C}}{\sqrt{gH}} = 0.6\sqrt{\xi}$

slope

● 1:3

+ 1:5

△ 1:7

$\overline{C}_{run-up}/\sqrt{gH_s}$

$\xi = S/\sqrt{H_s/L_\infty}$

Figure 6-36 The average velocity of the wave run-up edge, $\overline{C}_{run-up}$, normalized to $\sqrt{gH}$, where H is the wave height at the toe of the slope, as a function of the Iribarren number. [Adapted from Characteristics of Flow in Runup from Periodic Waves, A. Roos and J. A. Battjes, *Proceedings of the 15th Coastal Engineering Conference,* 1976. Reproduced with permission from the American Society of Civil Engineers.]

$$\frac{t_{run-up}}{T} = 0.7 \frac{1}{\sqrt{\xi_p}} \tag{6.30}$$

Roos and Battjes showed that when these two relationships are combined to yield an expression for the total run-up distance, the results are consistent with the simple formulation of Equation (6.22).

In a series of laboratory wave-channel experiments examining swash dynamics, Miller and Zeigler (1958) timed floats over measured distances to obtain average swash and backwash velocities. Similar measurements were obtained by Wright (1976) in the field on a beach with a 3° slope, by measuring the distance traveled by foam in successive movie frames. Water depths of the swash were also measured relative to poles placed on the beach face. At the time of the measurements the waves were small, approximately 30-cm high. As expected, the pattern is generally one of decreasing velocities during the progress of the swash up the beach face, accompanied by abrupt increases in water depths. The backwash is characterized by a progressive decrease in water depths, with an initial acceleration of the water to a maximum velocity, followed by a decrease of the velocity as the depth reduces to the order of 1 cm. Based on these measurements, Wright calculated values for the Froude number, $Fr = u/\sqrt{gh}$, which is of interest in that the values demonstrate that the backwash can become supercritical ($Fr > 1$). The dominance of supercritical flow would account for the plane sand bed (upper flow regime) of the beach face, as opposed to being covered by sediment ripple marks that develop under subcritical flows. Another result is the common development of a hydraulic jump when the supercritical flow of the backwash collides with the next incoming wave bore, which may be temporarily halted in its advance by the opposing flow. Depending on the combination of the size of the incoming bore and the Froude number of the backwash, the resulting hydraulic jump can be characterized by either an abrupt increase in the water depth at the bore front or an undular jump can form consisting of a series of secondary waves (Peregrine, 1966). An example of an undular hydraulic jump, temporarily halted in position, is seen in Figure 6-37(a) obtained on a low-sloping dissipative beach in Oregon. Figure 6-37(b) was taken a few seconds later and shows that the water motions associated with the undulations of the hydraulic jump produced a series of backwash ripples on the beach face, accentuated by the concentration of black heavy minerals where erosion was locally greatest. Broome and Komar (1979) analyzed the formation of backwash ripples by this process, including the development of numerical models to calculate the patterns of sediment transport beneath the undulations.

Matsunaga and Honji (1980, 1983) reported on laboratory observations of the internal velocity field of a breaking wave interacting with the backwash of the previous wave, utilizing flow-visualization techniques (the movement of particles that are slightly denser than water). In addition to measuring the internal velocities of the wave bore and backwash, they documented the formation of a vortex near the bottom at the edge of the swash zone where the backwash meets an incoming wave bore. It was found that this vortex can be important in the localized erosion of the sediment bed and can lead to the formation of a step in the beach profile, a commonly observed feature on steep reflective beaches. Larson and Sunamura (1993) have experimented further with this mechanism for the formation of a profile step, including more documentation of its dimensions and how they depend on the wave and sediment conditions.

Figure 6-37   Successive photographs about 10–15 sec apart, showing an undular hydraulic jump with five wavelets formed by the collision of the backwash and an incoming wave bore (a), and later the backwash ripples on the sandy beach face produced by the wavelets (b). [Used with permission of University of East Anglia, from R. Broome and P. D. Komar, Undular Hydraulic Jumps and the Formation of Backwash Ripples on Beaches, *Sedimentology 26,* p. 545. Copyright © 1979 University of East Anglia.]

The full description of wave run-up on the beach face requires the development of numerical models that predict the time and spatial variations of the water-surface profile and associated velocities of water movement. This involves solutions of the fundamental equations that govern the flow of fluids, in this application generally the depth-averaged nonlinear equations of continuity (water conservation) and momentum. Shen and Meyer (1963a, 1963b) were early contributors to the development of models to describe swash oscillations on a sloping beach, yielding simple predictions of the pattern of decreasing velocity as the swash ascends the beach and of the swash height as a function of the initial velocity following bore collapse. The predictions have been tested by Hughes (1992) using data obtained on Australian beaches dominated by incident waves, the measurements showing, for example, that the swash height on average reaches to only 65 percent of the predicted swash height. Although this result is likely due to the noninclusion of frictional drag and percolation in the theoretical analysis of the swash flow, Hughes could not detect any dependence in the measurements on the beach sediment grain size, which ranged from 0.3 to 2.0 mm. More detailed analyses of swash motions are provided by the models formulated by Hibberd and Peregrine (1979) and Kobayashi, DeSilva, and Watson (1989). These latter models have been tested by Raubenheimer et al. (1995) in an interesting series of field experiments where the run-up was measured using a vertical stack of five wires supported parallel to and above the beach face at elevations of 5, 10, 15, 20, and 25 cm, the series of wires stretching 80 m across the beach. With such an array, they were able to measure the changing profile of a wave bore as it is transformed into the wedge of the run-up surge. Their measurements showed good agreement with the theoretical model for wave run-up developed by Kobayashi and co-workers (1989).

INFRAGRAVITY MOTIONS AND EDGE WAVES IN THE NEARSHORE

In the discussion of run-up on beaches, it was seen that much of the energy of the swash oscillations is within the infragravity range, that is, it has periods in excess of 20 sec, beyond the range of normal wind-generated waves on the ocean. This was observed in the results of Guza and Thornton (1982) (Fig. 6-32), derived from swash measurements on a dissipative beach in California. At that site, an increase in the size of the offshore waves did not result in a corresponding increase in the swash energy associated with the incident waves themselves, but did produce an increase in the energy and swash-excursion elevations of the infragravity component. The implication is that the energy of the infragravity motions must ultimately be derived from the wind-generated waves reaching the coast, but somehow there has been a shift in the dominant periods of the energy, from less than to greater than 20 sec. Questions arise then as to the basic nature of this infragravity motion within the nearshore, the mechanism of energy transfer from the normal ocean waves, and whether this infragravity energy has any measurable importance within the nearshore. In answer to the last question, studies of sediment transport on beaches (Chapter 9) demonstrate that infragravity motions can be important to the suspension of sediment, particularly within the inner surf zone, and that water motions associated with the infragravity energy, specifically with edge waves, can produce a net transport of sediment that affects the overall beach morphology, at times forming beach cusps or offshore crescentic bars (Chapter 11).

Guza and Thornton (1982) and Holman and Sallenger (1985) measured the infragravity motions within the swash run-up, while other investigators have measured it at various positions across the surf zone, generally by using a series of electromagnetic current meters [e.g., Huntley (1976) and Huntley, Guza, and Thornton (1981)]. As with the run-up measurements, spectral analyses of the currents within the surf zone can separate the energy portions due to the incident waves versus the longer-period infragravity motions. The product of such an analysis is illustrated by the results in Figure 6-38, from Wright and Short (1983), showing the relative portions of the incident and infragravity components along a cross-shore profile and also a mean-flow contribution that is primarily a longshore current. The results are relative to the

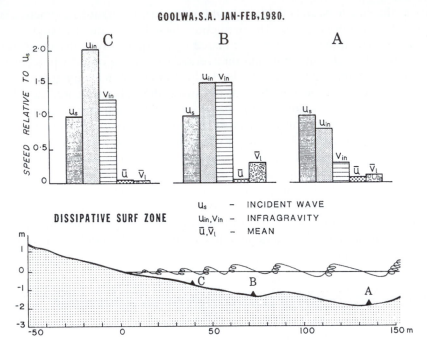

Figure 6-38   The relative magnitudes of longshore (v) and cross-shore (u) velocities measured with current meters of the incident waves, infragravity motions, and mean currents. At each cross-shore position the magnitudes are scaled relative to the velocity amplitude of the incident waves. The measurements are from a highly dissipative beach at Goolwa, south Australia. [Adapted with permission of CRC Press, from L. D. Wright and A. D. Short, Morphodynamics of Beach and Surf Zones in Australia, *Handbook of Coastal Processes and Erosion*, p. 47. Copyright © 1983 CRC Press.]

incident-wave contribution (set at $u_s = 1.0$) rather than being given as absolute energy levels. Both longshore and cross-shore components are included. Within the inner surf zone, position C on the profile, the infragravity component is substantially greater than the incident-wave energy. This would correspond to the observations of strong infragravity motions in the wave swash as observed by Guza and Thornton (1982). In addition, it can be seen in Figure 6-38 that there is a strong representation of the infragravity motions in the longshore direction as well as in the cross-shore, suggesting that the infragravity component is more complex than the simple run-up and run-down of the swash. At progressively greater depths across the beach profile, positions B and A, the importance of the infragravity energy decreases relative to the incident waves. It can be seen, however, that on the dissipative beach where these measurements were obtained by Wright and Short, the incident waves do not actually dominate until the outer surf zone. This is in keeping with the expected patterns of wave-energy losses on a dissipative beach, such that the breaking of more and more waves toward the inner surf zone and the progressive dissipation of individual bores during their travel will reduce the energy level of the incident waves. This progressive loss of incident wave energy toward the shore could account for the observed variations in incident-wave energy versus infragravity energy, as seen in Figure 6-38, even if the energy level of the infragravity component is uniform across the entire width of the surf zone. Direct measurements of the energy level of the infragravity motions show, however, that the energy is generally a maximum at the shoreline within the swash and decreases toward the offshore [e.g., Huntley and Bowen (1973), Huntley (1976), Huntley, Guza, and Thornton (1981), and Oltman-Shay and Guza (1987)].

Studies during the past two decades have established that in many instances, perhaps most, this pattern of cross-shore variation in infragravity energy is accounted for by the presence of edge waves within the nearshore (Holman, 1983). *Edge waves* are so termed because they are trapped at the edge of the ocean by the sloping beach. In a sense, they are held there by wave refraction—when waves are reflected from the shoreline at an angle (Fig. 6-39), they first travel seaward but refract as they go and eventually turn back toward the shore to be reflected once more to repeat the process. The net movement of such a trapped edge wave is in the longshore direction, a progressive wave. The amplitude varies sinusoidally along the shore and diminishes rapidly seaward from the shoreline where it is a maximum. The mathematics of such unusual motions for edge waves were first discovered in 1846 by Sir George Stokes, and for a long time, they were considered to be a mathematical oddity, having no consequence to the real world. It was not until the 1960s that their importance in nearshore processes was fully recognized, subsequent to which they have been the focus of numerous investigations.

Figure 6-40 depicts the movement of one type of progressive edge wave along the length of a sloping beach. The wave form is largest at the shoreline, with an exponential decrease in the amplitude toward the offshore. In the longshore direction there is a pattern of alternating crests and troughs, which on the sloping beach could be observed as positions of maximum run-up and run-down. Mathematically the amplitude variations of this specific edge-wave form are given by

$$\eta(x,y,t) = Ae^{-kx}\cos(ky - \sigma t) \tag{6.31}$$

where A is the amplitude at the shoreline, x and y are offshore and longshore coordinates, and $k = 2\pi/L_e$ and $\sigma = 2\pi/T_e$ are the wave number and radian frequency of the edge wave, L_e being its length as measured in the longshore direction. The components of the variation are apparent in this equation: the maximum amplitude A is at the shoreline ($x = 0$), there is an exponential offshore decrease in the amplitude governed by the e^{-kx} factor, and a cosine dependence containing y that controls the longshore rhythmicity of the edge-wave form. It is this mathematical relationship that was used to calculate the edge wave surface motions graphed in Figure 6-40 (Holman, 1983). This is but one possible form of edge waves, the simplest where there is an exponential offshore decay. A series of edge wave "modes" can actually exist, having different forms of the cross-shore amplitude variations, as graphed in Figure 6-41. This series is denoted by a modal number $n = 0, 1, 2, \ldots$ with the form of the amplitude variations progressively being more complicated. The mode number n actually signifies the number of nodal points where the surface crosses the x axis, the still-water level, with the edge-wave

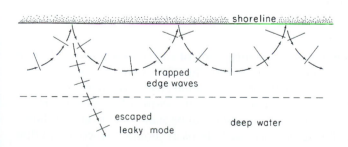

Figure 6-39   Edge waves are trapped to the sloping beach, in effect undergoing reflection from the shoreline but then move seaward at such an angle that they bend around due to refraction and return to shore to repeat the process. The result is that edge waves progress in the longshore direction and can be defined in terms of a longshore wave length. There can also be leaky modes that escape from the nearshore, refraction being insufficient to turn their energy back toward the shoreline.

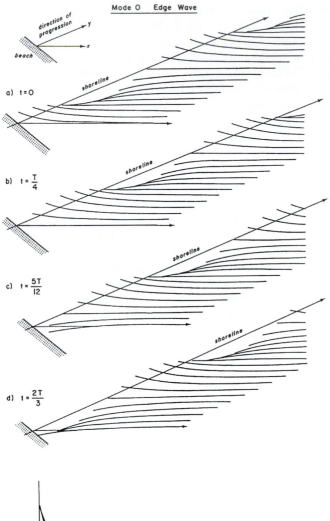

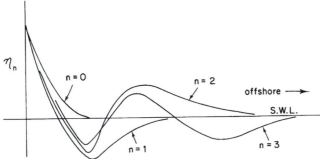

Figure 6-40 A progressive, mode $n = 0$ edge wave viewed obliquely from the offshore. The time progression of the wave is shown by the four figures, each being later in time than the one above it. The edge wave is sinusoidal in the longshore direction, while at each position there is an offshore exponential decrease in the amplitude. [Used with permission of CRC Press, from R. A. Holman, Edge Waves and the Configuration of the Shoreline, *Handbook of Coastal Processes and Erosion,* p. 23. Copyright © 1983 CRC Press.]

Figure 6-41 Cross-shore variations in the amplitudes of the series of edge waves of modes $n = 0$–3, with the mode number being equal to the number of cross-over node positions along the SWL.

surface moving vertically in opposite directions to either side of each cross-shore node. With the higher mode numbers, there is also a greater portion of the edge-wave energy shifted toward the offshore away from the shoreline. With the $n = 0$ mode edge wave, depicted in Figure 6-40, almost all of the amplitude variation occurs close to the shore, whereas in the mode $n = 3$ edge wave in Figure 6-42, some amplitude variation can be seen away from the shoreline with a highly complex pattern of water-surface oscillations. It should be recognized that such variations of the water surface due to edge waves would generally not be observable on

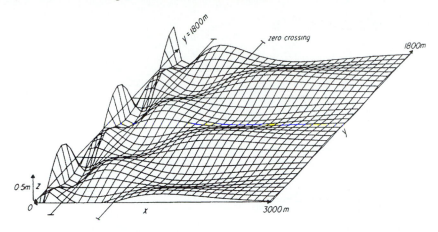

Figure 6-42   The mode $n = 3$ edge wave at an instant in time, showing the sinusoidal variations in the longshore, but a complex pattern of cross-shore variations in the water-surface elevations. [From *Marine Geology* 32, L. D. Wright, J. Chappell, B. G. Thom, M. P. Bradshaw and P. Cowell, Morphodynamics of Reflective and Dissipative Beach and Inshore Systems: Southeastern Australia, 1979 with kind permission of Elsevier Science-NL, Sara Burgerhartstratt 25, 1055 KV Amsterdam, The Netherlands.]

a beach, as they are small and would be obscured by the superimposed incident waves. The best chance for direct observations of edge waves is in the swash on the beach face where their amplitudes are greatest while the energy of the incident waves is at its lowest.

Of particular interest is the length of the edge waves in the longshore direction, which can be seen in the graphical depictions of Figures 6-40 and 6-42. For a beach of uniform slope angle β, Ursell (1952) derived the relationship

$$L_e = \frac{gT_e^2}{2\pi} \sin[(2n + 1)\beta] \tag{6.32}$$

which shows that the longshore length as well as the cross-shore amplitude variations depend on the mode number n. The length of the edge wave is also strongly dependent on its period T_e, and to a smaller degree on the beach slope. Because of the presence of the modal number n in the relationship, there are several possible values of L_e for each combination of period and beach slope. Equation (6.32) was derived by Ursell for edge waves on a beach of uniform slope; Eckart (1951) derived a slightly different form of the relationship but again by assuming that the beach has a uniform slope. Ball (1967) analyzed edge-wave motions on concave beach profiles that show better agreement with natural beach profiles. Unfortunately, the resulting equations derived by Ball for describing edge waves are substantially more complex than the uniform-slope solutions such as Equation (6.32). Holman and Bowen (1979) have shown that the use of Equation (6.32) on real beach profiles can lead to serious errors in estimating edge-wave lengths, but they provide a rule-of-thumb analysis for the choice of an effective beach slope that will yield improved evaluations of the wave length from Equation (6.32) or from the Eckart solution. Holman and Bowen further discuss the variations in edge-wave dimensions as tides change the water levels and the effective slope beneath a generally concave profile.

Figure 6-40 depicts the movement of a progressive edge wave along the length of a beach. It is possible that pairs of progressive edge waves can form but moving in opposite directions along the shoreline. If they have the same period and mode number, a pattern

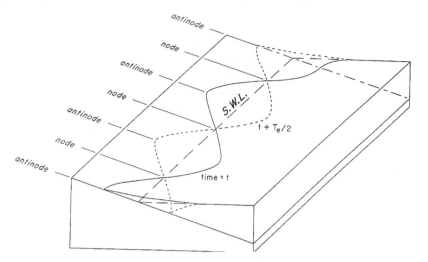

Figure 6-43 The pattern of a standing edge wave of mode $n = 0$, having fixed alternating nodes and antinodes in the longshore direction, respectively, where the amplitude of the edge wave and thus of its run-up are zero (node) and a maximum (antinode).

of *standing edge waves* will develop, as shown in Figure 6-43, characterized by alternating longshore antinodes and nodes where the amplitudes of the water motions are, respectively, a maximum and zero. With such a system, the contribution of the standing edge waves to run-up on the beach would vary systematically in the longshore direction, with the total excursion of the water due to the edge waves being greatest at the antinode positions, with zero contribution at the nodes. Again, this pattern would be largely obscured by the run-up of the incident waves, but documentation of a systematic pattern in the run-up along the length of a beach has been one mechanism for the identification of the existence of edge waves within the nearshore [e.g., Katoh (1981) and Holman and Bowen (1984)]. It is possible that standing edge waves are common in that the generation mechanisms are unlikely to prefer one direction versus the opposite direction along the beach, and therefore would give rise simultaneously to progressive edge waves moving in opposite directions and thus to a pattern of standing edge waves. Furthermore, edge waves can reflect from a headland or jetty extending across the nearshore or from the bounding wall of a laboratory wave basin. This reflection gives rise to progressive edge waves moving in opposite directions and again to a standing pattern of edge waves, as depicted in Figure 6-43. In such an instance, it is necessary that the headland or jetty be at an antinode position within the pattern of standing edge-wave motions, the node positions being characterized by strong longshore components of the orbital water motions, an impossibility where there is a headland or other obstacle.

   If the beach is bounded by headlands, as in the case of a pocket beach, a field of standing edge waves must have antinodes at each end of the beach. This is also the condition in a laboratory wave basin with bounding walls. In this situation, the total length of the beach must be some multiple of half of the edge-wave length, giving rise to the condition

$$L_e = \frac{2\Lambda}{m} \tag{6.33}$$

where Λ is the longshore length of the beach between headlands, and m is the longshore modal number that may have integral values $1, 2, 3, \ldots$. This establishes the potential condition for resonant forcing of edge-wave growth, where the incident wave conditions and beach slope are ideal to generate edge waves of period and a mode number that yields a longshore wave length L_e according to Equation (6.32), which also corresponds to the value given by Equation (6.33) for the fixed distance between headlands. Such an existence of a resonant condition for edge-wave growth was demonstrated in the wave basin studies of Bowen and Inman (1969), undertaken in connection with their investigation to examine the role of edge waves in the formation of rip currents (Chapter 8).

The mechanisms important to the generation of edge waves differ on reflective versus dissipative beaches, and the resulting periods and mode numbers also tend to differ on these two beach types. Guza and Davis (1974) and Guza and Bowen (1975) have analyzed edge wave generation on steep reflective beaches. Important is the reflection of the incident waves to form standing waves in the cross-shore direction, as the edge-wave generation involves a nonlinear transfer from that pattern of standing incident waves to edge waves that propagate in the longshore direction. Of particular significance, the analyses indicate that on reflective beaches, the dominant edge waves formed will have a period exactly twice that of the incident waves ($T_e = 2T_i$) and will be mode $n = 0$; these are termed *subharmonic edge waves*. There is strong evidence that standing subharmonic edge waves are important to the formation of beach cusps, and Equation (6.32) for the edge-wave length allows the prediction of cusp spacings when $T_e = 2T_i$ and $n = 0$ (Chapter 11). On low-sloping dissipative beaches, the generation of edge waves is linked to the modulation or beating of the incident waves, the slow variation of breaker heights resulting from the interaction of different wave trains having somewhat different periods. This is the variation that has given rise to the notion that every seventh wave is the largest (Chapter 5). This mechanism for the generation of edge waves was first proposed by Gallagher (1971) and was later expanded and tested in laboratory experiments by Bowen and Guza (1978). The theory was further expanded by Symonds, Huntley, and Bowen (1982), who focused on the time variations in the width of the surf zone and in the elevation of the set-up in response to wave modulations. These combined studies have established that the incident-wave modulation could, under the right conditions, transfer energy to edge waves having the same period as the modulation, which of course is much longer than the periods of the incident waves, and therefore yields edge waves that have periods in the infragravity range. In this case, unfortunately, the exact periods and modes of the generated edge waves cannot be predicted, and it is possible to simultaneously generate several edge waves of different periods and mode numbers. The theories predict that a continuum of frequencies and modes could be forced within the infragravity range, so the resulting superposition would be extremely complex. This has been confirmed by measurements of the infragravity energy within the surf zone, spectra typically showing several energy peaks whose relative importance depends on where the measurements are obtained in the cross-shore direction [e.g., Holman (1981) and Huntley, Guza, and Thornton (1981)].

An important attribute of edge waves is that the energy is trapped within the nearshore, with minimal leakage to the offshore. This means that even a weak forcing mechanism could result in the progressive buildup of significant levels of energy in the form of edge waves having infragravity periods. This would account for the observed growth of infragravity motions within the inner surf zone, particularly in the energy of the swash (Guza and Thornton, 1982; Holman and Sallenger, 1985). If the incident waves are depth limited so they are energy saturated within the inner surf zone, as depicted in Figure 6-13, then during a storm one would expect to see a significant increase in infragravity edge-wave energy in the swash but with little increase in the swash energy of the incident waves themselves. This is precisely what is depicted in the graph in Figure 6-32, derived from the run-up measurements of Guza and Thornton on a dissipative beach.

Although the infragravity dominance of swash motions suggests that edge waves may be an important energy component within the nearshore, a more conclusive demonstration of their existence and importance proved to be difficult. This proof came about primarily through measurements of their orbital motions with electromagnetic current meters, distinguishing the complex orbital water motions associated with progressive or standing edge waves versus the orbital motions of incident waves (Huntley and Bowen, 1973, 1975; Wright et al., 1979; Huntley et al., 1981; Holman, 1981; Oltman-Shay and Guza, 1987). Spectral analyses of the measured currents in the nearshore can distinguish between the incident-wave orbital velocities and those due to infragravity motions, again on the basis of their different ranges of periods, but more critical for establishing that edge waves are present is documenting that the cross-shore variations in the amplitudes of the motions correspond to those expected from edge waves and that there are the expected phase differences between the longshore-directed and cross-shore velocities.

Historically, the first study to obtain field measurements that firmly established the presence of edge waves on a beach was that of Huntley and Bowen (1973, 1975) on Slapton Beach, South Devon, England. Detection of the presence of edge waves was accomplished with a single electromagnetic current meter; thanks to tidal variations, this one meter obtained measurements from various cross-shore positions during 4 hours of data collection. A representative spectrum derived from the measured cross-shore component of the nearshore currents is shown in Figure 6-44. The principal peak in the spectrum, that containing the most energy, is centered at a 10–sec period, which is twice the period of the incident waves whose peak in the spectrum at 5 sec is less well defined due to the dissipation of the incident waves within the nearshore. From this alone, it would appear that a subharmonic edge wave ($T_e = 2T_i$) was present on this reflective beach at Slapton, in agreement with the theoretical development of Guza and Davis (1974) for subharmonic edge-wave generation on reflective beaches. However, this observation of a subharmonic energy peak in the spectrum, is not sufficient proof that this energy represents an edge wave. More positive identification involved a demonstration that the energy matched the predicted offshore decay of an edge wave; this

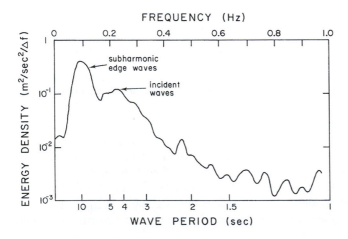

Figure 6-44 A spectrum derived from cross-shore water motions measured with an electromagnetic current meter at Slapton Beach, England. The incident waves are centered at a 5-sec period, but the greatest energy is in the 10-sec peak that was established to be a subharmonic standing edge wave. [Adapted with permission from *Nature*, 243: p. 161. Copyright © 1973 Macmillan Magazines Limited.]

is shown in Figure 6-45, which also established that this is the mode $n = 0$ edge wave, again in agreement with the theoretical predictions of Guza and Davis. Figure 6-45 shows the expected offshore decay of both the cross-shore and longshore components of the currents associated with the edge wave, and further analysis established a zero-phase relationship between the onshore and longshore velocities, which is consistent only with standing edge waves rather than with a progressive edge wave. This combination of observations was clear proof of the existence of a standing subharmonic edge wave, and the size of its spectral peak in Figure 6-44 further indicated that it contained a substantial amount of energy. Even with this amount of energy, Huntley and Bowen (1973, 1975) noted that its contribution to the run-up was hidden by the swash of the incident waves and did not form beach cusps, largely because the incident waves broke at an angle to the shore and generated a strong longshore current that acted to prevent cusp formation (Chapter 11).

The demonstration of the presence of edge waves within the infragravity motions on dissipative beaches proved to be much more difficult. In this instance one could not anticipate

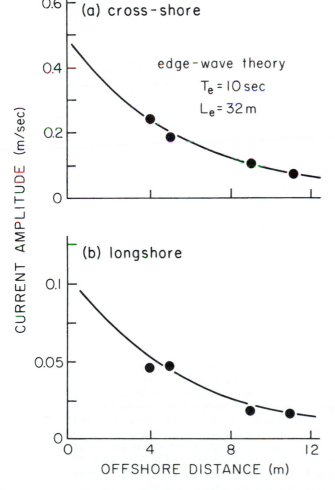

Figure 6-45 Graphs of the along-profile variations of the cross-shore (a) and longshore (b) amplitudes of the 10–sec energy component (Fig. 6-44) as measured with an electromagnetic current meter at Slapton Beach, South Devon, England. The measurements show good agreement with the expected offshore decay of a mode $n = 0$ edge wave having a $L_e = 32$ m wave length. The phase difference between the cross-shore and longshore velocities established that this is a standing subharmonic edge wave. [Adapted with permission from *Nature,* 243, p. 161. Copyright © 1973 Macmillan Magazines Limited.]

the presence of a specific edge wave like the subharmonic, and it is further expected that several edge waves could exist simultaneously, having different periods and modes, so that unraveling their separate existence from the measured water-current velocities would be difficult. A number of early investigators presented data from the infragravity band that were consistent with an edge-wave form but were not entirely conclusive (Holman, 1981; Sasaki, Horikawa, and Hotta 1976; Wright et al., 1979). The difficulty turned out to be the similarity in cross-shore amplitude variations of edge wave modes versus standing incident waves produced by wave reflection from the beach, such as those measured by Suhayda (1974). The analyses by Huntley (1976) of data collected on the dissipative beach within Hell's Mouth Bay, North Wales, England, provided the most convincing case of establishing the existence of infragravity edge waves using data solely from a cross-shore transit. Figure 6-46 shows measurements obtained with three electromagnetic current meters placed at different positions along a beach profile. The measured currents farthest offshore (84 m) are dominated by the on/offshore orbital motions of the incident waves having a 9-sec period. At the 24-m position and especially at the 4-m location closer to shore, the water motions are in the long-period in-

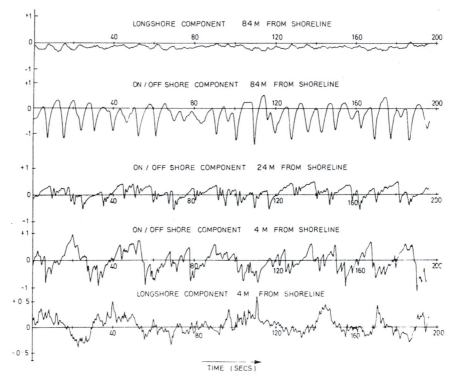

Figure 6-46 Measurements with electromagnetic current meters across the beach profile within Hell's Mouth Bay, North Wales, England. The measurements at the 84–m offshore position are dominated by incident waves of 9 sec, while at the more shoreward positions the water motions are dominated by infragravity motions, which appear in both the longshore and cross-shore components of the velocity measurements. These infragravity motions were identified as being due to a series of progressive edge waves in the period range 20–70 sec. [Used with permission of American Geophysical Union, from D. A. Huntley, Long Period Waves on a Natural Beach, *Journal of Geophysical Research 81*, p. 6444. Copyright © 1976 American Geophysical Union.]

fragravity range and affect both the on/offshore and longshore components of the measured currents. The phase differences between the longshore and cross-shore velocities again suggested that the infragravity motions were due to the presence of progressive edge waves.

The study by Huntley, Guza, and Thornton (1981) at Torrey Pines Beach, California, is generally considered to be the first to provide absolute evidence that infragravity motions can be in the form of edge waves. This could be established only when the longshore variation of the wave form was included in the data collection and analyses. Huntley and co-workers (1981) accomplished this by using an array of nineteen electromagnetic current meters, extending along 500 m of beach length, as well as in the cross-shore direction. Analysis of the immense data set was tedious but necessary to establish the presence of edge waves and to separate them as to periods and mode numbers. Cross-spectra were calculated between all instrument pairs in the longshore array. For each frequency band, the cospectral energy was then plotted versus the longshore lag, and a maximum-likelihood spectrum was then computed to determine the longshore wave length. The result is a two-dimensional wave number versus frequency spectrum, Figure 6-47 being an example. Of significance is that most of the energy lines up along the theoretical dispersion curves based on Equation (6.32), a result that proves that most of the infragravity energy measured in the longshore-directed currents by the array of electromagnetic current meters was in the form of a series of low-mode edge waves. The mode $n = 0$ dominates in the frequency range from about 0.005 Hz to 0.014 Hz (periods 200–70 sec), while mode $n = 1$ dominates above 0.014 Hz ($T < 70$ sec). This separation of the modes into different frequency ranges may be due to their generation by narrow-beam incident waves or may be a more general response of the nearshore zone to edge wave forcing (Huntley, Guza, and Thornton, 1981). In addition to establishing that much of the infragravity energy followed the edge-wave dispersion relationship, Huntley and co-workers (1981) also showed that the cross-shore variations in the amplitudes of the different modes, measured using 10 sensors along a profile, were in good agreement with that predicted by edge-wave theory. This full documentation of agreement with the dispersion relationship of Equation (6.32) and demonstration of the expected cross-shore variations in the amplitudes constituted a full proof of the existence on ocean beaches of edge waves having periods in the 20–300 sec range, and further established that the edge waves may contain significant levels of energy in comparison with the incident waves.

Oltman-Shay and Guza (1987) have undertaken many more analyses like those of Huntley and co-workers (1981), with data collected from two California beaches (Torrey Pines in San Diego and Leadbetter Beach in Santa Barbara). Again, the analyses were based on arrays of electromagnetic current meters extending along and across the beaches. Figure 6-48 is an example of a wave number versus frequency diagram like that in Figure 6-47, but here the energy is separated into eastward- versus westward-directed progressive edge waves (Leadbetter Beach in Santa Barbara has a nearly east-west trending shoreline).

Identification of the infragravity energy associated with edge waves is best established by measurements of the longshore-directed water motions, seen in Figure 6-48(a), while measurements of the cross-shore currents [Fig. 6-48(b)] show less energy that coincides directly with the edge-wave dispersion lines. This is because a greater portion of the longshore-directed energy is in the form of edge waves, while much of the cross-shore energy is in the form of "leaky modes," which can escape from the nearshore rather than be trapped as are edge waves (see Fig. 6-39 for the distinction). Oltman-Shay and Guza found that, on average, low mode ($n < 2$) edge waves constitute 69 percent of the variance in the longshore infragravity velocities, while in the cross-shore, they constitute only 17 percent. In fifteen-days of measurements at the two sites, the longshore infragravity energy in the form of low-mode edge

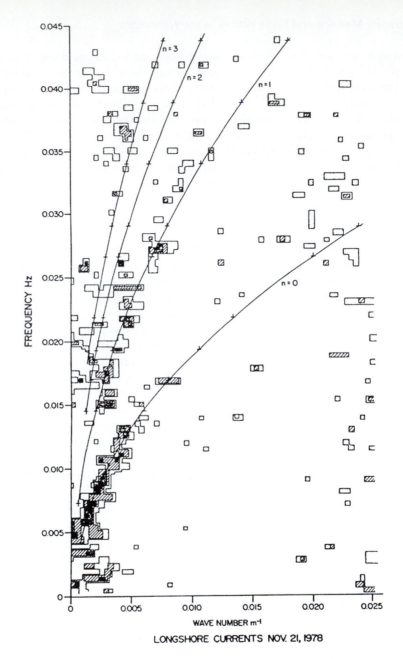

LONGSHORE CURRENTS NOV. 21, 1978

Figure 6-47 A two-dimensional spectrum derived from analyses of current measurements from an array of electromagnetic current meters that extended along 500 m of beach. The squares show where the measured energy existed within the frequency versus longshore wave-length domain. The series of curves are derived from Equation (6.32), the dispersion relationship for edge waves, calculated for the beach slope $\beta = 0.023$. Much of the measured energy is seen to line up with the mode $n = 0$ and 1 dispersion lines, indicating that in excess of 30 percent of the infragravity energy is in the form of progressive edge waves. [Used with permission of American Geophysical Union, from D. A. Huntley, R. T. Guza and D. B. Thornton, Field Observations of Surf Beat, 1. Progressive Edge Waves, *Journal of Geophysical Research 83*, p. 6458. Copyright © 1981 American Geophysical Union.]

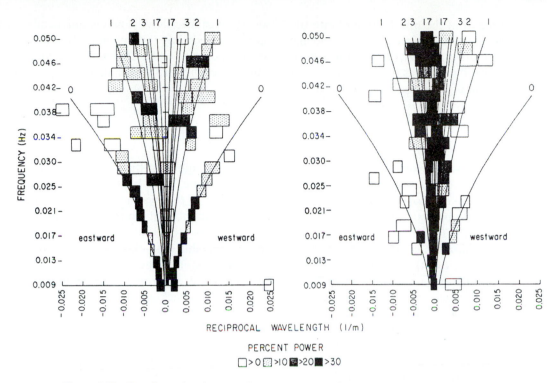

Figure 6-48 Two-dimensional spectra like that of Figure 6-47 but demonstrating that some of the infragravity energy is propagating eastward and part westward along the east-west trending shoreline at Santa Barbara, California. Spectrum (a) is based on measurements of longshore-directed velocities, while spectrum (b) is from cross-shore velocities. The longshore velocities better reveal the presence of mode $n = 0$ and 1 edge waves, while much of the infragravity energy in the cross-shore velocities is in the form of leaky modes that can escape offshore from the beach, not agreeing with the dispersion lines for edge waves. [Used with permission of American Meteorological Society, from J. Oltman-Shay and R. T. Guza, Infragravity Edge Wave Observations on Two California Beaches, *Journal of Physical Oceanography 17*, p. 655. Copyright © 1987 American Meteorological Society.]

waves was found to be in the 42–88 percent range. They also found that there are days when there is a factor 3 asymmetry in the edge-wave energy of a specific mode moving up and down the coast, but the ratio of up and down-coast energy is usually within 1.0 ± 0.2. Simultaneous measurements of the run-up on the beach face established that as much as 50 percent of its variance is contributed by low-mode edge waves. Therefore, the study of Oltman-Shay and Guza firmly established that low-mode edge waves contribute significantly to both water velocities within the nearshore and to the run-up components of the nearshore infragravity motions.

When it was first proposed back in the 1960s that much of the infragravity energy observed within the nearshore might be in the form of edge waves that could appreciably affect the morphology of the beach, some skepticism was voiced. At that time, there had been no direct measurements of edge waves on natural beaches, their possible importance being inferred only from the observed beach morphology, particularly the presence of offshore crescentic bars whose longshore wave lengths require edge waves having periods in the 20–300 sec range, according to the mechanism of bar generation proposed by Bowen and Inman (1971). It has been seen, in the above review, that direct measurements of edge waves did not come until the mid 1970s and into the 1980s. Many studies have now unequivocally proven the existence of edge waves and have demonstrated that they commonly contain large amounts of energy, often dominating over the incident waves. In Chapter 11 we will examine the probable roles of edge waves in affecting the beach morphology, specifically in the generation of offshore bars and beach cusps.

SUMMARY

A variety of surf-zone processes have been examined in this chapter, beginning with wave breaking, which defines the extent of the surf zone. Wave breaking represents the initial dissipation of wave energy in the nearshore, energy that was acquired from storms that may have been thousands of kilometers out to sea. A number of models have been developed to analyze the progressive dissipation of waves as they cross the surf zone, and we saw that the dissipation gives rise to a wave set-up in the nearshore, an elevation of the mean water surface above the still-water level of the sea. During storms, the set-up moves the effective shoreline upward and horizontally toward the land and, accordingly, can be a factor in producing beach erosion and the destruction of properties backing the beach. The other factor important to property erosion is the elevation achieved by the run-up of the wave swash beyond the effective mean shoreline established by the set-up. Distinct differences were found in the run-up on dissipative versus reflective beaches, with the unexpected discovery that on dissipative beaches the swash of the incident waves does not respond to changes in the offshore wave conditions, while there can be a remarkable increase in the long-period infragravity motions within the swash during storms. Later, it was established that much of this infragravity energy occurs in the form of edge waves trapped to the sloping beach. One of the most important areas of research during the past 30 years has focused on the physics of edge-wave generation within the nearshore, on measurements of the unusual properties of these trapped waves, and on their importance to the generation of nearshore currents, to patterns of sediment transport, and the role they play in affecting the nearshore morphology. These latter topics will be covered in subsequent chapters.

An underlying theme of this chapter has been the important distinction between reflective and dissipative beaches (and, as will be seen in Chapter 11, the intermediate stages between those two extreme beach types). This classification was first presented in Chapter 3 in a simplified form, mainly in terms of the contrasting beach slopes. In the present chapter the distinction has been more in terms of the surf-zone processes, there tending to be different types of wave breaking (spilling on dissipative beaches versus plunging to surging on reflective beaches), differences in the patterns of energy dissipation within the surf zone, and decidedly different natures of the run-up with regard to the importance of infragravity motions versus incident waves. Nearly all of these surf-zone properties were related to the dimensionless Iribarren number, which includes the slope of the beach and steepness of the waves. The governing importance of the Iribarren number as a surf-zone similarity parameter has been documented particularly by Battjes (1974), who also included considerations of wave reflection from the beach and the number of waves within the surf zone. In general, the value of the Iribarren number can be used to define a dissipative versus a reflective beach; a low value represents a beach of low slope and steeper waves, yielding a dissipative beach, while a high value of the Iribarren number would be indicative of a reflective beach of high slope and where the waves have low steepness values. Such distinctions will be explored more fully in Chapter 11, where considerations of the differences in nearshore water circulation patterns and three-dimensional beach morphologies are discussed.

PROBLEMS

1. For a beach slope $S = 0.05$, construct a graph of H_∞ versus T for combinations that yield spilling, plunging, and surging breakers, based on the ξ_∞ Iribarren number classification derived by Battjes.

2. For deep-water waves of height $H_\infty = 2$ m and period $T = 12$ sec, use Equations (6.3), (6.4), and (6.5) to calculate, respectively, the corresponding wave breaker heights by solitary wave, Airy wave, and empirical analyses.

3. For a breaking-wave height plus water depth total of $Y_b = 2.5$ m, use the results in Figure 6-5 to calculate water-particle velocities internal to the wave. Consider the forces acting on an individual diving into the face of this breaking wave, in particular, the safety of diving horizontally versus hitting the wave low toward its base.

4. For the breaking wave in Problem 3 (above), how far will the wave move forward toward the shore during breaking before the plunge crest touches down onto the water surface? If the celerity is given by $C_b = \sqrt{g(h_b + H_b)} = \sqrt{gY_b}$, approximately how much time is involved in this process? In that diving into a wave generally takes 1–2 sec, is there any danger in the individual being thrust downward by the plunging wave crest?

5. Because of offshore wave refraction, there is a resulting variation in wave heights along the 250–m length of a beach that in effect ranges from equivalent deep-water heights H_∞ of 1–3 m, while the period is uniform at $T = 13$ sec. Calculate the corresponding longshore variation in the maximum wave set-up, $\bar{\eta}_{max}$ and the longshore pressure gradient of the sloping water surface over this 250-m length, a pressure force that is important in the generation of longshore currents (Chapter 8).

6. Sea-cliff erosion occurred at Brookings, Oregon, during December 1975, when a storm generated deep-water waves having a significant wave height $H_\infty = 5.5$ m and $T = 14$ sec, and the high-tide level was +1.1 m relative to mean sea level (Komar, McDougal, and Ruggiero, 1996). The beach slope at the site is $S = 0.08$. Calculate the maximum set-up, $\bar{\eta}_{max}$, the $R_{2\%}$ run-up elevation above that set-up level, and the total water elevation (tide plus set-up and run-up) achieved during the storm. Compare this total water elevation with the surveyed elevation of 3.5 m for the junction between the beach and the base of the sea cliff, to judge whether cliff erosion would have been severe. Based on projections of wave statistics, it is estimated that the 100-year extreme storm would generate a deep-water significant wave height of 8.6 m and period 18 sec. For the same tide level of +1.1 m MSL, calculate the run-up and total water elevation for this potential event. How susceptible is the sea cliff to erosion during such extreme conditions?

7. Infragravity motions on dissipative beaches typically have periods ranging from 50 to 200 sec. If this energy is in the form of edge waves, what would be the ranges in wave lengths L_e for modes $n = 0$, 1, and 2 on a beach having a slope $S = 0.02$?

8. Incident waves reaching a beach typically range from 3 to 15 sec. For this range, calculate the possible wave lengths of subharmonic edge waves of mode $n = 0$ that might occur on a reflective beach of slope $S = 0.2$. There is evidence that subharmonic edge waves may be responsible for the formation of beach cusps, with two cusps forming within one edge-wave length (Chapter 11). If so, what would be the expected range of beach-cusp spacings in this example?

9. To achieve resonant conditions for the maximum growth of edge waves within a pocket beach of longshore length $\Lambda = 800$ m and slope $S = 0.03$, Equations (6.32) and (6.33) must be obeyed simultaneously. For a mode $n = 1$ infragravity edge wave, what edge-wave periods T_e would be required for resonant conditions for m = 1 through 5?

10. Program Equation (6.31) for the computation of the time variations (i.e., set $y = 0$) in the cross-shore amplitudes of mode $n = 0$ edge waves, with input values for A and T_e and for the beach slope, with L_e computed using Equation (6.32). Run the program with $A = 0.5$ m and $T_e = 24$ sec on a beach slope $S = 0.2$, a reasonable combination for subharmonic edge waves on a reflective beach.

REFERENCES

AAGAARD, T. and J. HOLM (1989). Digitization of Wave Runup Using Video Records. *Journal of Coastal Research* 5: 547–551.

ADEYEMO, M. D. (1970). Velocity Fields in the Wave Breaker Zone. *Proceedings of the 12th Coastal Engineering Conference, Amer. Soc. Civil Engrs.,* pp. 435–460.

AHRENS, J. P. (1981). *Irregular Wave Runup on Smooth Slopes.* Technical Aid No. 81–17, Coastal Engineering Research Center, Waterways Experiment Station, Vicksburg, Mississippi.

BAGNOLD, R. A. (1940). Beach Formation by Waves: Some Model Experiments in a Wave Tank. *Journal of the Institute of Civil Engineers* 15: 27–52.

BALL, F. K. (1967). Edge Waves in an Ocean of Finite Depth. *Deep Sea Research* 14: 79–88.

BATTJES, J. A. (1971). Runup Distributions of Waves Breaking on Slopes. *Journal of Waterways, Harbors, and Coastal Engineering Division, Amer. Soc. Civil Engrs.* 97(WW1): 91–114.

BATTJES, J. A. (1974). Surf Similarity. *Proceedings of the 14th Coastal Engineering Conference, Amer. Soc. Civil Engrs.,* pp. 466–479.

BATTJES, J. A. and J. P. F. M. JANSSEN (1978). Energy Loss and Set-Up Due to Breaking of Random Waves. *Proceedings of the 16th Coastal Engineering Conference, Amer. Soc. Civil Engrs.,* pp. 569–587.

BATTJES, J. A. and M. J. F. STIVE (1985). Calibration and Verification of a Dissipation Model for Random Breaking Waves. *Journal of Geophysical Research* 90: 9159–9167.

BENJAMIN, T. B. and M. J. LIGHTHILL (1954). On Cnoidal Waves and Bores. *Proceedings of the Royal Society of London,* Series A, 224: 448–460.

BIESEL, F. (1951). Study of Wave Propagation in Water of Gradually Varying Depth. *Gravity Waves,* pp. 243–253. Circular of the National Bureau of Standards, Gaithersburg, Maryland.

BOWEN, A. J. (1969). Rip Currents, 1: Theoretical Investigations. *Journal of Geophysical Research* 74: 5468–5478.

BOWEN, A. J., D. L. INMAN, and V. P. SIMMONS (1968). Wave "Set-Down" and Wave Set-Up. *Journal of Geophysical Research* 73: 2569–2577.

BOWEN, A. J. and D. L. INMAN (1969). Rip Currents, 2: Laboratory and Field Observations. *Journal of Geophysical Research* 74: 5479–5490.

BOWEN, A. J. and D. L. INMAN (1971). Edge Waves and Crescentic Bars. *Journal of Geophysical Research* 76: 8662–8671.

BOWEN, A. J. and R. T. GUZA (1978). Edge Waves and Surf Beat. *Journal of Geophysical Research* 83(C4): 1913–1920.

BROOME, R. and P. D. KOMAR (1979). Undular Hydraulic Jumps and the Formation of Backwash Ripples on Beaches. *Sedimentology* 26: 543–559.

CHESTER, W. (1966). A Model of the Undular Bore on a Viscous Fluid. *Journal of Fluid Mechanics* 24: 367–377.

COKELET, E. D. (1977). Steep Gravity Waves in Water of Arbitrary Uniform Depth. *Philosophical Transactions of the Royal Society of London,* Series A, 286: 183–230.

DALLY, W. R. (1990). Random Breaking Waves: A Closed-Form Solution for Planar Beaches. *Coastal Engineering* 14: 233–263.

DALLY, W. R. (1992). Random Breaking Waves: Field Verification of a Wave-by-Wave Algorithm for Engineering Application. *Coastal Engineering* 16: 369–397.

DALLY, W. R. R. G. DEAN, and R. A. DALRYMPLE (1985). Wave Height Variation Across Beaches of Arbitrary Profile. *Journal of Geophysical Research* 90: 11,917–11,927.

DALLY, W. R. and R. G. DEAN (1986). Transformation of Random Breaking Waves on Surf Beat. *Proceedings of the 20th Coastal Engineering Conference, Amer. Soc. Civil Engrs.,* pp. 109–123.

DIVOKY, D., B. LE MÉHAUTÉ and A. LIN (1970). Breaking Waves on Gentle Slopes. *Journal of Geophysical Research* 75: 1681–1692.

DYHR-NIELSEN, M. and T. SØRENSEN (1970). Sand Transport Phenomena on Coasts with Bars. *Proceedings of the 12th Coastal Engineering Conference, Amer. Soc. Civil Engrs.,* pp. 855–866.

EBERSOLE, B. A., M. A. CIALONE, and M. D. PRATER (1986). *Regional Coastal Processes Numerical Modeling System—Report 1, RCPWAVE-A Linear Wave Propagation Model for Engineering Use.* Technical Report CERC-86–4, Coastal Engineering Research Center, U.S. Army Corps of Engineers.

ECKART, C. (1951). *Surface Waves on Water of Variable Depth.* Scripps Institution of Oceanography, Wave Report 100, Reference 51–12.

GALLAGHER, B. (1971). Generation of Surf Beat by Nonlinear Wave Interactions. *Journal of Fluid Mechanics* 49: 1–20.

GALVIN, C. J. (1968). Breaker Type Classification on Three Laboratory Beaches. *Journal of Geophysical Research* 73: 3651–3659.

GALVIN, C. J. (1969). Breaker Travel and Choice of Design Wave Height. *Journal of the Waterways and Harbors Division, Amer. Soc. Civil Engrs.* 95: 175–200.

GAUGHAN, M. K. and P. D. KOMAR (1975). The Theory of Wave Propagation in Water of Gradually Varying Depth and the Prediction of Breaker Type and Height. *Journal of Geophysical Research* 80: 2991–2996.

GREATED, C. A., D. J. SKYNER, and T. BRUCE (1992). Particle Image Velocimetry (PIV) in the Coastal Engineering Laboratory. *Proceedings of the 23rd Coastal Engineering Conference, Amer. Soc. Civil Engrs.,* pp. 212–225.

GREENWOOD, B. and P. D. OSBORNE (1990). Vertical and Horizontal Structure in Cross-Shore Flows. An Example of Undertow and Wave Set-Up on a Barred Beach. *Coastal Engineering* 14: 543–580.

GRÜNE, J. (1982). Wave Run-Up Caused by Natural Storm Surge. *Proceedings of the 18th Coastal Engineering Conference, Amer. Soc. Civil Engrs.,* pp. 785–803.

GUZA, R. T. and R. E. DAVIS (1974). Excitation of Edge Waves by Waves Incident on a Beach. *Journal of Geophysical Research* 79: 1285–1291.

GUZA, R. T. and A. J. BOWEN (1975). The Resonant Instabilities of Long Waves Obliquely Incident on a Beach. *Journal of Geophysical Research* 80: 4529–4534.

GUZA, R. T. and E. B. THORNTON (1981). Wave Set-Up on a Natural Beach. *Journal of Geophysical Research* 86(C5): 4133–4137.

GUZA, R. T. and E. B. THORNTON (1982). Swash Oscillations on a Natural Beach. *Journal of Geophysical Research* 87(C1): 483–491.

HANSEN, J. B. and I. A. SVENDSEN (1984). A Theoretical and Experimental Study of Undertow. *Proceedings of 19th Coastal Engineering Conference, Amer. Soc. Civil Engrs.,* pp. 2246–2262.

HANSLOW, D. and P. NIELSEN (1993). Shoreline Set-Up On Natural Beaches. *Journal of Coastal Research* 15: 1–10.

HEDGES, T. S. and M. S. KIRKGOZ (1981). An Experimental Study of the Transformation Zone of Plunging Breakers. *Coastal Engineering* 4: 319–333.

HIBBERD, S. and D. H. PEREGRINE (1979). Surf and Run-Up on a Beach. *Journal of Fluid Mechanics* 95: 323–345.

HOLLAND, K. T. and R. A. HOLMAN (1993). The Statistical Distribution of Swash Maxima on Natural Beaches. *Journal of Geophysical Research* 98(C6): 10,271–10,278.

HOLMAN, R. A. (1981). Infragravity Energy in the Surf Zone. *Journal of Geophysical Research* 84: 6442–6450.

HOLMAN, R. A. (1983). Edge Waves and the Configuration of the Shoreline. In *CRC Handbook of Coastal Processes and Erosion,* P. D. Komar (editor), pp. 21–34, Boca Raton, FL: CRC Press.

HOLMAN, R. A. (1986). Extreme Value Statistics for Wave Run-Up on a Natural Beach. *Coastal Engineering* 9: 527–544.

HOLMAN, R. A. and A. J. BOWEN (1979). Edge Waves on Complex Beach Profiles. *Journal of Geophysical Research* 84(C10): 6339–6346.

HOLMAN, R. A. and A. J. BOWEN (1984). Longshore Structure of Infragravity Wave Motions. *Journal of Geophysical Research* 89: 6446–6452.

HOLMAN, R. A. and R. T. GUZA (1984). Measuring Run-Up on a Natural Beach. *Coastal Engineering* 8: 129–140.

HOLMAN, R. A. and A. H. SALLENGER (1985). Set-Up and Swash on a Natural Beach. *Journal of Geophysical Research* 90(C1): 945–953.

HORIKAWA, K. and C.-T. KUO (1966). A Study of Wave Transformation Inside the Surf Zone. *Proceedings of the 10th Coastal Engineering Conference, Amer. Soc. Civil Engrs.*, pp. 217–233.

HOTTA, S. and M. MIZUGUCHI (1980). A Field Study of Waves in the Surf Zone. *Coastal Engineering in Japan* 23: 79–89.

HUGHES, M. G. (1992). Application of a Non-Linear Shallow Water Theory to Swash Following Bore Collapse on a Sandy Beach. *Journal of Coastal Research* 8: 562–578.

HUNT, I. A. (1959). Design of Seawalls and Breakwaters. *Journal of the Waterways and Harbors Division, Amer. Soc. Civil Engrs.* 85(WW3): 123–152.

HUNTLEY, D. A. (1976). Long Period Waves on a Natural Beach. *Journal of Geophysical Research* 81(36): 6441–6449.

HUNTLEY, D. A. and A. J. BOWEN (1973). Field Observations of Edge Waves. *Nature* 243: 160–161.

HUNTLEY, D. A. and A. J. BOWEN (1975). Field Observations of Edge Waves and Their Effect on Beach Material. *Journal of the Geological Society of London* 131: 68–81.

HUNTLEY, D. A., R. T. GUZA, and D. B. THORNTON (1981). Field Observations of Surf Beat, 1: Progressive Edge Waves. *Journal of Geophysical Research* 83: 1913–1920.

IPPEN, A. T. and G. KULIN (1954). The Shoaling and Breaking of the Solitary Wave. *Proceedings of the 5th Coastal Engineering Conference, Amer. Soc. Civil Engrs.*, pp. 27–49.

IVERSEN, H. W. (1952). Studies of Wave Transformation in Shoaling Water, Including Breaking. In *Gravity Waves*. National Bureau of Standards Circular No. 521, pp. 9–32.

JAMES, I. D. (1974). Nonlinear Waves in the Nearshore Region: Shoaling and Set-Up. *Estuarine and Coastal Marine Sciences* 2: 207–234.

KAMINSKY, G. and N. C. KRAUS (1993). Evaluation of Depth-Limited Wave Breaking Criteria. *Waves '93, Amer. Soc. Civil Engrs.*, pp. 180–193.

KATOH, K. (1981). Analysis of Edge Waves by Means of Empirical Eigenfunctions. *Report of the Port and Harbor Institute* 20: 5–51.

KELLER, H. B., D. A. LEVINE, and G. B. WHITAM (1960). Motion of a Bore over a Sloping Beach. *Journal of Fluid Mechanics* 7: 302–316.

KISHI, T. and H. SAEKI (1966). The Shoaling, Breaking and Runup of the Solitary Wave on Impermeable Rough Slopes. *Proceedings of the 10th Coastal Engineering Conference, Amer. Soc. Civil Engrs.*, pp. 322–348.

KOBAYASHI, N., G. S. DESILVA, and K. D. WATSON (1989). Wave Transformation and Swash Oscillation on Gentle and Steep Slopes. *Journal of Geophysical Research* 94: 951–966.

KOBAYASHI, D. T. COX, and N., A. WURJANTO (1991). Permeability Effects on Irregular Wave Runup and Reflection. *Journal of Coastal Research* 7: 127–136.

KOMAR, P. D. and M. K. GAUGHAN (1972). Airy Wave Theory and Breaker Height Prediction. *Proceedings of the 13th Coastal Engineering Conference, Amer. Soc. Civil Engrs.*, pp. 405–418.

KOMAR, P. D., W. G. MCDOUGAL, and P. RUGGIERO (1996). Beach Erosion at Brookings, Oregon—Causes and Mitigation. *Shore & Beach* 64: 17–27.

LARSON, M. (1995). Model for Decay of Random Waves in Surf Zone. *Journal of Waterway, Port, Coastal, and Ocean Engineering, Amer. Soc. Civil Engrs.* 121: 1–12.

LARSON, M. and T. SUNAMURA (1993). Laboratory Experiment on Flow Characteristics at a Beach Step. *Journal of Sedimentary Petrology* 63: 495–500.

LEMÉHAUTÉ, B. and R. C. Y. KOH (1967). On the Breaking of Waves Arriving At an Angle to the Shore. *Journal of Hydraulic Research* 5: 541–549.

LIPPMANN, T. C. and R. A. HOLMAN (1991). Phase Speed and Angle of Breaking Waves Measured with Video Techniques. *Coastal Sediments '91, Amer. Soc. Civil Engrs.*, pp. 542–556.

LONGUET-HIGGINS, M. S. (1952). On the Joint Distribution of the Periods and Amplitudes of Sea Waves. *Journal of Geophysical Research* 80: 2688–2694.

LONGUET-HIGGINS, M. S. and R. W. STEWART (1963). A Note on Wave Set-Up. *Journal of Marine Research* 75: 4–10.

LONGUET-HIGGINS, M. S. and R. W. STEWART (1964). Radiation Stress in Water Waves, a Physical Discussion with Application. *Deep-Sea Research* 11: 529–563.

LONGUET-HIGGINS, M. S. and E. D. COKELET (1976). The Deformation of Steep Surface Waves in Water: I. A Numerical Method of Computation. *Proceedings of the Royal Society of London,* Series A, 350: 1–26.

MATSUNAGA, N. and H. HONJI (1980). The Backwash Vortex. *Journal of Fluid Mechanics* 99: 813–815.

MATSUNAGA, N. and H. HONJI (1983). The Steady and Unsteady Backwash Vortices. *Journal of Fluid Mechanics* 135: 189–197.

MATSUNAGA, N., K. TAKEHARA, and Y. AWAYA (1988). Coherent Eddies Induced by Breakers on a Sloping Beach. *Proceedings of the 21st Coastal Engineering Conference, Amer. Soc. Civil Engrs.*, pp. 234–245.

MAYER, R. H. and D. L. KRIEBEL (1994). Wave Runup on Composite-Slope and Concave Beaches. *Proceedings of the 24th Conference on Coastal Engineering, Amer. Soc. Civil Engrs.*, pp. 2325–2339.

MILLER, R. L. and J. M. ZEIGLER (1958). A Model Relating Dynamics and Sediment Pattern in Equilibrium in the Region of Shoaling Waves, Breaker Zone, and Foreshore. *Journal of Geology* 66: 417–441.

MILLER, R. L. and J. M. ZEIGLER (1964). The Internal Velocity Field in Breaking Waves. *Proceedings of the 9th Coastal Engineering Conference, Amer. Soc. Civil Engrs.*, pp. 103–122.

MIZUGUCHI, M. (1980). A Heuristic Model of Wave Height Distribution in Surf Zone. *Proceedings of the 17th Coastal Engineering Conference, Amer. Soc. Civil Engrs.*, pp. 278–289.

MIZUGUCHI, M. (1986). Experimental Study on Kinematics and Dynamics of Wave Breaking. *Proceedings of the 20th Coastal Engineering Conference, Amer. Soc. Civil Engrs.*, pp. 589–603.

MUNK, W. H. (1949). The Solitary Wave Theory and Its Applications to Surf Problems. *New York Academy of Science Annals* 51: 376–401.

NIELSEN, P. (1988). Wave Set-Up: A Field Study. *Journal of Geophysical Research* 93(C3): 15,643–15,652.

NIELSEN, P. and D. J. HANSLOW (1991). Wave Runup Distributions on Natural Beaches. *Journal of Coastal Research* 7: 1139–1152.

OKAYASU, A., T. SHIBAYAMA, and N. MIMURA (1986). Velocity Field Under Plunging Waves. *Proceedings of the 20th Coastal Engineering Conference, Amer. Soc. Civil Engrs.*, pp. 660–674.

OKAZAKI, S. and T. SUNAMURA (1991). Re-Examination of Breaker-Type Classification on Uniformly Inclined Laboratory Beaches. *Journal of Coastal Research* 7: 559–564.

OLTMAN-SHAY, J. and R. T. GUZA (1987). Infragravity Edge Wave Observations on Two California Beaches. *Journal of Physical Oceanography* 17: 644–663.

PEREGRINE, D. H. (1966). Calculations of the Development of an Undual Bore. *Journal of Fluid Mechanics* 25: 321–330.

RAUBENHEIMER, B., R. T. GUZA, S. ELGAR, and N. KOBAYASHI (1995). Swash on a Gently Sloping Beach. *Journal of Geophysical Research* 100(C5): 8751–8760.

ROOS, A. and J. A. BATTJES (1976). Characteristics of Flow in Runup from Periodic Waves. *Proceedings of the 15th Coastal Engineering Conference, Amer. Soc. Civil Engrs.*, pp. 781–795.

RUGGIERO, P., P. D. KOMAR, W. G. MCDOUGAL, and R. A. BEACH (1996). Extreme Water Levels, Wave Runup and Coastal Erosion. *Proceedings of the 25th Coastal Engineering Conference, Amer. Soc. Civil Engrs.*, pp. 2793–2805.

SASAKI, T., K. HORIKAWA, and S. HOTTA (1976). Nearshore Currents on a Gently Sloping Beach. *Proceedings of the 15th Coastal Engineering Conference, Amer. Soc. Civil Engrs.,* pp. 626–644.

SAVILLE, T. (1957). Scale Effects in Two-Dimensional Beach Studies. *Proceedings of the 7th Meeting of the International Association of Hydraulic Research,* July 24–31, 1957, Lisbon, Portugal.

SAVILLE, T. (1958). Wave Run-Up on Composite Slopes. *Proceedings of the 6th Coastal Engineering Conference, Amer. Soc. Civil Engrs.,* pp. 691–699.

SAVILLE, T. (1961). Experimental Determination of Wave Set-Up. *Proceedings of the 2nd Technical Conference on Hurricanes,* Beach Erosion Board.

SAWARAGI, T. and K. IWATA (1974). On Wave Deformation after Breaking. *Proceedings of the 14th Coastal Engineering Conference, Amer. Soc. Civil Engrs.,* pp. 481–498.

SHEN, M. C. and R. E. MEYER (1963a). Climb of a Bore on a Beach, 2. Nonuniform Beach Slope, 3. Run-Up. *Journal of Fluid Mechanics* 16: 108–112.

SHEN, M. C. and R. E. MEYER (1963b). Climb of a Bore on a Beach, 3. Run-Up. *Journal of Fluid Mechanics* 16: 113–125.

SMITH, E. R. and N. C. KRAUS (1991). Laboratory Study of Wave-Breaking over Bars and Artificial Reefs. *Journal of Waterways, Port, Coastal, and Ocean Engineering, Amer. Soc. Civil Engrs.* 117: 307–325.

SMITH, J. M., I. A. SVENDSEN, and U. PUTREVU (1992). Vertical Structure of the Nearshore Current at DELILAH: Measured and Modeled. *Proceedings of the 23rd Coastal Engineering Conference, Amer. Soc. Civil Engrs.,* pp. 2825–2838.

STIVE, M. J. F. (1980). VELOCITY AND PRESSURE FIELD OF SPILLING BREAKERS. *Proceedings of the 17th Coastal Engineering Conference, Amer. Soc. Civil Engrs.,* pp. 547–566.

STIVE, M. J. F. and H. J. DE VRIEND (1987). Quasi-3D Nearshore Current Modelling: Wave-Induced Secondary Current. *Proceedings of the Coastal Hydrodynamics, Amer. Soc. Civil Engrs.,* pp. 356–370.

STIVE, M. J. F. and H. G. WIND (1982). A Study of Radiation Stress and Set-Up in the Nearshore Region. *Coastal Engineering* 6: 1–25.

STIVE, M. J. F. and H. G. WIND (1986). Cross-Shore Mean Flow in the Surf Zone. *Coastal Engineering* 10: 325–340.

STOKER, J. J. (1948). The Formation of Breakers and Bores. *Communications Applied Mathematics* 1: 1–87.

SUHAYDA, J. H. (1974). Standing Waves on Beaches. *Journal of Geophysical Research* 79: 3065–3071.

SUHAYDA, J. N. and N. R. PETTIGREW (1977). Observations of Wave Height and Wave Celerity in the Surf Zone. *Journal of Geophysical Research* 82: 1419–1424.

SVENDSEN, I. A. (1984a). Wave Heights and Set-Up in a Surf Zone. *Coastal Engineering* 8: 303–329.

SVENDSEN, I. A. (1984b). Mass Flux and Undertow in a Surf Zone. *Coastal Engineering* 8: 347–365.

SVENDSEN, I. A., P. A. MADSEN, and J. B. HANSEN (1978). Wave Characteristics in the Surf Zone. *Proceedings of the 16th Coastal Engineering Conference, Amer. Soc. Civil Engrs.,* pp. 520–539.

SVENDSEN, I. A. and J. B. HANSEN (1988). Cross-Shore Currents in Surf-Zone Modelling. *Coastal Engineering* 12: 23–42.

SYMONDS, G., D. A. HUNTLEY, and A. J. BOWEN (1982). Two-Dimensional Surf Beat: Long Wave Generation by a Time-Varying Breakpoint. *Journal of Geophysical Research* 87(C7): 492–498.

THORNTON, E. B. and R. T. GUZA (1982). Energy Saturation and Phase Speeds Measured on Natural Beaches. *Journal of Geophysical Research* 87: 9499–9508.

THORNTON, E. B. and R. T. GUZA (1983). Transformation of Wave Height Distribution. *Journal of Geophysical Research* 88: 5925–5938.

URSELL, F. (1952). Edge Waves on a Sloping Beach. *Proceedings of the Royal Society of London,* Series A, 214: 79–97.

VAN DER MEER, J. W. and C.-J.M. STAM (1992). Wave Runup on Smooth and Rock Slopes of Coastal Structures. *Journal of Waterways, Port, Coastal, and Ocean Engineering* 118: 534–550.

VAN DORN, W. G. (1976). Set-Up and Run-Up in Shoaling Breakers. *Proceedings of the 15th Coastal Engineering Conference, Amer. Soc. Civil Engrs.,* pp. 738–751.

WASSING, F. (1957). Model Investigations of Wave Run-Up on Dikes Carried Out in the Netherlands During the Past Twenty Years. *Proceedings of the 6th Coastal Engineering Conference, Amer. Soc. Civil Engrs.,* pp. 700–714.

WEGGEL, J. R. (1972). Maximum Breaker Height. *Journal of the Waterways, Harbors, and Coastal Engineering Division, Amer. Soc. Civil Engrs.* 98: 529–548.

WEISHAR, L. L. and R. J. BYRNE (1978). Field Study of Breaking Wave Characteristics. *Proceedings of the 16th Coastal Engineering Conference, Amer. Soc. Civil Engrs.,* pp. 487–506.

WRIGHT, P. (1976). A Cine-Camera Technique for Process Measurement on a Ridge and Runnel Beach. *Sedimentology* 23: 705–712.

WRIGHT, L. D. and A. D. SHORT (1983). Morphdynamics of Beaches and Surf Zones in Australia. In *CRC Handbook of Coastal Processes and Erosion,* P. D. Komar (editor). Pp. 35–64. Boca Raton, FL: CRC Press.

WRIGHT, L. D., J. CHAPPELL, B. G. THOM, M. P. BRADSHAW, and P. COWELL (1979). Morphodynamics of Reflective and Dissipative Beach and Inshore Systems: Southeastern Australia. *Marine Geology* 32: 105–140.

7

Beach Profiles and Cross-Shore Sediment Transport

Although nature begins with the cause and ends with the experience, we must follow the opposite course, namely, begin with the experience and by the means of it investigate the cause.

Leonardo da Vinci, 1452–1519
Notebooks

An important aspect of a beach is its dynamic personality—the loose granular sediment continuously responds to the ever-changing waves and nearshore currents. The beach may achieve an equilibrium morphology in a laboratory wave tank where waves of constant height and period are maintained, but on natural beaches the changing wave and tide conditions give rise to an elusive equilibrium that the beach attempts to achieve but seldom does. However, the best way to understand beach profiles is in terms of a quasi-equilibrium and how it is determined by the waves, tides, and beach sediment. Ultimately, a full understanding depends on our knowledge of the processes of cross-shore sediment transport and on our ability to model the changing beach morphology using that knowledge.

In this chapter we will examine what is known about beach profiles: How sand shifts from the dry beach to the offshore and back again; why gravel beaches are steeper than sand beaches; what governs the number of offshore bars; and what are the effects of tides or a storm surge. Models and experimental studies of cross-shore sediment transport will also be reviewed in an attempt to understand the processes involved in beach profile development. For the most part, this chapter takes a two-dimensional view of the beach, the simple profile and the cross-shore sediment transport responsible for profile shifts in response to changing wave conditions. Although there will be some consideration of the third dimension of the beach—longshore variations in profile shapes—these will be considered in more detail in Chapter 11, where we examine rhythmic shoreline features such as crescentic bars and beach cusps.

The beach profile is important in that it can be viewed as a natural mechanism that causes waves to break and dissipate their energy. Waves can reach the coast with a tremendous amount of power, having the potential to bring havoc to communities. Faced with increased waves, the beach responds by reducing its overall slope, shifting the breaker zone farther offshore, thereby enhancing the dissipation of the waves before they reach the shore. This flattening of the beach under storm attack must not be confused with erosion, which implies a permanent loss of foreshore material. Instead, with such an altered slope, the beach can better act as a buffer, protecting sea cliffs and coastal properties from the intense wave attack and thereby preventing true erosion. The ability of a beach to adjust itself to the prevailing forces makes it an effective method of coastal defense.

THE MEASUREMENT OF BEACH PROFILES

Measuring beach profiles is more challenging than it might at first seem. This is particularly the case if the profile is to extend out through the breaker zone and into the deep water of the offshore. A variety of approaches have been taken to obtain complete profiles across the nearshore zone.

The measurement of most beach profiles involves standard practices of surveying with a rod and transit. This is a routine operation, except perhaps for the person holding the rod who must wade out into the breakers—the approach is obviously limited to the depth to which a tall student can safely wade. A low-tech variation involves use of the so-called "Emery boards," first devised by K. O. Emery (1961). As depicted in Figure 7-1, the apparatus consists of two flat boards connected by a rope having a standard length such as 5 or 10 m, which determines the measurement interval along the profile. Each board has a scale running its length, with 0 at the top. If the beach is sloping downward toward the sea, the procedure is to sight across the top of the seaward board to the level of the horizon, thereby determining the distance **a** from the top of the landward board (Fig. 7-1). If the beach is locally sloping upward in the offshore direction, then **a** is measured on the seaward board and the sighting is with the horizon over the top of the landward board. In either case, the measured distance **a** is equal to the distance **b** that the beach has either dropped or risen within the horizontal distance between the boards (the rope length). Although this approach may seem crude, it can provide a reasonably accurate measure of profile elevations. Its advantages are that the equipment is inexpensive, light, and easily carried to distant survey sites, and the profiles can be obtained very quickly. Some studies have required the collection of a large number of profiles along the length of a beach, all obtained during the hour close to low tide; Emery boards make this possible, as their use is significantly faster than employing a rod and transit.

The portion of the profile in water depths too deep to wade in can be measured with a fathometer on a boat. However, there is usually a considerable degree of inaccuracy in the measured depths due to the combined effects of waves, tides, and other factors that may alter water levels. There are additional problems in accurately establishing the horizontal position of the boat at all times during the profiling operation.

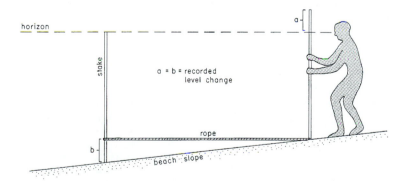

Figure 7-1 The "Emery board" apparatus for measuring beach profiles, consisting of two wooden boards marked from the top with rulers to measure the distance **a,** which is equal to the beach elevation change **b.** The length of the rope controls the measurement interval.

Some success has been achieved in developing vehicles that can obtain beach profiles that extend across the entire nearshore and into deep water. Seymour, Higgins, and Bothman (1978) and Dally, Johnson, and Osiecki (1994) have designed remotely controlled tractors. Figure 7-2 shows the Surf Rover of Dally and co-workers, which can be folded up and transported by truck. The tractors are run by electric motors with onshore sources of power delivered through umbilical cords. The horizontal position as well as vertical profiles are obtained from on-board instrumentation, all self-contained under water, and the data are immediately transmitted back to shore through the cable. Sallenger et al. (1983) have developed a sled that is towed across the surf zone with a winch and system of lines (Fig. 7-3). The sled is large and massive—5.5 m long, 4 m wide, and weighs approximately 800 kg (1800 lbs) including ballast. With this mass and size, the sled can be used under high wave conditions. The triangular line system makes it possible to tow the sled with the winch in either the onshore

Figure 7-2   The amphibious Surf Rover that can be used to survey the morphology of a beach. The vehicle is 7-m long, 5-m wide, and 1.5-m high, and its nominal dry weight is 1,360 kg (3,000 tons). [Used with permission of Marine Technology Society, from W. R. Dally, M. A. Johnson and D. A. Osiecki, Initial Development of an Amphibious ROV for Use in Big Surf, *Marine Technology Society Journal 28.* Copyright © 1994 Marine Technology Society.]

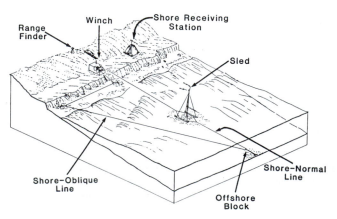

Figure 7-3   The use of a massive sled to measure beach profiles, the system of lines allowing it to be moved in either direction across the beach. [Reprinted from *Marine Geology* 51, A. H. Sallenger, P. C. Howard, C. H. Fletcher, and P. A. Howd, A System For Measuring Bottom Profile, Waves and Currents In the High-energy Nearshore Environment, p. 65, 1983 with kind permission of Elsevier Science-NL, Sara Burgerhartstratt 25, 1055 KV Amsterdam, The Netherlands.]

or offshore directions. The sled contains a steel mast with three optical prisms mounted at the top. An infrared range finder on the dry beach measures the slope distance to these prisms, which yields the horizontal and vertical distances in the cross-shore profile.

An innovative vehicle for measuring beach profiles is the Coastal Research Amphibious Buggy (CRAB) used at the Field Research Facility (FRF), Duck, North Carolina (Birkemeier and Mason, 1978). The CRAB, (Fig. 7-4) is a 10.6-m high motorized tripod that can survey the surf zone out to depths of 9 m. Power is supplied by a Volkswagen engine, and its top speed is 2.5 mph (4 km/hr) on land and somewhat less in water. Although it looks top heavy, the liquid-filled tires and wide wheel base provide excellent stability (the design overturning angle is 45°). Surveying utilizes an automated tracking device that aims at a prism cluster mounted on the CRAB and then measures the distance to the CRAB together with the horizontal and vertical angles; a micro-processor calculates the X, Y, and Z coordinates and stores them in memory. Based on repetitive measurements of the same profile, the vertical and horizontal accuracy are placed at ±3 cm. The operation requires only two people (vehicle and

Figure 7-4 The CRAB, consisting of a 10.6-m tall tripod run with a Volkswagen engine, used to measure beach profiles at the Field Research Facility of the Coastal Engineering Research Center at Duck, North Carolina. [Courtesy of R. Holman]

instrument operators), a considerable savings in personnel and time compared with standard surveying techniques. The only significant drawback of the CRAB is its lack of portability.

An unusual "remote sensing" technique has been developed by Holman and Lippmann (1986) and Lippmann and Holman (1989) to investigate beach profiles, in particular, the locations and forms of offshore bars. The technique is based on the patterns of wave breaking in the nearshore, which are determined mainly by the shallower water of the bars. As illustrated in Figure 7-5(a), a snapshot of wave breaking at some instant in time is suggestive of the presence of a sand bar, but the poor spatial coverage provided by the breaking wave crests and the statistical uncertainty associated with time variations in the sizes of the waves make the details of the bar unclear. However, a 10-minute time exposure [Figure 7-5(b)] yields a much clearer view of the pattern of wave dissipation and hence of the bar form. The long exposure time averages out the fluctuations of the individual waves and gives a statistically stable image of the wave dissipation and bar topography. The record for such an analysis can be obtained with an ordinary camera set for a time exposure, but more advanced techniques utilized by Holman and Lippmann employ a record from a video camera that is analyzed with a computer. A time exposure is obtained by summing successive images, and it is also possible to use differencing (the subtraction of consecutive frames) to remove residual foam from the images.

Figure 7-5 Oblique video images of the nearshore at the Field Research Facility, Duck, North Carolina. (a) A snapshot shows the instantaneous patterns of wave breaking that are suggestive of the beach bathymetry, particularly the presence of an offshore bar. (b) A 10-minute time exposure yields a white band in the offshore from many occurrences of wave breaking, showing the outlines of the bar and a zone of wave dissipation due to swash on the beach face. [Used with permission of American Geophysical Union, from T. C. Lippmann and R. A. Holman, Quantification of Sand Bar Morphology: A Video Technique Based on Wave Dissipation, *Journal of Geophysical Research 94*, p. 996. Copyright © 1989 American Geophysical Union.]

Lippmann and Holman (1989) tested their imaging technique at the FRF in Duck, North Carolina. In that specific experiment, a Panasonic black-and-white video camera was mounted on top of a 40-m high tower, erected on the dune crest immediately behind the beach. The recorded wave dissipation due to breaking was used to infer the bar position and form, which was then compared with actual beach profiles measured with the CRAB. An example is shown in Figure 7-6 where the intensity values determined from the video are non-dimensional with absolute magnitudes that reflect the bathymetry of the beach. There is a clear maximum in the intensity distribution that corresponds to the location and general form of the bar. A second maximum in the intensity occurs on the beach face due to the swash turbulence; the position of the peak intensity was found to coincide with the shoreline, defined as the intersection of the still-water level with the beach face. Figure 7-7 demonstrates

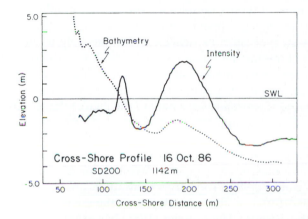

Figure 7-6 The correspondence between the cross-shore intensity of wave dissipation measured with the video-imaging technique and the bathymetry of the beach profile. [Used with permission of American Geophysical Union, from T. C. Lippmann and R. A. Holman, Quantification of Sand Bar Morphology: A Video Technique Based on Wave Dissipation, *Journal of Geophysical Research 94*, p. 1003. Copyright © 1989 American Geophysical Union.]

Figure 7-7 Crescentic bars outlined by the patterns of breaking waves, recorded with a 10-minute time exposure using the video-imaging system. [Used with permission of American Geophysical Union, from T. C. Lippmann and R. A. Holman, The Spatial and Temporal Variability of Sand Bar Morphology, *Journal of Geophysical Research 95*, p. 491. Copyright © 1990 American Geophysical Union.]

the ability of the technique to record the spatial patterns of bars, in this case an example of well-developed crescentic bars (Chapter 11). Lippmann and Holman (1990) obtained such images daily for 2 years at FRF in order to investigate how the bars respond to storms. It is under such conditions that their imaging technique is of greatest value, as the bar evolution is rapid and traditional survey methods and even the CRAB cannot be used during major storms. Their results will be examined in Chapter 11 where we will consider the three-dimensionality of beach morphology and how it changes under varying wave conditions.

GENERAL BEACH MORPHOLOGY AND VARIATIONS

Figure 7-8 contains examples of beach profiles from various coastal sites as well as from a laboratory wave channel. In most of these examples the profiles are graphed with extreme vertical exaggerations that make the bars appear more pronounced than they actually are and the beach face substantially steeper. Such vertical exaggerations help to emphasize the subtle variations in profiles but care must be taken to not be misled by the distortion. The series in Figure 7-8 illustrates the usual components of beach profiles, including the sloping beach face, the possible presence of a berm, and the common existence of one or more bars with intervening troughs. There is already some hint of the factors that control the profile morphology, such as the sediment grain size determining the beach-face slope—these factors will be explored below.

The collection of repeated surveys at a site, spanning months to years, generally reveals an envelope of profile variations. This is illustrated by Figure 7-9 containing profiles obtained at the FRF in North Carolina (Birkemeier, 1984). The vertical variations are greatest within the nearshore, particularly where bar-trough systems develop, with the envelope of change thinning and pinching out toward the offshore. Hallermeier (1981) has examined this zonation and attempted to relate it to the annual wave climate. The *closure depth, h_c,* the limit of the zone of extreme bottom changes for quartz-sand beaches, was found to be given approximately by

$$h_c = 2.28 H_e - 68.5 \left(\frac{H_e^2}{g T_e^2} \right) \tag{7.1}$$

where H_e is the nearshore storm-wave height that is exceeded only 12 hours per year, and T_e is the associated wave period. This in effect is a dependence on the wave height, with an adjustment for the wave steepness. Birkemeier (1985) has compared this relationship with profile variations at the FRF and found that a replacement of the coefficients in Equation (7.1) with the values 1.75 and 57.9 yields a better fit to the data. Birkemeier also noted that the simple proportionality $h_c = 1.57 H_e$ provides a satisfactory prediction of the closure depth.

Although beach profiles can be complex due to a series of bars and troughs, in general they are steepest at the shore and have a progressively decreasing slope as the water depth increases in the offshore direction. This regularity has inspired attempts to develop mathematical expressions to describe the profile shape. These formulations can then be used in analyses of wave dynamics during shoaling, in examinations of the generation of nearshore currents and sediment transport, and in computer models of beach morphology evolution and shoreline change. The beach profile expression that has seen most common use in engineering applications is that derived by Bruun (1954) and Dean (1976, 1977, 1991),

$$h = Ax^{2/3} \tag{7.2}$$

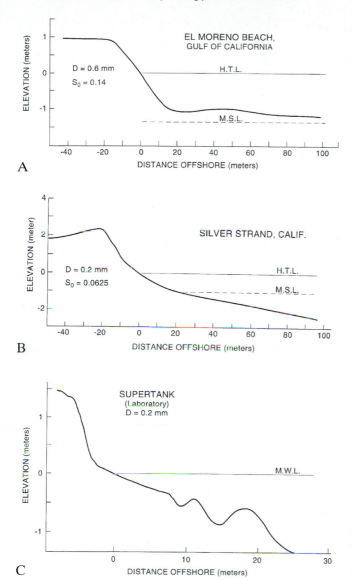

Figure 7-8 Examples of beach profiles from various coastal sites and also from a laboratory wave channel. All profiles are exaggerated in the vertical.

where h is the still-water depth at a horizontal distance x from the shoreline, and A is a dimensional shape parameter (length$^{1/3}$). An example of this relationship compared with a measured profile is shown in Figure 7-10. The original application by Bruun was limited to profiles seaward of the breaker zone, and he demonstrated its approximate congruence with profiles on the Danish North Sea coast and in California. Dean extended the application through the surf zone to the shoreline and showed that the relationship reasonably represents the 504 beach profiles compiled by Hayden et al. (1975) along the U.S. east and Gulf coasts. The exponent in Equation (7.2) determines the degree of profile concavity, and although Dean adopted a mean value of 2/3 for the exponent to be used as a functional constant, values for individual profiles within the 504 data base ranged from about 0.2 to 1.2.

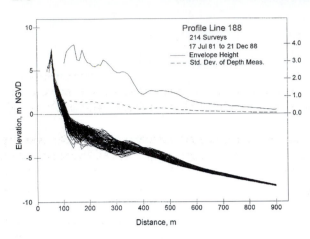

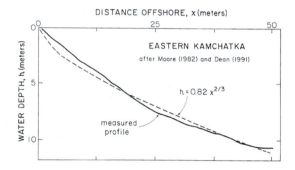

Figure 7-9 The envelope of a large number of beach profiles surveyed at the Field Research Facility, Duck, North Carolina. The upper graph shows the maximum vertical change in the depth and the standard deviation of the depth variations at each cross-shore position. [Courtesy of W. Birkemeier]

Figure 7-10 An example of the Bruun/Dean $x^{2/3}$ profile of Equation (7.2), compared with a measured beach profile. [Adapted with permission of Journal of Coastal Research from R. G. Dean, Equilibrium Beach Profiles: Characteristics and Applications, *Journal of Coastal Research 7*, p. 56. Copyright © 1991 Journal of Coastal Research.]

Justification for selecting $2/3 = 0.67$ was based in part on the apparent Gaussian distribution of values and the observed central tendency of this $2/3$ exponent; this value also corresponds to a uniform energy dissipation per unit volume of the waves as they travel across the beach profile. In a study of beaches on islands in the Caribbean, Boon and Green (1988) found an average exponent of approximately $1/2$ and concluded that this value resulted from the more reflective nature of Caribbean beaches with a greater concavity than typical of the continental quartz-sand beaches in the United States analyzed by Dean (1977). The study of Boon and Green established that, since the concavity of beaches varies as they range from reflective to intermediate to dissipative, the exponent will necessarily change rather than always being $2/3$.

The proportionality coefficient A in Equation (7.2) has been empirically related to the mean grain diameter of the beach sediment (Moore, 1982) and to the corresponding grain-settling velocity (Dean, 1987). It can be seen in Figure 7-11 that A increases with sediment diameter or settling velocity. This trend results because the value of A governs the overall slope of the beach profile from Equation (7.2), and as will be discussed below, there is a strong relationship between beach slope and sediment size on natural beaches. The beach slope also depends on the wave conditions, the wave steepness, and the absolute wave height or energy, implying that A should also have such a dependence. In their study of Caribbean beaches, Boon and Green (1988) particularly focused on how A varies with the grain size and wave conditions. They noted that A is a scaling parameter that is numerically equal to the depth at

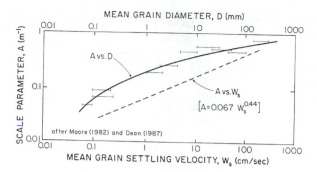

Figure 7-11 The dependence of the A proportionality coefficient in the Bruun/Dean profile Equation (7.2) on the mean grain diameter or settling velocity of the beach sand. [Adapted from Coastal Sediment Processes: Toward Engineering Solutions, R. G. Dean, *Coastal Sediments '87*, 1987. Reproduced with permission from the American Society of Civil Engineers.]

a unit distance from the shore, making it something of a surrogate measure of the beach slope but one having dimensions. However, Boon and Green did not find a significant relationship between measured beach-face slopes and the empirical A values. Although their data are scattered, they did establish some dependence of A on H_b^2/gDT^2, where H_b is the breaker height, and D is the beach-sediment grain size.

Pruszak (1993) demonstrated that there can be long-term variations in A, based on measurements at two coastal sites. Using 27 years of measured beach profiles at a location on the Baltic Sea, Pruszak found that A ranged from 0.05 to 0.10 m$^{1/3}$ in a regular cycle having a period of about 25–30 years. At the second location on the Black Sea, monthly profiles were available from 1972 to 1978, and the measurements showed a distinct annual cycle in A values due to changes in the seasonal wave climate, as well as revealing a long-term trend wherein the mean value for A increased from about 0.15 to 0.25 m$^{1/3}$ during the 6 year interval. The changes in A again show that the value depends on the wave conditions as well as on the sediment grain size or settling velocity and that there can be long-term trends due to decadal changes in the beach morphology.

Work and Dean (1991) have explored models where A varies with the offshore distance x, possibly due to varying sediment sizes along the profile, but found that these more complex models do not significantly improve the accuracy beyond the simple model of Equation (7.2) with a fixed A. Larson (1991) also examined alternative models that would account for cross-shore variations in grain sizes, taking a theoretical approach that focuses on wave-energy dissipation. In comparisons with measured beach profiles at three coastal locations, Larson found that the modified profile expression provides a better prediction than does the original Bruun/Dean model.

In formulating Equation (7.2) with application to the offshore, Bruun (1954) showed that it could be derived from an assumption of an equal bottom stress exerted by the waves on the seabed. Dean (1976, 1977) provided a theoretical justification based on an assumption that there is a uniform dissipation of wave energy flux or power per unit volume within the surf zone, $(dP/dx)/h = D_e$, where D_e is the equilibrium value of wave-power dissipation, and h is the local water depth. Employing Airy-wave theory to evaluate P in shallow water and assuming $H = \gamma h$ for the depth control on the local wave height (Chapter 6), Dean derived Equation (7.2) for the profile shape where

$$A = \left(\frac{24D_e}{5\rho g^{3/2}\gamma^2} \right)^{2/3}$$

(7.3)

suggesting that A is more aptly related physically to the wave power dissipation than simply to the beach-sediment grain size or settling velocity. Larson and Kraus (1989) provide a slightly modified derivation based on $D_e = K(P - P_{stable})/h$ from Dally, Dean, and Dalrymple (1985), which compares the local wave energy flux or power to a stable value expected at that water depth (Chapter 6). Integration of the energy dissipation equation then yields the profile form

$$x = \frac{h}{S_0} + \left(\frac{h}{A_*}\right)^{3/2} \tag{7.4}$$

where the offshore distance x is a function of the water depth. Unfortunately, the relationship cannot be inverted to make the depth the dependent variable, which would make it more directly usable. In this derivation, A_* has the same dependence on D_e as given by Equation (7.3), but here A_* is clearly different from the original A proportionality coefficient in Equation (7.2).

A different theoretical justification for the original Bruun/Dean profile relationship of Equation (7.2) was established by Bowen (1980), using an energetics model of suspended sediment transport based on wave-orbital motions and a "perturbation" or drift velocity within the nearshore. The first-order solution, assuming symmetrical wave-orbital motions, yielded a beach profile having a $x^{2/3}$ depth variation, as was found empirically by Dean (1977). However, when Bowen included the higher harmonic terms to simulate the more realistic asymmetrical wave-orbital motions as found in Stokes-type waves having strong onshore oribital velocities under the wave crests, the analysis then yielded

$$h \propto g^{1/5}\left(\frac{w_s x}{\sigma^2}\right)^{2/5} \tag{7.5}$$

where w_s is the grain-settling velocity of the beach sediment, and $\sigma = 2\pi/T$ is the wave radian frequency. This relationship suggests that $h \propto x^{2/5}$, which is closer to the 1/2 exponent found by Boon and Green (1988) for Caribbean beaches than to the ⅔ exponent established empirically by Dean as a mean value. Equation (7.5) includes a dependence on the grain-settling velocity, much as found by Dean (1977) for the proportionality coefficient A in Equation (7.2), but also indicates that there should be a strong dependence on the wave period.

The Bruun/Dean beach profile of Equation (7.2) accounts for the general increase in water depth and decrease in bottom slope with distance offshore, but cannot account for the presence of bars or troughs. Inman, Elwany, and Jenkins (1993) used a compound form, as illustrated in Figure 7-12, with each segment being fitted to a Bruun/Dean profile having the form $h = Ax^m$, where the exponent m is used as a variable in matching the equation to the profile segments. Inman and co-workers determined that $m = 0.4$ is nearly the same for the two segments in the profiles they examined. Inman and co-workers also shift the origins for the curves, indicated by the + symbols in Figure 7-12, locating them above the level of the beach so as to avoid infinite beach slopes (discussed below). The inshore portion of the profile is termed the "bar-berm segment," while the offshore portion is the "shorerise segment." They justify this division on the basis of different processes acting within the two segments, with shoaling waves occurring over the shorerise, and breaking waves and bores acting over the bar-berm segment. The use of multiple segments as proposed by Inman and co-workers can partially account for the presence of a bar, especially one that is the outer edge of a terrace within the surf zone; however, this curve-fitting approach cannot account for the presence of a longshore trough that gives a true positive relief to the offshore bar. Of special interest in the study of Inman and co-workers is their analysis of seasonal beach changes (summer versus winter) and how the observed variations are reflected in the profile segments.

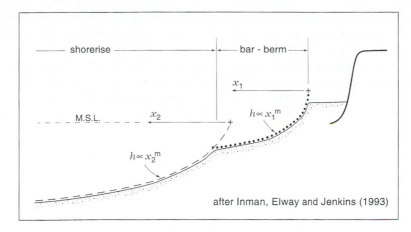

Figure 7-12   The compound profile of Inman and co-workers (1993) where the generalized Bruun/Dean profile $h = Ax^m$ is fitted separately to the inshore bar-berm and offshore shorerise segments, in part giving rise to an offshore bar where the segments join. The origins for the fitted curves, indicated by crosses, are elevated above the beach so as to avoid infinite slopes at the shore and on the bar. [Adapted with permission of American Geophysical Union, from D. L. Inman, M. H. S. Elwany and S. A. Jenkins, Shorerise and Bar-berm Profiles on Ocean Beaches, *Journal of Geophysical Research 98*, pp. 18,184. Copyright © 1993 American Geophysical Union.]

A shortcoming of the Bruun/Dean beach profile relationship of Equation (7.2) is its prediction of an infinite slope at the shore. The derivative of the relationship yields the variation in the beach slope S:

$$S = \frac{dh}{dx} = \frac{2A}{3x^{1/3}} \tag{7.6}$$

which becomes infinite when $x = 0$ at the shore. This is true for any model having the form of Equation (7.2) so long as the exponent is less than 1 (which it is for concave profiles). Accordingly, it can be expected that the simple Bruun/Dean model might agree satisfactorily with measured profiles in the offshore, as first undertaken by Bruun (1954), but will progressively and then substantially fail as the shore is approached. The modified profile form of Equation (7.4) does not suffer from this problem. The derivative of that relationship gives dx/dh, which can then be inverted to yield the beach slope

$$S = \frac{S_0}{1 + \frac{3}{2}\left(\frac{S_0}{A_*^{3/2}}\right)h^{1/2}} \tag{7.7}$$

at the offshore water depth h, which corresponds to the offshore distance x as calculated with Equation (7.4). At the shore itself, $h = 0$, so that Equation (7.7) gives S_0 as the beach-face slope.

An alternative to the Bruun/Dean beach-profile formulation is the exponential profile that has been developed by Bodge (1992) and Komar and McDougal (1994). Bodge explored the use of an exponential model having the relationship

$$h = B(1 - e^{-kx}) \tag{7.8}$$

where B and k are empirical coefficients, k determines the concavity, and according to Bodge k has a range 3×10^{-5} to 1.16×10^{-3} m^{-1} for the Hayden et al. (1975) data set. The proportionality coefficient B is the offshore depth to which the sloping profile of Equation (7.8) approaches asymptotically. Bodge demonstrated that this exponential profile agrees more closely with the measured profiles of the Hayden et al. data set than does the form of the Bruun/Dean profile of Equation (7.2), even when one allows the exponent to empirically depart from the mean 2/3 value suggested by Dean. The slope associated with the relationship of Equation (7.8) is given by

$$S = \frac{dh}{dx} = kBe^{-kx} \tag{7.9}$$

At the shore ($x = 0$) the beach-face slope is $S_0 = kB$, a finite value but one that combines the two empirical coefficients. This suggests the alternate form of the exponential profile developed by Komar and McDougal (1994):

$$h = \frac{S_0}{k}(1 - e^{-kx}) \tag{7.10}$$

This form contains only the coefficient k to be used in fitting the relationship to the observed profile, since a number of studies have documented how the beach-face slope S_0 at the shore varies with grain size and wave conditions (to be discussed in the following section). The variation in bottom slope along the profile is now given by

$$S = \frac{dh}{dx} = S_0 e^{-kx} \tag{7.11a}$$

an exponential decrease with distance offshore from the S_0 value at the shoreline, or by

$$S = S_0 - kh \tag{7.11b}$$

a linear offshore decrease with depth h. The one free coefficient, k, can be based on a best-fit to the overall profile or simply calculated with one offshore coordinate such as the closure depth of profile changes analyzed by Hallermeier (1981) and Birkemeier (1985). Figure 7-13 shows the relationship of Equation (7.10) in dimensionless terms of kx_c versus $(h_c/x_c)/S_0$, with the profile governed by the offshore depth coordinates (x_c, h_c), which could be the closure-depth coordinates. Simplifying approximations are also shown in Figure 7-13, one involving $e^{-kx} < 1$ and a Taylor series expansion solution suitable for use when $(h_c/x_c)/S_0 > 0.8$. The ratio $(h_c/x_c)/S_0$ represents the average beach slope out to the depth coordinates (x_c, h_c), compared with the steeper slope S_0 at the shore. In general, the beach-face slope S_0 will depend on the sediment grain size and somewhat on the daily wave conditions, while the closure depth will be a function of the annual wave climate.

An example of exponential profiles calculated with Equation (7.10) is shown in Figure 7-14, an example where S_0 was calculated using the empirical formula developed by Sunamura (1984), to be examined later [Equation (7.15) and Fig. 7-20]. With S_0 fixed mainly by the sediment diameter D, and the profile necessarily passing through the offshore closure depth calculated with $h_c = 1.57H_e$ from Birkemeier (1985), the values of k progressively decrease in profiles A to C, with the beach becoming steeper and more reflective. In the profile series presented in Figure 7-14, each exponential profile converges to a fixed offshore slope

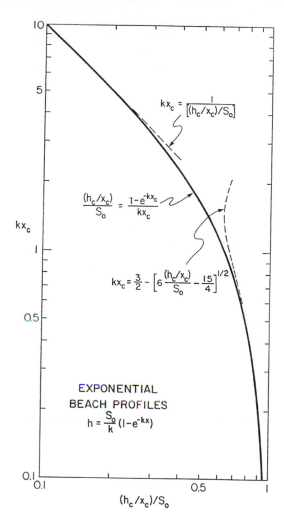

Within the figure:

$$kx_c = \frac{1}{[(h_c/x_c)/S_0]}$$

$$\frac{(h_c/x_c)}{S_0} = \frac{1-e^{-kx_c}}{kx_c}$$

$$kx_c = \frac{3}{2} - \left[6\frac{(h_c/x_c)}{S_0} - \frac{15}{4}\right]^{1/2}$$

kx_c

EXPONENTIAL
BEACH PROFILES
$$h = \frac{S_0}{k}(1-e^{-kx})$$

$(h_c/x_c)/S_0$

Figure 7-13 The evaluation of the coefficient k in the exponential beach-profile expressions of Equations (7. 8) through (7.11), based on the value of the beach-face slope S_0 and an offshore depth coordinate (x_c,h_c) such as the closure depth of profile variations. [Used with permission of Journal of Coastal Research from P. D. Komar and W. G. McDougal, The Analysis of Beach Profiles and Nearshore Processes Using the Exponential Beach Profile Form, *Journal of Coastal Research 10*, p. 61. Copyright © 1994 Journal of Coastal Research.]

$S_c = 0.001$ at and beyond the closure depth, a slope that is assumed to correspond to the gradient of the continental shelf beyond the nearshore zone. Figure 7-15 from Komar and McDougal (1994) provides a comparison with a beach profile from the Nile Delta of Egypt. The measured slope at the shoreline is $S_0 = 0.0289$, and k was evaluated using the graph of Figure 7-13, based on a closure depth 300 m offshore where $h_c = 3.5$ m. Shown in Figure 7-15 is the Bruun/Dean $x^{2/3}$ profile expression of Equation (7.2), which is seen to depart significantly from the measured profile. Komar and McDougal also analyzed this profile from the Nile Delta in terms of slope variations, comparing the measured variations with the exponential-profile predictions of Equation (7.11). Good agreement was found, while as expected, disagreement with the Bruun/Dean profile was still more profound.

The exponential beach-profile relationship of Equation (7.10) is superior to the Bruun/Dean profile of Equation (7.2) in providing better agreement with measured beach

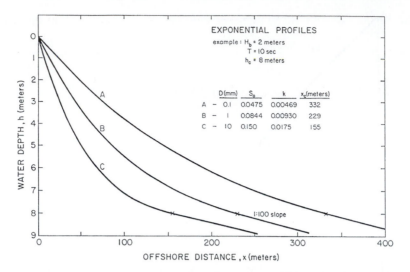

Figure 7-14 Examples of exponential beach profiles of Equation (7.10) for a series of sediment sizes $D = 0.1, 1,$ and 10 mm with the corresponding beach-face slope S_0 calculated with Equation (7.15) from Sunamura (1984). [Used with permission of Journal of Coastal Research from P. D. Komar and W. G. McDougal, The Analysis of Beach Profiles and Nearshore Processes Using the Exponential Beach Profile Form, *Journal of Coastal Research 10,* p. 63. Copyright © 1994 Journal of Coastal Research.]

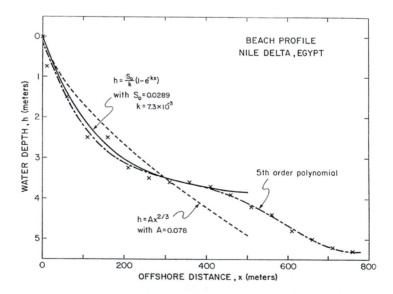

Figure 7-15 A measured beach profile from the Nile Delta of Egypt compared with the exponential profile of Equation (7.10) and with the Bruun/Dean $x^{2/3}$ profile. [Used with permission of Journal of Coastal Research from P. D. Komar and W. G. McDougal, The Analysis of Beach Profiles and Nearshore Processes Using the Exponential Beach Profile Form, *Journal of Coastal Research 10,* p. 68. Copyright © 1994 Journal of Coastal Research.]

profiles and in accounting for the expected range of profiles from dissipative to reflective beaches. Both of these expressions for the mean profile are deficient in not accounting for the presence of bars and troughs that are typical of most natural beaches.

THE SLOPE OF THE BEACH FACE AT THE SHORE

The slope of the beach face is governed by the asymmetry in the intensity of the wave-swash uprush versus the return backwash and the resulting asymmetrical cross-shore sediment transport. Because of water percolation into the beach and frictional drag on the swash, the return backwash tends to be weaker than the shoreward uprush. This asymmetry moves sediment onshore until a slope is built up over which gravity supports the backwash and enhances the offshore sediment transport. When the same amount of sediment is transported seaward as is moved landward, the beach-face slope becomes effectively constant and is in a state of dynamic equilibrium.

The slope of this equilibrium beach face depends in part on the quantity of water lost through percolation into the beach. The percolation is governed principally by the grain size of the beach sediment—water percolates much more rapidly into gravel than into a sand beach. The result is that on the gravel beach the return backwash is weakened, and therefore the slope is greater than on beaches composed of sand.

Numerous field studies have quantitatively demonstrated that beaches composed of coarse particles have steeper slopes (Bascom, 1951; Wiegel, 1964; McLean and Kirk, 1969;

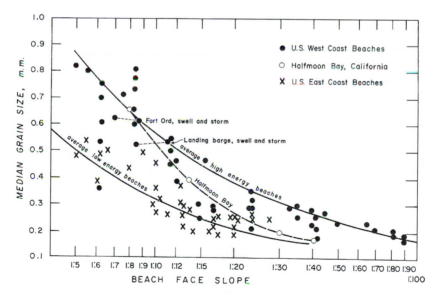

Figure 7-16 The beach-face slope as a function of the median grain size of the beach sediment. Also shown is the difference between U.S. west coast and east coast beaches, which reflects the importance of the overall wave-energy level as a control on the beach slope. [Adapted with permission of American Geophysical Union, from W. H. Bascom, The Relationship Between Sand Size and Beach Face Slope, Transactions American Geophysical Union 32, p. 872. Copyright © 1951 American Geophysical Union and with permission of Prentice-Hall Inc., from R. L. Wiegel, *Oceanographical Engineering.* Copyright © 1964 Prentice Hall.]

Dubois, 1972), and this has been shown in laboratory wave-tank experiments as well (Bagnold, 1940; Rector, 1954; van Hijum, 1974). Figure 7-16, after Bascom (1951) and Wiegel (1964), relates the slope to the grain size at a "reference point," the part of the beach face subjected to wave action at mid-tide. The use of a reference point standardizes the sampling location and thereby removes some of the scatter from such a plot due to cross-shore variations in grain sizes and slopes. As expected, Figure 7-16 reveals that the coarser the sand, the steeper the beach, but also shows a dependence on the wave energy.

Bascom (1951) and Wiegel (1964) included data only from sand beaches. Figure 7-17, based on the compilation of Shepard (1963), carries the relationship up through pebble beaches where slopes achieve 15°. On cobble beaches, the slopes can reach 25°, which is close to the angle of repose (approximately 32°), the limit to which most noncohesive grains can be piled.

The rate of percolation is affected by the degree of sediment sorting as well as by the mean grain size. Sorting therefore should also have an effect on the beach slope. This was demonstrated by Krumbein and Graybill (1965), who found that well-sorted coarse-sand beaches have steeper slopes than poorly sorted coarse-sand beaches. The importance of the degree of sorting has been further investigated by McLean and Kirk (1969) on mixed sand-shingle beaches in New Zealand. Their curve is compared with the Shepard (1963) curve in Figure 7-17. Because of the overall poorer sorting, the New Zealand beaches have lower slopes than those measured by Shepard, even for the same median grain size. Also note that the curve of McLean and Kirk is wavy. This is due to the nature of the sources of sediments to the New Zealand beaches, sources that yield beaches that consist of pebbles with diameters of 4–16 mm or of sand having a median diameter of 0.5 mm or mixtures of the two. When the individual modes of pebbles or sand occur alone, the sediment sorting is good and the resulting beach is steeper; when the modes are mixed, the sorting is poorer and water percolation and the beach slope are reduced. Accordingly, the curve of McLean and Kirk in Figure 7-17 rises (higher slope) for sands of 0.5 mm and for pebbles, but lowers (reduced slope) for intermediate grain sizes.

Bascom's (1951) data in Figure 7-16 were derived from high-energy exposed beaches on the Pacific coast of the United States. Wiegel (1964) added data from the lower-energy beaches of New Jersey, North Carolina, and Florida. For a given mean grain size, these lower-energy beaches have greater slopes than the exposed Pacific beaches. The terms "exposed" and "protected" also are used to describe the large differences in the amount of wave energy reaching a beach. Included in Figure 7-16 is a series of points from Halfmoon Bay, California. This bay is partially protected by a headland (Fig. 7-18), such that the energy is lowest in close proximity to the headland and progressively increases to the south as the sheltering of the

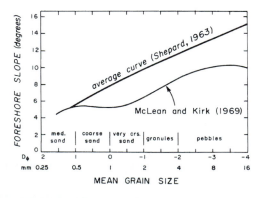

Figure 7-17 Beach-face slope versus the grain size according to Shepard (1963) and McLean and Kirk (1969). The wavy nature of the curve from McLean and Kirk demonstrates the importance of the degree of sediment sorting on the beach slope. [Adapted with permission of SIR Publishing, from R. F. McLean and R. M. Kirk, Relationship Between Grain Size, Size-Sorting, and Foreshore Slope on Mixed Sand-Shingle Beaches, *New Zealand Journal of Geology and Geophysics,* p. 153. Copyright © 1969 SIR Publishing.]

headland is lost. In response to the increasing energy level, the beaches farthest from the headland are coarsest—there is a systematic increase in grain size in the longshore direction. The beach face therefore increases its slope, the steepest being the exposed beaches to the south (Fig. 7-18). From the trend of the Halfmoon Bay data points in Figure 7-16, it is apparent that this increase in slope to the south is not entirely due to the increasing grain size but also reflects the wave-energy level due to sheltering by the headland. Close to the headland, the energy level is reduced to that of the "low energy" curve of Wiegel (1964), otherwise established by Atlantic coast beaches. [Note: Bascom's (1951) measurements at Halfmoon Bay were collected prior to construction of the breakwater in the 1960s, which has substantially changed the wave exposure, beach-sand sorting, and profiles.]

A number of studies have attempted to develop empirical relationships for the beach-face slope in terms of the wave steepness, suggesting a dependence on the wave period as well as on the height or energy. In experiments in a wave channel containing sand of 0.22-mm diameter, Rector (1954) found that the foreshore slope above the still-water level depends on the deep-water wave steepness according to the relationship

$$S_0 = 0.30 \left(\frac{H_\infty}{L_\infty} \right)^{-0.30} \tag{7.12}$$

This indicates that the greater the wave steepness, the lower the beach slope. Utilizing regression analysis techniques on field data, Harrison (1969) similarly found an inverse relationship between the beach-face slope and wave steepness. As will be discussed later, this change is one aspect of beach profile variations between calm weather conditions and storms that have higher wave steepnesses. This dependence on wave steepness has the effect of adding scatter to the diagram of Figure 7-16. This is demonstrated by the pairs of points from Fort Ord and Landing Barge, both in California. The ranges in slopes at these two locations

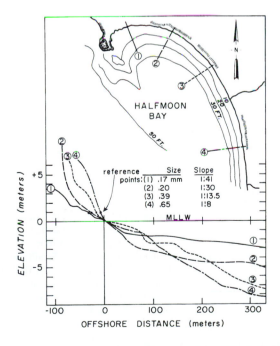

Figure 7-18 Systematic changes in beach slopes along the length of Halfmoon Bay, California, due to longshore variations in beach sediment grain sizes and exposure to the waves. [Adapted with permission of American Geophysical Union, from W. H. Bascom, The Relationship Between Sand Size and Beach Face Slope, *Transactions American Geophysical Union 32*, p. 872. Copyright © 1951 American Geophysical Union.]

reflect the varying wave steepness between the calm summer months and winter storms, with the lower slopes corresponding to the winter months of steeper waves.

   Although the dependence on the wave steepness can help explain part of the observed variation in beach slopes, it is not the full answer. The wave steepness may be the same for both ocean waves and in a laboratory wave tank or small lake, yet the beach slope will be higher in the low-energy wave tank. In addition to the effects of the wave steepness, there must also be a dependence on the actual wave-energy level or wave height.

   Attempts have been made to relate the beach-face slope to a variety of dimensionless parameters that incorporate measures of wave heights and periods as well as sediment sizes. Dalrymple and Thompson (1976) employed the dimensionless settling velocity or Dean number, $H_\infty/w_s T$, where w_s is the mean grain settling velocity of the beach sediment. From the physical arguments of Dean (1973), this ratio is a measure of whether a sediment particle lifted into suspension by a passing wave can fall to the bottom during the time when its net displacement is shoreward, due to the horizontal water orbital flow. Figure 7-19, from Dalrymple and Thompson, combines data from several laboratory studies, showing a general decrease in beach slope with increasing $H_\infty/w_s T$. The H_∞/T portion of the Dean number is nearly equivalent to the deep-water wave steepness, so the trend found in Figure 7-17 agrees with the basic dependence of Equation (7.12) found by Rector (1954). The dependence on the Dean number yields an inverse relationship to w_s; an increase in sediment size and w_s reduces the value of the Dean number and predicts an increase in beach slope (Fig. 7-17). This of course is the dependence observed by numerous field and laboratory investigations, but it remains to be established whether it is the grain settling velocity *per se* that is important in governing the beach slope, or whether it is acting as a surrogate for the grain diameter that determines the permeability and surface roughness of the beach sediment. Measurements by Dubois (1972) of slopes on beaches enriched in heavy minerals revealed trends of increasing slope with decreasing grain size, the opposite found by others and expected if the control is mainly by water percolation. The strongest correlation found by Dubois was between the beach slope and the percentage of heavy minerals. This suggests a direct dependence on the grain settling velocity as proposed by Dean (1973), the higher densities of the heavy minerals yielding the higher settling rates and slopes in the beaches studied by Dubois. It is appar-

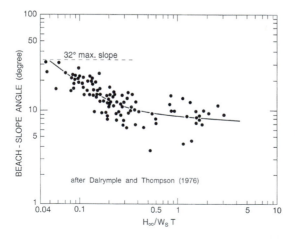

Figure 7-19 A compilation of laboratory data to relate the beach-face slope to $H_\infty/w_s T$, where w_s is the settling velocity of the beach sand. [Adapted from Study of Equilibrium Beach Profiles, R. A. Dalrymple and W. W. Thompson, *Proceedings of the 15th Coastal Engineering Conference*, 1976. Reproduced with permission from the American Society of Civil Engineers.]

ent, however, from studies such as those of McLean and Kirk (1969) on New Zealand beaches, where sediment sorting affects the permeability and thus the beach slope, that the grain size and its direct effects on permeability must be important, in addition to the possible control by the grain settling velocity.

Kemp and Plinston (1968) related the beach slope to $H_b/D^{1/2}T$. Sunamura (1984) employed the same basic ratio, expressed in dimensionless form as $H_b/g^{1/2}D^{1/2}T$. Sunamura compiled a large quantity of laboratory and field data to establish empirical relationships for the beach slope. The laboratory data alone yield the empirical relationship

$$S_0 = \frac{0.013}{(H_b/g^{1/2}D^{1/2}T)^2} + 0.15 \tag{7.13}$$

for the slope S_0 at the shore. The field data are highly scatter but can be described by the equation

$$S_0 = \frac{0.12}{(H_b/g^{1/2}D^{1/2}T)^{1/2}} \tag{7.14}$$

According to these empirical results, the slopes of laboratory beaches are on average greater than those of natural beaches for the same value of $H_b/g^{1/2}D^{1/2}T$. The reason for this difference is uncertain, but it may be due to the general absence of infragravity energy on the laboratory beaches, the presence of which on natural beaches would act to reduce the beach slope. Using $L_\infty = (g/2\pi)T^2$ and the Komar and Gaughan (1972) relationship for the breaker height as a function of the deep-water wave parameters [Equation (6.6)], the above relationships can be converted into dependencies on D/H_∞ and H_∞/L_∞. Equation (7.14) for the field data thereby becomes

$$S_0 = 0.25\left(\frac{D}{H_\infty}\right)^{0.25}\left(\frac{H_\infty}{L_\infty}\right)^{-0.15} \tag{7.15}$$

showing an inverse dependence on the deep-water wave steepness similar to that found by Rector (1954) in Equation (7.12), and also depending on the sediment grain size and overall wave height or energy level. Equation (7.15) has been used to generate the series of curves in Figure 7-20.

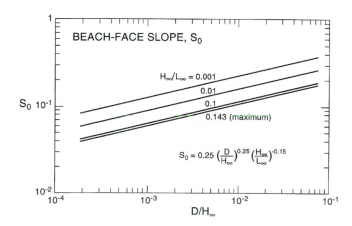

Figure 7-20 The beach-face slope as a function of D/H_∞ and the wave steepness H_∞/L_∞ according to Equation (7.15) as modified from Equation (7.14), obtained by Sunamura (1984) with field data.

The general dependencies of beach slopes on sediment sizes and wave conditions are clear from the trends of the field and laboratory data. However, there has not been full agreement as to the choice of which dimensionless ratios are most appropriate for establishing predictive relationships or accounting for the physical processes that control the beach slope. There is some correspondence between the different forms of dimensionless ratios used by the various investigators. This was seen above in the conversion of Equation (7.14), based on the analysis of Sunamura (1984) in terms of $H_b/g^{1/2}D^{1/2}T$, into the equivalent Equation (7.15) dependent on D/H_∞ and the wave steepness H_∞/L_∞. The $H_b/g^{1/2}D^{1/2}T$ ratio also has a close affinity to the Dean number H_∞/w_sT, since the settling velocity $w_s \propto (gD)^{1/2}$ for sand sizes and coarser particles. The choice should be governed by the dominant physical processes controlling the beach slope. If the slope is established mainly by cross-shore transport of suspended sediments, as envisioned in the heuristic model of Dean (1973), then a ratio containing w_s would be appropriate. On the other hand, if percolation into and out of the beach face is more important, then the grain diameter (and range of sizes) would be more relevant. If both processes of percolation and cross-shore sediment transport are important, then the beach-face slope would be expected to depend on both the settling velocity and median grain size and also on the degree of sorting, which in part controls the rate of percolation. At this stage, in order to resolve such questions, it is more important to focus on the processes rather than on further empirical analyses employing various dimensionless ratios.

BERM FORMATION

The berm is the nearly horizontal portion of the exposed beach, the parcel most familiar to the sunbather. It is formed by sediment brought ashore during low wave conditions. The presence of a berm is not always apparent on a fine-sand beach that is already nearly horizontal. A distinct berm with a marked difference from the sloping beach face is better developed on medium- to coarse-sand beaches and may be best developed on gravel beaches where the separation from the steep beach face is pronounced.

In laboratory wave-tank studies, Bagnold (1940) demonstrated that the berm elevation coincides with the wave run-up height on the sloping beach face. The run-up distance in part depends on the wave height H (Chapter 6), and accordingly Bagnold found the simple proportionality

$$\text{Berm elevation} = bH \tag{7.16}$$

where

$$b = \begin{cases} 1.68 & \text{for } D = 0.7 \text{ cm} \\ 1.78 & \text{for } D = 0.3 \text{ cm} \\ 1.8 & \text{for } D = 0.05 \text{ cm} \end{cases}$$

The factor b is seen to have a small dependence on the grain size of the beach sediment, being smaller for the coarser sizes since percolation is greater.

Takeda and Sunamura (1982) [see Sunamura (1989)] further explored the role of run-up in controlling the elevation of the berm. Using the results of Hunt (1959) for the run-up height as a function of the wave conditions, Takeda and Sunamura obtained a relationship for the berm elevation B_h:

$$B_h = 0.125 H_b^{5/8}(gT^2)^{3/8} \qquad (7.17)$$

The proportionality coefficient is based on a comparison with laboratory and field measurements, as seen in Figure 7-21. The field data were obtained from beaches along the Pacific coast of Japan where the mean tidal range is approximately 1 m—the berm height was measured relative to mean sea level. Sediment sizes ranged 0.2–1.3 mm without showing differences in berm elevations.

A correspondence between wave height and berm elevation also had been noted by Bascom (1953) on ocean beaches. He provides a good description of berm development, having based his observations on the Pacific coast of the United States. After a wave breaks, the water rushes forward up the beach face, carrying sand with it, losing velocity as it goes because it is opposed by gravity and friction and because of water losses through percolation. As the beach builds seaward, it leaves a nearly horizontal berm that corresponds to the elevation to which sand had been carried by the swash run-up. If the waves were uniformly the same height, a rapidly growing berm could have undulations reflecting the tide levels; this was found by Strahler (1966), who showed that the berm elevation is raised in response to the high waters of spring tides.

Bascom (1953) noted that the upward growth of the berm depends mainly on the largest waves that reach the beach, since their swash passes completely over the crest and deposits the bulk of the sand atop the berm. The upward growth is greatest at high tides, and the combination of tides with large waves sometimes produces a landward-sloping berm if the crest grows higher as it builds seaward. The swash runs landward down this inclined berm and collects in a "lagoon" at the back of the beach. At times enough water gathers that it is able to break through the berm crest at a low point and then flows back out to sea.

High storm waves cut back the berm and tend to destroy it. Bascom (1953) pointed out the interesting paradox that storm waves also build up a berm to higher elevations, due to the greater wave heights and run-up. The uprush of large waves adds sand to the top of the berm, even while the beach face is being eroded and the extent of the berm is reduced. A storm may leave a high, narrow berm that can survive until removed by subaerial erosion or a larger storm. A prograding beach may have a series of abandoned storm berms on its landward side, reflecting the history of extreme storms.

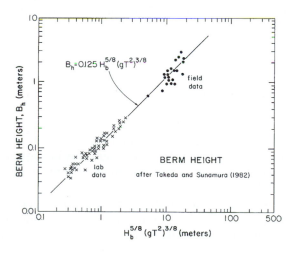

Figure 7-21 Measurements of berm elevations that confirm Equation (7.17), based on considerations of wave run-up levels. [Adapted from *Coastal Modeling*, T. Sunamura, Sandy Beach Geomorphology Elucidated by Laboratory Modeling, p. 172, 1989 with kind permission of Elsevier Science-NL, Sara Burgerhartstratt 25, 1055 KV Amsterdam, The Netherlands.]

LONGSHORE BARS AND TROUGHS

When sediment is shifted offshore, having been eroded from the beach face and berm, it generally is deposited to form a longshore bar with a trough on its shoreward side. Such bars may have a considerable longshore extent; for example, Evans (1940) traced bars for up to 50 km along the shore of Lake Michigan. Longshore bars are often broken by gaps due to rip currents but are otherwise continuous along the length of beach. The bars also may take on a rhythmical crescentic shape, a form whose origin will be considered in Chapter 11. Here we focus on longshore bars that are essentially two dimensional and parallel to the shoreline.

A good summary of the early literature on the characteristics of longshore bars and troughs can be found in Johnson (1919). Some of the earliest investigations were made by German scientists along the coasts of the Baltic and North Sea. The first description is that given by Hagan (1863), who explained bar formation by a seaward-flowing undertow meeting shoreward-moving waves, an explanation that is not much different from our present view of the origin of break-point bars. Otto (1912) studied a series of bars on the Baltic Sea for a period of 5 years, demonstrating that most changes occur during violent storms and concluding that sand moves offshore from the trough during a storm and is deposited to form the bar.

The papers by Evans (1940), Keulegan (1948), King and Williams (1949), and Shepard (1950a) were particularly important early contributions to our understanding of the characteristics and variability of longshore bars and troughs. The study of Keulegan involved the controlled conditions of a laboratory wave channel; the field investigations of Evans (Lake Michigan) and King and Williams (Mediterranean Sea) included tideless areas, while the study of Shepard on the California coast reflected the effects of tides. All of these studies recognized that breaking waves comprise an important element in longshore bar formation, because the breakers often determine the offshore positions of bars, their sizes, and depths of occurrence. These studies all agree that the larger the waves, the deeper the resulting bars and troughs. In his controlled wave-tank investigations, Keulegan found that the bar position is governed by the wave height and wave steepness—an increase in wave height moves the bar seaward into deeper water, while holding the wave height constant and increasing the steepness (decreasing the wave period) moves the bar shoreward. Keulegan found that starting with a smooth beach face, the bar initially forms just shoreward of the breaker position but then migrates landward as it grows, the breaker position shifting with it. The fully developed bar is appreciably closer to shore than the initial breaker position. In addition to shifting shoreward with the bar, the wave breakers also change character, tending to become more of the plunging type than spilling. The bar is effective in dissipating the wave energy. Keulegan found that with bars present, the breaking waves are able to impart a lesser amount of energy to the reformed waves than would be the case if the bars were absent [also see Carter and Balsillie (1983)].

Shepard (1950a) also noted that plunging breakers appear to be more conducive to the development of bars and troughs than are spilling breakers. He found that with spilling breakers, which accompany short-period storm waves, the bar-trough system may be entirely eliminated, primarily because of the absence of a well-defined breaker zone under those conditions.

As part of their effort to model cross-shore sediment transport and profile changes, Larson, Kraus, and Sunamura (1988) developed relationships for the factors controlling the component parts of the profile, including dimensions of the offshore bar. They based their analyses on laboratory wave-channel data but restricted the inclusion to measurements from large channels that provide near prototype conditions. Figure 7-22 shows their compilation of data for the volume increase of the bar with time, demonstrating that it can take many hours for the bar size to achieve equilibrium. The final equilibrium volume of the bar, V_{eq}, was found empirically to depend on the relationship

$$\frac{V_{eq}}{L_\infty^2} = 0.028\left(\frac{H_\infty}{w_s T}\right)^{1.32}\left(\frac{H_\infty}{L_\infty}\right)^{1.05} \tag{7.18}$$

This equation explained 70 percent of the variation in the data. Larson and Kraus (1989) present a number of such relationships in which other geometric components of the profile are related to wave and sediment properties.

The generation of longshore troughs that accompany bars can also be related to the processes of wave breaking. Based on an extensive series of wave-tank experiments, Miller (1976) found that the trough is excavated by the turbulence associated with breaking waves, with the eroded sand moving offshore to form the bar. The excavation of the trough is greatest under plunging waves, due to their downward thrust with the dissipation confined to a narrow zone, but excavation still occurs with spilling breakers, although it is not as deep. Based on these observations, Sunamura (1985, 1989) developed a model that correlates the distance from the trough position to the location of the breaking waves. Experimental results from three laboratory wave channels ranging from small to medium to prototype scale were consistent (Fig. 7-23), yielding

$$\frac{\ell_t}{L_\infty} = \frac{10}{S}\left(\frac{H_b}{gT^2}\right)^{4/3} \tag{7.19}$$

for the distance ℓ_t between the breaking wave position and the center of the resulting trough, where S is the average slope of the beach. The distance ℓ_c to the bar crest was found to be

$$\frac{\ell_c}{H_b} = 0.18\left(\frac{\ell_t}{H_b}\right)^{3/2} \tag{7.20}$$

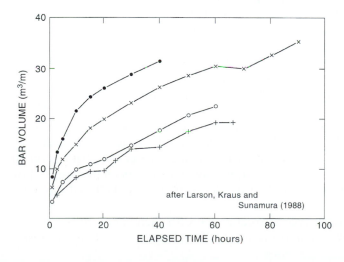

after Larson, Kraus and
Sunamura (1988)

Figure 7-22 The increase with time of volumes of sand within offshore bars, approaching equilibrium volumes that are related to sediment and wave parameters by Equation (7.18). [Adapted from Beach Profile Change: Morphology, Transport Rate and Numerical Simulation, M. Larson, N. C. Kraus, and T. Sunamura, *Proceedings of the 21st Coastal Engineering Conference*, 1988. Reproduced with permission from the American Society of Civil Engineers.]

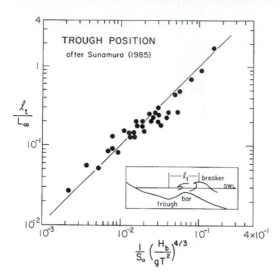

Figure 7-23 Measurements of the cross-shore distance from the breaking wave position to the eroded longshore trough, yielding Equation (7.19). [Adapted from *Coastal Modeling,* T. Sunamura, Sandy Beach Geomorphology Elucidated by Laboratory Modeling, p. 181, 1989 with kind permission of Elsevier Science-NL, Sara Burgerhartstratt 25, 1055 KV Amsterdam, The Netherlands.]

   In addition to the direct action of breaking waves in generating a longshore bar and trough, several investigations have suggested the importance of the wave-induced undertow (Chapter 6). Its possible role in bar formation was first described qualitatively by Dhyr-Nielsen and Sorensen (1970). Quantitative models for the undertow, the resulting sediment transport, and the beach-profile evolution have been presented by Dally and Dean (1984) for regular waves and by Stive and Battjes (1984) and Stive (1986) for random waves. Roelvink and Stive (1989) conducted laboratory wave-tank experiments to investigate the role of undertow in generating bars and to test formulae that are responsible for the cross-shore sediment transport. The results indicate that on dissipative beaches, the undertow induced by modulated waves breaking at different water depths can have an important role in bar formation within the surf zone.

   Some beaches have more than one longshore bar separated by troughs. Figure 7-24 shows an extreme example from the U.S. Gulf coast where there are nearly a dozen bars that are essentially parallel. The number of bars depends mainly on the overall slope of the nearshore, the more gradual the slope the greater the number. Bars are commonly absent on steep beaches. At the other extreme, Kindle (1936) and more recently Dolan and Dean (1985) documented the presence of multiple bars at various sites within Chesapeake Bay, the large number in part being due to the gentleness of the slope within the bay. Dolan and Dean found four to seventeen bars at various locations, all being within about 550 m from shore and in water depths generally less than 2 m. Spacings between bar crests ranged from 12–70 m, generally increasing with distance offshore. Evans (1940) found an average of three bars on Lake Michigan beaches, rarely more than four, and in a few places where the water deepens rapidly, only one. He found that the size of the bar increases with distance from shore, as does the distance between adjacent bars, the depths of the outer bars being greater than bars closer to shore in spite of their larger sizes. A similar field of multiple bars has been investigated by Boczar-Karakiewicz and Davidson-Arnott (1987) in Georgian Bay, a sheltered bay connected to Lake Huron.

   Multiple bars may result from multiple breaker zones, with each bar corresponding to the average breaking position of waves of a certain size. The deepest bars therefore result

Figure 7-24 Multiple bars within Mississippi Sound on the Gulf of Mexico coast of Mississippi. Beaches of the Sound are partially sheltered by a system of small barrier islands. [Courtesy of Dag Nummedal]

from the largest waves. Evans (1940) demonstrated in a wave tank that an inner and outer bar could be formed by a succession of large waves followed by smaller waves. He first used large waves to generate the outer bar, while the subsequent lower waves passed over this outer bar with little effect to themselves or to the bar and broke closer to the shore, forming an inner bar. King and Williams (1949) found that the outer bar at Sidi Ferruch, North Africa, owed its position to the predominant break point of the storm waves, which are comparatively rare on the Mediterranean Sea but are of great energy, while the inner bar depends on the predominant calm-weather waves.

On a low-sloping beach, once waves break they may reform as they continue across the surf zone (Chapter 6). These reformed waves may break for a second time and perhaps even a third time before they finally reach the shore. An outer bar may form in association with the initial zone of breaking, while inner bars develop where the reformed waves break, giving a multiple-bar system. Since the reformed waves are reduced in size, the topographic relief of the inner bar and trough is smaller. Such a development is shown in Figure 7-25, derived from the laboratory experiments of Sunamura and Maruyama (1987). Dolan and Dean (1985) extended this argument in a conceptual model for the formation of multiple bars in Chesapeake Bay. They argue that, due to the low slope, the waves are able to repeatedly reform after breaking, so that, with time, the field of bars grows in the shoreward direction.

Some multiple-bar systems might be explained by the varying water level and breaker positions resulting from the cycles of tides. There may be one set of bars for the high-tide water level and a second set for the low-tide position. King and Williams (1949) have shown in wave-tank experiments that a falling tide will tend to destroy the shallower bars. However, on ocean beaches the inner bars exposed at low tide are generally able to persist, though modified and flattened by the varying water levels (Shepard, 1950a).

During storms, waves may break on all bars in a multiple-bar system, the largest waves breaking over the deepest bar and progressively smaller breakers occurring on the inner bars. The waves that finally reach the shore are much reduced in size. During times of smaller waves, the waves pass over the deep outer bars without appreciable affect and do not break until they reach the comparatively shallow water over the inner bar. During the 14 months of his study in Lake Michigan, Evans (1940) found little or no movement of the outer two bars, while the shallowest inner bar showed considerable migration and variability. He found that the inner bar, built during storm conditions, may be driven inshore by moderately sized waves. These migrations of the inner bar on Lake Michigan were studied further by Davis and Fox (1972) and Davis et al. (1972). Figure 7-26 shows the onshore movement of the inner bar, its initial increase in height as it constricts the landward trough during the migration, and its eventual welding onto the beach face. This landward migration of the bar was the primary mode of onshore sediment transport back to the beach face during times of low waves following a storm that had shifted sand offshore and formed the bar.

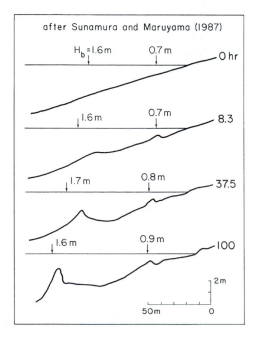

Figure 7-25   Laboratory wave-channel experiments showing the development of two bars, the outer associated with the initial breaking of waves and an inner bar resulting from the second breaking of the reformed waves. [Adapted from Wave-induced Geomorphic Response of Eroding Beaches, with Special Reference to Seaward Migrating Bars, T. Sunamura and K. Maruyama, *Coastal Sediments '87*, 1987. Reproduced with permission from the American Society of Civil Engineers.]

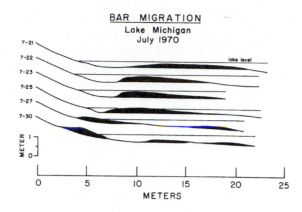

Figure 7-26 The onshore migration of the inner bar on a Lake Michigan beach. [Adapted with permission of Society for Sedimentary Geology, from R. A. Davis et al., Comparison of Ridge and Runnel Systems in Tidal and Nontidal Environments, *Journal of Sedimentary Petrology 42,* p. 416. Copyright © 1972 Society for Sedimentary Geology.]

Such movements of inner bars on ocean beaches have been documented by Hayes and Boothroyd (1969), Hayes (1972), Davis et al. (1972), and Owens and Frobel (1977). They applied the term *ridge and runnel systems* to rapidly migrating inner bars, an unfortunate choice since that term had been used earlier by King and Williams (1949) for systems of nearly stationary bars found on profiles affected by a large tidal range (Orford and Wright, 1978). The migrating inner bars of Davis and Fox (1972) on Lake Michigan and of Hayes and Boothroyd (1969) and others on ocean beaches are more akin to the *swash bars* described by King and Williams (1949, pp. 81–82).

The systems of true ridges and runnels, originally described by King and Williams (1949) at Blackpool Beach, England, by Gresswell (1937, 1953) on the Lancashire coast of England, and by Pugh (1953) between Dieppe, France, and Ostend, Belgium, are found on the low-tide terrace of beach profiles where there is a significant tidal range. They are similar to the offshore bars found on most beaches but are smaller and lower in profile. The ridges are cut by numerous gaps, formed by water escaping seaward from the runnels as the tide falls. The gaps also allow the water of the rising tide to penetrate quietly into the runnels while the sea is still breaking on the seaward face of the ridge. With a rising or falling water level from the tides, the breaker position jumps from one ridge to the next, with little wave action occurring within the runnels. At Blackpool Beach, studied by King and Williams, the ridges and associated runnels tended to increase in size toward the low-tide level. They found no systematic movement of the ridges over several years of observation. Gresswell (1937) noted that, unlike longshore bars, the ridges build up during calm weather conditions of moderate swell and are reduced in height during storms. King and Williams indicated that the factors that appear to be important to the development of a ridge and runnel system are: (1) a large tidal range, such that an extensive low-tide terrace is developed with a low offshore slope; (2) low wave energy, such systems forming best in protected areas sheltered from the full ocean fetch; and (3) an abundance of sand.

Most of the bars discussed thus far, whether single or multiple, are in sufficiently shallow water so that waves can break over the bar and could in many instances account for the bar formation. This mode of origin generally is referred to as the *break-point hypothesis.* In some multiple-bar systems, however, the outer bars are in water too deep to be associated with breaking waves (e.g., Scott, 1954; O'Hare and Davies, 1990), so another mode of formation is required. Although these deep bars demonstrate the need for explanations beyond the break-point hypothesis, these alternative hypotheses might also explain the origin of bars within the surf zone that are otherwise affected by wave-breaking processes.

The hypothesis developed by Boczar-Karakiewicz and Davidson-Arnott (1987) relates the formation of shore-parallel bars directly to the transformations of shoaling waves as they approach the shore, there being a regular pattern of energy transfer between the peak frequency of the waves and the first higher harmonic. Free second-harmonic waves have been observed to accompany progressive waves in flumes, and Bijker, van Hijum, and Vellinga (1976) found that the resultant pattern of near-bottom water particle velocities can cause sand bars to form at regular intervals along the flume. Boczar-Karakiewicz and Davidson-Arnott developed a model that predicts the numbers and spacings of bars, starting from an initially planar slope, and found that the predicted pattern correlates well with wave transformations and observed bars in Georgian Bay, Lake Huron. Further analyses have been undertaken by Boczar-Karakiewicz, Forbes, and Drapeau (1995), with comparisons between the model and bars found along beaches within the Gulf of St. Lawrence.

Another theory of multiple-bar formation is associated with the presence of standing waves produced by the reflection of surface waves from the beach. Lettau (1932) was apparently the first to suggest that sediment would tend to be scoured from beneath the nodes of standing waves and accumulate under the antinodes, possibly resulting in the formation of longshore bars. This pattern of sediment movement corresponds to the water-particle velocities under standing waves, the horizontal velocities being a maximum under nodes and decreasing to zero at the antinode positions. Carter, Liu, and Mei (1973) and Lau and Travis (1973) investigated the formation of multiple bars under standing waves but focused on the offshore distribution of the mass transport associated with the waves (Chapter 5). With simple progressive waves approaching the shore, the mass transport near the sediment bed is directed onshore. If there is significant wave reflection from the beach, then as illustrated in Figure 7-27, a set of standing waves is established and there are reversals in the near-bottom mass transport of water. The residual circulation now consists of quarter-wavelength cells in which the trapped fluid slowly rotates. This residual motion is upward beneath the antinodes of the standing waves and downward beneath the nodes. Hence, close to the bed the current direction is from nodes to antinode positions, and one would expect that sediment carried

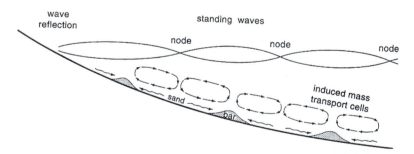

Figure 7-27 Schematic diagram illustrating bar formation associated with standing waves due to wave reflection from the beach. The standing waves form an envelope of surface-amplitude variations, with zero amplitudes at the nodes and maximum amplitudes at the antinodes. The mass transport of water associated with the standing waves forms a series of cells, with the bottom current flowing from node to antinode positions. Sediment carried by the mass-transport currents can be expected to accumulate as bars at the antinodes where the currents converge.

along by the current would tend to accumulate beneath the antinodes to form a system of bars spaced at half the surface wavelength of the standing waves (Fig. 7-27).

This model attributing multiple bars to standing reflected waves was developed theoretically by Carter, Liu, and Mei (1973), while Lau and Travis (1973) extended the theory to account for a sloping bottom. The theoretical development was partially confirmed by Carter and co-workers in laboratory wave-tank experiments. Systems of bars were formed in the flat portion of the tank at the toe of a reflecting beach slope. The theory predicts that the number of bars is likely to increase when the bottom gradient decreases and that the spacings between bars increase in the offshore direction. Both predictions generally conform to what is observed in nature. Lau and Travis showed good agreement between theory and observations from sites in Lake Michigan, the Black Sea, and in Escambia Bay, Florida. According to the theory, for good bar development, wave reflection from the beach should be on the order of 40 percent (the percentage of wave energy reflected). Studies of wave reflection suggest that this would require a beach slope of about 17°, unusually steep for a natural beach. However, subsequent investigations by Davies and Heathershaw (1984) and O'Hare and Davies (1990) demonstrated that bars can develop when less than 20 percent of the wave energy is reflected.

Heathershaw and Davies (1985) have shown theoretically and experimentally that after one or two bars are formed, their presence can increase the wave reflection that results in the formation of additional bars seaward from the original bars. The theoretical analysis indicates the existence of a resonant Bragg-type interaction between the surface waves and the bedforms, which gives rise to a substantial back-reflected wave. Wave-flume experiments by Heathershaw and Davies and by O'Hare and Davies (1990) have demonstrated the existence of this resonant reflection and that it can lead to the formation of additional offshore bars. Heathershaw and Davies documented the initial formation of bars due to wave reflection in flume experiments where a thin layer of sand was placed over a uniformly sloping bottom. The experiments of O'Hare and Davies were more extensive and involved a deep sediment substrate rather than a thin layer of sand. One set of experiments used fine glass spheres that were easily transported in suspension by the waves. In this case, the sediment was carried from positions beneath the nodes of the standing-wave envelope toward positions beneath the antinodes, just as diagrammed in Figure 7-27. The second series of experiments involved the use of a natural sand that was coarser than the glass spheres and had a greater spread of diameters. In those experiments, more of the sand was transported as bedload, less in suspension. The bars were then observed to form approximately midway between the nodes and antinodes of the standing waves. This was explained in terms of the mechanics of bedload transport, where asymmetries in wave orbital motions are important to the cross-shore sand transport, in addition to the effects of the directions of the imposed mass transport. The results indicate that bar crests may form at any position relative to the nodes and antinodes of the standing waves, depending on the proportions of coarse and fine grains present in the sediment.

In some of the experiments completed by O'Hare and Davies (1990), large bars developed, leading to a reduction in the local water depth over the bar crests and resulting in wave breaking over the bars. This caused large regions of turbulence and led to the rapid destruction of the bars. The results of one such experiment are given in Figure 7-28. The initial set of bars took about 30 minutes to develop, during which time the amplitude of the reflected waves increased and the reflection coefficient evaluated from the offshore waves increased from about 0.1 to more than 0.5, demonstrating the importance of the bars in enhancing the reflection (note in Fig. 7-28 that there was little increase in reflection from the beach face).

A   Incident & reflected wave amplitudes

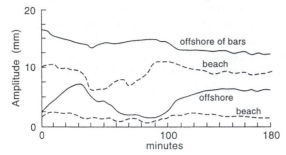

B   Reflection Coefficients

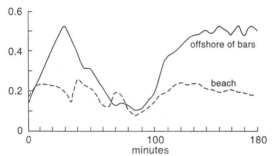

Figure 7-28 (a) Time variations in the amplitudes of the incident and reflected waves measured offshore from the developing bars and also in the immediate offshore directly in front of the sloping beach face but shoreward of the bars. (b) Reflection coefficients at the beach face and for the total reflection offshore from the developing bars. The time variations resulted from the initial development of the bars during the first 30 minutes of the experiment, their subsequent destruction by wave breaking over the bar crests and then the development of a second set of bars after 90 minutes into the experiment. (c) The final equilibrium bed profile, showing the presence of thirteen bars. [Adapted with permission of Journal of Coastal Research, from T. J. O'Hare and A. G. Davies, A Laboratory Study of Sand Bar Evolution, *Journal of Coastal Research 6,* p. 541. Copyright © 1990 Journal of Coastal Research.]

C   Profile of final bars

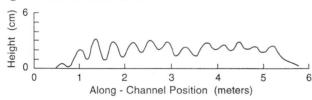

About 30 minutes into the experiment, waves began to break over the bars, leading to their erosion and a parallel decrease in the amplitude of the reflected wave and reflection coefficient. The bed and wave conditions essentially returned to those at the beginning of the experiment. Once the original bars had been removed, the waves again became stable and at 90 minutes into the run a second set of bars began to form, again leading to higher reflection coefficients offshore from the bars. Curiously, although the development of this second set of bars did lead to some wave breaking, it did not result in their destruction; the final bed profile is included in Figure 7-28(c), showing a patch of thirteen bars. From this, O'Hare and Davies concluded that there must be a very fine dividing line between stable conditions in which bars form and exist in equilibrium with the surface waves, and unstable conditions in which the bars are eroded by breaking waves. They further suggested that the original set of bars formed in the experiment were unstable because insufficient quantities of coarse grains had migrated to the bar crests, whereas when the bars later reformed, the accumulated coarse sand stabilized the bars even after wave breaking began.

The above studies provide strong support for the formation of multiple bars by wave reflection and indicate that once bars have formed in the nearshore, by whatever mechanism, the resonant interaction with incoming waves may result in enhanced wave reflection that can induce further bar formation. However, reflected incident waves have nodes every 10–20 m in the cross shore, too short in most locations to match the offshore distance to the first bar or the spacings between multiple bars. Short (1975a, 1975b) and Bowen (1980) extended the short-wave theory of Carter and co-workers (1973) and Lau and Travis (1973) by proposing that the bar formation is in response to long-period infragravity motions that are known to contain considerable energy within the surf zone (Chapter 6). Short examined the role of infragravity motions in the generation of multiple bars in the Chukchi Sea, north Alaska, and was able to show that measured infragravity wave spectra contained significant amounts of energy at the correct frequencies for the observed bar spacings (periods of 75–100 sec).

It has been proposed that edge waves within the infragravity range may be responsible for the formation of nearshore bars. As will be discussed in Chapter 11 where we consider various rhythmic shoreline forms, long-period (infragravity) standing edge waves almost certainly account for crescentic bars, as proposed by Bowen and Inman (1971). Holman and Bowen (1982) suggested that progressive edge waves may similarly give rise to multiple parallel bars. The mechanism is analogous to that described above for standing reflected waves, with bars tending to form beneath the antinodes of progressive edge waves where the cross-shore velocities are a minimum. It was seen in Chapter 6 that field studies have documented the simultaneous existence of several edge waves, progressive and standing, with different periods and modes. Therefore, it is possible that a complex pattern of bars could be generated, as demonstrated in the computer models developed by Holman and Bowen. Support for an edge-wave origin of multiple bars has come from the work of Aagaard (1991) on the coast of Denmark. During an intense storm, progressive edge waves developed, while a less intense storm produced standing edge waves. In both cases a single edge-wave mode dominated the measured energy spectra. It was shown by Aagaard that the scales of the offshore linear bars corresponded with those expected from the progressive edge waves developed during major storms but would otherwise remain stationary during low-wave conditions. In contrast, the standing edge waves that developed during a less intense storm resulted in the formation of crescentic bars, as hypothesized by Bowen and Inman.

One of the complicating factors in understanding the generation of nearshore bars is the feedback of the progressively deformed bottom. The presence of a sand bar leads to a concentration in wave breaking over its crest. The infragravity model is also affected in several ways. Most importantly, Kirby, Dalrymple, and Liu (1981) have shown that when sand bars of finite amplitude are present, they perturb the cross-shore structure of the standing waves to displace the velocity nodes toward the bar positions, thereby reinforcing the bar-generation mechanism. In addition, Symonds and Bowen (1984) suggested that the presence of a sand bar could lead to a preferred forcing of long waves that would fit the scale of the bar. Thus, the presence of a small bar would be expected to alter the wave field in several ways that could encourage further bar growth.

Howd et al. (1991) have proposed a mechanism for bar generation wherein a strong longshore current has the same effect on the standing infragravity waves as the presence of a sand bar. Even though the bottom may have an essentially uniform slope, the longshore-current interaction produces an "effective bar" that could have the same feedback effect as a real bar. In a comparison with measurements at the FRF, Duck, North Carolina, Howd et al. found that the storm adjustment of bar positions was always toward the effective bar location but with a lag time of roughly 10 hours. Therefore, the proposed mechanism of Howd et al. might account for bar formation when a strong longshore current is present.

As recounted here, several hypotheses have been offered to explain the origin of bars within the nearshore. There is evidence both pro and con for each of these hypotheses with no consensus having emerged. The break-point hypothesis is valid at least for some bars that develop within the surf zone and is most easily demonstrated with monochromatic waves in laboratory wave channels or with very narrow spectra of regular swell waves on natural beaches. There is a problem in the application of the break-point hypothesis with random waves having a range of heights, a common situation on natural beaches, as bars are observed to form within the wave-saturated portion of the surf zone where waves are breaking at all depths (Sallenger and Howd, 1989; Holman and Sallenger, 1993). The hypotheses involving bar development due to standing waves formed by wave reflection from the beach, or by sediment movement associated with progressive or standing edge waves, or due to the existence of higher harmonics in shoaling waves, all appear theoretically reasonable and have received support from field observations. It is possible that, depending on the circumstances, these various hypotheses could at different times account for the formation of multiple bars. In a review of the research that has been undertaken at the FRF, Holman and Sallenger (1993) concluded that bar generation at that site is not simply a result of break-point generation or of sediment transport toward the nodes of a single standing infragravity wave, but instead results from a hybrid of mechanisms. They note that sorting out the respective contributions of the different mechanisms will require the implementation of complex numerical models combined with extensive programs of field-data collection. This is expected to be the focus of a great deal of research in the future.

PROFILE CHANGES DUE TO STORMS

Observed Profile Responses

Thus far we have focused on the components of beach profiles—the slope of the beach face, the elevation of the berm, and the number of offshore bars. Now we turn to an examination of how the overall profile responds to changes in the energy level of the waves, to water elevation variations associated with tides, and to the direct effects of winds. These responses are revealed when repeated profiles of the beach are obtained at some fixed location on the coast, and a reference bench mark is used so that long-term systematic changes can be determined as well as short-term fluctuations.

The main systematic variation observed in the series of beach profiles would most likely be in response to changing wave conditions, a variation that could result from a single storm of unusual magnitude or could be spread out over the year due to the general seasonality of wave intensity experienced on most coasts. The pattern of this beach-profile response to wave intensity is illustrated schematically in Figure 7-29. Such shifts in profiles were first documented by Shepard (1950b) and Bascom (1953) on the west coast of the United States where storm waves dominate the winter and lower energy waves occur throughout the summer. Based on this marked seasonality, Shepard referred to the two profile types as *summer profiles* versus *winter profiles*. In a wave tank study that documented a similar response to the wave energy, Johnson (1949) applied the terms "storm profile" and "normal profile," which eliminates the seasonality. In the previous edition of this text, the terms "storm profile" and

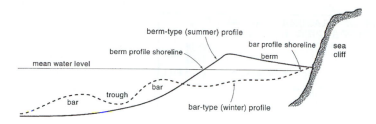

Figure 7-29   The bar-type profile that forms during a storm versus the berm-type profile that occurs under lower-energy wave conditions.

"swell profile" were used, also eliminating any seasonal implications, since in many parts of the world the response is to individual storms such as hurricanes. Here I opt for *bar profile* and *berm profile*, following the use by Larson, Kraus, and Sunamura (1988), reflecting the usual presence of bars versus an accretional berm in the two profile types. Strictly speaking, this still is not entirely satisfactory as bars do not always form during high-wave conditions, and a distinct berm may not develop on a fine-sand beach. Whatever terms are used, the cycle itself is important because it shows a tendency for beach profiles to change morphology in response to the intensity of the wave conditions, as illustrated in Figure 7-29. With low-energy waves, the profile typically is characterized by a wide berm and by a smooth offshore profile with little or no development of bars except perhaps in relatively deep water. In contrast, under higher wave conditions, the berm is destroyed by the intensified wave swash and the sediment is shifted offshore to form one or more bars that extend parallel to the shoreline. The volume of sand involved remains relatively constant so long as there has been minimal longshore movement—the areas under the two profiles of Figure 7-29 are about the same, representing the volume of sand per unit shoreline length.

Shepard (1950b) obtained repeated profiles for several years along Scripps Pier, La Jolla, California, that show the seasonal profile shifts illustrated schematically in Figure 7-29. The profile series (Fig. 7-30) demonstrates the tendency for the beach to shift from more evenly sloping berm profiles during the summer to fall, to a bar profile during the winter to spring, and back again. The change in profile geometry is brought about by a cyclical cut and fill, clearly representing a cross-shore sediment transport and exchange between the offshore bar, which grows during winter storms, and the upper beach which increases in elevation during the summer. Figure 7-31 from Bascom (1953) shows approximately monthly beach profiles obtained during 1946–1947 at Carmel, California. On this coarse-sand beach, a well-defined, nearly horizontal to landward-sloping berm develops during the summer, shown by the accretional series of profiles from 4-21-46 to 9-4-46. With the return of high wave conditions during the winter, the beach face and berm are cut back, seen in the dashed profiles of 12-10-46 and 2-21-47. The sand removed from the berm presumably moved offshore to form bars, but the profiles do not extend into water sufficiently deep to reveal the bar growth. Also noteworthy in these two examples of profile cycles from California beaches are the marked contrasts in the elevation changes of the berms experienced between summer and winter at these two locations. On Scripps Beach studied by Shepard (1950b), the cut-and-fill cycle changes the elevation of the berm by on the order of 1–1.5 m (Fig. 7-32), while at Carmel the change was 2–3 m. This may result in part from differences in the wave-energy levels at the two locations, but undoubtedly a more significant factor is the contrasting beach-sediment

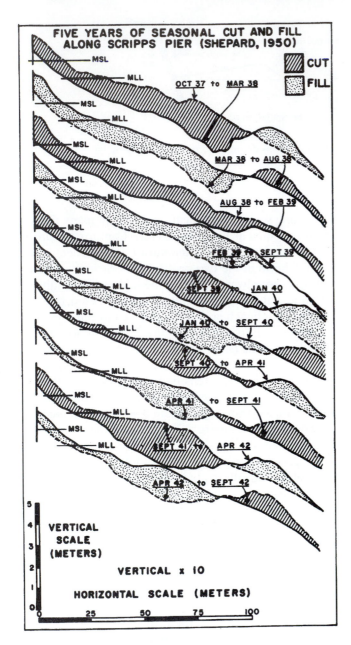

Figure 7-30 Profile changes along Scripps Pier, La Jolla, California, showing the tendency to shift from a more evenly sloping berm-type profile in the summer to a bar-type profile in the winter and back again, the seasonal cycle of profiles found on many beaches. [From F. P. Shepard, *Beach Cycles in Southern California,* Beach Erosion Board Technical Memo No. 20, U.S. Army Corps of Engineers, 1950.]

grain sizes, with Scripps Beach being a fine- to medium-sand while Carmel is a coarse-sand beach. The general observation is that, at least within the range of sandy beaches, the coarser-sand beaches show greater elevation changes over the seasonal cycle and also in response to individual storms. This has been established by profile series on the Oregon coast where the beaches otherwise have the same wave climate (Shih and Komar, 1994). Those studies demonstrate that, due to the large vertical changes of the coarse-sand beach profiles and also due to their rapid cut-back by storm waves, coastal properties backing the coarse-sand beaches are

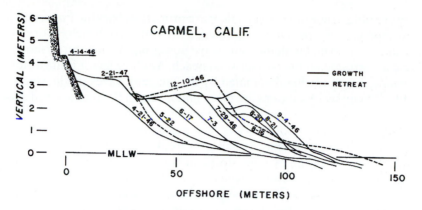

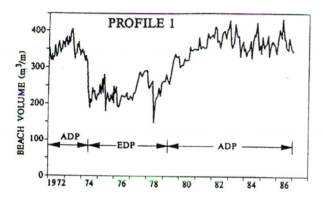

Figure 7-31 The progressive growth and retreat of the berm at Carmel, California, during 1946–1947. [Adapted from Characteristics of Natural Beaches, W. H. Bascom, *Proceedings of the 4th Coastal Engineering Conference,* 1953. Reproduced with permission from American Society of Civil Engineers.]

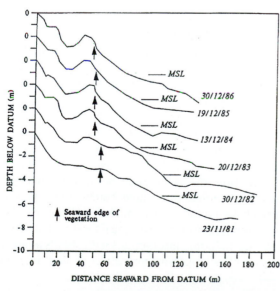

Figure 7-32 (a) Beach volume changes in a profile time series obtained at Moruya, a pocket beach on the New South Wales coast of Australia. Two distinct accretion-dominated periods (ADP) are separated by an erosion-dominated period (EDP) that occurred during a major storm. (b) Profiles from Moruya, showing the growth of a foredune during the accretion-dominated period, which developed to such an extent that the dunes represented some 10 percent of the dry sand on the beach. [Used with permission of John Wiley & Sons, Inc., from B. G. Thom and W. Hall, Behavior of Beach Profiles During Accretion and Erosion Dominated Periods, *Earth Surface Processes and Landforms 16,* pp. 116–120. Copyright © 1991 John Wiley & Sons, Inc.]

more susceptible to episodic erosion than are properties backing finer-sand beaches. A factor in this response is that the coarser-grained, more reflective beaches are steeper and the surf zone is narrower, so that with the same wave power input from the offshore, the wave energy is more concentrated on the coarser-sand beach. Another factor is the difference in the wave swash characteristics, being dominated by the incident waves on the steep reflective beach, while infragravity swash motions are more important on the fine-grained dissipative beaches, which tend to filter out the incident-wave energy (Chapter 6).

While a beach may erode very rapidly during the one or two days of a major storm, its subsequent recovery phase may extend over weeks or years. In a study of beaches in southern California following an extreme storm on January 16–18, 1988, Egense (1989) found that, during the first phase of recovery, which covered the first three weeks following the storm, the berm regained between 45 and 65 percent of the sand volume lost during the storm. The second phase was marked by a much lower rate of recovery and was still continuing nearly a year later when the study was completed. In a study on the New Jersey coast, Birkemeir (1979) similarly found an initially rapid rate of beach recovery following a storm, with over half of the eroded volume returning within two days.

The study of Thom and Hall (1991) on the coast of Australia has particularly documented the cycles of beach erosion and recovery, based on monthly beach profiles collected for 16 years (1972–1988). The study was centered on a 5.5-km crescent-shaped pocket beach, which ensured the conservation of the total volume of sand so that any long-term trends of erosion or accretion did not result from losses or additions of sand to the system. Figure 7-32(a) shows a representative time series of sand volumes on beach profiles above a low-tide datum, expressed as the volume of sand per meter longshore length. The series is clearly divided into two accretion-dominated periods (ADP) and a single erosion-dominated period (EDP). The EDP interval was initiated by a major storm in 1974 that eroded the beach, while a second storm in 1978 cut back the beach farther. Subsequent to that 1978 storm, the beach volumes have recovered with the initial rate of increase (0.269 m^3/m/day) being the most rapid, and with subsequent rates progressively declining, just as found by other studies. It can be seen in Figure 7-32(a) that this accretion-dominated period extended over several years, with most of the volume recovery having been completed within 5 years. Thom and Hall showed that much of the volume gain, especially since 1981, has been in the growth of a foredune at the back of the beach. This dune growth can be seen in the series of profiles in Figure 7-32(b). With time, the foredune shifted landward as well as growing upward and became increasingly asymmetric. The foredunes contain some 10 percent of the total volume of subaerial sand above the datum, which serves as a reserve and buffer during major erosion events.

A similar study by Morton, Paine, and Gibeaut (1994) documented the beach recovery on the Gulf of Mexico coast of Texas, following Hurricane Alicia that eroded the beaches during August 1983. Their observations are based on a 10-year monitoring program following the storm. They identified four time-dependent stages of recovery: (1) rapid forebeach accretion following the storm, (2) slower backbeach aggradation, (3) dune formation, and (4) dune expansion and vegetation recolonization. With time, the natural beaches passed through all four stages, but beaches backed by developments reached only stage 2 since additional recovery was prevented by the presence of houses backing the beach which restricted dune-sand accumulation. Morton and co-workers also found that there were local factors that affected the recovery of the beaches following the storm, including interactions with shoals at an adjacent tidal inlet, the adverse effects of jetties, and the longshore migration of rhythmic shoreline features (Chapter 11).

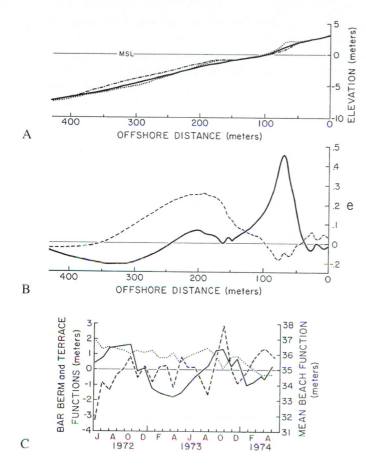

A

B

C

Figure 7-33 Eigenfunction analysis of beach profiles from Torrey Pines Beach, California. (a) The dash-dot profile was measured during April 1972, while the dotted profile is from October 1972. The largest eigenfunction obtained in the analysis of many profiles corresponds to the "mean beach function," the solid profile in the diagram, representing an average profile for the years. (b) The second eigenfunction, the "bar-berm function" (solid curve), has a large maximum at the location of the summer berm and a minimum at the location of the winter bar; it is related to the cross-shore movement of sand resulting from seasonal variations in wave energy. The third eigenfunction is the "terrace function" (dashed curve). (c) The magnitude and sign of the bar-berm function (solid line) varies with the season as the profile shifts from a summer berm-type profile to a winter bar-type profile, while the terrace function (dashed) shows no systematic variation. [Adapted with permission of American Geophysical Union, from C. D. Winant, D. L. Inman and C. E. Nordstrom, Description of Seasonal Beach Changes Using Empirical Eigenfunctions, *Journal of Geophysical Research 80*, p. 1985. Copyright © 1975 American Geophysical Union.]

Eigenfunction Analysis of Profile Variations

The spatial and temporal variations of beach profiles due to changing waves and other processes can be statistically analyzed by using empirical eigenfunctions. This approach was first applied to beach profiles by Winant, Inman, and Nordstrom (1975), with additional verification of the approach provided by Winant and Aubrey (1976). Eigenfunction analysis expresses the profiles in terms of the principal components and determines the major modes by which changes occur in the profile form. The principal components represent the dominant modes of change along the profile, simplifying the examination of the main trends and setting up classifications of beach-profile data.

Winant, Inman, and Nordstrom (1975) analyzed a 2-year series of profiles collected from Torrey Pines Beach, California, and found that most of the variation in profile configuration can be accounted for by three eigenfunctions. An example is presented in Figure 7-33. The dash-dot profile of Figure 7-33(a) was obtained during April 1972, while the dotted profile is from October 1972. The largest eigenfunction obtained in the analysis of many profiles corresponded to the "mean beach function," which represents an average profile; this is the solid profile shown in Figure 7-33(a). The second eigenfunction, the "bar-berm function," has a large maximum at the location of the summer berm and a minimum at the location of the winter bar [Fig. 7-33(b)]; it is related to the cross-shore movement of sand resulting from seasonal variations in wave energy. The magnitude and sign of this bar-berm function is seen in Figure 7-33(c) to vary systematically with the season as the profile shifts from a summer berm-type profile to a winter bar-type profile. The third eigenfunction obtained in the analysis is termed the "terrace function" as its position of maximum importance [Figure 7-33(b)], corresponds to a terrace located in the measured profiles between the changing beach face and offshore bar. It does not vary systematically with the season [Fig. 7-33(c)]. The results demonstrate a high degree of sensitivity in the eigenfunction analysis of beach-profile data and in its ability to detect relatively small changes in the profile configuration.

Aubrey (1979) demonstrated the usefulness of eigenfunction analysis in determining patterns of profile changes and cross-shore sediment transport. Based on 5 years of monthly profiles obtained at Torrey Pines Beach, California, the bar-berm function suggested two seasonal pivotal points separating eroding and accreting regions; one is the expected pivot point between the berm and offshore bar, while the second is at a water depth of 6 m and represents an exchange of sand between the bar and the offshore. The results from the eigenfunction analysis are confirmed by profile shifts monitored by surveys and by accurate rod measurements of sand elevations in the offshore. The results indicate that during the initial winter storms, sand is eroded from both the foreshore and from depths of 6–10 m and is deposited in water depths from 2 to 6 m to form the bar. During less energetic periods, sediment migrates both shoreward to the beach face and seaward from the bar. A sediment budget for the seasonal cross-shore transport establishes that the exchange of sand between the bar and foreshore through the 3-m pivotal point involves 85 m^3/m, while the offshore exchange through the 6-m pivotal point is 15 m^3/m. Utilizing the same data set, Aubrey, Inman, and Winant (1980) have shown that daily and weekly beach changes, expressed by the eigenfunction analysis, are correlated with and predictable from the wave-energy level.

Felder, Hayden, and Dolan (1979) employed eigenfunction analysis to compare the inshore profiles of the shoreface with profiles that extend across the inner continental shelf. The results reflected differences in the history of equilibrium slope evolution and also reflected wave climates, tidal ranges, and sediment characteristics. Bowman (1981) provides an example of the use of eigenfunction analysis to classify beach profiles and their changes, based on a 1-year set of profiles from the Mediterranean coast of Israel. The second eigenfunction, which accounted for 51–84 percent of the residual variance from the mean beach profile (the first eigenfunction), revealed patterns of change in three subenvironments of the exposed beaches—the upper foreshore, a stable segment than may contain a longshore trough, and the upper backshore.

Prediction of Profile Changes

Earlier in this chapter we reviewed the attempts that have been made to correlate the geometry of the components of the beach profile, such as the beach-face slope, to the wave and sediment parameters. In that changes in the slope and numbers and sizes of bars are also a part of the overall shift from the berm profile to the bar profile with a changing wave-energy level, one would expect the same types of correlations to predict the overall shift. The earliest attempts to predict the form of the profile focused on the wave steepness H_∞/L_∞, the ratio of the deep-water wave height to the deep-water wave length, which is related to the wave period by $L_\infty = (g/2\pi)T^2$. Storm waves have high steepness values because of their greater heights and shorter periods, while long swell waves have low steepness values. The wave steepness can be increased either by an increase in the wave height or by a decrease in the wave period. In his wave-channel experiments, Johnson (1949) found profile changes like those described above, which he associated with the magnitude of the wave steepness. He determined that with a wave steepness greater than 0.03, an offshore bar always forms (bar profile), whereas if the steepness is less than 0.025, an offshore bar is never formed (berm profile). Rector (1954) and Watts (1954) both found that a wave steepness of 0.016 is critical for developing a berm profile versus a bar profile. King and Williams (1949) concluded that a wave steepness of 0.012 is important in governing whether sand moves onshore or offshore within the surf zone and therefore determines the profile type. The differences in the exact value of this critical wave steepness, found by various studies, is due in part to a dependence on the grain size of the beach sediment (Rector, 1954). Some of the variation may also result from the scale of the experiments. Results with large waves in a large wave tank or on real ocean beaches might be different than with small waves in a small wave tank. This is indicated by the results of Saville (1957) in a large-scale wave channel, obtaining a critical wave steepness of 0.0064 that is much lower than those found in studies utilizing small-scale wave channels. The grain size and scale effects on the beach profile type have been considered by Iwagaki and Noda (1962). Their plot in Figure 7-34 relates the profile type to the wave steepness and the ratio of the deep-water wave height to the mean sediment diameter, H_∞/D, demonstrating that the critical wave steepness does depend on the wave height and sediment grain size.

Dean (1973) presented a model for the shift from a berm-type to a bar-type profile, based on a consideration of the trajectory of a suspended sand particle during its fall to the bottom, acted upon at the same time by the horizontal water-particle velocities of the waves. If the grain fall requires a short time relative to the wave period, then the particle will be acted upon predominantly by onshore velocities. On the other hand, if the fall velocity is low, then the grains will tend to shift offshore. Based on this model, Dean established that the

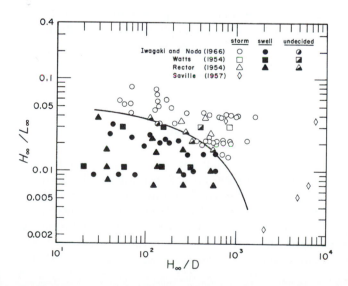

Figure 7-34 The critical wave steepness H_∞/L_∞ at which the beach profile shifts from a bar-type to a berm-type profile; D is the median sediment diameter. [Adapted from Laboratory Study of Scale Effects in Two-Dimensional Beach Processes, Y. Iwagaki and H. Noda, *Proceedings of the 8th Coastal Engineering Conference*, 1962. Reproduced with permission from the American Society of Civil Engineers.]

dimensionless ratio w_s/gT is important in governing the critical wave steepness. Using various laboratory data sets, Dean obtained the relationship

$$critical \frac{H_\infty}{L_\infty} = \frac{1.7\pi w_s}{gT} \tag{7.21}$$

for the critical wave steepness. For those data, the relationship correctly predicted whether the transport is onshore or offshore in 87.5 percent of the cases, that is, whether there is a berm profile or a bar profile. Allen (1985) applied this relationship to beaches on Sandy Hook, New Jersey, and found that, with a proportionality coefficient of 2.0 and using the breaker height rather than the deep-water wave height, the model correctly predicted 98 percent of the erosion events, the shift to a bar profile. On the other hand, it predicted only 45 percent of the shifts to a depositional berm profile. Allen attributed this poorer prediction of the accretional phase to the considerable lag time that can exist on natural beaches for profile shifts in that direction. Kriebel, Dally, and Dean (1986) carried out a number of additional laboratory experiments and found that using large-scale tests yielded proportionality coefficients between 4 and 5 in Equation (7.21), rather than the 1.7 value found by Dean. Also utilizing only large-scale wave channel data, Kraus and Larson (1988) and Larson and Kraus (1989) found a 5.5 coefficient but concluded that a better separation between bar-type profiles and berm profiles is provided by the relationship

$$critical \frac{H_\infty}{L_\infty} = 115\left(\frac{\pi w_s}{gT}\right)^{3/2} \tag{7.22}$$

Kraus and Larson (1988) and Larson, Kraus, and Sunamura (1988) also compared the critical wave steepness for the transition from bar profiles to berm profiles with the dimensionless ratio $H_\infty/w_s T$, utilizing data from large wave channels. The results (Fig. 7-35) yielded

$$critical \frac{H_\infty}{L_\infty} = 0.00070\left(\frac{H_\infty}{w_s T}\right)^3 \tag{7.23}$$

as the line separating the two profile types. As discussed by Dalrymple (1992), this relationship is potentially confusing since for a fixed value of $H_\infty/w_s T$, increasing the wave steepness leads to a shift from a bar-type (storm) profile to a berm profile, opposite to the expected dependence and trend established by other studies. Dalrymple demonstrated that this results from a direct relationship between H_∞/L_∞ and $H_\infty/w_s T$. He suggested that the relationship be rearranged into a single dimensionless number

$$\frac{gH_\infty^2}{w_s^3 T} = \frac{2\pi}{0.0007} \approx 9.0 \times 10^3 \tag{7.24}$$

and further noted that Equation (7.21) can be altered to the same form with a constant of 1.04×10^4.

An understanding of the critical wave conditions that governs the shift from the berm profile (summer profile) to the bar profile (winter profile) is still incomplete. The above relationships are based almost entirely on laboratory data and their applicability to field conditions remains uncertain. Although there are many field studies that demonstrate an offshore shift of beach sand during storms to form bars, and a shoreward return during low-wave conditions (Shepard and LaFond, 1940; Shepard, 1950a, 1950b; Bascom, 1953; Strahler, 1966; Gorsline, 1966), none of those studies obtained satisfactory wave measurements that

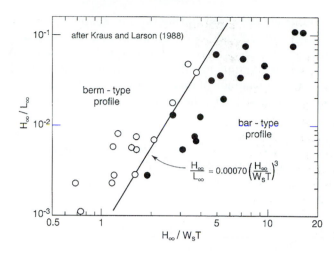

Figure 7-35 The critical wave steepness H_∞/L_∞ at which the beach profile shifts from a berm-type (summer) profile to a bar (winter) profile as a function of $H_\infty/w_s T$. [Adapted from N. D. Kraus and M. Larson, *Beach Profile Change Measured in the Tank for Large Waves, 1956–1957 and 1962*. Technical Report CERC-88-6, U.S. Army Corps of Engineers, 1988.]

could determine a critical wave steepness. The studies all demonstrate that an increase in the wave height during storm conditions clearly leads to a bar profile with an offshore shift of sand. The dependence on the wave period is less clear. According to the hypothesis of a critical wave steepness, a shift to a bar profile could be accomplished by decreasing the wave period while maintaining the wave height constant. This is indicated by pairs of profiles obtained in a wave tank by Iwagaki and Noda (1962) but has not been established clearly on ocean beaches. Shepard and LaFond (1940) indicated that longer-period waves tend to shift sand offshore into bars, opposite to that implied by a critical wave steepness, although the effect of the wave period is not as great as the wave height. Their evidence is not conclusive, but it points to the need for additional field studies that carefully tie the cross-shore shifts in beach profiles to the nature of the waves.

A complicating aspect in the two-dimensional consideration of beach profiles presented here is that profile variations on natural beaches are commonly three dimensional. Sonu and Russell (1966) pointed out that when crescentic bars are present (e.g., Fig. 7-7), a profile could be either a bar profile or a berm profile lacking bars, depending on where the profile is obtained within the alongshore-varying crescentic bar. If repeated profiles are obtained along a single transit line, longshore migrations of the crescentic bars could result in alternating profile types, even though there are no accompanying change in wave conditions. Clarke and Eliot (1988) further demonstrated that due to the presence of rip-current cells and the re lated rhythmic topography, spatially restricted surveys may not yield a reliable representation of the total morphology of a beach and the changes it undergoes. It is clear that, in field studies of beach profiles, variations in the longshore direction must be monitored.

PROFILE CHANGES DUE TO TIDES

In addition to the response of beach profiles to wave conditions, repeated profiles over a month or longer might show alterations that correspond to the tides. This involves hourly changes resulting from the rising and falling water levels and also longer-term effects due to differences in the ranges of spring versus neap tides.

Although not pronounced, the profile changes from spring tides to neap tides are interesting. Thompson and Thompson (1919) and LaFond (1939) demonstrated that on southern California beaches, the sand surface a few meters above the mean tide level reaches its minimum elevation a few days after spring tide and its maximum elevation following neap tide. Grunion (the fish *Leuresthes tenuis*) take advantage of this cycle by laying their eggs just after spring tides. The eggs are then buried by sand during the neap tide when deposition occurs. During the next spring tide, the sand is again removed and the eggs are ready to hatch. This spawning takes place from March to August, when the wave conditions are steadiest, so the tide-induced changes are most effective. At other times of the year profile changes due to varying wave conditions are greater, so the tidal influence is masked. This is also true at other beaches—cycles due to spring and neap tides are usually small compared with wave effects.

Several studies have investigated the hourly profile changes that result from varying water levels of the daily tidal cycle. Strahler (1966) made half-hourly observations of the foreshore changes during one tidal cycle at Sandy Hook, New Jersey. Sand scoured from the swash zone was deposited at the upper limit of the swash and in the step under the breaker zone. As the tide rose, at any fixed location on the beach face, a small amount of deposition occurred, followed by erosion as the site came under the intense swash, in turn followed by deposition as the breaker zone passed over. Just the reverse occurred with the falling tide. A similar pattern of sediment movement was demonstrated by Otvos (1965) and Schwartz (1967) through the use of fluorescent sand tracers. Schwartz determined that the finer fraction of the sediment was sorted out and deposited at the top of the swash or was carried seaward of the breaker zone, while the coarser material accumulated in building the step beneath the breaker zone. Duncan (1964) investigated the effects of the water table on this pattern of profile readjustment to a daily tidal cycle. During a flood tide the water level rises faster than the water table within the beach, so the seaward edge of the water table slopes shoreward as shown in Figure 7-36 (Emery and Foster, 1948; Grant, 1948; Harrison, Fang, and Wang, 1971). During ebb tide the water table slopes seaward. Because of this, during a flood tide the water from the swash run-up is lost to percolation into the beach, and the return backwash is weakened. Just the opposite occurs during the ebb tide, since water is added to the backwash. Duncan (1964) found that most of the sediment transported up the beach face by the swash during flood tide is deposited at the top of the swash limit, due to the enhanced percolation (Fig. 7-36). During ebb tide the stronger backwash removes this sand and deposits it on the shoreward side, where the backwash collides with the incoming surf and loses its transporting capacity.

The importance of the interaction of the swash and the water table has been further demonstrated by Harrison (1969). Using regression analysis statistical techniques, Harrison found that most of the variability in the quantity of sand eroded from or deposited on the foreshore over the interval from one low water to the next can be explained by (1) the steepness of the breaking waves, (2) the hydraulic head between the water table within the beach and the swash run-up level, and (3) the angle of wave approach to the shoreline. He provides an empirical equation for this net-volume change, as well as equations for the resulting advance or retreat of the shoreline and the mean slope of the foreshore.

In areas where the tidal range is large, the beach profiles tend to be characterized by a relatively steep beach face, terminated abruptly at its base by a wide low-tide terrace (Inman and Filloux, 1960). Generally, this high-tide beach face is composed of sediments that are coarser than those of the terrace. In the example of Figure 7-37 from the coast of Wales, the steep shingle upper beach is distinct from the sandy terrace. In addition, low-tide terraces

Flood Tide

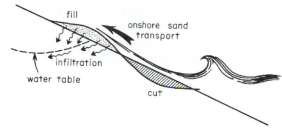

A

Ebb Tide

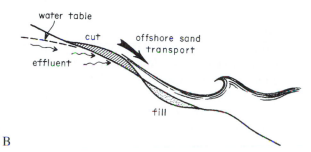

B

Figure 7-36 Water table effects on the cut and fill of the beach profile during (a) flood tides and (b) ebb tides. [Adapted from *Marine Geology* 2, R. Duncan, The Effects of Water Table and Tidal Cycle on Swash-Backwash Sediment Distribution and Beach Profile Development, 1964 with kind permission of Elsevier Science-NL, Sara Burgerhartstratt 25, 1055 KV Amsterdam, The Netherlands.]

Figure 7-37 Newgale Sands, Pembroke, Wales, showing an example of a high-tide beach consisting of shingle, fronted by a fine-sand tidal flat that is exposed during low tide.

are characterized by poorly sorted finer-grained sediments, well-formed and complex ripple marks, and abundant organisms living on and within the sand. There may be a "low-tide beach" seaward of the terrace, where the slope is slightly greater than the terrace slope, and the sand is slightly coarser and better sorted. Because a brief still stand in the water level occurs at high tide and at low tide, the energy of the waves is concentrated at those levels. The water level sweeps rather quickly over the low-sloping terrace, and the wave energy tends to be dissipated by the low water depths. The concentration of energy at the high- and low-water levels accounts for the differences between the terrace sediments and the characteristics of the high- and low-tide beaches.

PROFILE CHANGES DUE TO WINDS

An onshore wind produces a landward movement of the surface water that must be compensated for by a seaward current at depth. Just the reverse occurs with an offshore wind, the near-bottom flow being onshore. These currents can be important to the cross-shore transport of sediments and therefore are a factor in the response of the beach profile.

The effects of winds are illustrated by the laboratory experiments of King and Williams (1949), undertaken to investigate cross-shore sand transport in a wave tank both shoreward and seaward of the breakers. One set of experiments involved the superposition of an onshore wind on the wave action. With an onshore wind, it was found that within the surf zone the sand was transported offshore, even though for the wave steepness used, a berm-type profile should have developed from onshore transport. Outside the breaker zone the offshore near-bottom current generated by the onshore wind counteracted the landward mass movement of water due to the waves (Chapter 5). The former onshore sand transport due to the waves alone was converted into a slight seaward transport by the presence of the onshore wind.

On natural ocean beaches, separating the effects of any wind-induced currents from the sediment movement due to waves alone is more difficult than in laboratory experiments. With strong onshore winds, steep waves may be locally generated that have a destructive influence on the beach profile in addition to the reaction of the beach to the wind-induced currents that also tend to cut back the beach. With offshore winds the advancing waves tend to be reduced in height by the head winds, so the waves reaching the shore are of lower steepness. Any constructive effects the near-bottom onshore wind-induced currents might have, are enhanced by the reduced wave steepness.

Shepard and LaFond (1940) found in their beach profiles from southern California that a cutting of the beach corresponded with high onshore wind velocities. The maximum amount of cutting for 1 year occurred at a time when the wind velocity reached a maximum, whereas the waves were only the fifth largest occurrence for that year. Observations by King (1953) on Marsden Bay, County Durham, England, also demonstrated such effects of onshore versus offshore winds on the profile. With an onshore wind, erosion took place thirteen times on the upper foreshore above the mean-tide level, while accretion occurred only four times. With an offshore wind, accretion occurred thirteen times and erosion only three times. On the lower foreshore, just the reverse took place.

Observations such as these confirm that the strength and direction of coastal winds can be an important factor in beach-profile development. In any study of beach profiles and cross-shore sediment transport, it is important to monitor the wind and to assess its influence.

CROSS-SHORE SEDIMENT TRANSPORT AND PROFILE MODELS

The changes in two-dimensional beach profiles documented throughout this chapter are brought about by the cross-shore transport of sediment. This is particularly apparent in the change from a berm-type profile (summer profile) to a bar profile (winter profile) with an increase in the wave-energy level. As diagrammed in Figure 7-38, the profile change itself permits one to infer the pattern of net cross-shore sediment transport. When there is an offshore transport due to higher wave conditions, the beach face and berm experience erosion, and the eroded sediment moves offshore where deposition occurs to form a bar. There is a nodal point, a line in the third dimension, that separates the zones of erosion and deposition. Although there has been no vertical change in the profile at the nodal point, it is apparent that there must have been a large quantity of sediment transported offshore past that point. Indeed, the net cross-shore transport rate would have been a maximum at the nodal point, denoted by q_{sm} on the dashed curve in Figure 7-38 for the cross-shore variations in the sediment transport. As one moves seaward from the shore, more and more sand has been eroded from the beach face, so there would have been a progressively increasing transport up to the maximum at the node. Beyond the node, just the opposite trend occurs, with the progressive deposition resulting in a decreasing quantity of transported sediment. The dashed curve for the cross-shore transport decreases to zero at the offshore position where the profile changes cease (i.e., at the closure depth). The lower diagram of Figure 7-38 shows the inverse condition brought about by a net onshore sediment transport during low-wave conditions. The maximum transport again occurs at the node.

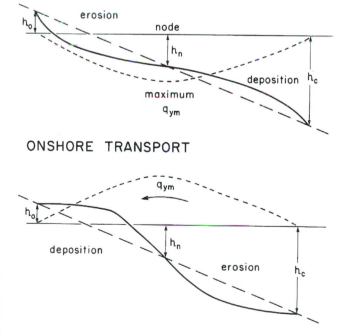

OFFSHORE TRANSPORT

ONSHORE TRANSPORT

Figure 7-38 Distributions of cross-shore sediment transport rates (dashed curves) with maximum q_{sm}, derived from erosion-accretion patterns of beach-profile changes that occur during offshore or onshore transport.

The amount of erosion or accretion at any point in the profile, that is, the elevation change in the profile at that point, is not simply related to the quantity of the transported sediment. Instead, the amount of erosion or deposition depends on the cross-shore gradient of the transport, that is on dq_s/dx, where q_s is the volume of the cross-shore transport per unit shoreline length, and x is the cross-shore coordinate (units for q_s might be cubic meters of sand transported per second per meter of shoreline length). This relationship is illustrated in Figure 7-39, centered on the zone of profile erosion. If q_{in} and q_{out} are the respective rates of transport in and out of the small segment of profile, then $\Delta q_s = q_{out} - q_{in}$ is the difference, an increase that accounts for the erosion of the profile. If we consider an elapsed time Δt, then $\Delta q_s \Delta t$ is the quantity of sediment removed from that segment of profile per unit longshore length. But that quantity will also be given by $\Delta h \Delta x$, where Δh is the elevation change of the profile, and Δx is the cross-shore length of the small segment under consideration (Fig. 7-39). Equating the two volumes, we have $\Delta q_s \Delta t = -\Delta h \Delta x$, or rearranging

$$\frac{\Delta q_s}{\Delta x} = -\frac{\Delta h}{\Delta t} \tag{7.25a}$$

Shrinking this to infinitesimals yields

$$\frac{dq_s}{dx} = -\frac{dh}{dt} \tag{7.25b}$$

a relationship between the rate of change in the beach elevation, dh/dt, versus the cross-shore gradient of the sediment transport rate, dq_s/dx. The reason for the presence of the negative sign now becomes apparent. If by convention we signify that $dh/dt = -$ represents erosion, then to produce that erosion we must have $dq_s/dx = +$, an increase in the rate of sediment

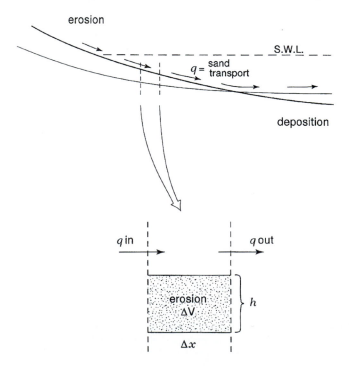

Figure 7-39 The derivation of Equation (7.25) for the conservation of sediment within a small segment of the beach profile where the elevation change h of the beach is related to the quantities of sediment brought in (q_{in}) or transported out (q_{out}) of the segment.

transport. In the opposite sense, deposition represented by $dh/dt = +$ corresponds to $dq_s/dx = -$, a decreasing rate of transport. The relationships of Equation (7.25) represent the conservation of sediment as it is being redistributed along the profile by the cross-shore transport. Sediment must be conserved within a profile as long as there is no net longshore movement that redistributes the quantity of sand along the length of the beach.

Equation (7.25) provides a useful tool for converting observed patterns of profile changes in the form of erosion versus deposition, into the corresponding patterns of the cross-shore transport responsible for those changes, much as shown schematically in Figure 7-39. To obtain the actual patterns of the transport rates, that is, the dashed curves for q_s, the dq_s/dx gradients derived from Equation (7.25) must be integrated in the cross-shore direction. The laboratory wave-channel study of Sawaragi and Deguchi (1980) provides an example of this analysis approach to determine the transport patterns responsible for the observed profile changes. Many of the profile shifts are more complex than the simple cases diagrammed in Figure 7-38, produced by an onshore transport in some areas of the profile, while at the same time there is an offshore transport in other areas. Sawaragi and Deguchi analyzed a large number of experiments, focusing on how the maximum transport rate in the distribution, q_{sm}, and the component parts of the profile such as the water depth at the node, depend on wave parameters and beach-sediment grain sizes. They documented how these parameters change with time, particularly that q_{sm} decreases exponentially with time as the beach initially changes at a rapid rate from an even slope but subsequently changes more slowly as an equilibrium profile is approached. Watanabe, Riho, and Horikawa (1980) have undertaken similar analyses of cross-shore sediment transport as reflected in profile shifts and compared the results with the general formulation of Madsen and Grant (1976) for sediment transport under waves.

Based on such analyses, models have been developed with the objective of predicting beach-profile changes in response to varying wave conditions and changing water elevations due to tides or a storm surge. One such numerical model in common use is SBEACH, which was fully developed in the report by Larson and Kraus (1989), with a published summary presented by Larson, Kraus, and Sunamura (1988). The model simulates the macroscale profile changes, including the growth and movement of break-point bars and the berm. The overall direction of the transport (onshore versus offshore) is evaluated on the basis of empirical relationships between the wave steepness and the Dean number, H/w_sT, criteria reviewed earlier in this chapter. Empirical relationships also are used for the speed of migration of the center of mass of the bar and its growth with time (Fig. 7-22) to the equilibrium volume V_{eq} given by Equation (7.18). The sediment-transport computations are separated into the four zones shown in Figure 7-40, the divisions being based in part on the expected contrasting modes of sediment movement. In Zone I, the offshore zone of transforming unbroken waves, the transport rate is given by

$$q = q_b e^{-\lambda x} \tag{7.26a}$$

where x is the cross-shore coordinate originating at the break point, q_b is the transport rate at the break point, and λ is the spatial decay coefficient given by

$$\lambda = 10.3(D/H_b)^{0.47} \tag{7.26b}$$

where D is the sediment grain size. The transport rate in Zone II (breaker zone) is described by a similar relationship but with a smaller decay coefficient ($\lambda = 0.21$). A transport relationship similar to that used by Kriebel and Dean (1985) is applied in the region of fully broken

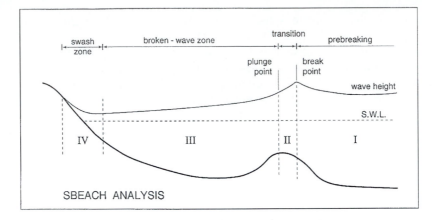

Figure 7-40   The four zones of computation included in the SBEACH analysis of pro-
file changes. [From M. Larson and N. C. Kraus, SBEACH: *Numerical Model for
Simulating Storm Induced Beach Change*, Technical Report CERC-89-9, U.S. Army
Corps of Engineers]

waves (Zone III). In an analysis, sand is exchanged between the four zones and a conservation of sediment relationship similar to Equation (7.25) is applied to maintain a sand balance
within the evolving profile.

The SBEACH model was calibrated initially with data from large near prototype wave
channels, although monochromatic waves were used in the tests. An example is shown in Figure 7-41, where the initial profile (dashed), consisting of a uniform sand slope, was eroded
into a bared profile (measured). Also shown are the series of computed profiles at 1 through
59 hours derived from the SBEACH analysis, the final profile being the one that should be compared with the measured profile. Also shown is the measured versus calculated distributions
of wave heights across the width of the surf zone. The agreement between the calculated and
measured profiles generally is good, particularly in the correspondence between the sizes and
shapes of the offshore bar and in the erosion of the scarp at the shoreline. The main disagreement occurs within the inshore, where a second bar has formed on the measured profile, a feature not produced by the SBEACH analysis that apparently does not fully account for
wave reformation and rebreaking. Figure 7-42 shows a comparison with measured profiles at
the FRF, North Carolina, a severe test of SBEACH due to the changing wave conditions and

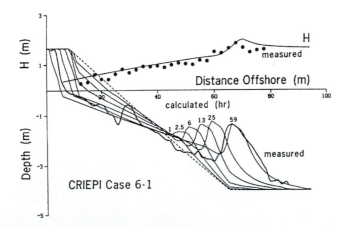

Figure 7-41   A laboratory example of profile
variations analyzed by SBEACH, with the final
computed profile compared with a measured
profile from a large wave-channel experiment.
[From M. Larson and N. C. Kraus, SBEACH: *Numerical Model for Simulating Storm Induced
Beach Change*, Technical Report CERC-89-9,
U.S. Army Corps of Engineers]

tide levels. Recalibration of the model was required, as it was found that the laboratory-based coefficients yielded bar migration rates that were too rapid and bars that were too pronounced in relief. The overall movement of the bar was fairly well predicted by SBEACH (Fig. 7-42), although the amount of material moved was underestimated and the trough was not as pronounced as in the measured profile. The results look promising, especially considering that several environmental parameters were varying, while at this stage in its development, the SBEACH model has only one adjustable parameter in the calibration.

Many aspects of the SBEACH analyses are empirical, based on observed profile changes in large wave tanks and on natural beaches. The development of more process-based models is a challenging undertaking. Such advanced models are provided by Hedegaard, Deigaard, and Fredsøe (1991), in the papers by Southgate and Nairn (1993) and Nairn and Southgate (1993), and by Dibajnia, Shimizu, and Watanabe (1994) and Watanabe and Shiba (1994). These models variously incorporate analyses of wave transformations in the offshore, wave breaking and decay within the surf zone, wave set-up effects on the mean water level in the presence of waves, the generation of undertow, which can be important to cross-shore sediment movements, and the transport of the suspended and bedload sediments are independently evaluated. The models developed by several investigators differ primarily in how they evaluate these processes and in the computational procedures used in their models. Figure 7-43 shows the results of the Hedegaard and co-workers model compared with the evolution of a beach

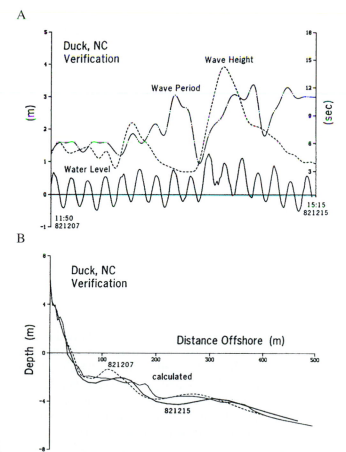

Figure 7-42 Comparison of the SBEACH analysis with profile shifts measured at the FRF, North Carolina. [From M. Larson and N. C. Kraus, SBEACH: *Numerical Model for Simulating Storm Induced Beach Change,* Technical Report CERC-89-9, U.S. Army Corps of Engineers]

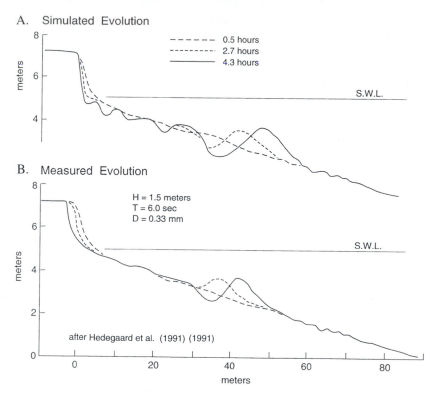

A. Simulated Evolution

B. Measured Evolution

Figure 7-43   Model results of Hedegaard and co-workers (1991) simulating changes (a) in the beach profile, compared with measured profiles (b) in a large wave channel. [Adapted from Onshore/Offshore Sediment Transport and Morphological Modelling of Coastal Profiles, I. B. Hedegaard, R. Deigaard and J. Fredsøe, *Coastal Sediments '91*, 1991. Reproduced with permission from the American Society of Civil Engineers.]

profile in a large-scale wave channel. The model has been successful in duplicating the details of the measured profiles and their evolution with time from an initially uniform slope. Figure 7-44 includes comparisons of the Nairn and Southgate model with profiles in large wave channels, including an example with multiple bars. Southgate (1991) provides additional comparisons and sensitivity tests of the model components. Again, the computed and measured profiles are in close agreement. Nairn and Southgate concluded that the processes related to the incident waves can be predicted to a level of accuracy that is sufficient to permit a confident estimate of the evolution of the beach-profile morphology, but the description of the long-wave (infragravity) structure and its influence on the sediment transport is still inadequate. Nairn and Southgate also include an analysis of longshore-directed processes (the S_{xy} radiation stress component, etc.), which allows them to model longshore currents and the cross-shore distribution of the longshore sediment transport; the later evaluation was shown to agree with the measurements of Kraus, Gingerich, and Rosati (1989) at the FRF, North Carolina. This inclusion permits the development of three-dimensional analyses of surf-zone processes and beach morphology evolution. The foremost value in advanced models of this type rests with their ability to identify links between observed beach changes and the underlying coastal processes, leading to improvements in the models as our understanding of those processes improve.

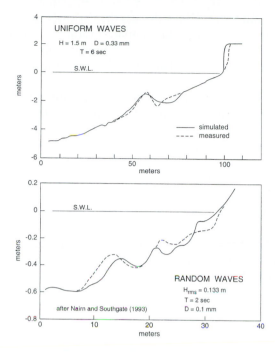

Figure 7-44 Model results of Nairn and Southgate (1993) compared with measured beach-profile changes in large wave channels using uniform waves and random waves. [Adapted from *Coastal Engineering, 19,* R. B. Nairn and H. N. Southgate, Deterministic Profile Modelling of Nearshore Process, 2. Sediment Transport and Beach Profile Development, 1993 with kind permission of Elsevier Science-NL, Sara Burgerhartstratt 25, 1055 KV Amsterdam, The Netherlands.]

The models developed by Dibajnia, Shimizu, and Watanabe (1994) and Watanabe and Shiba (1994) for the cross-shore sediment transport and accompanying profile changes are conceptually similar to those reviewed above, differing in details as to how processes such as wave dissipation in the surf zone are modeled and in the approach to evaluate sediment transport rates. The model predictions again have been compared with laboratory data obtained in large wave channels and show good agreement for conditions where the waves break either by plunging or by spilling.

Refinements of the cross-shore sediment transport analyses are still needed. Hedegaard et al. (1992) and Roelvink and Hedegaard (1993) compared several proposed models against measured beach-profile changes in large wave channels. They found significant differences in predictions of the models and varying degrees of agreement with the measured profiles. The models generally underestimated the initially rapid offshore sediment transport, leading again to the conclusion that the models inadequately account for wave reflection from the beach and infragravity long waves. The models did not succeed in accounting for swash processes leading to dune erosion, a conclusion also arrived at by Nairn (1991) in a more extensive series of analyses where his model was directed toward dune-erosion assessments.

Laboratory and field studies have been undertaken and models developed specifically with the objective of assessing foredune erosion when attacked by storm waves and elevated water levels due to an accompanying storm surge. Fisher, Overton, and Chisholm (1986), Overton, Fisher, and Fenaish (1987) and Overton et al. (1993) conducted a series of laboratory and field tests using artificially constructed sand dunes to measure erosion rates due to the attack of the wave swash. They succeeded in deriving a relationship between the rate of dune erosion and the force of the wave uprush, evaluated as the product of the square of the bore velocity and its depth. Overton and Fisher (1988) discuss the various aspects involved in applying the results to dune-erosion evaluations and compare predicted versus measured erosion rates at the FRF. Nairn (1991) suggested the incorporation of this approach for evaluating dune erosion into the models discussed above that otherwise mainly attempt to account for beach-profile responses.

Models have been formulated that specifically attempt to evaluate dune erosion during storms. The Dutch, in particular, have concentrated on these models due to their reliance on the presence of large sand dunes along their coast to prevent inundation by North Sea storms. Edelman (1968, 1972) formulated a model for dune erosion based on numerous surveys of pre- and post-storm beach profiles. It was observed that during a storm, the beach change is mainly a result of sand transport perpendicular to the shore, with erosion of the berm and dune and deposition within the outer surf zone to the depth of wave breaking in the nearshore. Following this observation, Edelman developed the model shown schematically in Figure 7-45, where sand eroded from the dunes is deposited as a wedge on the beach. Using actual and idealized pre-storm profiles, Edelman established the post-storm profile relative to the level of the peak storm surge and then shifted the profile landward to conserve sand volumes. He concluded that the equilibrium profile relative to the instantaneous water level is approximately constant and is represented by a uniform slope that depends on sediment characteristics. Assuming an idealized profile in which the dunes are represented by a vertical face and uniform crest elevation, Edelman (1972) developed generalized graphs that relate dune recession to the storm-tide water-level change. This analysis approach has been improved by Graaff (1977, 1986) and Vellinga (1982), based on results from large-scale wave tank tests. While remaining conceptually similar to the Edelman model, Graaff and Vellinga developed revised relationships for evaluating the shapes of beach profiles and for establishing depths of effective sediment motion (closure depths). Figure 7-46 shows the results of a representative test of the model, an example that employed a time-varying storm surge level that simulated the extreme surge experienced on the coast of the Netherlands during 1953.

Kriebel and Dean (1985) and Kriebel (1986, 1990) developed models for dune erosion and beach-profile adjustments during storms that provide a better assessment of the lag time of the erosion behind the causative processes. Their models include assessments of the cross-shore sediment transport due to the disequilibrium of wave-energy dissipation produced by the storm. The transport equation together with a relationship for sediment continuity are solved numerically to predict the time-dependent two-dimensional beach and dune erosion. Since the analysis considers sediment-transport processes, the models can account for time variations in wave heights and water levels, and therefore can be used to examine response

DUNE - EROSION MODEL
(Edelman, 1968, 1970)

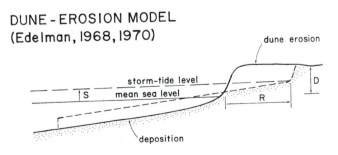

Figure 7-45 Diagram of the Edelman model that evaluates dune erosion resulting from elevated water levels of a storm surge, with the eroded sand distributed across the beach profile. [Adapted from Dune Erosion During Storm Conditions, T. Edelman, *Proceedings of the 11th Conference on Coastal Engineering*, 1968. Reproduced with permission from the American Society of Civil Engineers.]

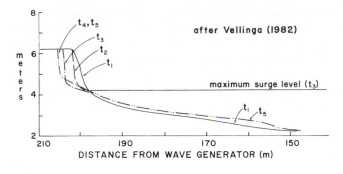

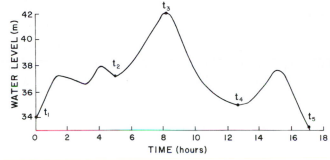

Figure 7-46 Model simulation of dune erosion and beach-profile development that resulted from the extreme storm surge experienced during 1953 along the coast of the Netherlands. [Adapted from *Coastal Engineering* 6, P. Vellinga, Beach and Dune Erosion During Storms, p. 376, 1982 with kind permission of Elsevier Science-NL, Sara Burgerhartstratt 25, 1055 KV Amsterdam, The Netherlands.]

times of beaches including the time lag of the erosion behind the processes. The results indicate that time scales of natural beaches may be on the order of 10–100 hours for storm conditions and on the order of 1,000–10,000 hours when the effective limit of sediment transport is far offshore. The lag of the profile response, therefore, can be considerable and in general results in the actual erosion during a storm being only some 15–30 percent of the potential erosion predicted by equilibrium models based on simple shifts of beach profiles. Figure 7-47 compares the profile changes during Hurricane Eloise in 1975 on the Florida coast with those predicted by the model. The agreement is good, the main difference being the presence of a small offshore bar in the measured post-storm profile, a recovery feature that may actually have formed following the storm but before profiling was undertaken.

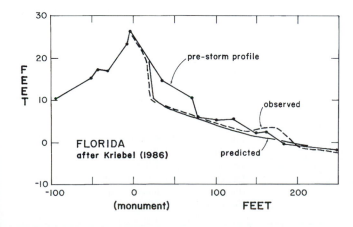

Figure 7-47 Dune erosion and profile development predicted by the model of Kriebel and Dean (1985), compared with erosion experienced on the Florida coast during Hurricane Eloise. [From D. L. Kriebel, Verification Study of a Dune Erosion Model, *Shore & Beach 54*, 1986.]

Analyses of the cross-shore sediment transport that give rise to profile changes and redistribute sand derived from dune erosion require equations that relate the rate of sediment transport to the local wave conditions and any superimposed current. For example, the study of Watanabe, Riho and Horikawa (1980) noted above employed the sediment-transport formulations of Madsen and Grant (1976), while Hedegaard et al. (1991) relied on the approach of Engelund and Fredsøe (1976). In their evaluations of cross-shore and longshore sediment transport rates, Nairn and Southgate (1993) employed the transport formulations of Bagnold (1963), as modified and extended by Bowen (1980) and Bailard and Inman (1981). Laboratory tests of the Bagnold transport relationships have been undertaken by Pruszak (1989). The Bagnold transport equations include the effects of the downslope component of gravity, which is important in cross-shore sediment transport and acts to determine the overall slope of the equilibrium profile. Bowen analyzed the simple case in which the asymmetry of the wave-orbital velocities, with stronger onshore flows under the wave crests compared with weaker offshore velocities under the troughs, result in an onshore sediment transport that continues until balanced by the offshore slope of the bottom, at which time an equilibrium is achieved. For this case, Bowen derived Equation (7.5), where $h \propto x^{2/5}$ for the variation in water depth with distance offshore. Bowen also used the modified Bagnold equations to model the formation of multiple bars by reflected incident waves or by low-mode edge waves. The study of Holman and Bowen (1982), to be discussed in Chapter 11, provides more complex analyses that model the three-dimensional development of crescentic bars and transverse bars that trend obliquely to the shoreline, again using the sediment-transport formulations of Bagnold (1963) as modified by Bowen (1980) and Bailard and Inman (1981).

Another aspect of cross-shore sediment transport and the development of beach profiles is the sorting of the sediment according to their grain sizes. In Chapter 3 it was noted that, in the simplest case, beaches tend to have the coarsest grain sizes at the shore with a progressive decrease in grain size toward the offshore. In more complicated cases, there may be a coarsening of sediment over offshore bars due to the concentrated wave-energy expenditure and again in the swash zone of intense wave run-up. This cross-shore sediment sorting is an important attribute of beach profiles, one that has not been adequately accounted for in the models.

Part of the cross-shore sorting of sediment can be attributed to the patterns of orbital motions under waves as they shoal. The orbital velocities of Stokes-type waves (Chapter 5) in shallow water are diagrammed in Figure 7-48, consisting of a forward orbital motion under the wave crest that is short in duration but high in velocity, while the return flow under the trough is slower but of longer duration. This is due to the steepening of waves

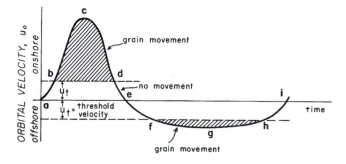

Figure 7-48 The Stokes-wave asymmetry of orbital motions, with a strong onshore velocity versus a weaker offshore velocity but with a longer duration, resulting in an asymmetry in the cross-shore sediment transport and a net shoreward transport of coarse sediments.

in shallow water, with isolated crests separated by wide troughs. Cornish (1898) suggested that the shoreward component of such waves may be more effective in moving coarser sediment particles, since a shorter-lived current of high velocity could transport material that is too large to be moved by the longer-duration but weaker seaward current. On the other hand, silt and sand will readily move a nearly equal distance in both directions. In this way the waves may selectively drive pebbles and cobbles toward the beach. Johnson (1919) summarizes some dramatic examples of the onshore transport of coarse debris. Murray (1853) demonstrated that shingle and chalk ballast dropped into the sea off Sunderland, England, at a distance of 10–15 km from land where water depths are 20–40 m, were eventually thrown onshore by storm waves. Gaillard (1904) reported on pig lead washed up at Madras, India, during a violent storm, the lead having come from a vessel wrecked more than 2 km offshore.

The selective-sorting model of different grain sizes proposed by Cornish (1898) was verified by Bagnold (1940, pp. 32–33) in a series of wave-channel experiments. He observed that the largest particles are moved only during the most violent part of the orbital motion and progressively creep toward the shore, the larger the particle the more pronounced the creep. The biggest particles eventually collected high on the beach face. This process has been modeled by Horn (1992), including evaluations of sediment threshold under Stokes second-order waves as they transform during their approach to the shore. The model generally was successful in predicting the broad pattern of increasing grain sizes in the onshore direction, with more success in accounting for the width of the sediment size distribution than in providing an accurate prediction of the mean grain sizes across the beach profile. The model did not calculate sediment transport rates, nor did it account for the sloping sea floor.

In 1898 Cornaglia, an Italian engineer, proposed an expansion of the selective-transport model due to stronger onshore orbital motions, an expansion that recognized the fundamental importance of (1) the influence of waves in trying to move the grains shoreward, and (2) gravity trying to pull the particles offshore in the downslope direction. Like Cornish (1898), Cornaglia recognized the existence of a threshold water-particle velocity, denoted by u_t in Figure 7-48, necessary to move a given grain size, and established that, depending on the equality or inequality of the areas b-c-d and f-g-h in Figure 7-48, the particles will either be in equilibrium or will tend to have a net onshore motion. As the water depth increases, the difference between the trough and crest velocities becomes less, until there is not enough difference to move the particle a net distance either onshore or off-shore. According to Cornaglia, the influence of gravity becomes increasingly important at these depths in tending to pull sediment particles seaward. At some unique depth, all of these forces balance and the particle simply moves back and forth. This position of dynamic equilibrium is the so-called *null point*, the location where there is zero net movement for that particular grain size along a line parallel to the wave direction. Cornaglia recognized that the null point will vary in position for each grain size on the beach and for different wave conditions. The coarser the particle, the shallower its null-point position, a prediction that corresponds to the general observation that sediment grain sizes tend to coarsen toward shore as the water depth decreases. The model also predicts that if a certain grain size is seaward of its null point, its net movement will be offshore, while that grain size placed landward of the null point will tend to move onshore; therefore, although the null-point represents an equilibrium position for a certain grain size, it would be an unstable equilibrium. These aspects of the null-point model have been placed into mathematical terms by Ippen and Eagleson (1955), Eagleson, Dean, and Peralta (1958) and Eagleson and Dean (1961), who also conducted laboratory wave-channel experiments that in part verified the theory. This included the use of a fixed-bottom slope artificially roughened by sand glued to its surface. Single sediment particles were then introduced and their movements under the waves were followed. Some grains moved onshore at all depths less than the depth at which threshold was achieved. Other grain sizes and densities demonstrated the existence of null points, with seaward transport offshore and forward transport shoreward of the null-point position. The predictive equations for the null-point conditions derived from those laboratory studies were tested in the field by Miller and Zeigler (1958, 1964) but with mixed results.

The null-point model also predicts that if many different grain sizes are present at a certain location, only one of the sizes can be at its null point. The model indicates that the larger grains will move offshore and the finer ones will move onshore. This prediction by the null-point model is demonstrably incorrect, as shown by the use of sediment tracers. For example, Zenkovitch (1946) tagged different grain sizes with different fluorescent colors and released the mixture into the offshore. He found that the coarsest particles moved onshore and the fine ones offshore, while those having a grain size close to the *in situ* grains oscillated back and forth in a sort of null-point position of equilibrium.

Although there are attractive aspects of the null-point hypothesis as proposed by Cornaglia and placed in to mathematical terms by Ippen and Eagleson (1955), many of its predictions are erroneous and the model has to be rejected. In discussing Cornaglia's hypothesis, Munch-Peterson (1938) held that it was not valid because the influence of gravity is overestimated relative to wave forces. However, the hypothesis was correct in focusing on the importance of asymmetries in the wave-orbital motions versus the downslope component of gravity. These factors are also inherent in the Bagnold (1963) sediment-transport relationships, as revised and applied by Bowen (1980) and Bailard and Inman (1981). Bowen noted in passing that, when the equations are used to analyze cross-shore sediment transport and equilibrium profiles, the equations correctly predict that the coarser sediment moves onshore while finer sediment drifts offshore. Unfortunately, this has not been pursued further, so it is not yet possible to integrate the processes of sediment sorting into the models of cross-shore sediment transport and profile change.

SUMMARY

This chapter has provided a detailed consideration of beach profiles, first examining the components such as the development of the berm, the slope of the beach face, and the number of offshore bars and processes that may be important to their formation. It was documented how beach profiles change in response to varying wave conditions and water levels, and the chapter ended with an examination of the cross-shore sediment transport that ultimately is responsible for those changes.

The chapter has been limited to a two-dimensional consideration of beach morphology, the simple profile and cross-shore sediment transport. Most beaches are more three dimensional, many having rhythmic features such as crescentic bars and beach cusps, or have bars and troughs that run transverse to the shoreline and vary in their development in the longshore direction. The generation of longshore currents and rip currents and the accompanying longshore sediment transport are important in the development of the three-dimensional beach morphology, processes that are examined in Chapters 8 and 9. In Chapters 10 and 11 we return to beach morphology to consider its full three dimensionality, first to examine models of the overall curvature of the shoreline governed by sediment sources and transport processes, and then to the variety of rhythmic shoreline forms and their origins.

PROBLEMS

1. The storm-wave conditions at a coastal site are represented by a deep-water significant wave height of 7 m and period of 18 sec. Assuming these parameters are appropriate for application in evaluating closure depths of profile variations, apply Equation (7.1) from Hallermeier (1981) and the modified relationships of Birkemeier (1985) to evaluate the closure depths. There is no predicted dependence on the beach-sediment grain size. Is this reasonble, or how might one expect the closure depth to differ for a coarse-sand beach versus a fine-sand beach?

2. With $A = 0.1$ m^{-1}, which corresponds to a beach composed of sand of diameter 1 mm according to Figure 7-11, compute and graph the Bruun/Dean profile given by Equation (7.2), and also graph the cross-shore variations in the corresponding beach slopes, S.

3. While Dean (1977) established a median value of 2/3 for the exponent in the Bruun/Dean profile relationship of Equation (7.2), he found that exponents of individual profiles ranged from 0.2 to 1.2. With $A = 0.1$ m^{-1}, calculate and graph the profiles that correspond to these extremes in the possible exponents.

4. The slope at the shoreline of a beach composed of 1-mm diameter sand is about $S_0 = 0.08$. Compute and graph the exponential beach profile of Equation (7.10), where the profile passes through an offshore closure depth of 8 m at an offshore distance 230 m from the shoreline.

5. The slope of a beach at the shoreline is $S_0 = 0.05$, which progressively decreases to a slope of 0.001 about 300 m offshore. Match the exponential profile of Equation (7.10) to these conditions and graph the cross-shore variations in profile slopes and depths.

6. Various relationships have been proposed for factors controlling the slope of the beach face, while the graphs in Figures (7-16), (7-17), (7-19), and (7-20) provide alternative evaluations. For a beach composed of sand of diameter $D = 0.5$ mm, which corresponds to a settling velocity $w_s = 8$ cm/sec, evaluate the expected beach-face slopes from this series of graphs for the deep-water wave height 3 m and period 12 sec. Compare the results and discuss why a range of predicted values is obtained.

7. The development of Equation (7.17) to predict the elevation of the berm is based on the run-up equation of Hunt (1959). From the review in Chapter 6, it is apparent that recent advances in measuring and analyzing run-up elevations have outdated the Hunt formula, particularly in showing that the run-up depends on the beach slope as well as the wave height and period. Reformulate the predictive relationship for the run-up, based on equations given in Chapter 6. Compare the results with Equation (7.17) for the deep-water wave conditions $H_\infty = 3$ m and $T = 15$ sec (it will be necessary to evaluate the corresponding approximate breaker height), and for beach slopes ranging $S_0 = 0.02, 0.05$, and 0.1.

8. Predictions of the berm elevation, as in Problem 7, account for the wave run-up alone, and it is noted that the test of Equation (7.17) in Figure 7-23 was with data from effectively tideless coasts. Discuss how one could include tides in the analysis, providing predictions of berm elevations as the expected joint occurrence of extreme tides and storm-wave conditions.

9. Assuming that the exponent for $H_\infty/w_s T$ in Equation (7.18) for the equilibrium volume of the offshore bar is actually $1.33 = 4/3$, algebraically derive the equivalent dimensional form of the relationship for V_{eq} as a direct function of the deep-water wave height, period, and grain-settling velocity. Graph the resulting relationship as V_{eq} versus H_∞, with a fixed period $T = 12$ sec as representative, with separate curves for $w_s = 20$ cm/sec (coarse sand) and 0.8 cm/sec (fine sand). Are

the resulting dependencies reasonable, in particular, the relative bar volumes for coarse- versus fine-sand beaches?

10. For a coarse-sand beach ($D = 1.5$ mm; $w_s = 20$ cm/sec) and fine-sand beach ($D = 0.1$ mm; $w_s = 0.8$ cm/sec) and assuming $T = 12$ sec as a representative wave period, use the relationships and graphs presented in the text to evaluate the deep-water significant wave heights required for the transition between berm-type and bar-type beach profiles. Compare the results and discuss why a range of predicted values is obtained.

REFERENCES

AAGAARD, T. (1991). Multiple-Bar Morphodynamics and Its Relation to Low-Frequency Edge Waves. *Journal of Coastal Research* 7: 801–813.

ALLEN, J. R. (1985). Field Evaluation of Beach Profile Response to Wave Steepness as Predicted by the Dean Model. *Coastal Engineering* 9: 71–80.

AUBREY, D. G. (1979). Seasonal Patterns of Onshore/Offshore Sediment Movement. *Journal of Geophysical Research* 84: 6347–6354.

AUBREY, D. G., D. L. INMAN, and C. D. WINANT (1980). The Statistical Prediction of Beach Changes in Southern California. *Journal of Geophysical Research* 85: 3264–3276.

BAGNOLD, R. A. (1940). Beach Formation by Waves: Some Model Experiments in a Wave Tank. *Journal Institution of Civil Engineers* 15: 27–52.

BAGNOLD, R. A. (1963). Mechanism of Marine Sedimentation. In *The Sea,* M. N. Hill (editor). Vol. 3, pp. 507–528. New York: Wiley-Interscience.

BAILARD, J. A. and D. L. INMAN (1981). An Energetics Bedload Model for a Plane Sloping Beach: Local transport. *Journal of Geophysical Research* 86(C3): 2035–2043.

BASCOM, W. H. (1951). The Relationship Between Sand Size and Beach Face Slope. *Transactions American Geophysical Union* 32: 866–874.

BASCOM, W. H. (1953). Characteristics of Natural Beaches. *Proceedings of the 4th Coastal Engineering Conference, Amer. Soc. Civil Engrs.,* pp. 163–180.

BIJKER, E. W., E. VAN HIJUM, and P. VELLINGA (1976). Sand Transport by Waves. *Proceedings of the 15th Coastal Engineering Conference, Amer. Soc. Civil Engrs.,* pp. 1149–1167.

BIRKEMEIER, W. A. (1979). The Effects of the 19 December 1977 Coastal Storm on Beaches in North Carolina and New Jersey. *Shore & Beach* 47(1): 7–15.

BIRKEMEIER, W. A. (1984). Time Scales of Nearshore Profile Change. *Proceedings of the 19th Coastal Engineering Conference, Amer. Soc. Civil Engrs.,* pp. 1507–1521.

BIRKEMEIER, W. A. (1985). Field Data on Seaward Limit of Profile Change. *Journal of Waterway, Port, Coastal, and Ocean Engineering, Amer. Soc. Civil Engrs.,* 111(3), 598–602.

BIRKEMEIER, W. A. and C. MASON (1978). The CRAB: A Unique Nearshore Surveying Vehicle. *Journal of Survey Engineering* 110(1): 1–7.

BOCZAR-KARAKIEWICZ, B. and R. G. D. DAVIDSON-ARNOTT (1987). Nearshore Bar Formation by Non-Linear Wave Processes—A Comparison of Model Results and Field Data. *Marine Geology* 77: 287–304.

BOCZAR-KARAKIEWICZ, B., D. L. FORBES, and G. DRAPEAU (1995). Nearshore Bar Development in Southern Gulf of St. Lawrence. *Journal of Waterway, Port, Coastal, and Ocean Engineering, Amer. Soc. Civil Engrs.* 121: 49–60.

BODGE, K. R. (1992). Representing Equilibrium Beach Profiles with an Exponential Expression. *Journal of Coastal Research* 8: 47–55.

BOON, J. D. and M. O. GREEN (1988). Caribbean Beach-Face Slopes and Beach Equilibrium Profiles. *Proceedings of the 21st Coastal Engineering Conference, Amer. Soc. Civil Engrs.,* pp. 1618–1630.

Bowen, A. J. (1980). Simple Models of Nearshore Sedimentation. Beach Profiles and Longshore Bars. *The Coastline of Canada,* by S. B. McCann (editor). Paper 80–10, pp. 1–11. Geological Survey of Canada, Ottawa, Canada.

Bowen, A. J. and D. L. Inman (1971). Edge Waves and Crescentic Bars. *Journal of Geophysical Research* 76(36): 8662–8671.

Bowman, D. (1981), Efficiency of Eigenfunctions for Discriminate Analysis of Subaerial Non-Tidal Beach Profiles. *Marine Geology* 39: 243–258.

Bruun, P. (1954). *Coast Erosion and the Development of Beach Profiles.* Beach Erosion Board, Technical Memo 44.

Carter, R. W. G. and J. H. Balsillie (1983). A Note on the Amount of Wave Energy Transmitted over Nearshore Sand Bars. *Earth Surface Processes and Landforms* 8: 213–222.

Carter, T. G., P. L.-F. Liu, and C. C. Mei (1973). Mass Transport by Waves and Offshore Sand Bedforms. *Journal Waterways, Harbors, and Coastal Engineering Division, Amer. Soc. Civil Engrs.* 99(WW2): 165–184.

Clarke, D. J. and I. G. Eliot (1988). Low-Frequency Changes of Sediment Volume on the Beachface at Warilla Beach, New South Wales, 1975–1985: *Marine Geology* 79: 189–211.

Cornish, V. (1898). On Sea Beaches and Sand Banks. *Geographical Journal* 11: 528–559, 628–647.

Dally, W. R. and R. G. Dean (1984). Suspended Sediment Transport and Beach Profile Evolution. *Journal Waterway, Port, Coastal, and Ocean Engineering, Amer. Soc. Civil Engrs.* 110(1): 15–33.

Dally, W. R., R. G. Dean, and R. A. Dalrymple (1985). Wave Height Variation Across Beaches of Arbitrary Profile. *Journal of Geophysical Research* 90: 11,917–11,927.

Dally, W. R., M. A. Johnson, and D. A. Osiecki (1994). Initial Development of an Amphibious ROV for Use in Big Surf. *Marine Technology Society Journal* 28(1): 3–10.

Dalrymple, R. A. (1992). Prediction of Storm/Normal Beach Profiles. *Journal of Waterway, Port, Coastal, and Ocean Engineering, Amer. Soc. Civil Engrs.* 118 (2): 193–200.

Dalrymple, R. A. and W. W. Thompson (1976). Study of Equilibrium Beach Profiles. *Proceedings of the 15th Coastal Engineering Conference, Amer. Soc. Civil Engrs.,* pp. 1277–1296.

Davies, A. G. and A. D. Heathershaw (1984). Surface Wave Propagation Over Sinusoidally Varying Wave Topography. *Journal of Fluid Mechanics* 144: 419–443.

Davis, R. A. and W. T. Fox (1972). Coastal Processes and Nearshore Sand Bars. *Journal of Sedimentary Petrology* 42: 401–412.

Davis, R. A., W. T. Fox, M. O. Hayes, and Boothroyd (1972). Comparison of Ridge and Runnel Systems in Tidal and Non-Tidal Environments. *Journal of Sedimentary Petrology* 42: 413–421.

Dean, R. G. (1973). Heuristic Models of Sand Trans in the Surf Zone. *Proceedings of the 1st Australian Conference on Coastal Engineering,* Engineering Dynamics in the Surf Zone, Sydney, Australia, pp. 209–214.

Dean, R. G. (1976). Beach Erosion: Causes, Processes and Remedial Measures. *CRC Critical Review of Envir. Control* 6(3): 259–296.

Dean, R. G. (1977). *Equilibrium Beach Profiles: U.S. Atlantic and Gulf Coasts.* Dept. of Civil Engr., Ocean Engineering Technical Report No. 12, University of Delaware, Newark.

Dean, R. G. (1987). Coastal Sediment Processes: Toward Engineering Solutions. *Coastal Sediments '87, Amer. Soc. Civil Engrs.,* pp. 1–24.

Dean, R. G. (1991) Equilibrium Beach Profiles: Characteristics and Applications. *Journal of Coastal Research* 7: 53–84.

Dhyr-Nielsen, M. and T. Sorensen (1970). Sand Transport Phenomena on Coasts with Bars. *Proceedings of the 12th Coastal Engineering Conference, Amer. Soc. Civil Engrs.,* pp. 855–866.

Dibanjnia, M., T. Shimizu, and A. Watanabe (1994). Profile Change of a Sheet Flow Dominated Beach. *Proceedings of the 24th Conference on Coastal Engineering, Amer. Soc. Civil Engrs.,* pp. 1946–1960.

Dolan, T. J. and R. G. Dean (1985). Multiple Longshore Sand Bars in the Upper Chesapeake Bay. *Estuarine Coastal Shelf Science* 21: 727–743.

Dubois, R. N. (1972). Inverse Relation Between Foreshore Slope and Mean Grain Size as a Function of the Heavy Mineral Content. *Geological Society of America Bulletin* 83: 871–875.

Duncan, R. (1964). The Effects of Water Table and Tidal Cycle on Swash-Backwash Sediment Distribution and Beach Profile Development. *Marine Geology* 2: 186–197.

Eagleson, P. S., R. G. Dean, and L. A. Peralta (1958). *The Mechanics of the Motion of Discrete Spherical Bottom Sediment Particles Due to Shoaling Waves.* U.S. Army Corps of Engineers, Beach Erosion Board Technical Memo No. 104.

Eagleson, P. S. and R. G. Dean (1961). Wave-Induced Motion of Bottom Sediment Particles. *Transactions American Society of Civil Engineers* 126: 1162–89.

Edelman, T. (1968). Dune Erosion During Storm Conditions. *Proceedings of the 11th Conference on Coastal Engineering, Amer. Soc. Civil Engrs.*, pp. 719–722.

Edelman, T. (1972). Dune Erosion During Storm Conditions. *Proceedings of the 13th Conference on Coastal Engineering, Amer. Soc. Civil Engrs.*, pp. 1305–1312.

Egense, A. K. (1989). Southern California Beach Changes in Response to Extraordinary Storm. *Shore & Beach* 57(4): 14–17.

Emery, K. O. (1961). A Simple Method for Measuring Beach Profiles. *Limnology and Oceanography* 6: 90–93.

Emery, K. O. and J. F. Foster (1948). Water Tables in Marine Beaches. *Journal of Marine Research* 7: 644–654.

Engelund, F. and J. Fredsøe (1976). A Sediment Transport Model for Straight Alluvial Channels. *Nordic Hydrology* 7: 1–20.

Evans, O. F. (1940). The Low and Ball of the East Shore of Lake Michigan. *Journal of Geology* 48: 476–511.

Felder, W., B. Hayden, and R. Dolan (1979). Analysis of Inshore and Offshore Profiles. *Journal of Geology* 87: 455–461.

Fisher, J. S., M. F. Overton, and T. Chisholm (1986). Field Measurements of Dune Erosion. *Proceedings of the 20th Coastal Engineering Conference, Amer. Soc. Civil Engrs.*, pp. 1107–1115.

Gaillard, D. D. (1904). *Wave Action in Relation to Engineering Structures.* U.S. Army Corps of Engineers Prof. Paper No. 31.

Gorsline, D. S. (1966). Dynamic Characteristics of West Florida Gulf Coast Beaches. *Marine Geology* 4: 187–206.

Graaff, J. van de (1977). Probabilistic Design of Dunes. An Example from the Netherlands. *Coastal Engineering* 9: 479–500.

Graaff, J. van de (1986). Dune Erosion During a Storm Surge. *Coastal Engineering* 1: 99–134.

Grant, U. S. (1948). Influence of the Water Table on Beach Aggradation and Degradation. *Journal of Marine Research* 7: 655–660.

Gresswell, R. K. (1937). The Geomorphology of the South-West Lincolnshire Coastline. *Geographical Journal* 90: 335–349.

Gresswell, R. K. (1953). *Sandy Shores of South Lincolnshire:* Liverpool, England: University of Liverpool Press.

Hagan, G. (1863). *Handbuch der Wasserbaukunst*: 3. Berlin, Germany: Teil-Das Meer.

Hallermeier, R. J. (1981). A Profile Zonation for Seasonal Sand Beaches from Wave Climate. *Coastal Engineering* 4: 253–277.

Harrison, W. (1969). Empirical Equations for Foreshore Changes over a Tidal Cycle. *Marine Geology* 7: 529–551.

Harrison, W., C. S. Fang, and S. N. Wang (1971), Groundwater Flow in a Sandy Tidal Beach, 1: One-Dimensional Finite Element Analysis. *Water Resources Research* 7: 1313–1322.

Hayden, B., W. Felder, J. Fisher, D. Resio, L. Vincent, and R. Dolan (1975). *Systematic Variations in Inshore Bathymetry.* Technical Report No. 10, Department of Environmental Sciences, University of Virginia, Charlottesville.

HAYES, M. O. (1972). Forms of Sediment Accumulation in the Beach Zone. In *Waves on Beaches,* R. E. Meyer (editor). Pp. 297–356. New York: Academic Press.

HAYES, M. O. and J. C. BOOTHROYD (1969). Storms as Modifying Agents in the Coastal Environment. In *Coastal Environments: NE Massachusetts,* M. O. Hayes (editor). Pp. 290–315. Department of Geology, University of Massachusetts, Amherst, MA.

HEATHERSHAW, A. D. and A. G. DAVIES (1985). Resonant Wave Reflection by Transverse Bedforms and Its Relation to Beaches and Offshore Bars. *Marine Geology* 62: 321–338.

HEDEGAARD, I. B., R. DEIGAARD, and J. FREDSØE (1991). Onshore/Offshore Sediment Transport and Morphological Modelling of Coastal Profiles. *Coastal Sediments '91, Amer. Soc. Civil Engrs.,* pp. 643–657.

HEDEGAARD, I. B., J. A. ROELVINK, H. SOUTHGATE, P. PECHON. J. NICHOLSON, and L. HAMM (1992). Intercomparison of Coastal Profile Models. *Proceedings of the 23rd Coastal Engineering Conference, Amer. Soc. Civil Engrs.,* pp. 2108–2121.

HOLMAN, R. A. and A. J. BOWEN (1982). Bars, Bumps and Holes: Models for the Generation of Complex Beach Topography. *Journal of Geophysical Research* 87: 457–468.

HOLMAN, R. A. and A. H. SALLENGER (1993). Sand Bar Generation: A Discussion of the Duck Experiment Series. *Journal of Coastal Research,* A. D. Short (editor). Special Issue No. 15, pp. 76–92.

HOLMAN, R. A. and T. C. LIPPMANN (1986). Remote Sensing of Nearshore Bar Systems—Making Morphology Visible. *Proceedings of the 20th Coastal Engineering Conference, Amer. Soc. Civil Engrs.,* pp. 929–944.

HORN, D. P. (1992). A Numerical Model for Shore-Normal Sediment Size Variation on a Macrotidal Beach. *Earth Surface Processes and Landforms* 17: 755–773.

HOWD, P. A., A. J. BOWEN, R. A. HOLMAN, and J. OLTMAN-SHAY (1991). Infragravity Waves, Longshore Currents, and Linear Sand Bar Formation. *Coastal Sediments '91, Amer. Soc. Civil Engrs.,* pp. 72–84.

HUNT, I. A. (1959). Design of Seawalls and Breakwaters. *Proceedings of the American Society of Civil Engineers* 85(WW3): 123–152.

INMAN, D. L. and J. FILLOUX (1960). Beach Cycles Related to Tide and Local Wind Wave Regime. *Journal of Geology* 68: 225–231.

INMAN, D. L., M. H. S. ELWANY, and S. A. JENKINS (1993). Shorerise and Bar-Berm Profiles on Ocean Beaches. *Journal of Geophysical Research* 98(C10): 18,181–18,199.

IPPEN, A. T. and P. S. EAGLESON (1955). *A Study of Sediment Sorting by Waves Shoaling on a Plane Beach.* U.S. Army Corps of Engineers, Beach Erosion Board, Technical Memo No. 63.

IWAGAKI, Y. and H. NODA (1962). Laboratory Study of Scale Effects in Two-Dimensional Beach Processes. *Proceedings of the 8th Coastal Engineering Conference, Amer. Soc. Civil Engrs.,* pp. 194–210.

JOHNSON, D. W. (1919). *Shore Processes and Shoreline Development:* New York: John Wiley & Sons. [facsimile edition, New York: Hafner].

JOHNSON, J. W. (1949). Scale Effects in Hydraulic Models Involving Wave Motion. *Transactions American Geophysical Union* 30: 517–525.

KEMP, P. H. and D. T. PLINSTON (1968). Beaches Produced by Waves of Low Phase Difference. *Journal of the Hydraulics Division, Amer. Soc. Civil Engrs.* 94: 1183–1195.

KEULEGAN, G. H. (1948). *An Experimental Study of Submarine Sand Bars.* U.S. Army Corps of Engineers, Beach Erosion Board, Technical Report No. 5.

KINDLE, E. M. (1936). Notes on Shallow-Water Sand Structures. *Journal of Geology* 44: 861–869.

KING, C. A. M. (1953). The Relationship Between Wave Incidence, Wind Direction, and Beach Changes at Marsden Bay, C. Durham: *Transactions of the Institute of British Geographers* 19: 13–23.

KING, C. A. M. and W. W. WILLIAMS (1949). The Formation and Movement of Sand Bars by Wave Action. *Geographical Journal* 113: 70–85.

KIRBY, J. T., R. A. DALRYMPLE, and L. F. LIU (1981). Modification of Edge Waves by Barred-Beach Topography. *Coastal Engineering* 5: 35–49.

KOMAR, P. D. and M. K. GAUGHAN (1972). Airy Wave Theory and Breaker Height Prediction. *Proceedings of the 13th Coastal Engineering Conference, Amer. Soc. Civil Engrs.,* pp. 405–418.

KOMAR, P. D. and W. G. McDOUGAL (1994). The Analysis of Beach Profiles and Nearshore Processes Using the Exponential Beach Profile Form. *Journal of Coastal Research* 10: 59–69.

KRAUS, N. C. and M. LARSON (1988). *Beach Profile Change Measured in the Tank for Large Waves, 1956–1957 and 1962*. Technical Report CERC-88-6, U.S. Army Corps of Engineers.

KRAUS, N. C. , K. J. GINGERICH, and J. ROSATI (1989). *DUCK85 Surf Zone Sand Transport Experiment*. Technical Report CERC-89-5, U.S. Army Corps of Engineers.

KRIEBEL, D. L. (1986). Verification Study of a Dune Erosion Model. *Shore & Beach* 54: 13–21.

KRIEBEL, D. L. (1990). Advances in Numerical Modelling of Dune Erosion. *Proceedings of the 22nd Coastal Engineeering Conference, Amer. Soc. Civil Engrs.*, pp. 2304–2317.

KRIEBEL, D. L. and R. G. DEAN (1985). Numerical Simulation of Time-Dependent Beach and Dune Erosion. *Coastal Engineering* 9: 221–245.

KRIEBEL, D. L., W. R. DALLY, and R. G. DEAN (1986). Undistorted Froude Scale Model for Surf Zone Sediment Transport. *Proceedings of the 20th Coastal Engineering Conference, Amer. Soc. Civil Engrs.*, pp. 1296–1310.

KRUMBEIN, W. C. and F. A. GRAYBILL (1965). *An Introduction to Statistical Models in Geology*. New York: McGraw-Hill.

LaFOND, E. C. (1939). Sand Movement near the Beach in Relation to Tides and Waves. *Proceedings of the 6th Pacific Science Congress*, pp. 795–799.

LARSON, M. (1991). Equilibrium Profile of a Beach with Varying Grain Size. *Coastal Sediments '91, Amer. Soc. Civil Engrs.*, pp. 905–919.

LARSON, M., N. C. KRAUS, and T. SUNAMURA (1988). Beach Profile Change: Morphology, Transport Rate, and Numerical Simulation. *Proceedings of the 21st Coastal Engineering Conference, Amer. Soc. Civil Engrs.*, pp. 1295–1309.

LARSON, M. and N. C. KRAUS (1989). *SBEACH: Numerical Model for Simulating Storm Induced Beach Change*. Technical Report CERC-89-9, U.S. Army Corps of Engineers.

LAU, J. and B. TRAVIS (1973). Slowly Varying Stokes Waves and Submarine Longshore Bars. *Journal of Geophysical Research* 78(21): 4489–4497.

LETTAU, H. (1932). Stehende Wellen als Ursache und gestaltender Vorgänge in Seen. *Ann. Hydrogr. Mar. Meteorol.* 60: 385.

LIPPMANN, T. C. and R. A. HOLMAN (1989). Quantification of Sand Bar Morphology: A Video Technique Based on Wave Dissipation. *Journal of Geophysical Research* 94(C1): 995–1011.

LIPPMANN, T. C. and R. A. HOLMAN (1990). The Spatial and Temporal Variability of Sand Bar Morphology. *Journal of Geophysical Research* 95(C7): 11,575–11,590.

MADSEN, O. S. and W. D. GRANT (1976). Qualitative Description of Sediment Transport by Waves. *Proceedings of the 15th Coastal Engineering Conference, Amer. Soc. Civil Engrs.*, pp. 1093–1112.

McLEAN, R. F. and R. M. KIRK (1969). Relationship between Grain Size, Size-Sorting, and Foreshore Slope on Mixed Sand-Shingle Beaches. *New Zealand Journal of Geology and Geophysics* 12: 138–155.

MILLER, R. L. (1976). Role of Vortices in Surf Zone Prediction: Sedimentation and Wave Forces. In *Beach and Nearshore Sedimentation*, R. A. Davis and R. L. Ethington (editors). Special Publication No. 24, pp. 92–114. Soc. Econ. Paleontologists and Mineralogists, Tulsa, Oklahoma.

MILLER, R. L. and J. M. ZEIGLER (1958). A Model Relating Dynamics and Sediment Pattern in Equilibrium in the Region of Shoaling Waves, Breaker Zone, and Foreshore. *Journal of Geology* 66: 417–441.

MILLER, R. L. and J. M. ZEIGLER (1964). A Study of Sediment Distribution in the Zone of Shoaling Waves over Complicated Bottom Topography. In *Papers in Marine Geology*, R. L. Miller (editor), pp. 133–153, New York: MacMillan.

MUNCH-PETERSON (1938). Littoral Drift Formula. U.S. Army Corps of Engineers, *Beach Erosion Board Bulletin No.* 4 (1950): 1–36.

MOORE, B. D. (1982). *Beach Profile Evolution in Response to Changes in Water Level and Wave Height*. Masters thesis, Department of Civil Engineering, University of Delaware, Newark.

MORTON, R. A., J. G. PAINE, and J. C. GIBEAUT (1994). Stages and Durations of Post-Storm Beach Recovery, Southeastern Texas Coast, U.S.A. *Journal of Coastal Research* 10: 884–908.

MURRAY, J. (1853). On Movement of Shingle in Deep Water. *Minutes of the Proceedings of the Institute of Civil Engineering* 12: 551.

NAIRN, R. B. (1991). Problems Associated with Deterministic Modelling of Extreme Beach Erosion Events. *Coastal Sediments '91, Amer. Soc. Civil Engrs.*, pp. 588–602.

NAIRN, R. B. and H. N. SOUTHGATE (1993). Deterministic Profile Modelling of Nearshore Processes, 2: Sediment Transport and Beach Profile Development. *Coastal Engineering* 19: 57–96.

O'HARE, T. J. and A. G. DAVIES (1990). A Laboratory Study of Sand Bar Evolution. *Journal of Coastal Research* 6: 531–544.

ORFORD, J. D. and P. WRIGHT (1978). What's in a Name?—Descriptive or Genetic Implications of "Ridge and Runnel" Topography. *Marine Geology* 28: M1–M8.

OTTO, T. (1912). Der Darss und Zingst. *Jahrb. d. Geo. Gesell. zu Greifswald* 13: 393–403.

OTVOS, E. G. (1965). Sedimentation-Erosion Cycles of Single Tidal Periods on Long Island Sound Beaches. *Journal of Sedimentary Petrology* 35: 604–609.

OVERTON, M. F., J. S. FISHER, and T. FENAISH (1987). Numerical Analysis of Swash Forces on Dunes. *Coastal Sediments '87, Amer. Soc. Civil Engrs.*, pp. 632–641.

OVERTON, M. F. and J. S. FISHER (1988). Simulation Modeling of Dune Erosion. *Proceedings of the 21st Coastal Engineering Conference, Amer. Soc. Civil Engrs.*, pp. 1857–1867.

OVERTON, M. F., W. A. PRATIKTO, J. C. LU, and J. S. FISHER (1993). Laboratory Investigation of Dune Erosion as a Function of Sand Grain Size and Dune Density. *Coastal Engineering* 23: 151–165.

OWENS, E. H. and D. H. FROBEL (1977). Ridge and Runnel Systems in the Magdalen Islands, Quebec. *Journal of Sedimentary Petrology* 47: 193–198.

PRUSZAK, Z. (1989). On-Offshore Bed-Load Sediment Transport in the Coastal Zone. *Coastal Engineering* 13: 273–292.

PRUSZAK, Z. (1993). The Analysis of Beach Profile Changes Using Dean's Method and Empirical Orthogonal Functions. *Coastal Engineering* 19: 245–261.

PUGH, D. C. (1953). Etudes Minéralogique dees Plages Picardes et Flamandes. *Bull. d'Inf. Com. Cent. d'Oceanog. et d'Etudes des Cotes* 5(6): 245–276.

RECTOR, R. L. (1954). *Laboratory Study of the Equilibrium Profiles of Beaches.* U.S. Army Corps of Engineers, Beach Erosion Board, Technical Memo No. 41.

ROELVINK, J. A. and M. J. F. STIVE (1989). Bar-Generating Cross-Shore Flow Mechanisms on a Beach. *Journal of Geophysical Research* 94(C4): 4785–4800.

ROELVINK, J. A. and I. B. HEDEGAARD (1993). Cross-Shore Profile Models. *Coastal Engineering* 21: 163–191.

SALLENGER, A. H., P. C. HOWARD, C. H. FLETCHER, and P. A. HOWD (1983). A System For Measuring Bottom Profile, Waves and Currents in the High-Energy Nearshore Environment. *Marine Geology* 51: 63–76.

SALLENGER, A. H. and P. A. HOWD (1989). Nearshore Bars and the Break-Point Hypothesis. *Coastal Engineering* 12: 301–313.

SAVILLE, T. (1957). Scale Effects in Two Dimensional Beach Studies. *International Association of Hydraulic Research,* Lisbon, Portugal.

SAWARAGI, T. and I. DEGUCHI (1980). On-Offshore Sediment Transport Rate in the Surf Zone. *Proceedings of the 17th Coastal Engineering Conference, Amer. Soc. Civil Engrs.*, pp. 1194–1214.

SCHWARTZ, M. L. (1967). Littoral Zone Tidal Cycle Sedimentation. *Journal of Sedimentary Petrology* 37: 677–683.

SCOTT, T. (1954). *Sand Movement by Waves.* U.S. Army Corps of Engineers, Beach Erosion Board, Technical Memo No. 48.

SEYMOUR, R. J., A. L. HIGGINS, and D. P. BOTHMAN (1978). Tracked Vehicle for Continuous Nearshore Profiles. *Proceedings of the 16th Coastal Engineering Conference, Amer. Soc. Civil Engrs.*, pp. 1542–1554.

SHEPARD, F. P. (1950a). *Longshore Bars and Longshore Troughs.* U.S. Army Corps of Engineers, Beach Erosion Board, Technical Memo No. 15.

SHEPARD, F. P. (1950b). *Beach Cycles in Southern California.* U.S. Army Corps of Engineers, Beach Erosion Board, Technical Memo No. 20.

SHEPARD, F. P. (1963). *Submarine Geology.* New York: Harper & Row. [see also 3rd edition, 1973].

SHEPARD, F. P. and E. C. LaFOND (1940). Sand Movements near the Beach in Relation to Tides and Waves. *American Journal of Science* 238: 272–285.

SHIH, S.-M. and P. D. KOMAR (1994). Sediments, Beach Morphology and Sea Cliff Erosion Within an Oregon Coast Littoral Cell. *Journal of Coastal Research* 10: 144–157.

SHORT, A. D. (1975a). Multiple Offshore Bars and Standing Waves. *Journal of Geophysical Research* 80: 3838–3840.

SHORT, A. D. (1975b). Offshore Bars Along the Alaskan Arctic Coast. *Journal of Geology* 83: 209–221.

SONU, C. J. and R. J. RUSSELL (1966). Topographic Changes in the Surf Zone Profiles. *Proceedings of the 10th Coastal Engineering Conference, Amer. Soc. Civil Engrs.,* pp. 502–524.

SOUTHGATE, H. N. (1991). Beach Profile Modelling: Flume Data Comparisons and Sensitivity Tests. *Coastal Sediments '91, Amer. Soc. Civil Engrs.,* pp. 1829–1841

SOUTHGATE, H. N. and R. B. NAIRN (1993). Deterministic Profile Modelling of Nearshore Processes, 1: Waves and Currents. *Coastal Engineering* 19: 27–56.

STIVE, M. J. F. (1986). A model for Cross-Shore Sediment Transport. *Proceedings of the 20th Coastal Engineering Conference, Amer. Soc. Civil Engrs.,* pp. 1550–1564.

STIVE, M. J. F. and J. A. BATTJES (1984). A Model for Offshore Sediment Transport. *Proceedings of the 19th Coastal Engineering Conference, Amer. Soc. Civil Engrs.,* pp. 1420–1436.

STRAHLER, A. N. (1966). Tidal Cycle of Changes on an Equilbrium Beach. *Journal of Geology* 74: 247–268.

SUNAMURA, T. (1984). Quantitative Predictions of Beach-Face Slopes. *Geological Society of America Bulletin* 95: 242–245.

SUNAMURA, T. (1985). Predictive Relationships for Position and Size of Longshore Bars. *Proceedings of the 32nd Japan Conference on Coastal Engineering,* pp. 331–335 [in Japanese].

SUNAMURA, T. (1989). Sandy Beach Geomorphology Elucidated by Laboratory Modeling. In *Applications in Coastal Modeling,* V. C. Lakhan and A. S. Trenhail (editors), pp. 159–213. Amsterdam: Elsevier.

SUNAMURA, T. and K. MARUYAMA (1987). Wave-Induced Geomorphic Response of Eroding Beaches, with Special Reference to Seaward Migrating Bars. *Coastal Sediments 87, Amer. Soc. Civil Engrs.,* pp. 788–801.

SYMONDS, G. and A. J. BOWEN (1984). Interactions of Nearshore Bars with Incoming Wave Groups. *Journal of Geophysical Research* 89: 1953–1959.

TAKEDA, I. and T. SUNAMURA (1982). Formation and Height of Berms. *Transactions Japanese Geomorphological Union* 3: 145–157 [in Japanese].

THOM, B. G. and W. HALL (1991). Behaviour of Beach Profiles During Accretion and Erosion Dominated Periods. *Earth Surface Processes and Landforms* 16: 113–127.

THOMPSON, W. F. and J. B. THOMPSON (1919). *The Spawning of the Grunion.* California State Fish and Game Commission of Fish Bulletin No. 3.

VELLINGA, P. (1982). Beach and Dune Erosion During Storms. *Coastal Engineering* 6: 361–387.

VAN HIJUM, E. (1974). Equilibrium Profiles of Coarse Material Under Wave Attack. *Proceedings of the 14th Coastal Engineering Conference, Amer. Soc. Civil Engrs.,* pp. 939–957.

WATANABE, A., Y. RIHO, and K. HORIKAWA (1980). Beach Profiles and On-Offshore Sediment Transport. *Proceedings of the 17th Coastal Engineering Conference, Amer. Soc. Civil Engrs.,* pp. 1106–1121.

WATANABE, A. and K. SHIBA (1994). Numerical Modeling of Beach Profile Change Under Sheet-Flow Condition Using Nonlinear Wave Theory. *Proceedings of the 24th Coastal Engineering Conference, Amer. Soc. Civil Engrs.,* pp. 2785–2798.

WATTS, G. M. (1954). *Laboratory Study on the Effect of Varying Wave Periods on the Beach Profiles.* U.S. Army Corps of Engineers, Beach Erosion Board, Technical Memo No. 53.

WIEGEL, R. L. (1964). *Oceanographical Engineering.* Englewood Cliffs, NJ: Prentice-Hall.

WINANT, C. D., D. L. INMAN, and C. E. NORDSTROM (1975). Description of Seasonal Beach Changes Using Empirical Eigenfunctions. *Journal of Geophysical Research* 80: 1979–1986.

WINANT, C. D. and D. G. AUBREY (1976). Stability and Impulse Response of Empirical Eigenfunctions. *Proceedings of the 15th Coastal Engineering Conference, Amer. Soc. Civil Engrs.,* pp. 1312–1325.

WORK, P. A. and R. G. DEAN (1991). Effect of Varying Sediment Size on Equilibrium Beach Profiles. *Coastal Sediments '91, Amer. Soc. Civil Engrs.,* pp. 890–904.

ZENKOVITCH, V. P. (1946). On the Study of Shore Dynamics. *Transactions Inst. Okeanol., Akad. Nauk S.S.S.R.* 1: 99–112.

8

Wave-Generated Currents in the Nearshore

It [a rip current] is a sort of river, a stream showing on the surface deep and powerful, easily perceptible, running with the velocity of a mill race. So swift and powerful is it that a motorboat could not stem its sweeping current. It will carry brick, large rocks and even chunks of lead far out to sea. The most powerful swimmer will find himself helpless as a babe in its rushing grasp.

M. P. Hite
Science (1925)

When waves reach the coast and break on a sloping beach, they generate currents in the nearshore that take on a variety of forms depending on the wave and beach conditions (Basco, 1982, 1983; Komar and Oltman-Shay, 1990). These currents may achieve a considerable strength, becoming a hazard to swimmers and causing many drownings each year. However, the nearshore currents have their beneficial aspects, producing a flushing of the nearshore water, replacing it with generally cleaner offshore water (Inman, Tait, and Nordstrom, 1971). Nearshore currents are particularly important in that they combine with waves to transport beach sediment (Chapter 9) and accordingly are a significant factor in controlling the morphology of the beach (Chapter 11).

There are two wave-induced current systems in the nearshore that dominate water movements: (1) a cell-circulation system consisting of rip currents and associated longshore currents and (2) longshore currents generated by an oblique wave approach to the shoreline. These systems are illustrated, respectively, in Figures 8-1(a) and 8-1(c), and can be considered as end members or limiting cases within an otherwise continuum of flow-field types, with Figure 8-1(b) illustrating an intermediate pattern that takes on aspects of both. The current pattern that dominates the nearshore circulation depends in large part on the obliquity of wave approach to the shore. When the waves break with their crests effectively parallel to the average shoreline trend, the generated currents take the form of a cell circulation [Fig. 8-1(a)]. Rip currents are the most obvious portion of this circulation—strong, narrow currents that flow seaward from the surf zone. If the waves break at appreciable angles to the shoreline, the resulting longshore current flows parallel to the shore and is largely confined to the nearshore between the breakers and shoreline [Fig. 8-1(c)]. It is this type of current that is particularly significant in causing a longshore movement of beach sediment, a longshore transport that can involve hundreds of kilometers of sediment displacement along the coast (Chapter 9). The intermediate pattern [Fig. 8-1(b)] commonly occurs when waves break at small angles to the shoreline or where there is a strong control by the beach topography on the pattern of nearshore currents.

A   Cell Circulation ($\alpha_b \approx 0°$)

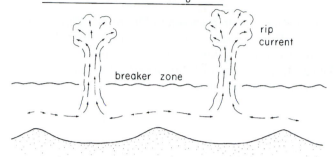

B   General Circulation (small α_b)

C   Oblique Wave Approach (large α_b)

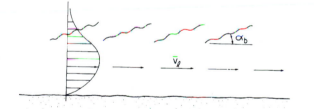

Figure 8-1   Patterns of currents observed within the nearshore, which depend in large part on the angle of wave breaking (α_b). With the wave crests parallel to the shore ($\alpha_b = 0$), a cell circulation tends to develop, dominated by seaward-flowing rip currents. With large α_b, the wave-generated currents flow parallel to the shoreline.

These currents give direction to the movement of beach sediments and as a result act to mold the nearshore topography. The beach morphology often reflects the water-current patterns, being markedly different for the two end members in Figure 8-1 (Chapter 11). The beach topography in turn becomes an important factor in controlling the current patterns, and at times it is uncertain which came first, the current or the topography.

The generation mechanisms of nearshore currents will be examined in this chapter. The focus will be primarily on the two limiting cases depicted in Figure 8-1 because they contrast in their basic modes of formation. Some consideration will be given to the interplay between the currents and beach morphology, but the effects on the morphology will be dealt with at greater length in Chapter 11.

RIP CURRENTS AND THE CELL CIRCULATION

A detailed depiction of the cell circulation system is shown in Figure 8-2. Rip currents are the most visible and important feature. They are strong, narrow currents that flow seaward through the surf zone, often carrying debris and sediment that gives the water a distinct color compared with the adjacent clear water. The seaward-flowing rip currents also affect the waves, so the presence of a rip is sometimes made apparent by the altered wave refraction pattern. The flow within a rip current can achieve a high magnitude, attested to by the quotation at the beginning of this chapter, which observes that the seaward directed flow can be too strong for even the ablest swimmer who becomes as "helpless as a babe." However, the knowledgeable beach user will first swim parallel to shore to escape the confines of the narrow rip current and only then attempts to swim back to shore.

Rip currents are fed by longshore-directed currents within the surf zone that increase from zero at a point between two neighboring rips, reaching a maximum just before turning seaward to form the rip (Fig. 8-2). The longshore currents are in turn fed by the slow shoreward transport of water into the surf zone from breaking waves. A nearshore circulation cell thus consists of longshore currents feeding the rips, the seaward flowing rip currents that extend through the breaker zone and spread out into rip heads, and a return onshore flow to replace the water moving offshore through the rips (Shepard and Inman, 1950a, 1950b). The idealized form of the circulation cell is one in which the rips are oriented perpendicular to the shoreline, with the longshore directed surf-zone currents feeding the rips equally from either side (Fig. 8-2). However, rip currents often cut diagonally through the surf zone [Fig. 8-1(b)], and the longshore feeder currents on either side differ both in longshore extent and in intensity.

The first scientific observations of rip currents were made by Shepard, Emery, and La-Fond (1941). They found that the velocity of rip currents and the distance they flow seaward are related to the heights of the incoming waves. They further recognized that the positioning of rip currents could be governed by offshore topography, the rip currents generally occurring away from areas of wave convergence, that is, away from the areas of highest waves. Shep-

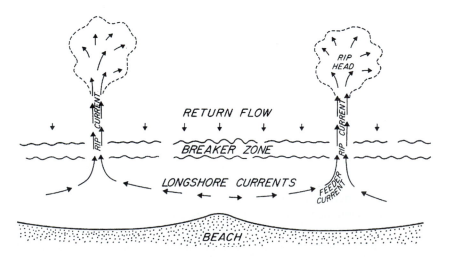

Figure 8-2   The nearshore cell circulation consists of (1) feeder longshore currents, (2) seaward-flowing rip currents, and (3) a return flow of water from the offshore into the surf zone.

ard and Inman (1950a, 1950b) obtained the first comprehensive series of field measurements of the complete cell circulation. It was recognized that although offshore topography and its effects on wave refraction may govern the positioning of rip currents along a beach, cells with rip currents can also exist on long, straight beaches with a regular bottom topography.

McKenzie (1958) demonstrated that each incident wave condition arriving at the beach forms a characteristic pattern of longshore currents and rip currents. He noted that with large waves, only a few strong rips are produced, whereas when the waves are smaller the rips are weaker and more numerous.

The first attempts to explain the generation of rip currents were based on the existence of an onshore mass transport of water associated with the waves, a Stokes drift (Chapter 5). It was hypothesized that this water would pile up on the beach and provide a head for the out-flowing rip currents. However, because of the difficulty of making quantitative estimates of the onshore mass transport near the breakers as a function of the wave parameters, this attempt to explain the occurrence of the cell circulation was never particularly successful (e.g., Putnam, Munk, and Traylor, 1949).

The modern explanation for the generation of the cell circulation was developed following the introduction of the concept of radiation stress by Longuet-Higgins and Stewart (1964). Radiation stress, the excess flow of momentum due to the presence of waves, is now the basic theoretical tool with which nearshore circulation cells are modeled. We saw in Chapter 6 that the shoreward component of the radiation stress produces a set-down immediately offshore of the breakers and a set-up within the surf zone. As first demonstrated and analyzed by Bowen (1969a), the cell circulation results from a longshore variation in wave heights, which in turn produce a longshore variation in set-up elevations. This is illustrated schematically in Figure 8-3. In simple terms, the longshore currents flow "downhill" from areas of greatest wave heights and set-up, to areas of lower waves and set-up where the converging longshore currents turn seaward to form a rip current. Bowen's analysis involved an examination of the longshore variations in wave set-down and set-up, as well as the longshore component (S_{yy}) of the radiation stress.

We saw in Chapter 6 that the set-up results from the S_{xx} onshore component of the radiation stress being balanced by the pressure gradient of the seaward-sloping water surface in the nearshore. Balancing those forces yields

$$\frac{\partial \overline{\eta}}{\partial x} = -\left(1 + \frac{8}{3\gamma^2}\right)\frac{\partial h}{\partial x} \tag{8.1}$$

for the cross-shore slope of the set-up denoted by $\overline{\eta}$. There is a direct proportionality with the beach slope $S_0 = \partial h/\partial x$, but not a direct dependence on the wave height. However, the larger waves break in deeper water than smaller waves, and the set-up therefore begins farther seaward at longshore locations where the larger waves occur. This dependence of the set-up on the wave height is shown in Figure 8-4, based on the laboratory wave-channel experiments undertaken by Bowen (1969a). Although the cross-shore gradients of the set-up are approximately the same for both the large and small waves, as predicted by Equation (8.1), the set-up achieves higher elevations shoreward from the larger breakers. Inside the surf zone, the mean water level is higher shoreward of the larger breakers than it is shoreward of the small waves. A longshore pressure gradient therefore exists, which will drive a longshore current from positions of high waves and set-up to adjacent positions of low waves, as illustrated in Figure 8-3.

It would appear from Figure 8-4 that since the set-down associated with the larger waves is greater, there would be a similar tendency for water to flow from positions of low breakers

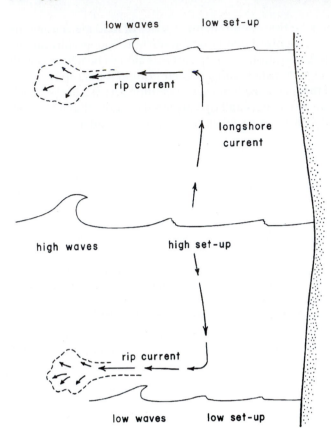

low waves   low set-up

rip current

longshore
current

high waves   high set-up

Figure 8-3 Schematic illustration of the generation of the cell circulation by a longshore
variation in the heights of breaking waves,
which produces a parallel variation in the elevation of the set-up within the surf zone. The longshore currents flow from positions of high waves
and set-up, to positions of low waves and set-up
where the converging currents turn seaward as
rip currents.

rip current

low waves   low set-up

to positions of high breakers just outside the break point, opposite in direction to the current
generated within the surf zone. It turns out, however, that the longshore y-component of the
radiation stress, which has not been included until now, prevents this latter current and at the
same time enhances the current within the surf zone. Recall from Chapter 5 that, in addition
to the S_{xx}-component of the radiation stress, there is a S_{yy}-component, a momentum flux acting parallel to the wave crests, in this case acting parallel to the shoreline. This component is
given by

$$S_{yy} = \frac{1}{8} \rho g H^2 \left[\frac{kh}{\sinh(2kh)} \right] \tag{8.2}$$

Since the wave height varies alongshore, S_{yy} will similarly vary and there will exist a longshore gradient

$$\frac{\partial S_{yy}}{\partial y} = \frac{1}{4} \rho g H \left[\frac{kh}{\sinh(2kh)} \right] \frac{\partial H}{\partial y} \tag{8.3}$$

As derived in Chapter 6, the set-down offshore from the breaker zone is given by

$$\overline{\eta} = -\frac{kH^2}{8 \sinh(2kh)} \tag{8.4}$$

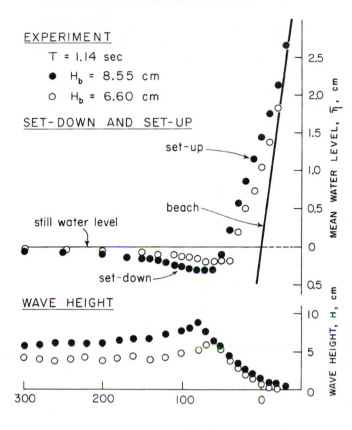

EXPERIMENT

T = 1.14 sec

● H_b = 8.55 cm

○ H_b = 6.60 cm

SET-DOWN AND SET-UP

set-up

beach

still water level

set-down

WAVE HEIGHT

MEAN WATER LEVEL, $\bar{\eta}$, cm

WAVE HEIGHT, H, cm

300   200   100   0

Figure 8-4 Wave set-down and set-up for two different wave breaker heights as measured in a laboratory wave channel. The larger wave height produces a greater set-down, and although the water slope of the set-up is approximately the same for both wave heights, the water level is higher shoreward of the larger breakers since the set-up slope is initiated farther offshore. [Adapted with permission of American Geophysical Union, from A. J. Bowen, Rip Currents, 1. Theoretical Investigations, *Journal of Geophysical Research* 74, p. 5472. Copyright © 1969 American Geophysical Union.]

The longshore pressure gradient resulting from variations in the set-down is therefore

$$\rho g(\bar{\eta} + h)\frac{\partial \bar{\eta}}{\partial y} = \rho g(\bar{\eta} + h)\frac{kH}{4 \sinh(2kh)}\frac{\partial H}{\partial y}$$

$$\approx -\frac{1}{4}\rho g H\left[\frac{kh}{\sinh(2kh)}\right]\frac{\partial H}{\partial y}$$

(8.5)

having used $\bar{\eta} << h$ outside the surf zone. Comparing Equations (8.3) and (8.5), we see that they differ only in sign, that is, in the longshore direction in which they act. Therefore, as determined by Bowen (1969a), the longshore variation in the wave set-down outside the breaker zone is balanced by the gradient in the y-directed radiation stress ($\partial S_{yy}/\partial y$), so that no net force exists outside the breaker zone to produce the circulation. It is a different matter within the surf zone where the longshore variation in the set-up and the longshore gradient in the radiation stress both act in the same direction, away from positions of high waves and toward areas of low waves. Within the surf zone the two forces combine to produce the observed flow of water away from regions of high waves and toward positions of low waves. The flow then turns seaward as a rip current where the waves and the set-up are lowest and the longshore currents converge.

Bowen (1969a) formulated his analysis by using the equations of continuity and momentum (which includes the radiation stress terms), along with general assumptions about the flow conditions: (1) the currents are steady; (2) velocities are depth integrated; (3) the water is homogeneous and incompressible; (4) the pressure is hydrostatic; (5) the Coriolis force can be neglected; and (6) the currents are sufficiently small that their interactions with waves are negligible. An example of the numerical solutions obtained by Bowen is shown in Figure 8-5, where the spacings of the streamlines reveal strong longshore currents within the surf zone, a narrow rip current of high velocity in the region of low waves, weak currents outside the surf zone, and a low-velocity onshore flow in the region of high waves. Bowen's inclusion of the nonlinear terms in the analysis, as first proposed by Arthur (1962), produced this narrowing of the rip current and strengthening of its velocities. Even with the above simplifying assumptions, the calculated cell circulation derived in the numerical analysis of Bowen bears a strong resemblance to the flow patterns of the cell circulation observed in nature. Of the assumptions made by Bowen, (2) and (6) are the most questionable and have been the focus for subsequent analyses of the cell circulation and improvements in the model (e.g., LeBlond and Tang, 1974; Dalrymple and Lozano, 1978; Longuet-Higgins, 1983; Haines, 1984).

The cell circulation therefore depends primarily on the existence of variations in wave heights along the shore. The most obvious way to produce this variation is by wave refraction, which can concentrate the wave rays in one area of the beach, causing high waves, and at the same time spread the rays in an adjacent area of beach and so produce low waves (Chapter 5). The positions of the rip currents and the overall cell circulation will then be governed by wave refraction and hence the offshore topography. The classic example of this situation is found on the beach at La Jolla, California, near the Scripps Institution of Oceanography, shown in Figure 8-6. The nearshore circulation was initially described by Shepard and Inman (1950a, 1950b) but was further discussed by Bowen and Inman (1969) in terms of the analyses of Bowen (1969a). At this location, wave refraction over submarine canyons (Chapter 5,

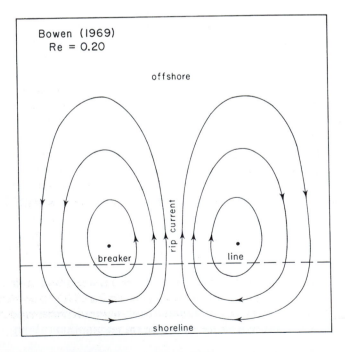

Figure 8-5 The pattern of streamlines for the cell circulation obtained in a numerical solution by Bowen (1969a). The closer the spacings of the streamlines, the faster the velocity of the flow. [Adapted with permission of American Geophysical Union, from A. J. Bowen, Rip Currents, 1. Theoretical Investigations, *Journal of Geophysical Research 74*, p. 5476. Copyright © 1969 American Geophysical Union.]

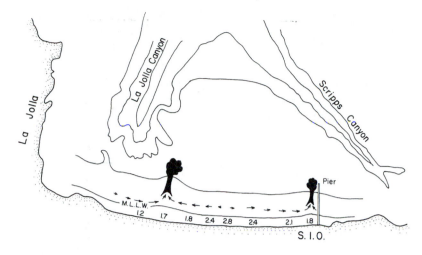

Figure 8-6 Rip currents and longshore currents at La Jolla, California, produced by a longshore variation in wave breaker heights caused by wave refraction over offshore submarine canyons. The numbers along the shore are measured values of breaker heights in meters. [Adapted from Nearshore Circulation, F. P. Shepard and D. L. Inman, *Proceedings of the 1st Coastal Engineering Conference,* 1950. Reproduced with permission from the American Society of Civil Engineers.]

Fig. 5-44) yields pronounced longshore variations in breaker heights with low waves in the lee of the canyon and larger waves to either side. Figure 8-6 shows the resulting longshore variation in breaker heights as measured by Shepard and Inman (1950b) and the resulting nearshore cell circulation with rip currents located in positions of lowest waves. A strong rip current is always present in the canyon lee, shifting position only slightly as waves arrive from different deep-water directions.

Headlands, breakwaters, and jetties can affect the incident waves by the partial sheltering of the shore and thereby produce significant longshore variations in wave heights and set-up (Gourley, 1974, 1976; Sasaki, 1975; Mei and Liu, 1977). Wave refraction and diffraction produce alongshore gradients with lower waves and set-up in the lee of the headland or breakwater, which in turn generate longshore currents flowing inward toward the sheltered region. In some situations this process can account for the development of strong rip currents adjacent to jetties and breakwaters. Gourley (1974, 1976) noted a number of field examples of longshore currents generated by wave-sheltering effects and undertook wave-basin experiments to examine the flow patterns and to relate the magnitudes of the currents to the longshore gradient in wave-breaker heights. The experimental arrangement and example results are shown in Figure 8-7. Sheltering by the breakwater produced a systematic longshore variation in the local breaker height H_b in reference to the unsheltered breaker height H_b^*. The wave set-up was measured by a system of piezometers located at fixed points within the concrete bed; the results are graphed in Figure 8-7 as $\overline{\eta}/\overline{\eta}^*$, which shows the longshore variation with the lowest elevations being within the sheltered region behind the breakwater. The longshore current was analyzed from photographs that followed the movements of a large number of neutrally buoyant floats. As expected, the flows were inward toward the sheltered zone where the measured set-up elevations are lowest, the magnitude of the longshore current V_m depending on the deep-water wave height H_∞, as graphed in Figure 8-7. The longshore current was deflected seaward by the breakwater, in effect producing a rip current.

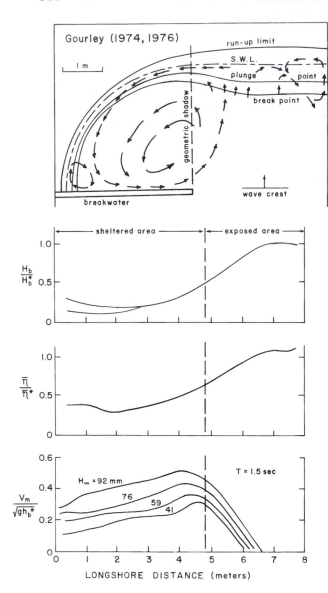

Figure 8-7 Experimental arrangement and results of Gourley (1974, 1976) for nearshore currents generated in the sheltered region of a breakwater. H_b is the wave breaker height, $\overline{\eta}$ is the wave set-up, and V_m is the maximum longshore current. The asterisk (*) parameters represent values in the exposed area distant from the immediate effects of the breakwater. [Adapted from Wave Set-up and Wave Generated Currents in the Lee of a Breakwater or Headland, M. R. Gourley, *Proceedings of the 14th Coastal Engineering Conference*, 1974. Reproduced with permission from the American Society of Civil Engineers.]

Shepard and Inman (1950a, 1950b) noted that circulation cells with rip currents can exist on long, straight beaches with regular offshore bottom topography, so clearly are not the product of wave refraction or local sheltering as described above. Bowen and Inman (1969) have shown both theoretically and experimentally in a wave basin that the ordinary incident swell waves can generate standing edge waves on the beach that have the same period as the incoming waves (i.e., synchronous edge waves). The interaction or summation of the incident waves and edge waves produces alternating high and low breakers along the shoreline and therefore gives rise to a regular pattern of circulation cells with evenly spaced rip currents.

As discussed in Chapter 6, with standing edge waves there will be alternate positions of nodes, where there is no observable up-and-down motion of the water surface due to the edge waves, and antinodes where the full edge-wave height is observed as the up-and-down motion (Fig. 6-43). Important to the generation of the cell circulation is that the edge waves have the same period as the incident waves so they can be systematically in phase and out of phase along the length of the shore. This is illustrated in Figure 8-8 from Bowen and Inman (1969), showing uniform incident waves having a constant height alongshore, superimposed on edge waves having the same period. At the instant of wave breaking, the standing edge wave at one antinode may be in phase with the incoming wave, so they add to enhance the height of the breaking wave. At the same time, at the next antinode position along the beach, the incoming wave and edge wave would be out of phase and therefore would subtract to yield a lower breaker height. Only at the positions of the nodes, where the edge wave makes no contribution to the breaker, would the true height of the incoming wave be observed. Since the input waves and edge waves have the same period, the large or small breakers persistently remain at fixed antinode positions, every other antinode being the site of large breakers and the alternate antinodes being the locations of small breakers. The result is a consistent longshore variation in the breaker heights, with a regular alternation of high and low breakers. This in turn produces a regular pattern of rip currents and cell circulation, as illustrated in Figure 8-9, the rips being found in the positions of low breakers, that is, at every other antinode of the causative edge waves. Since rip currents occur at every other antinode, the rip spacing will be equal to the edge-wave length, L_e, given by

$$L_e = \frac{gT_e^2}{2\pi} \sin(2n + 1)\beta \tag{8.6}$$

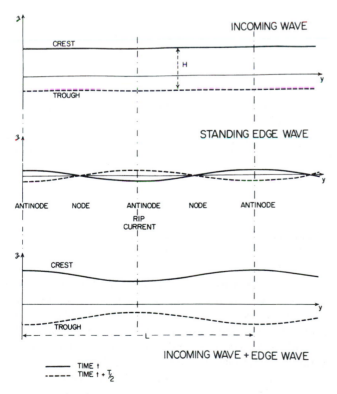

Figure 8-8 text labels: INCOMING WAVE, CREST, H, TROUGH, STANDING EDGE WAVE, ANTINODE, NODE, ANTINODE, NODE, ANTINODE, RIP CURRENT, CREST, TROUGH, L, INCOMING WAVE + EDGE WAVE, TIME t, TIME t + T/2

Figure 8-8 The addition of an incoming ocean wave and a standing edge wave at the breaker position to give a longshore variation in observed breaker heights. The net height is greatest where the edge wave and incoming wave are in phase and lowest where they are 180° out of phase. [Used with permission of American Geophysical Union, from A. J. Bowen and D. L. Inman, Rip Currents, 2. Laboratory and Field Observations, *Journal of Geophysical Research 74*, p. 5483. Copyright © 1969 American Geophysical Union.]

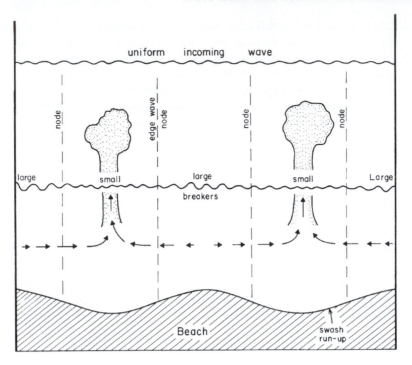

Figure 8-9 Positioning of the rip currents where the breaker height is smallest, that is, where the edge wave and incoming wave are 180° out of phase.

where T_e is the edge-wave period (and that of the incident waves), β is the local beach-slope angle, and n is the mode number of the edge waves (Chapter 6). The rip spacing is mainly dependent on the wave period and, to a smaller extent, on the beach slope. Because of the presence of the modal number n in the relationship, there are several possible values of L_e and rip spacings for each combination of wave period and beach slope.

Bowen and Inman (1969) demonstrated the basic validity of this mechanism in a series of wave-basin experiments. Extra boundary conditions are imposed on the edge-wave system in the laboratory wave basin, the end walls necessarily being at antinodes, requiring that

$$L_e = \frac{2\Lambda}{m} \tag{8.7}$$

where Λ is the longshore length of the beach (the distance between the walls), and m is the longshore modal number that may have integral values $1, 2, 3, \ldots$. In the laboratory the edge waves therefore have both cross-shore and longshore modal numbers. If the period of the generating waves in the basin is such that it yields a wave length L_e from Equation (8.6) that corresponds to an L_e value that is possible according to Equation (8.7), then the edge waves are at a resonating period and their interactions with the incoming waves produce a simple but pronounced pattern of rip currents. If the L_e values of Equations (8.6) and (8.7) do not agree for the input wave period, then they are non-resonating and the wave length that corresponds to the closest resonating T_e usually predominates. Under such circumstances an additional mode may simultaneously produce secondary rip currents and therefore a rather complex circulation.

Bowen and Inman (1969) also obtained field measurements of rip-current spacings that support their hypothesis that synchronous edge waves are important to the generation of the cell circulation. These data were obtained at El Moreno Beach on the Gulf of California, Mexico, a steep ($S_0 = 0.148$) reflective beach that extends uninterrupted in the longshore direction for several kilometers. On such a long beach, Equation (8.7) no longer applies, so the edge-wave length and rip-current spacings depend only on Equation (8.6) where a series of cross-shore mode numbers n are possible for a given incident wave period and beach slope. On a long beach the edge waves could also be progressive rather than standing, since there would be the same incident/edge-wave interaction as documented in the laboratory wave basin. Bowen and Inman obtained simultaneous measurements of rip-current spacings and spectra-derived measurements of peak periods of the incident waves. For several data sets obtained at various times and under different conditions, good agreement was found between the measured rip spacings and L_e calculated from Equation (8.6) using the incident-wave period and $n = 1$, supporting the synchronous edge-wave hypothesis.

This model proposed by Bowen and Inman (1969) requires the presence of synchronous edge waves, edge waves of the same period as the incident waves, a correspondence that is believed to be restricted to steep, reflective beaches (Guza and Inman, 1975; Bowen and Guza, 1978). In addition, although edge waves of other periods have been measured in the field (Chapter 6), the detection of synchronous edge waves is extremely difficult and they have been observed with certainty only in laboratory wave basins. As a result, we still lack positive field confirmation of this model.

Dalrymple (1975) proposed another model to account for regularly spaced rip currents on long, straight beaches, a mechanism that involves two wave trains having the same period, approaching the beach at different angles. The intersecting waves create regular longshore variations in the mean water level, producing a longshore variation in incident wave heights and set-up. Dalrymple's model predicts a rip-current spacing

$$\lambda = \frac{L_\infty}{\sin \alpha_\infty - \sin \theta_\infty} \tag{8.8}$$

where L_∞ is the deep-water wavelength of the two sets of incident waves, and α_∞ and θ_∞ are their deep-water approach angles off the normal to the beach. It is not only necessary that the incident waves in this model have the same period, they must also have a fixed phase relationship—their phases cannot randomly drift relative to one another because the longshore variations in mean water level would then average out over time. Although this mechanism may on occasion generate regularly spaced rip currents, its circumstances are unusual and will not normally account for observed cell circulation systems.

An instability model for the generation of cell circulation has been proposed by Hino (1974). The analysis considered both bottom boundary and hydrodynamic instability. Subsequent work has concentrated on pure hydrodynamic instability problems where the bottom is considered to be non-deformable (LeBlond and Tang, 1974; Iwata, 1976; Mizuguchi, 1976; Miller and Barcilon, 1978). The fundamental mechanism underlying these models is an instability in the basic nearshore state of uniform (longshore independent) set-up. Hydrodynamic instability is examined with perturbation expansions of the mean variables found in the momentum and energy conservation equations (i.e., incident-wave energy density, set-up, energy dissipation, and currents). The zero-order equations yield solutions for the longshore-invariant set-up. The cell-circulation currents appear in the solutions of the first-order equations where coupling between these currents and the incident waves is also included

(LeBlond and Tang, 1974). Differences in the various instability models occur in the form of their energy density and dissipation equations, as well as in the criteria used to identify favorable rip-current spacings. For instance, LeBlond and Tang suggest that rip-current spacings occur at values that minimize the rate of energy dissipation, normalized to the total kinetic energy in the cell-circulation system. Comparisons with field data indicate that their predicted spacings are too large. Predictions of spacings in the model of Miller and Barcilon (1978) met with more success, although these investigators note that the available data are insufficient to verify the predicted trends in their equations. Their model requires a balance between the dissipation of kinetic energy through friction and the release in infinitesimal perturbations of potential energy stored in the set-up.

The instability models for the formation of the cell circulation yield predictions of expected ratios between rip-current spacings and surf-zone widths. For example, the analysis by Hino (1974) indicated that the ratio should be on the order of 4. Direct measurements in the field yield a wide range of results from 1.5 to 8 (Short, 1985; Huntley and Short, 1992). Those field studies find statistically significant correlations (but with lots of scatter) between the rip-current spacings and surf-zone widths, which Huntley and Short suggest may be because the surf-zone width governs which edge wave mode n dominates, which in turn has a large effect on the value of L_e from Equation (8.6). This would also account for the observation that wider rip-current spacings occur on a beach under higher wave conditions (McKenzie, 1958), since the heights of the waves would largely determine the width of the surf zone.

The cell circulation may redistribute beach sediments and therefore have a marked effect on the beach topography (Chapter 11). This becomes a complicating factor with regard to the generation of the cell circulation in that the troughs scoured by rip currents may act to stabilize their positions. The cell circulation is then strongly affected by the beach topography and is not completely free to respond to changing conditions of incident waves and edge waves. A number of studies have found shoreward-moving currents over bars or shoals, a longshore current confined to a trough extending along the length of the beach, and narrow seaward-flowing rip currents passing through troughs that cut across the bar [see, for example, McKenzie (1958)]. The bottom configuration and accompanying cell circulation change markedly only during high wave conditions of a storm.

Sonu (1972) conducted an especially thorough study of the nearshore circulation in an area of irregular bottom topography at Seagrove, Florida. The bathymetry at Seagrove consists of shoals and seaward-trending troughs, alternating along the length of the beach. Sonu's measurements revealed that the shoreward currents in the circulation cell typically occur over the shoals, while the rip currents are positioned within the troughs (Fig. 8-10). The outflowing rips attained velocities as great as 2 m/sec. The interesting discovery was that the breaker heights were uniform along the beach, so one would not at first expect a cell circulation. However, over the shoals the waves broke by spilling and tended to maintain the breaking process through the surf zone, while those entering the rip troughs broke by plunging in a narrow strip near the bar and traveled the remainder of the surf zone with relatively unbroken crests. Measurements revealed that shoreward of the shoals and the spilling breakers, there was a higher set-up than shoreward of the plunging breakers at the rip-current positions. Taking a similar approach to Bowen (1969a) but assuming a constant wave breaker height alongshore and an undulatory bottom within the surf zone, Sonu demonstrated theoretically that cell circulation can be produced. Noda (1974) later developed detailed numerical models for the control of the cell circulation by bottom topography as proposed by Sonu.

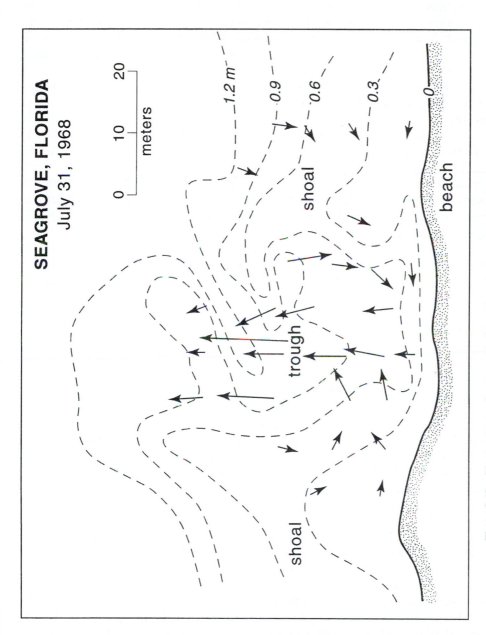

Figure 8-10 Observations at Seagrove, Florida, of longshore currents that flow from bar positions to the trough where the water turns seaward as a rip current. Measurements showed that the circulation was maintained by differences in set-up over the bars versus over the trough, in spite of the uniformity of breaking wave heights along the beach. $H_b = 39.5$ cm, $T = 5$ sec, and $\alpha_b = 0$. [Adapted with permission of American Geophysical Union, from C. J. Sonu, Field Observation of Nearshore Circulation and Meandering Currents, *Journal of Geophysical Research* 77, p. 3234. Copyright © 1972 American Geophysical Union.]

The observations of Sonu (1972) at Seagrove, Florida, demonstrate that rip currents can be maintained even when the breaking waves are uniform in height alongshore, so long as there are undulations in the nearshore bottom consisting of alternating bars and seaward-trending troughs. Based on such observations, some investigators have argued that the irregular topography always exerts a dominant control over the water circulation, that is, the topography came first and generated the circulation. However, as pointed out by Noda (1974), there remains the basic question of how the original bottom deformation was formed, if not by a cell circulation. In addition, rip currents can develop on smooth beaches without any influence of bottom irregularities. Whatever the cause of the cell circulation—whether wave refraction effects on the distribution of breaker heights, synchronous edge waves, or perhaps some instability mechanism—the rip currents most probably come first, causing sediment transport and producing the irregular bottom topography. At some later stage the irregularities of the bottom may become sufficient to provide the principal control over the nearshore circulation.

It is apparent from this discussion of the generation of cell circulation that any one or a combination of mechanisms could be responsible for its formation. It has been difficult to adequately test the relative contributions of the different theoretical models. Circulation cells are three dimensional and therefore require an extensive deployment of current meters and wave sensors, followed by complex analyses. Cell circulation patterns are known to be dynamic and constantly changing in their spatial configurations, and at times the entire system is observed to slowly migrate alongshore.

LONGSHORE CURRENTS DUE TO OBLIQUELY BREAKING WAVES

There has been considerable interest in longshore currents generated by waves breaking at angles to the shoreline, due to their importance in moving sediment along the beach (Chapter 9). Furthermore, this current system is less complex than the cell circulation system, and has been more amenable to mathematical analyses and to field measurement programs. As a result, we have a fairly good understanding of its generation and a reasonable ability to predict velocities. However, much remains to be learned about the details of the generation of these longshore currents.

Observations on natural beaches as well as in laboratory wave basins have confirmed that this longshore current is largely confined to the nearshore, rapidly decreasing in velocity beyond the surf zone [Fig. 8-1(c)]. Such observations confirm that the current is driven by forces associated with waves breaking on a sloping beach and that ocean currents, wind-driven currents, and tides do not in general play significant roles (as will be discussed later, in some cases these factors can modify the current that is otherwise generated by waves).

Over the years a number of theories have been proposed to account for longshore currents generated by obliquely breaking waves. These early theories have been reviewed by Galvin (1967) and Komar (1976a, p. 183–188) and for the most part can be abandoned. The modern analysis of longshore-current generation originated with the papers by Bowen (1969b), Longuet-Higgins (1970a, 1970b), and Thornton (1970). Each of these studies employed radiation stress to describe the flux of momentum associated with the waves (Chapter 5), associating the generation of the longshore current with the longshore-directed component of the radiation stress of the obliquely breaking waves. The three investigations differ primarily in their formulations of the frictional drag on the resulting current and in their modeling of the horizontal mixing across the surf-zone width. The analyses of Longuet-Higgins (1970a, 1970b) provided the simplest solutions and have been the main point of departure for most subsequent studies. Therefore, we will concentrate mainly on the analyses he developed, first for the mean longshore current and then for the distribution of that current across the width of the surf zone.

When waves break parallel to the shore, as was seen in Chapter 5, there is an onshore-directed radiation stress (S_{xx}), a momentum flux associated with the waves, and a longshore-directed radiation stress (S_{yy}) resulting from the effects of the wave motions on hydrodynamic pressures. When the waves approach the coast at an angle with respect to the shoreline, each of these two portions of the radiation stress have longshore components, which combine to yield

$$S_{xy} = En \sin \alpha \cos \alpha \qquad (8.9a)$$

$$= (ECn \cos \alpha)\left(\frac{\sin \alpha}{C}\right) \qquad (8.9b)$$

where E is the wave-energy density, n is the ratio of the wave group and phase velocities, and α is the angle the wave crests make with the shoreline. As such, S_{xy} is the longshore-directed (y-component) of radiation stress that is moving toward the shoreline (in the x-direction). As seen in Equation (8.9b), S_{xy} can be written as the product of the wave energy flux per unit shoreline length, $(ECn)\cos \alpha$, times $\sin \alpha/C$, which is constant according to Snell's law if the depth contours are parallel to the shoreline (Chapter 5). Therefore, if bottom friction is negligible, the energy flux is constant and it follows that $S_{xy} = $ constant. Thus, with a wave train arriving from deep water, S_{xy} reaches the nearshore relatively unaltered and is expended when the waves break on the beach. This dissipation of S_{xy} within the nearshore is the direct cause for the generation of longshore currents and is so analyzed in the studies of Bowen (1969b), Longuet-Higgins (1970a, 1970b), Thornton (1970), and in essentially all subsequent studies dealing with these currents. Note that when $\alpha = 0, S_{xy} = 0$ according to Equation (8.9) and there will be no driving force for a longshore current.

The actual "thrust" that generates the longshore current is the onshore gradient $\partial S_{xy}/\partial x$, that is, the local dissipation of S_{xy} as the waves progress across the beach. Employing linear-wave theory to evaluate the wave-energy density ($E = \rho g H^2/8$), the derivative of Equation (8.9) becomes

$$\frac{\partial S_{xy}}{\partial x} = \frac{5}{4} \varsigma\rho g h S \sin \alpha \cos \alpha \qquad (8.10a)$$

where h is the local water depth, and $S = dh/dx$ is the beach slope (in taking the derivative, it has been assumed that $\cos\alpha \approx$ constant). The relationship can also be expressed in terms of the maximum horizontal orbital velocity of the waves, u_m, which by linear-wave theory in shallow water is $u_m = \sqrt{gh}$, so that

$$\frac{\partial S_{xy}}{\partial x} = \frac{5}{4} \varsigma\rho u_m^2 S \sin \alpha \cos \alpha \qquad (8.10b)$$

In both relationships $\varsigma = 1/[1 + (3\gamma^2/8)]$, where γ is the ratio of the wave height to the local water depth, and ς accounts for wave set-up modifications of the water depth compared with the still-water depth. In the analysis, γ and hence ς are assumed to remain approximately constant as the waves break and travel as bores across the beach. The water depth h and the orbital velocity u_m will both be reduced as the waves cross the beach and approach the shore. Because of refraction of the bores, $\sin\alpha$ also progressively decreases, and therefore the local thrust $\partial S_{xy}/\partial x$ of Equation (8.10) decreases across the beach, being a maximum at the breaker zone and 0 at the shoreline.

The simplest longshore-current model balances this thrust against the frictional drag on the resulting flow. Longuet-Higgins (1970a) formulated this drag as

$$\langle R_y \rangle = \frac{2}{\pi} C_f \rho u_m v_\ell \tag{8.11}$$

where v_ℓ is the velocity of the longshore current, and C_f is a dimensionless frictional-drag coefficient. Equating the two opposing forces of Equations (8.10b) and (8.11) for a steady current, and solving for v_ℓ yields

$$v_\ell = \frac{5\pi}{8} \frac{sS}{C_f} u_m \sin\alpha_b \cos\alpha_b \tag{8.12a}$$

or

$$v_\ell = \frac{5\pi}{8} \frac{sS}{C_f} \sqrt{gh_b} \sin\alpha_b \cos\alpha_b \tag{8.12b}$$

for the magnitude of the longshore current. Evaluations of the wave parameters have been placed at the breaker zone for convenience, with h_b being the depth at wave breaking and α_b the wave-breaker angle; u_m is similarly evaluated at the breaker zone.

The simple relationship of Equation (8.12a), derived theoretically by Longuet-Higgins (1970a), implies that the longshore current is proportional to $u_m \sin\alpha_b \cos\alpha_b$, with the beach slope S and drag coefficient C_f potentially being other factors governing the magnitude of the flow. Komar and Inman (1970) independently obtained a similar relationship,

$$v_\ell = 2.7 u_m \sin\alpha_b \cos\alpha_b \tag{8.13}$$

based on a comparison of equations utilized to evaluate rates of longshore sand transport (Chapter 9) but also empirically tested using longshore-current data. Here v_ℓ is specifically the longshore current measured at the mid-surf position, halfway between the breaker zone and the shoreline, while u_m and α_b are evaluated at the breaker position. Longshore currents most commonly have been measured at approximately the mid-surf position, and this magnitude is also roughly the maximum value of the longshore current across the surf-zone width. As was subsequently shown by Komar (1979), since linear-wave theory gives $u_m = \sqrt{gh_b} = \sqrt{gH_b}/\gamma$, it is preferable to express the dependence as

$$\bar{v}_\ell = 1.17\sqrt{gH_{br}} \sin\alpha_b \cos\alpha_b \tag{8.14a}$$

in terms of the root-mean-square wave breaker height, H_{br}, or the heights of uniform waves in laboratory experiments. A comparison with field and laboratory data is given in Figure 8-11 [a more complete comparison between individual data sets is provided by Komar (1979)]. When dealing with random waves, the relationship to the significant wave-breaker height H_{bs} can be taken as approximately

$$\bar{v}_\ell = 1.0\sqrt{gH_{bs}} \sin\alpha_b \cos\alpha_b \tag{8.14b}$$

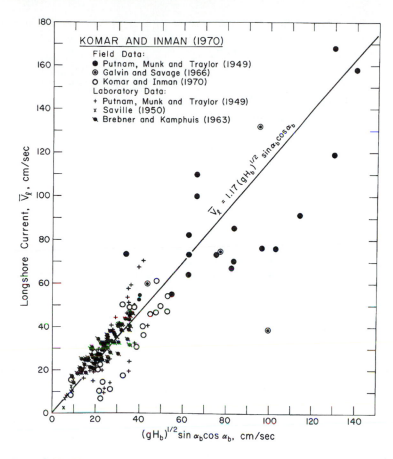

Figure 8-11 Field and laboratory data for the longshore current generated by obliquely breaking waves, used to test Equation (8.14a). [Used with permission of Gulf Publishing Company, from P. D. Komar and J. Oltman-Shay, Nearshore Currents, *Handbook on Coastal and Ocean Engineering* v2, p. 655. Copyright © 1990 Gulf Publishing Company.]

Of the many relationships that have been proposed for the prediction of longshore currents on beaches due to an oblique-wave approach, the equivalent Equations (8.13) and (8.14) provide the best predictions of the current as measured at the mid-surf position. The agreement with the measured velocities of longshore currents is good up to breaker angles as large as 45° (Komar, 1975).

The *Shore Protection Manual* (CERC, 1984) has recommended use of a similar formula

$$\bar{v}_\ell = 41.4S\sqrt{gH_b}\,\sin\alpha_b\cos\alpha_b \tag{8.15}$$

differing from Equation (8.14) in its dependence on the beach slope S. This relationship is based on the Longuet-Higgins solution, [Equation (8.12)] but with an assumed constant value $C_f = 0.01$ for the drag coefficient, while retaining the beach-slope dependence. The only support offered in the *Shore Protection Manual* for this relationship was the field data of Putnam, Munk, and Traylor (1949) and the laboratory measurements of Galvin and Eagleson (1965). It can be seen in Figure 8-12 that those two data sets actually offer minimal support

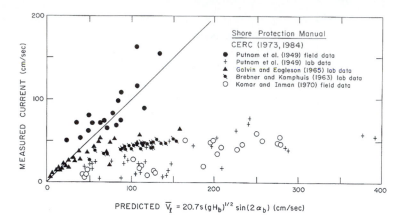

Figure 8-12   Field and laboratory data for longshore currents used to test Equation (8.15) from the *Shore Protection Manual* (CERC, 1984). The poor agreement results from the formula's direct dependence on the beach slope *s*. [Used with permission of Gulf Publishing Company, from P. D. Komar and J. Oltman-Shay, Nearshore Currents, *Handbook on Coastal and Ocean Engineering* v2, p. 656. Copyright © 1990 Gulf Publishing Company.]

for Equation (8.15), the measurements of Galvin and Eagleson in particular having lower velocities than predicted. Komar (1979) demonstrated the inadequacy of Equation (8.15) when tested against additional data sets, apparent in Figure 8-12 where the laboratory data of Putnam and co-workers (1949) and Brebner and Kamphuis (1963) and the field data of Komar and Inman (1970) have been added. The departure of the measurements from predictions by Equation (8.15) is directly attributable to its inclusion of the beach slope. This is demonstrated by data sets such as the laboratory measurements of Putnam and co-workers, where a large range of beach slopes were employed; the higher the beach slope, the greater the error in using Equation (8.15) to predict the longshore current (Komar, 1979). This also accounts for the large divergence of the field data of Komar and Inman (Fig. 8-12), in that those measurements were obtained on a relatively steep, coarse-sand beach. The conclusion is unmistakable—the longshore current is not directly proportional to the beach slope as given by Equation (8.15), and the use of that formula can lead to errors in excess of an order of magnitude.

The observed lack of dependence of the longshore current on the beach slope implies that $S/C_f \approx$ constant in Equation (8.12), yielding Equation (8.13) and its equivalent Equation (8.14). This might seem reasonable from the expected behavior of the drag coefficient and beach slope with a change in grain size of beach sediment; as the grain size increases, both S and C_f will increase such that their ratio might remain approximately constant. However, as pointed out by Huntley (1976), this does not explain the agreement with the laboratory measurements where the beaches are usually solid with a fixed beach slope. Huntley instead explains the apparent constancy of S/C_f in terms of beach-slope effects on the turbulence generated by wave breaking. An increase in beach slope will produce a narrower surf zone over which turbulent breaking occurs and hence a greater level of turbulence per unit surface area. This increases the horizontal mixing, which results in a decreased longshore current and hence a greater apparent value for C_f.

The above-mentioned longshore current predictions are for the magnitude of the flow at the mid-surf position. Most applications require calculations of the complete velocity distribution, the profile of longshore-current variations across the surf-zone width. Bowen

(1969b), Longuet-Higgins (1970b), and Thornton (1970) analyzed such variations by considering the local values of the $\partial S_{xy}/\partial x$ thrust and frictional drag, the variations of these forces with x, the cross-shore coordinate. Their analyses also include processes of horizontal mixing, which act to smooth the resulting cross-shore profile of the longshore current. This latter process is often envisioned as the cross-shore transfer of momentum by horizontal eddies, which develop primarily by wave breaking, but as will be seen later, can also be accomplished by waves of various heights breaking in different water depths.

The various derivations of the velocity profile differ in the way in which the horizontal eddy coefficient is evaluated. Bowen (1969b) assumed that this eddy coefficient is constant, while Longuet-Higgins (1970b) utilized a form equivalent to $\rho x u_m$, arguing that its length scale will depend on the offshore distance x, which provides a limit to the eddy size, while the velocity scale is represented by u_m. This form has been used in many subsequent studies [for example, by Madsen, Ostendorf, and Reyman (1978) and Kraus and Sasaki (1979)]. In a detailed analysis of surf-zone turbulence, Battjes (1975) argued that the characteristic size of eddies is limited by the local water depth, not by the distance from the shore. He further evaluated the velocity scale from the local rate of wave-energy dissipation per unit bottom area. However, for a uniformly sloping beach this leads to a relationship that is essentially the same as that derived by Longuet-Higgins (1970b), differing mainly in the inclusion of the beach slope in the horizontal mixing term; the difference is more important on non-uniform beaches. On the basis of a comparison with laboratory data, Madsen, Ostendorf, and Reyman (1978) concluded that there should be a beach-slope dependence as derived by Battjes. This is also indicated by the field measurements of Huntley (1976), but his simultaneous evaluations of the drag coefficient indicate a more complex model in which lateral mixing results from the turbulence of wave breaking and is not distributed evenly through the water column. Thornton (1970) and Jonsson, Skovgaard, and Jacobsen (1974) evaluated the eddy coefficient on the basis of a different rationale, placing it in terms of wave-orbital motions. Battjes (1975) criticized this approach, pointing out that the main source of turbulent energy within the surf zone is the intense dissipation of wave energy by breaking and that the orbital motions are directly accounted for by the radiation stresses.

The solution obtained by Longuet-Higgins (1970b) for the profile of the longshore current is

$$V = \begin{cases} B_1 X^{p_1} + AX & \text{for } 0 < X < 1 \\ B_2 X^{p_2} & \text{for } 1 < X < \infty \end{cases} \tag{8.16a}$$

where $X = x/X_b$, with X_b being the distance from the shoreline to the breaker line, and $V = v/v_0$, where v is the magnitude of the longshore current at the offshore distance x, normalized to

$$v_0 = \frac{5\pi}{16} \gamma s^2 \frac{S}{C_f} (gH_b)^{1/2} \sin \alpha_b \cos \alpha_b \tag{8.16b}$$

which contains the dependence on the wave conditions and is similar in form to Equation (8.12) already tested. The various coefficients in the solution for the complete profile are

$$p_1 = -\frac{3}{4} + \left(\frac{9}{16} + \frac{1}{sP}\right)^{1/2} \tag{8.16c}$$

$$p_2 = -\frac{3}{4} - \left(\frac{9}{16} + \frac{1}{sP}\right)^{1/2} \tag{8.16d}$$

$$P = \frac{\pi N S}{\gamma C_f} \qquad (8.16e)$$

$$A = \frac{1}{[1 - (5sP/2)]} \qquad \left(P \neq \frac{2}{5s}\right) \qquad (8.16f)$$

$$B_1 = \frac{p_2 - 1}{p_1 - p_2} A \qquad (8.16g)$$

$$B_2 = \frac{p_2 - 1}{p_1 - p_2} A \qquad (8.16h)$$

This solution yields a family of longshore current profiles as graphed in Figure 8-13, one profile for each value of P. According to Equation (8.16e), P is a non-dimensional parameter that represents the relative importance of the horizontal mixing, accounted for by the coefficient N, compared with the frictional drag represented by C_f. If there is no horizontal eddy mixing, then $N = 0$ and $P = 0$, and the solution yields a saw-tooth velocity distribution with a discontinuity at the breaker line (Fig. 8-13). This distribution demonstrates that the current is driven by $\partial S_{xy}/\partial x$ only within the surf zone, there being no driving force seaward of the breaker line. With increasing horizontal mixing, producing an increase in P, the profile becomes smooth and more realistic; the maximum velocity of the profile shifts toward the shoreline and decreases in magnitude. Inclusion of horizontal mixing also couples the water seaward of the breaker zone to the flow within the surf zone, so now there is a longshore current outside the surf zone with no discontinuity at the breaker line.

The analyses of Longuet-Higgins (1970a, 1970b) included assumptions that the angle of wave approach is small and $v/u_m < 1$, that is, the magnitude of the generated longshore current is small in comparison with the wave-orbital velocities. The validity of the latter assumption can be evaluated approximately from Equation (8.13), which indicates that one must have

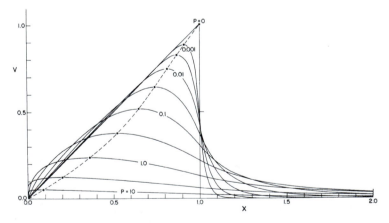

Figure 8-13 A family of longshore current profiles across the surf zone obtained with the solution of Equation (8.16), given in dimensionless form where $X = x/x_b$, and $V = v/v_0$, the local value of the current v divided by v_0 of Equation (8.16c). The larger the value of P [Equation (8.16e)], the greater the effect of horizontal mixing. [Adapted with permission of American Geophysical Union, from M. S. Longuet-Higgins, Longshore Currents Generated by Obliquely Incident Waves, *Journal of Geophysical Research 75*, p. 6793. Copyright © 1970 American Geophysical Union.]

$\alpha_b < 24°$ in order to ensure that $v/u_m < 1$. Breaker angles are usually smaller than this, and since Equation (8.13) was found to actually agree with data up to breaker angles approaching 45°, this 24° cut-off based on $v/u_m < 1$ would appear to be too conservative. However, the desire to eliminate such assumptions made by Longuet-Higgins has been the impetus for subsequent analyses. Liu and Dalrymple (1978) analyzed the drag in terms of the combined motions of the waves plus the longshore current but neglected lateral mixing, whereas Madsen, Ostendorf, and Reyman (1978) and Kraus and Sasaki (1979) included mixing as well. Another assumption made by Longuet-Higgins is that there is a constant γ, the ratio of the wave height to the local water depth. This is equivalent to assuming that, as the broken waves proceed as bores across the surf zone, their decay is linearly related to the water depth (Chapter 6). Smith and Kraus (1987) examined this assumption and demonstrated that the decay is not linear but takes the form of a power law $H = \gamma_b h_b (h/h_b)^f$, where the exponent f is empirically related to the beach slope, and γ_b is evaluated at the breaker line. Analytical solutions for the longshore-current profile showed that with mild beach slopes, values of f will be large and the current distribution is more peaked, with the maximum current position closer to the breaker line. As f approaches 1, the solutions become more like those obtained by Longuet-Higgins (1970b).

Most of the field data for longshore currents consist of measurements at the mid-surf position, the data used to establish Equation (8.14) in Figure 8-11. This corresponds to the velocity at $X = 0.5$ in the profile solution of Equation (8.16) and can be used to constrain that solution. This was done by Komar (1976b) by forcing the profile distribution to agree with the value predicted by Equations (8.13) and (8.14), thereby indirectly making the distribution agree with the available data at $X = 0.5$. This yields a relationship for S/C_f as a function of P or N (Fig. 8-14). Once a reasonable selection for P has been made to define the overall shape of the profile, the graph of Figure 8-14 provides a value for S/C_f and thus for C_f with a known beach slope. Examples of computed longshore current profiles using this approach are shown in Figure 8-15 for two values of P but with the same wave and beach conditions. Both distributions yield a longshore current magnitude of 64 cm/sec at the mid-surf position ($X = 0.5$), the velocity given by Equation (8.13) for the specified wave conditions.

Several investigators have compared theoretical velocity profiles with measurements obtained in laboratory wave basins. Based on comparisons with the measurements of Galvin

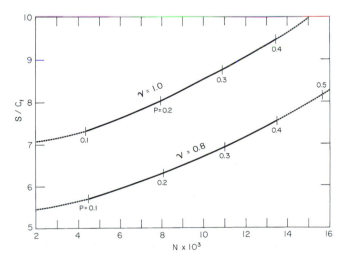

Figure 8-14 Relationship between S/C_f and N, with values of P also shown, obtained by equating the longshore current velocity of Equations (8.13) or (8.14) to the mid-surf ($X = 0.5$) solution of Equation (8.16). This calibrates the solution for the complete velocity distribution against the field and laboratory data for the magnitude of the current at the mid-surf position. [From Longshore Currents and Sand Transport on Beaches, P. D. Komar, *Ocean Engineering III*, 1976. Reproduced with permission from the American Society of Civil Engineers.]

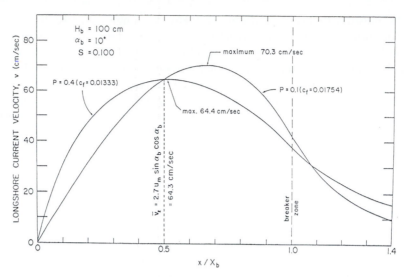

Figure 8-15 Example calculations for two values of P for cross-shore distributions of the longshore current, where the curves of Figure 8-14 have been used to evaluate the drag coefficient C_f such that the distributions agree with Equations (8.13) and (8.14a) at the mid-surf position. [From Longshore Currents and Sand Transport on Beaches, P. D. Komar, *Ocean Engineering III*, 1976. Reproduced with permission from the American Society of Civil Engineers.]

and Eagleson (1965), Longuet-Higgins (1970b) concluded that $P \approx 0.1$–0.4. Kraus and Sasaki (1979) compared the theoretical solutions with the laboratory data of Mizuguchi, Oshima, and Horikawa (1978). The results are shown in Figure 8-16, where the dashed theoretical distributions are calculated from Equation (8.16) obtained by Longuet-Higgins (1970b), while the solid curves are based on the revised solutions of Kraus and Sasaki. The agreement is good inside the surf zone, while there is some divergence offshore; as noted by Kraus and Sasaki, the disagreement in the offshore is likely due to inaccuracies in measuring the low-magnitude longshore currents in the presence of the stronger wave motions. Smith and Kraus (1987) similarly compared their theoretically derived profiles with the laboratory data of Mizuguchi and co-workers (1978).

The study of Kraus and Sasaki (1979) was the first to compare field measurements of the complete velocity profile with theoretical solutions, the data having been obtained at Urahama Beach in Japan. It can be seen in Figure 8-17 that agreement is good in the outer surf and breaker zones and in the offshore. This portion of the velocity profile is controlled by the primary breaker zone. The poor agreement within the inner surf zone results from the breakdown of assumptions in the analysis that the beach is planar and the wave-height decrease is linear. Instead, the step-like profile produced a zone of secondary breakers at approximately 20 m from the shoreline and a secondary zone of longshore-current generation.

Additional field measurements of longshore-current distributions were obtained as part of the Nearshore Sediment Transport Study (NSTS) at Santa Barbara, California. The beach there is relatively uniform in slope with straight and parallel bottom contours, conditions that correspond to the assumptions made in most of the theoretical analyses. Wu, Thornton, and Guza (1985) compared velocity-profile measurements collected during 5 days of low-wave activity, at which time the waves had narrow-banded frequency and directional spectra, permitting model assumptions of nearly unidirectional and monochromatic waves.

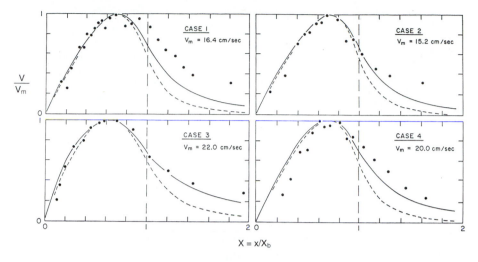

Figure 8-16 Comparisons of the solution by Longuet-Higgins (1970b) [Equation (8.16)] for the longshore current profile with the laboratory measurements of Mizuguchi, Oshima, and Horikawa (1978). [Adapted with permission of Marcel Dekker, Inc., from N. C. Kraus and T. O. Sasaki, Effect of Wave Angle and Lateral Mixing on the Longshore Current, *Marine Science Communications* 5. Copyright © 1979 Marcel Dekker, Inc.]

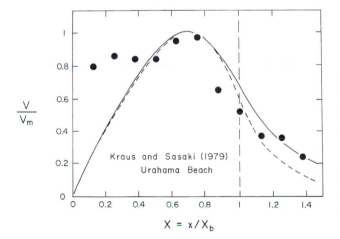

Figure 8-17 A comparison of the field measurements on Urahama Beach, Japan, of the longshore current profile with the solution of Longuet-Higgins (1970b) [Equation (8.16)]. The departure of the data from the theoretical distribution within the inner surf zone results from the non-uniform slope of the beach profile. [Adapted with permission of Marcel Dekker, Inc., from N. C. Kraus and T. O. Sasaki, Effect of Wave Angle and Lateral Mixing on the Longshore Current, *Marine Science Communications* 5. Copyright © 1979 Marcel Dekker, Inc.]

Their models include bottom friction, lateral mixing, and nonlinear convective accelerations, but this completeness requires numerical solutions. The solution predicts the root-mean-square (rms) wave height variation in the cross-shore direction as well as the longshore-current profile. An example of a calculated current profile is shown in Figure 8-18, compared with one set of measurements. The theoretical profiles are based on a match with the measurements that yields a minimum overall error, obtained by varying C_f and N until a best fit is achieved; in this particular example, $C_f = 0.012$ and $N = 0.006$ for the linear solution, values that agree reasonably well with those suggested by other investigators. Comparisons also were made between linear and non-linear solutions, and it can be seen in Figure 8-18 that the linear solution provided a better overall fit to the data.

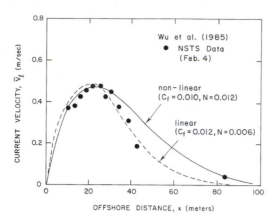

Figure 8-18 Measurements of the longshore current profile at Santa Barbara, California, obtained during the Nearshore Sediment Transport Study, compared with theoretical solutions. The linear solution is that of Longuet-Higgins (1970b) [Equation (8.16)].[Adapted with permission of American Geophysical Union, from C-S. Wu, E. B. Thornton and R. T. Guza, Waves and Longshore Currents: Comparison of a Numerical Model With Field Data, *Journal of Geophysical Research 90,* p. 4956. Copyright © 1985 American Geophysical Union.]

The above analyses of longshore-current distributions all assumed regular waves of uniform height, breaking at a fixed position on the profile. However, with a spectrum of waves, the larger the individual wave height the deeper the water depth in which it initially breaks. There is then a zone of breaking waves and bores, with the number of breakers increasing shoreward until almost all waves are breaking within the inner surf zone (Chapter 6). This pattern in turn affects the distribution of the longshore current. Battjes (1972) examined this effect, utilizing a wave-by-wave description of the irregular waves. The lateral momentum exchange due to turbulence was neglected, but in spite of this, smooth velocity profiles were obtained, similar in distribution to those in Figure 8-13 from Longuet-Higgins (1970b). This result demonstrates that irregular waves breaking in a range of water depths will have much the same effect as lateral mixing, and in the field it may be difficult to separate the two influences. This is demonstrated by the study of Thornton and Guza (1986), who analyzed the same NSTS measurements as Wu, Thornton, and Guza (1985). A narrow-band, random-wave transformation model was applied to describe the cross-shore spatial variations of the root-mean-square wave height, H_{rms}, and these variations served as the basis for the derivation of the longshore-current profile. An example is shown in Figure 8-19, a solution that assumes no horizontal mixing (i.e., $N = 0$)—the current distribution results solely from waves of different heights breaking in a range of water depths. For completeness, Thornton and Guza did include horizontal mixing in a few of their analyses, but optimal N values were small (average 0.002) and their inclusion did not significantly improve the fit to the data.

Thornton and Guza (1986) included a randomness associated with the spectra of wave heights but did not consider variations in wave directions. Their analyses utilized data only from days when there was a single wave train having a small angular distribution of wave directions. Battjes (1972) demonstrated that a large angular distribution can strongly affect the radiation stress that provides the driving force for the longshore current. He estimated that if the waves are treated as long-crested, then the total longshore thrust could be overestimated by as much as 100 percent.

Guza, Thornton, and Christensen (1986) examined situations where waves of two periods arrive at a beach from separate directional quadrants. A common example is swell from a distant storm arriving from one direction, while locally generated wind waves have a different direction. Their analyses again were based on the NSTS data from Santa Barbara where there tends to be two distinct quadrants of wave arrivals due to windows between offshore islands and the mainland coast. Their study therefore included complex sets of data that

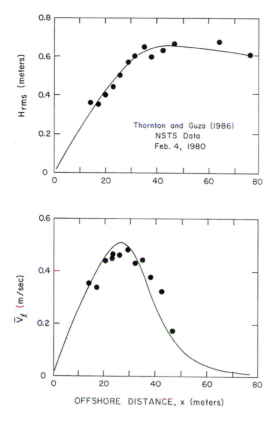

Figure 8-19 A model for the cross-shore distribution of root-mean-square wave heights and the distribution of the longshore current, based on the principle that different wave heights break in different water depths on the sloping beach, and this factor yields the smoothed distribution for the current. [Used with permission of American Meteorological Society, from E. B. Thornton and R. T. Guza, Surf Zone Longshore Currents and Random Waves, *Journal of Physical Oceanography* 16, p. 1175. Copyright © 1986 American Meteorological Society.]

were not utilized in the analyses of Wu and co-workers (1985) and Thornton and Guza (1986). Guza and co-workers considered the total radiation stress S_{xy}^T, which is the sum of individual S_{xy} contributions from distinct wave trains. One could envision cases where two wave trains arrive from opposite quadrants such that their individual S_{xy} cancel to yield $S_{xy}^T = 0$ whereas if they arrived from the same quadrant, they would reinforce to yield a larger S_{xy}^T and presumably a stronger longshore current. Both circumstances were found in the data analyzed by Guza and co-workers, and a direct correlation was found between the magnitude and direction of the longshore current and the net magnitude and direction of S_{xy}^T. Such a direct correlation was further justified by scaling arguments based on the balance between the $\partial S_{xy}^T / \partial x$ thrust and the mean frictional drag on the current, following the analysis of Longuet-Higgins (1970a).

Of particular interest in the study of Guza, Thornton, and Christensen (1986) is that the cross-shore profiles of the longshore currents are apparently insensitive to the structure of the individual S_{xy} making up S_{xy}^T. Again, one could envision two wave trains arriving from opposite quadrants but with one having lower wave heights than the second, and thus breaking in shallower depths on the beach profile. In such a case, would it be possible to have a reversal of the current direction between the inner and outer surf zone, with a strong shear between? Such structures were not found at the NSTS experiment site by Guza and co-workers, and they tentatively interpreted this as implying that shallow-water wave breaking occurs as a broad-band process that extracts energy and momentum simultaneously from many incident-wave frequency bands and directions, or alternately, that horizontal eddy mixing suppresses strong shear in the currents. The beach at the NSTS site has a uniform slope where the different wave trains could fully interact. Therefore, the possibility remains that, on other beaches having pronounced multiple bars, the individual S_{xy} would be expended by breaking on different bars and hence could drive currents in opposite directions.

Most of the theoretical analyses of longshore current distributions, including Equation (8.16) from Longuet-Higgins (1970b), assume a uniform beach slope. McDougal and Hudspeth (1983) obtained solutions for concave beach profiles in which the still-water depth is proportional to $x^{2/3}$, the Bruun/Dean profile, which approximates many beach profiles (Chapter 7). This particular beach profile produces marked changes in the velocity distributions of longshore currents from those given in Figure 8-13 for a uniform beach slope, shifting the maximum current to the shoreline position ($X = 0$). This unrealistic result occurs because the $x^{2/3}$ profile has an infinite slope at the shore, and the local slope is important in the solution for the longshore current distribution. McDougal and Hudspeth (1989) developed improved solutions for composite beach profiles consisting of a planar beach face joined to an $x^{2/3}$ profile in the offshore. Elimination of the infinite slope at the shore by the addition of the planar slope resulted in current profiles that are similar to the Longuet-Higgins solution in Figure 8-13 for a completely planar beach, but the concavity causes the maximum in the distribution to shift toward the shore.

Complex beach topography with multiple bars and troughs requires numerical rather than analytical solutions. A recent formulation to deal with complex beach profiles is the NMLONG model for simulating longshore currents as analyzed by Kraus and Larson (1991) and Larson and Kraus (1991). The approach calculates the wave heights, wave directions, and mean water elevations (set-down and set-up) across the complete beach profile, as well as the time-averaged longshore current. NMLONG also can account for the longshore component of wind stress and its effects on the distribution and magnitude of the longshore current. One objective of the approach was to simplify the computational techniques sufficiently so that NMLONG could be run on a desk-top computer. The model was compared with the laboratory wave-basin data of Visser (1991) and the field data of Kraus and Sasaki (1979) and Thornton and Guza (1986). The comparison with one experimental run of Visser with a uniform beach slope is shown in Figure 8-20; included is a comparison between computed versus measured changes in wave heights, mean longshore currents, and mean water surface elevations. The overall patterns are in good agreement but are shifted horizontally by a small degree that prevents congruency. The calculated distributions are based on the point of incipient wave breaking; Larson and Kraus suggest that agreement would be improved if the analyses are based on the plunge point shoreward of the break point (Chapter 6) but note that this would require an improved prediction of the plunge point for incorporation into NMLONG. Similar results were found in the comparison with the field data of Thornton and Guza (1986), obtained during the NSTS experiments at Santa Barbara, a beach having a fairly uniform slope. The comparison with the data of Kraus and Sasaki (1979), obtained on a beach having a significant bar and trough, is shown in Figure 8-21. The computed longshore current distribution predicts a maximum over the offshore bar and a second maximum in the swash zone over the beach face, with a sharp minimum over the trough of the profile. There again appears to be a systematic horizontal shift between the computed and measured distributions. Furthermore, the measured longshore current within the trough is stronger than the computed current. The computed distribution is based on the longshore current being driven by local wave breaking, and this accounts for the maximum being over the bar and in the

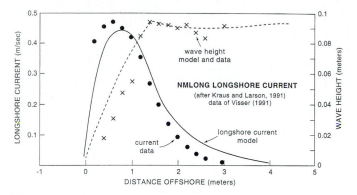

Figure 8-20 NMLONG analyses of the longshore current distribution, together with the distribution of wave heights across the nearshore, compared with the laboratory data of Visser (1991). [Adapted from Numerical Model of Longshore Current for Bar and Trough Beaches, M. Larson and N. C. Kraus, *Journal of Waterway, Port, Coastal and Ocean Engineering,* 1991. Reproduced with permission from the American Society of Civil Engineers.]

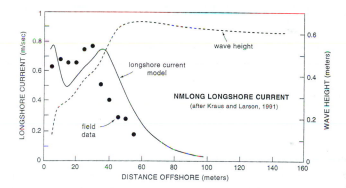

Figure 8-21 NMLONG analysis of the longshore current distribution compared with the field data of Kraus and Sasaki (1979) obtained on a beach with a bar and trough. [Reprinted from N. C. Kraus and M. Larson, *MNLONG: Numerical Model for Simulating the Longshore Current—Report 1, Model Development and Tests,* Technical Report DRP-91-1, U.S. Army Corps of Engineers.]

swash zone. Similar to the Longuet-Higgins (1970b) analysis, NMLONG developed by Larson and Kraus, includes a horizontal mixing term that produces some longshore current within the trough in spite of the minimal wave breaking in that zone. This is also shown by their analyses of longshore current distributions over an idealized bar-trough topography, obtaining results that are similar to the computed distributions of Symonds and Huntley (1980). It can be seen in Figure 8-22 from Symonds and Huntley that the longshore current concentrates over the bar with a minimum over the trough, the velocity in the trough increasing with increasing P since it is driven mainly by horizontal mixing. Symonds and Huntley also obtained field measurements under such conditions, suggesting $P \approx 0.1$–0.4, a more exact estimate being precluded by equipment failure. However, in contrast to Figure 8-22, field observations commonly show the maximum longshore current to be within the trough, rather than over the bar. Symonds and Huntley demonstrated that this is produced by longshore

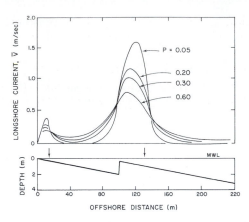

Figure 8-22 Theoretical distributions of long-shore currents generated over an idealized bar-trough system of a beach profile, such that there is a primary breaker line over the outer bar and a secondary line in the swash zone (indicated by the arrows). [Adapted with permission of National Research Council of Canada, from G. Symonds and D. A. Huntley, Waves and Currents over Nearshore Bar Systems, *Proceedings of the Canadian Coastal Conference.* Copyright © 1980 National Research Council of Canada.]

pressure gradients due to variations in wave set-up along the length of the beach, generally the situation for longshore currents associated with the cell circulation system discussed earlier. The disagreement between the NMLONG computed versus measured currents in the trough (Fig. 8-21), in part, points to our limited understanding of the processes responsible for horizontal mixing within the surf zone. However, another possibility is that the longshore current within the trough is driven in part by a longshore gradient in wave set-up, a factor that is not included in the NMLONG analyses.

CURRENTS DUE TO COMBINED OBLIQUE WAVES AND LONGSHORE VARIATIONS IN SET-UP

The most general condition of nearshore currents is where they are generated by a combination of the two end members discussed above: obliquely breaking waves and longshore variations in wave set-up. Such a condition was examined by O'Rourke and LeBlond (1972) in their study of longshore currents within a semicircular bay. Their analysis demonstrated that not only were oblique incidence and longshore variations in wave heights important driving forces for nearshore currents, but longshore variations in breaker angles can also be significant. Keeley and Bowen (1977) applied the analysis of O'Rourke and LeBlond to measurements obtained along a beach of more than 1 km in length. They found that the currents were dominated by the oblique wave approach, with the $\partial H_b/\partial y$ and $\partial \alpha_b/\partial y$ terms together contributing approximately 10 percent to the overall current strength. Superimposed on these large-scale currents (>600 m) were regular, small-scale circulation cells, probably produced by edge waves.

Keeley (1977) demonstrated a correspondence between the patterns of the longshore currents and the development of a large-scaled cuspate shoreline. This again points out the important interactions between nearshore currents and beach topography. Komar (1971) analyzed an unusual situation that developed in a wave-basin study of cuspate shorelines. A cell circulation with rip currents initially rearranged the beach sand into cusps and embayments. However, an equilibrium condition eventually was achieved in which the oblique-wave approach to the cusp flanks opposed the longshore variation in wave breaker heights and set-up that was driving the cell circulation. The two mechanisms for generating longshore currents therefore were in operation, and an equilibrium was achieved where they opposed and balanced one another such that no current developed.

These observations inspired a more general analysis by Komar (1975) of longshore currents that are driven by a combination of obliquely breaking waves and longshore variations in wave heights and set-up. The derived relationship for the velocity at the mid-surf position, modified slightly from that originally given, is

$$v_\ell = 1.17\sqrt{gH_b}\,\sin\alpha_b\cos\alpha_b - a\sqrt{gH_b}\,\frac{\partial H_b}{\partial y} \qquad (8.17)$$

where

$$a = \frac{\pi\sqrt{2}}{C_f\gamma^{5/2}}\left(1 + \frac{\gamma^2}{8}\right)$$

The first term on the right is equivalent to Equation (8.14) for the longshore current generated by the oblique wave approach (α_b), while the second term accounts for the longshore variation in wave-breaker heights ($\partial H_b/\partial y$), which determines the longshore gradient of the set-up elevations. With $\partial H_b/\partial y = -$, the longshore gradient supports the current generated by the oblique waves, while $\partial H_b/\partial y = +$ opposes and hence reduces the current due to the oblique-wave approach. In the case of the cuspate shoreline observed by Komar (1971) and described above, the two terms on the right of Equation (8.17) balanced so the resultant was $v_\ell \approx 0$, even though waves continued to break at an angle to the flank of the cusp and a $\partial H_b/\partial y$ gradient persisted.

Although there may be a balance at the mid-surf position of the forces generating longshore currents such that $v_\ell = 0$ in Equation (8.17), there still could be a non-balance at other positions across the surf-zone width. This was examined by Komar (1975) in a complete solution of the velocity distribution across the surf zone. This solution is an extension of that obtained by Longuet-Higgins (1970b), Equation (8.16), but now includes a forcing term due to the longshore variation in wave set-up, $\partial\overline{\eta}/\partial y$. Example profiles are shown in Figure 8-23. When $\partial\overline{\eta}/\partial y = 0$, the solution is the same as that of Longuet-Higgins (1970b), where the longshore current is due solely to an oblique-wave approach. With $\partial\overline{\eta}/\partial y = -0.0005$, the water slopes downward in the positive y-direction, the direction of the longshore component of the oblique waves, so the two forces combine to produce stronger longshore currents. For positive values of $\partial\overline{\eta}/\partial y$, the water slope opposes the oblique wave approach and the current is reduced. For $\partial\overline{\eta}/\partial y = 0.0025$ (Fig. 8-23), the forces are close to balancing and the resulting longshore current is small. The strongest residual current is close to the shoreline and is flowing in the opposite direction to the current beyond the mid-surf zone. The results indicate that a near balance can be achieved across the entire width of the surf zone, a condition observed on the cuspate shoreline described above (Komar, 1971). The analysis in Figure 8-23 is for a uniformly sloping beach. With a bar and trough topography where the waves break principally over the bar, the S_{xy} thrust of the obliquely breaking waves would be concentrated over the bars and in the swash zone, whereas the set-up gradient $\partial\overline{\eta}/\partial y$ would act mainly within the intervening trough. The balance between the driving forces will then be complex, and the final longshore current will depend on the degree of horizontal mixing.

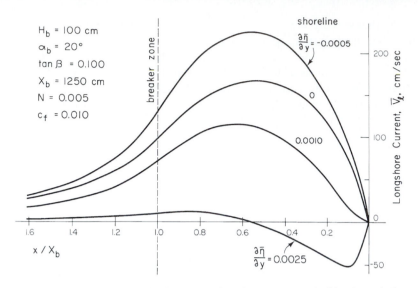

Figure 8-23 Solutions for distributions of longshore current velocities through the surf zone, affected by a longshore variation in the wave set-up, $\partial \bar{\eta} / \partial y$, as well as by an oblique-wave approach. With $\partial \bar{\eta} / \partial y = 0.0025$, the set-up slope in the longshore direction nearly opposes and balances the thrust due to the oblique-wave approach, resulting in greatly weakened currents. [Used with permission of John Wiley & Sons, Inc., from P. D. Komar, Nearshore Currents: Generation by Obliquely Incident Waves and Longshore Variations in Breaker Heights, *Nearshore Sediment Dynamics and Sedimentation*, p. 40. Copyright © 1975 John Wiley & Sons, Inc.]

WIND AND TIDE EFFECTS ON NEARSHORE CURRENTS

In addition to longshore currents being generated by an oblique-wave approach or by longshore variations in wave heights (or set-up), a wind blowing in the longshore direction and tidal water draining from the beach might also contribute to the observed flow. However, the importance of the wind and tides remains largely unevaluated.

Shepard and Inman (1950a, 1950b) pointed out the significance of the wind in generating longshore currents, although they noted that it is difficult to separate the currents generated directly by the wind stress from the currents associated with the wind-generated waves. With a wind stress acting in the longshore direction, the resulting current should be greater than the current predicted with Equation (8.14), which only accounts for the breaker heights and angles of the wind-generated waves.

Several investigators have attempted to apply multiregression analyses to the prediction of longshore currents (Harrison and Krumbein, 1964; Brebner and Kamphuis, 1964; Harrison, 1968; Allen, 1974; Nummedal and Finley, 1978), an approach that can include the effects of the wind as well as wave heights and angles. Such empirical analyses yield results that likely are applicable only to the beaches where the measurements were obtained but still should be indicative of the relative importance of the wind. The formulas derived from these several studies differ as to the relative importance of the various factors but most find the expected dependence on wave heights and breaker angles. Of interest is the study of Nummedal and Finley (1978) on the coast of South Carolina, an investigation that found that the longshore component of wind velocity accounted for most of the measured longshore current. In this instance, however, the strong dependence of the current on the wind likely resulted in part from the use of visual observations of the wave heights and angles in the analyses, which would have weakened the relationship between the current and those parameters. The importance of the wind is more clearly demonstrated by the study of Hubertz (1986) at the Field Research Facility (FRF), Duck, North Carolina. His analyses show a direct relationship between the measured longshore currents and the mean wind speed, firmly establishing that waves reaching the coast are not always the sole generation mechanism of nearshore currents.

As the tide ebbs, water drains from the beach and may enhance the rip currents and feeder longshore currents. This increase is particularly strong as low tide is approached and water is trapped within troughs running parallel to the shore, prevented from direct offshore flow by the bar that is now partly exposed. The only escape is for the water to flow alongshore within the trough, and then seaward through rip channels that cut across the bar. The effect may be enhanced by waves breaking over the bar and spilling into the trough, contributing to the total flow. Sonu (1972) found at Seagrove, Florida, that the tide level controlled the intensity of wave breaking and therefore the strength of the circulation. He noted that rip currents generally were stronger during low tide than at high tide. In addition, Sonu found that rip currents at high tide fluctuated with the period of the incoming swell, whereas the stronger rips at low tide tended to fluctuate with the beat of groups of waves having periods of 25–50 sec. Shepard and Inman (1950a) similarly found pulsations in longshore current velocities resulting from surf beat, due to alternating sets of large and small breaking waves.

The level of the tide will also directly affect water depths within the nearshore, and this in turn will affect the heights of the waves and hence the magnitude of the longshore current. This tide control on magnitudes of longshore currents has been investigated by Thornton and Kim (1993), based on 19 days of continuous measurements of water depths, wave heights, and longshore-current velocities at various positions within the nearshore at the FRF. The strongest tidal signatures in the longshore currents appear near the shore and over the bar, with a decrease in tidal signature in the trough and in the offshore. This pattern can be directly attributed to the control of wave heights over the bar by the tide altering water depths and a similar control within the inner surf zone where wave heights and energy are saturated due to the local water depth. The observed variations in wave heights and longshore currents were successfully modeled using the Thornton and Guza (1986) approach to analyzing the cross-shore distributions of wave heights and generated longshore currents.

SHEAR INSTABILITIES IN LONGSHORE CURRENTS

An unusual form of pulsating longshore currents, termed *shear waves* due to their mode of formation, first came to light during the 1986 SUPERDUCK experiments at the FRF. Figure 8-24 from Oltman-Shay, Howd, and Birkemeier (1989) shows a time series of measured alongshore and cross-shore components of the longshore current that reveals the strong pulsations of shear waves. During that data run, offshore significant wave heights progressively increased from 40 to 210 cm, and the southward mean longshore current in the surf zone increased dramatically from 10 to 160 cm/sec. Apparent in the time series is the development of strong oscillations, which had a period of about 400 sec near the middle of the run but decreased to about 200 sec toward the end. The oscillations are strongest in the longshore component of the current, being less dramatic in the cross-shore component where they are partly obscured by the wave-orbital motions. The changing magnitude of the longshore current due to the shear waves amounted to

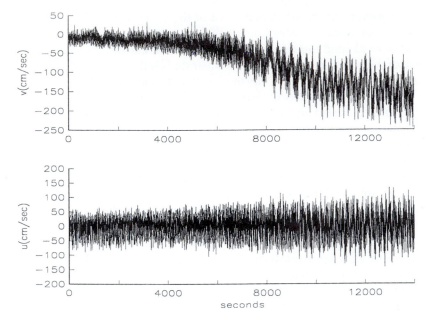

Figure 8-24   Time series of (a) longshore and (b) cross-shore water velocities measured within the surf zone at the FRF, North Carolina, on October 10, 1986. The strong oscillations about the mean longshore current are due to the presence of shear waves. [Used with permission of American Geophysical Union, from J. Oltman-Shay, P. A. Howd and W. A. Berkemeier, Shear Instabilities of the Mean Longshore Current: 2. Field Observations, *Journal of Geophysical Research 94*, p. 18,037. Copyright © 1989 American Geophysical Union.]

some 100 cm/sec within 5– to 10–minute intervals and was dramatic enough to be noted by investigators working in the surf at the time of data collection. Oltman-Shay and co-workers reported that the pulses "occasionally left us hanging onto the sensor poles as flags in high winds."

Oltman-Shay, Howd, and Birkemeier (1989) fully documented the properties of the shear-wave oscillations within the longshore current, while the accompanying paper by Bowen and Holman (1989) developed a theoretical explanation and presented mathematical models of the flows. The results of the combined studies established that the observed pulses represent a new class of nearshore waves that depends on instabilities in the horizontal shear of the longshore current, that is, on $V_x = dv/dx$, where x is in the cross-shore direction. Important is the presence of a maximum or minimum in the flow's vorticity, V_x/h, where h is the local water depth. Figure 8-25 provides an example distribution based on measurements at the FRF. It can be seen that the minimum in the vorticity depends on the combined cross-shore patterns of the shear in the longshore current and the changing water depth h of the beach profile. The results suggest that shear instabilities and the generation of shear waves will be more important on barred profiles than on beaches of uniform slope. The instability leading to shear waves develops in the velocity shear zone that is seaward of the minimum in V_x/h, giving rise to a horizontal oscillation that, when superimposed on the mean longshore current, appears as an alongshore progressive meandering current. This is seen in the model calculations of Bowen and Holman in Figure 8-26, which represents the velocity pattern for one wave length of the motion. The meandering wave form migrates in the direction of the mean longshore current, predicted by Bowen and Holman to progress at a rate of $V_0/3$, where

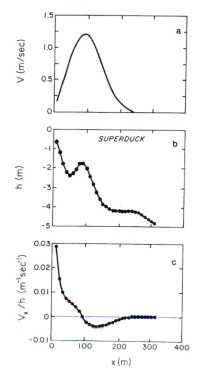

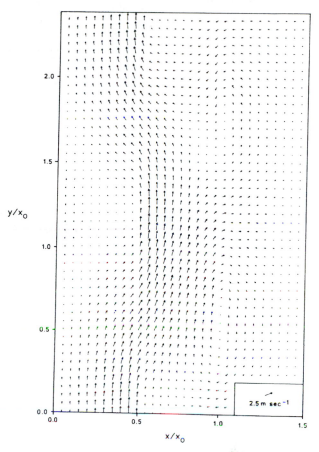

Figure 8-25 An example calculation of the cross-shore profile of the vorticity $V_x/h = (dV/dx)/h$ of the longshore current at the FRF, North Carolina, which depends on the profile of the longshore current V and depth h of the beach profile. Shear waves develop from instabilities within the zone of shear (dV/dx) seaward of the minimum in the vorticity. [Used with permission of American Geophysical Union, from A. J. Bowen and R. A. Holman, Shear Instabilities of the Mean Longshore Current: 1. Theory, *Journal of Geophysical Research 94*, p. 18,029. Copyright © 1989 American Geophysical Union.]

Figure 8-26 The total velocity pattern for one wave length of a shear wave superimposed on a mean longshore current, yielding a meandering flow. The motion is progressive in the direction of the mean current. [Used with permission of American Geophysical Union, from A. J. Bowen and R. A. Holman, Shear Instabilities of the Mean Longshore Current: 1. Theory, *Journal of Geophysical Research 94*, p. 18,030. Copyright © 1989 American Geophysical Union.]

V_0 is the mean longshore current, a prediction that is substantiated by field measurements presented in Oltman-Shay and co-workers.

The energy associated with shear waves appears in wavenumber-frequency spectra (Fig. 8-27) at combinations of low frequencies and high wavenumbers that are beyond the range of possible edge-wave modes. This places the energy in the far-infragravity range, that is, beyond the infragravity range of possible edge-wave modes (analogous to infrared versus far-infrared in the light spectrum). Being in the far-infragravity range, shear waves alternately have been referred to as FIG waves. It can also be seen in Figure 8-27 that there is a directionality associated with the energy of shear waves, in that example toward the south, resulting from the movement of the shear waves in the direction of the mean longshore current. The presence of a strong longshore current also affects the alignment of the edge-wave

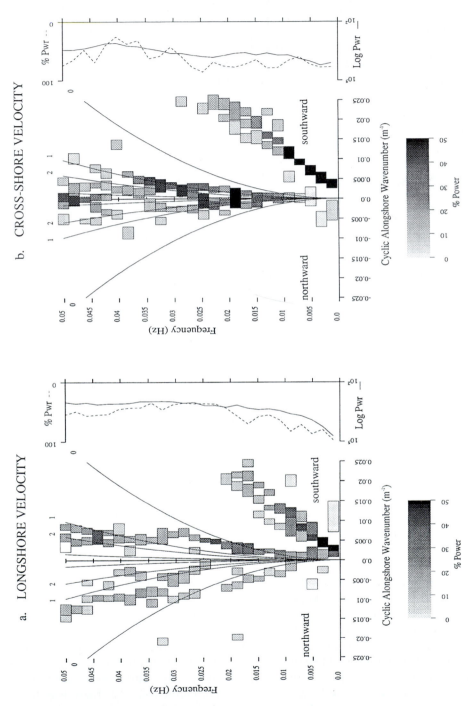

Figure 8-27 Wavenumber-frequency spectra of the (a) longshore and (b) cross-shore measured water velocities in the surf zone, revealing the presence of shear waves in the far infragravity. Positive wavenumbers indicate shear wave propagation toward the south along the north-south trending beach at the FRF. [Used with permission of American Geophysical Union, from J. Oltman-Shay, P. A. Howd and W. A. Berkemeier, Shear Instabilities of the Mean Longshore Current: 2. Field Observations, *Journal of Geophysical Research 94*, p. 18,030. Copyright © 1989 American Geophysical Union.]

energy in comparison with the series of dispersion curves for modes $n = 0, 1$, etc., producing a slight overall asymmetry. This effect is analogous to the Doppler shift in the frequency of sound waves emanating from a moving source (for example, the change in the sound of a train as it passes by). Howd, Oltman-Shay, and Holman (1991) developed techniques for the analysis of edge waves in this situation, as well as any accompanying shear waves that might be generated by instabilities within the mean longshore current.

The presence of shear waves superimposed on a mean longshore current can lead to strong pulsations in the observed flow, variations that in turn can affect other surf-zone processes and the transport of sediment. In particular, the cross-shore oscillations of shear waves are believed to be an important agent of mixing within the nearshore, having an effect on the cross-shore profile of the mean longshore current. This has been analyzed by Dodd, Oltman-Shay, and Thornton (1992) and Church, Thornton, and Oltman-Shay (1992) and has been investigated in laboratory wave-basin experiments by Deguchi, Sawaragi, and Ono (1992).

SUMMARY

There are two principal patterns of wave-generated currents within the nearshore. The cell circulation is dominated by strong seaward-flowing rip currents, which can be a hazard to swimmers, but also have the positive effect of flushing the nearshore water and replacing it with generally cleaner offshore water. The cell circulation can have a profound effect on the beach morphology, with the rip currents generally cutting through the offshore bar and eroding embayments into the beach berm, an erosion that sometimes reaches back to coastal properties (Chapter 11).

The cell circulation can be generated by longshore variations in wave-breaker heights, which produce a parallel variation in wave set-up elevations. The longshore currents flow parallel to shore from areas of high waves and set-up, to zones of low waves and set-up where the currents converge and turn seaward as rips. The required longshore variation in wave heights can be produced by wave refraction over irregular offshore topography such as submarine canyons, or by the local sheltering of headlands or engineering structures such as jetties and breakwaters. More uncertain is the formation of cell circulation systems over long uniform beaches. Models have been developed that demonstrate that synchronous edge waves can interact with the incident waves to produce longshore variations in breaker heights and thus generate a cell circulation. Instability models also have been suggested for the generation of the cell circulation. A problem in establishing the mode of formation of cell circulation systems on ocean beaches is the modification of the beach topography by the currents, as this modified topography can become dominant and control the circulation. Field studies have demonstrated that the presence of a seaward-trending trough can maintain the flow of a rip current due to differences in set-up elevations over the trough versus adjacent bars.

Longshore currents may also be produced by waves breaking at an angle to the shoreline. An understanding of the generation of this current has come from analyses of the longshore component of the radiation stress of the waves (S_{xy}), which exerts a thrust on the water within the nearshore. The best prediction for the magnitude of the current at the mid-surf position is provided by Equation (8.14), a simple dependence on the wave breaker height and angle. Formulations also are available for the complete distribution of the longshore current across the profile of the nearshore zone, the solution of Longuet-Higgins (1970b) being most commonly employed [Equation (8.16)]. A shortcoming of that solution is its assumption of a uniform beach slope. Longshore current distributions can be substantially different over profiles having significant bars and troughs. Fundamental problems remain in understanding the processes of horizontal mixing that smooth the velocity profile (the coefficients N and P in the Longuet-Higgins solution). More recent analyses account for the randomness of the waves, with waves of different heights breaking in different water depths. This randomness has been shown to yield smooth velocity profiles for the generated longshore current, without the necessity of assumptions regarding cross-shore mixing.

An understanding of the generation of nearshore currents has been considerably advanced in recent years. Much remains to be investigated, in particular the generation of nearshore currents produced by the combined forces of obliquely breaking waves and longshore variations in wave set-up, the currents generated by multiple wave trains, and especially the effects of the beach topography on the currents. More study also is needed of the properties and significance of shear waves that can give rise to strong pulsations in the velocities of longshore currents.

PROBLEMS

1. A cell circulation is generated by a longshore gradient in the set-up, $\partial \bar{\eta}/\partial y = 0.0025$, measured from mid-way between two rip currents that are 300 m apart. If the beach initially is uniform with a 5° slope, how much farther landward will the shoreline be displaced at the point midway between the rips where the set-up is a maximum, compared with the shoreline position at the rip position? Later, erosion by the seaward-flowing rip current lowers the level of the beach at its position by 1 m, while maintaining the 5° slope and not affecting the elevation of the profile midway between the rips. What will be the relative horizontal positions of the shoreline at this stage?

2. Waves of period 10 sec reach a beach having a slope angle $\beta = 5°$. According to the hypothesis of Bowen and Inman (1971) that attributes the formation of the cell circulation to synchronous edge waves, what are possible rip-current spacings?

3. Edge waves of period $T_e = 12$ sec and mode $n = 1$ are present on a beach of slope 7°. Groynes are built across the beach (Chapter 12) that segment it into a series of nearly self-contained pockets. What groyne spacing, Λ in Equation (8.7), might result in resonant conditions and therefore the development of adverse currents?

4. Two wave trains in deep water have periods of 12 sec and approach directions of $\alpha_\infty = 80°$ and $\theta_\infty = 75°$, measured with respect to a normal to the shoreline. According to Equation (8.8), based on the intersecting-wave model of Dalrymple (1975), what would be the resulting spacings between the rip currents?

5. Waves break on the beach with a significant wave height of 3 m and at an angle of 10° with the shoreline. Evaluate the magnitude of the resulting longshore current at the mid-surf position. If the slope of the beach is 10°, what is the estimated magnitude if Equation (8.15) from the *Shore Protection Manual* had mistakenly been used?

6. At which beach slope S does Equation (8.15) from the *Shore Protection Manual* yield the same calculated magnitude as Equation (8.14) for the longshore current at the mid-surf position?

7. Write a computer program for the calculation of the cross-shore distribution of the longshore current as evaluated using Equations (8.16) derived by Longuet-Higgins (1970b). Input will be the value of P, which governs the distribution as graphed in Figure 8-13. Test the program with various values of P; the results should be the same as graphed in Figure 8-13.

8. Modify the program of Problem 7 (above) so as to calculate longshore current profiles where the magnitude of the velocity will agree with Equation (8.14a) at the mid-surf position. Input values will be the root-mean-square wave breaker height and breaker angle, beach slope, and again an assumed value for P that would yield a reasonable velocity distribution according to Figure 8-13. Formulate a subroutine for the evaluation of the current at the mid-surf position according to Equation (8.14a) and use that result for the evaluation of the C_f friction coefficient by matching that magnitude with the velocity at the mid-surf position ($X = 0.5$) given by the profile solution of Equation (8.16). Test the program for the conditions given in Figure 8-15, resulting from the analysis of Komar (1976b) that is much the same as developed in this Problem.

REFERENCES

ALLEN, J. (1974). Empirical Models of Longshore Currents. *Geogr. Annal.,* Series A, 56: 238–240.

ARTHUR, R. S. (1962). A Note on the Dynamics of Rip Currents. *Journal of Geophysical Research* 67(7): 2778–2779.

BASCO, D. R. (1982). *Surf Zone Currents.* U.S. Army Corps of Engineers, Coastal Engineering Research Center, Misc. Report 82–7.

BASCO, D. R. (1983). Surf Zone Currents. *Coastal Engineering* 7: 331–355.

BATTJES, J. A. (1972). Radiation Stresses in Short-Crested Waves. *Journal of Marine Research* 30: 56–64.

BATTJES, J. A. (1975). A Note on Modeling of Turbulence in the Surf Zone. *Proceedings Symposium on Modeling Techniques, Amer. Soc. Civil Engrs.,* San Francisco, pp. 1050–1061.

BOWEN, A. J. (1969a). Rip Currents, 1: Theoretical Investigations. *Journal of Geophysical Research* 74: 5468–5478.

BOWEN, A. J. (1969b). The Generation of Longshore Currents on a Plane Beach. *Journal of Marine Research* 27: 206–215.

BOWEN, A. J. and D. L. INMAN (1969). Rip Currents, 2: Laboratory and Field Observations. *Journal of Geophysical Research* 74: 5479–5490.

BOWEN, A. J. and R. T. GUZA (1978). Edge Waves and Surf Beats. *Journal of Geophysical Research* 83: 1913–1920.

BOWEN, A. J. and R. A. HOLMAN (1989). Shear Instabilities of the Mean Longshore Current, 1: Theory. *Journal of Geophysical Research* 94(C12): 18,023–18,030.

BREBNER, A. and J. W. KAMPHUIS (1963). *Model Tests on the Relationship Between Deep-Water Wave Characteristics and Longshore Currents.* Queen's University, Kingston, Ontario, Canada, C.E. Research Report No. 31.

BREBNER, A. and J. W. KAMPHUIS (1964). Model Tests on the Relationship Between Deep-Water Wave Characteristics and Longshore Currents. *Proceedings of the 9th Coastal Engineering Conference, Amer. Soc. Civil Engrs.,* pp. 191–196.

CERC (1984). *Shore Protection Manual.* Coastal Engr. Res. Center, U.S. Army Corps of Engrs. Washington, D.C.: U.S. Govt. Printing Office.

CHURCH, J. C., E. B. THORNTON, and J. OLTMAN-SHAY (1992). Mixing by Shear Instabilities of the Longshore Current. *Proceedings of the 23rd Coastal Engineering Conference, Amer. Soc. Civil Engrs.,* pp. 2999–3011.

DALRYMPLE, R. A. (1975). A Mechanism for Rip Current Generation on an Open Coast. *Journal of Geophysical Research* 80: 3485–3487.

DALRYMPLE, R. A. and C. J. LOZANO (1978). Wave-Current Interaction Models for Rip Currents. *Journal of Geophysical Research* 83: 6063–6071.

DEGUCHI, I., T. SAWARAGI, and M. ONO (1992). Longshore Current and Lateral Mixing in the Surf Zone. *Proceedings of the 23rd Coastal Engineering Conference, Amer. Soc. Civil Engrs.,* pp. 2642–2654.

DODD, N., J. OLTMAN-SHAY, and E. B. THORNTON (1992). Shear Instabilities in the Longshore Current: A Comparison of Observation and Theory. *Journal of Physical Oceanography* 22(1): 62–82.

GALVIN, C. J. (1967). Longshore Current Velocity: A Review of Theory and Data. *Reviews of Geophysics* 5: 288–304.

GALVIN, C. J. and P. S. EAGLESON (1965). *Experimental Study of Longshore Currents on a Plane Beach.* U.S. Army Corps Engr., CERC Technical Memo No 10.

GALVIN, C. J. and R. P. SAVAGE (1966). *Longshore Currents at Nags Head, North Carolina.* U.S. Army Coastal Engineering Research Center Bulletin II, pp. 11–29.

GOURLEY, M. R. (1974). Wave Set-Up and Wave Generated Currents in the Lee of a Breakwater or Headland. *Proceedings of the 14th Coastal Engineering Conference, Amer. Soc. Civil Engrs.*, pp. 1976–1995.

GOURLEY, M. R. (1976). Non-Uniform Alongshore Currents. *Proceedings of the 15th Coastal Engineering Conference, Amer. Soc. Civil Engrs.*, pp. 701–720.

GUZA, R. T. and D. L. INMAN (1975). Edge Waves and Beach Cusps. *Journal of Geophysical Research* 80: 2998–3012.

GUZA, R. T., E. B. THORNTON, and N. CHRISTENSEN (1986). Observations of Steady Longshore Currents in the Surf Zone. *Journal of Physical Oceanography* 16: 1959–1969.

HAINES, J. W. (1984). Steady Flows in the Nearshore Zone. *Proceedings of the 19th Coastal Engineering Conference, Amer. Soc. Civil Engrs.*, pp. 2280–2292.

HARRISON, W. (1968). Empirical Equation for Longshore Current Velocity. *Journal of Geophysical Research* 73: 6929–6936.

HARRISON, W. and W. C. KRUMBEIN (1964). *Interactions of the Beach—Ocean—Atmosphere System at Virginia Beach, Virginia.* U.S. Army Coastal Engr. Res. Center, Technical Memo 7.

HINO, M. (1974). Theory on Formation of Rip Current and Cuspidal Coast. *Proceedings of the 14th Coastal Engineering Conference, Amer. Soc. Civil Engrs.*, pp. 901–919.

HOWD, P. A., J. OLTMAN-SHAY, and R. A. HOLMAN (1991). Wave Variance Partitioning in the Trough of a Barred Beach. *Journal of Geophysical Research* 96: 12,781–12,796.

HUBERTZ, J. M. (1986). Observations of Local Wind Effects on Longshore Currents. *Coastal Engineering* 10: 275–288.

HUNTLEY, D. A. (1976). Lateral and Bottom Forces on Longshore Currents. *Proceedings of the 15th Coastal Engineering Conference, Amer. Soc. Civil Engrs.*, pp. 645–659.

HUNTLEY, D. A. and A. D. SHORT (1992). On the Spacing Between Observed Rip Currents. *Coastal Engineering* 17: 211–225.

INMAN, D. L., R. T. TAIT, and C. E. NORDSTROM (1971). Mixing in the Surf Zone. *Journal of Geophysical Research* 76(15): 3493–3514.

IWATA, N. (1976). Rip Currents. *Journal Oceanographic Society of Japan* 32: 1–10.

JONSSON, I. G., O. SKOVGAARD, and T. S. JACOBSEN (1974). Computation of Longshore Currents. *Proceedings of the 14th Coastal Engineering Conference, Amer. Soc. Civil Engrs.*, pp. 699–714.

KEELEY, J. R. (1977). Nearshore Currents and Beach Topography, Martinque Beach, Nova Scotia. *Canadian Journal of Earth Sciences* 14: 1906–1915.

KEELEY, J. R. and A. J. BOWEN (1977). Longshore Variations in Longshore Currents. *Canadian Journal of Earth Sciences* 14: 1898–1905.

KOMAR, P. D. (1971). Nearshore Cell Circulation and the Formation of Giant Cusps. *Geological Society of America Bulletin* 82: 2643–2650.

KOMAR, P. D. (1975). Nearshore Currents: Generation by Obliquely Incident Waves and Longshore Variations in Breaker Heights. In *Nearshore Sediment Dynamics and Sedimentation*, J. Hails and A. Carr (editors), pp. 18–45. London, England, Wiley.

KOMAR, P. D. (1976a). *Beach Processes and Sedimentation:* Englewood Cliffs, NJ: Prentice-Hall.

KOMAR, P. D. (1976b). Longshore Currents and Sand Transport on Beaches. *Ocean Engineering III, Amer. Soc. Civil Engrs.*, pp. 333–354.

KOMAR, P. D. (1979). Beach-Slope Dependence of Longshore Currents. *Journal of Waterway, Port, Coastal, and Ocean Division, Amer. Soc. Civil Engrs.*, 105(WW4): 460–464.

KOMAR, P. D. and D. L. INMAN (1970). Longshore Sand Transport on Beaches. *Journal of Geophysical Research* 75: 5914–5927.

KOMAR, P. D. and J. OLTMAN-SHAY (1990). Nearshore Currents. In *Handbook on Coastal and Ocean Engineering,* J.B. Herbich (editor). Vol. 2, Chapt. 10, pp. 651–680, Houston, TX: Gulf Publishing Co.

KRAUS, N. C. and T. O. SASAKI (1979). Effect of Wave Angle and Lateral Mixing on the Longshore Current. *Marine Science Communications* 5: 91–126.

KRAUS, N. C. and M. LARSON (1991). *MNLONG: Numerical Model for Simulating the Longshore Current—Report 1, Model Development and Tests.* Technical Report DRP-91–1, Dredging Research Program. Washington, D.C.: U.S. Army Corps of Engrs.

LARSON, M. and N. C. KRAUS (1991). Numerical Model of Longshore Current for Bar and Trough Beaches. *Journal of Waterway, Port, Coastal, and Ocean Engineering, Amer. Soc. Civil Engrs.* 117: 326–347.

LEBLOND, P. H. and C. L. TANG (1974). On Energy Coupling Between Waves and Rip Currents. *Journal of Geophysical Research* 79: 811–816.

LIU, P. L.-F. and R. A. DALRYMPLE (1978). Bottom Frictional Stresses and Longshore Currents Due to Waves with Large Angles of Incidence. *Journal of Marine Research* 36: 357–375.

LONGUET-HIGGINS, M. S. (1970a). Longshore Currents Generated by Obliquely Incident Waves, 1. *Journal of Geophysical Research* 75: 6778–6789.

LONGUET-HIGGINS, M. S. (1970b). Longshore Currents Generated by Obliquely Incident Waves, 2. *Journal of Geophysical Research* 75: 6790–6801.

LONGUET-HIGGINS, M. S. (1983). Wave Set-Up, Percolation and Undertow in the Surf Zone. *Proceedings of the Royal Society of London,* Series A, 390: 283–291.

LONGUET-HIGGINS, M. S. and R. W. Stewart (1964). Radiation Stress in Water Waves, a Physical Discussion with Applications. *Deep-Sea Research* 11: 529–563.

MADSEN, O. S., D. W. OSTENDORF, and A. S. REYMAN (1978). A Longshore Current Model. *Proceedings of the Coastal Zone '78, Amer. Soc. Civil Engrs.,* pp. 2332–2341.

McDOUGAL, W. G. and R. T. HUDSPETH (1983). Wave Setup/Setdown and Longshore Current on Non-Planar Beaches. *Coastal Engineering* 7: 103–117.

McDOUGAL, W. G. and R. T. HUDSPETH (1989). Longshore Current and Sediment Transport on Composite Beach Profiles. *Coastal Engineering* 12: 315–338.

McKENZIE, R. (1958). Rip Current Systems. *Journal of Geology* 66: 103–113.

MEI, C. C. and P. L.-F. LIU (1977). Effects of Topography on the Circulation in and near the Surf Zone—Linear Theory. *Journal of Estuary Coastal Marine Sciences* 5: 25–37.

MILLER, C. and A. BARCILON (1978). Hydrodynamic Instability in the Surf Zone as a Mechanism for the Formation of Horizontal Gyres. *Journal of Geophysical Research* 83: 4108–4116.

MIZUGUCHI, M. (1976). Eigenvalue Problems for Rip Current Spacing. *Transactions of the American Society of Civil Engineers* 248: 83–88.

MIZUGUCHI, M., Y. OSHIMA, and K. HORIKAWA (1978). Laboratory Experiments on Longshore Currents. *Proceedings of the 25th Conference on Coastal Engineering in Japan,* Japan Soc. Civil Engr. [in Japanese].

NODA, E. K. (1974). Wave-Induced Nearshore Circulation. *Journal of Geophysical Research* 79(27): 4098–4106.

NUMMEDAL, D. and R. J. FINLEY (1978). Wind-Generated Longshore Currents. *Proceedings of the 16th Coastal Engineering Conference, Amer. Soc. Civil Engrs.,* pp. 1428–1438.

OLTMAN-SHAY, J., P. A. HOWD, and W. A. BIRKEMEIER (1989). Shear Instabilities of the Mean Longshore Current: 2. Field Observations. *Journal of Geophysical Research* 94 (C12): 18,031–18,042.

O'ROURKE, J. C. and P. H. LEBLOND (1972). Longshore Currents in a Semicircular Bay. *Journal of Geophysical Research* 77: 444–452.

PUTNAM, J. A., W. H. MUNK, and M. A. TRAYLOR (1949). The Predictions of Longshore Currents. *Transactions of the American Geophysical Union* 30: 338–345.

SASAKI, T. (1975). Simulation on Shoreline and Nearshore Current. *Proceedings of the Specialty Conference III, Civil Engineering in the Ocean, Amer. Soc. Civil Engrs.,* pp. 179–196.

SAVILLE, T. JR., (1950). Model Study of Sand Transport Along an Infinitely Long, Straight Beach. *Transactions American Geophysical Union* 31: 555–565.

SHEPARD, F. P., K. O. EMERY, and E. C. LAFOND (1941). Rip Currents: A Process of Geological Importance. *Journal of Geology* 49: 338–369.

SHEPARD, F. P. and D. L. INMAN (1950a). Nearshore Circulation Related to Bottom Topography and Wave Refraction. *Transactions American Geophysical Union* 31(4): 555–565.

SHEPARD, F. P. and D. L. INMAN (1950b). Nearshore Circulation. *Proceedings of the 1st Coastal Engineering Conference, Amer. Soc. Civil Engrs.*, pp. 50–59.

SHORT, A. D. (1985). Rip Current Type, Spacing and Persistence, Narrabeen Beach, Australia. *Marine Geology* 65: 47–71.

SMITH, J. M. and N. C. KRAUS (1987). Longshore Current Based on Power Law Wave Decay. *Coastal Hydrodynamics Conference, Amer. Soc. Civil Engrs.*, pp. 155–169.

SONU, C. J. (1972). Field Observation of Nearshore Circulation and Meandering Currents. *Journal of Geophysical Research* 77(18): 3232–3247.

SYMONDS, G. and D. A. HUNTLEY (1980). Waves and Currents over Nearshore Bar Systems. *Proceedings of the Canadian Coastal Conference,* Natl. Res. Council, Canada, pp. 64–78.

THORNTON, E. B. (1970). Variations of Longshore Current Across the Surf Zone. *Proceedings of the 12th Coastal Engineering Conference, Amer. Soc. Civil Engrs.,* pp. 291–308.

THORNTON, E. B. and R. T. GUZA (1986). Surf Zone Longshore Currents and Random Waves: Field Data and Models. *Journal of Physical Oceanography* 16: 1165–1178.

THORNTON, E. B. and C. S. KIM (1993). Longshore Current and Wave Height Modulation at Tidal Frequency Inside the Surf Zone. *Journal of Geophysical Research* 98(C9): 16,509–16,519.

VISSER, P. J. (1991). Laboratory Measurements of Uniform Longshore Currents. *Coastal Engineering* 15: 563–593.

WU, C.-S., E. B. THORNTON, and R. T. GUZA (1985). Waves and Longshore Currents: Comparison of a Numerical Model with Field Data. *Journal of Geophysical Research* 90(C3): 4951–4958.

9

The Longshore Transport of Sediments on Beaches

The marine engineer has no greater problem to deal with than this. The construction of harbours upon a sandy coast is always risky, resulting in no end of trouble and expense. . . . The interference with the natural sand-travel upon a coast cannot but be injurious; the breaking of any of Nature's laws has a detrimental effect.

Ernest R. Matthews
Coastal Erosion and Protection (1934)

Waves and currents in the nearshore combine to produce a transport of beach sediment. In some cases this is only a local rearrangement of sand into bars and troughs (Chapter 7) or into a series of rhythmic embayments cut into the beach by rip currents (Chapter 11). In other cases there are extensive longshore displacements of sediment, at times moving hundreds of thousands of cubic meters of sand along the coast each year. This along-coast movement is referred to as *littoral sediment transport* or as the *longshore sediment transport,* while the actual volumes of sediment involved in the movement are termed the *littoral drift.*

The longshore transport of sediment on beaches manifests itself whenever this natural movement is prevented by the construction of jetties, breakwaters, and groynes. Such structures act as dams to the moving sediment, causing a buildup of the beach on the updrift sides and simultaneous erosion in the downdrift direction. The construction of such structures, therefore, can have serious consequences in the erosion of coastal properties—in a few cases, entire communities have disappeared. In order to anticipate such impacts and to improve harbor design, it is desirable to have the ability to predict quantities of sediment transport as a function of the wave conditions and nearshore currents, and to assess how the transport is governed by environmental factors such as the sediment grain size and beach morphology. This desire has led to numerous research efforts, both on natural beaches and in laboratory wave basins. However, research has gone well beyond simple empirical attempts to predict total quantities of the littoral drift, and many investigations have focused on considerations of the transport processes. Such studies have examined the relative importance of suspension versus bedload transport and the distributions of the longshore transport across the surf-zone width.

The objective of the chapter is to examine the degree of success of those investigations in meeting the goal of attaining a better understanding of the transport processes, leading to improved predictions of quantities of sediment movement along the coast.

THE COASTAL RESPONSE TO ENGINEERING STRUCTURES

Many occurrences of destructive erosion have resulted from the construction of jetties, breakwaters, or other engineering structures that block longshore sediment transport. Some of the earliest examples occurred along the east coast of India, where there is a large net littoral transport of sediment toward the northeast. The longshore transport rate at Madras has been estimated to be about 500,000 m³/year, and the blockage of transport by construction of a breakwater for a harbor has led to a long history of coastal impacts (Vernon-Harcoart, 1881; Spring, 1919; Johnson, 1957; Cornick, 1969; Komar, 1983). The Madras area lacked the natural protection needed for a harbor, so a breakwater had to be developed on the open coast. The straight shoreline of 1876 that existed prior to breakwater construction (Fig. 9-1), was quickly disrupted with the now familiar pattern of sediment accretion on the updrift side of the structure and erosion in the downdrift direction. In the 36 years (1876–1912) following the initial breakwater construction, more than a million square meters of new land formed on the updrift side, while at the same time, rapid erosion occurred along some 5 km of shoreline in the downdrift direction to the north (Fig. 9-1). The placement of revetments and seawalls immediately downdrift from the breakwater, shifted the zone of maximum erosion farther to the north, where it continues. A suction dredge has been installed to pump sand past the harbor—bypassing the sand around the harbor by dredging has replaced the natural transport system where waves produced the sand movement.

The breakwater at Ceará, Brazil, provides another early example of accretion-erosion problems (Carey, 1903; Johnson, 1957). A detached breakwater was constructed in 1875, extending for approximately 430 m more or less parallel to the shoreline, but was not connected to the shore itself. The idea of using a detached breakwater was that the longshore sediment transport would continue to move uninterrupted by the presence of a structure built across the surf zone, the hope being that no accretion-erosion problem would result. This is a fallacy, since removing the wave action eliminates the energy required to transport the littoral sediments, and this loss of energy results in sediment accumulation within the sheltered region of the structure. Following construction of the detached breakwater at Ceará, a tombolo-shaped mass of sand grew outward from the original shoreline in the protected lee of the breakwater, finally reaching the structure and completely closing the passage. Sand continued to accumulate along the updirft side of the tombolo, eventually moving around the seaward side of the breakwater.

A similar detached breakwater was constructed at Santa Monica, California, in 1934 (Fig. 9-2). Like Ceará, sand began to deposit in the protected lee of the 600–m long break-

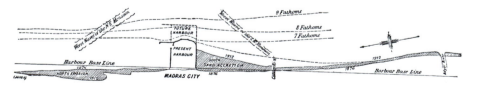

Figure 9-1   Shoreline changes due to the construction of a breakwater at Madras, India, showing beach accretion to the updrift side of the harbor and erosion in the downdrift direction. [Used with permission of the Institute of Civil Engineers, from F. J. E. Spring, Coastal Sand Travel Near Madras Harbor, *Minutes of the Proceedings of the Institute of Civil Engineers.* Copyright © 1919 Institute of Civil Engineers.]

water (Handin and Ludwick, 1950). The shoreline along the updrift side also advanced, while erosion occurred in the downdrift direction. Dredging prevented attachment of the tombolo to the breakwater and the complete closure of the harbor.

The Santa Barbara breakwater was constructed on the California coast beginning in 1927. At that stage it was to be a detached breakwater like that at Santa Monica, but in 1930 it was extended and connected to the shore in order to prevent shoaling that already was affecting use of the harbor. The predominant waves are from a westerly direction, causing a longshore sand transport to the east, estimated to average about 215,000 m^3/year (Johnson, 1953). The breakwater interrupted this longshore transport and again resulted in accretion along its updrift side (Figs. 9-3 and 9-4). Sand accumulated to the west of the breakwater until the entire area was filled; the sand then moved along the arm of the break- water and deposited as a tongue or spit of sand into the quiet water of the harbor. Without dredging, the spit eventually would have grown across the harbor mouth, attaching to the opposite shore and closing off the entrance. Had this been allowed to occur, the entire lit- toral transport of sand would then have passed around the breakwater, and a new equilib- rium shoreline would have been achieved. However, to prevent closure of the harbor, dredging of the spit was initiated and now operates nearly continuously. The dredged sand is dumped on the beach to the immediate east of the breakwater in order to replenish the sand lost by the blockage of the longshore transport and to prevent a continuation of the erosion that took place following breakwater construction (Fig. 9-5). A fully documented case his- tory of the problems at Santa Barbara can be found in Wiegel (1959, 1964).

Jetties have a comparable impact in blocking the longshore sediment transport, re- sulting in shoreline erosion. Construction of jetties in 1935 to stabilize the inlet south of Ocean City, Maryland, trapped a strong southerly transport of sand. The shoreline advanced to the north of the jetties, opposite Ocean City, but the jetty blockage induced massive ero- sion to the downdrift side along Assateague Island (Fig. 9-6). The shoreline retreated about 450 m in the first 20 years following jetty construction (Shepard and Wanless, 1971). By 1961 the south beach had eroded to the point where it actually separated from the landward end of the south jetty, leaving a gap of almost 240 m of open water. Storms in 1962 opened a breach over a kilometer wide, 2.5 km south of the jetties. Subsequent dredging and fill mended the breach and restored the fronting shore, but future storms can be expected to re- new the problem.

As noted for some of the above examples, in many instances sand bypassing systems have been installed in an attempt to maintain the sediment movement along the coast past

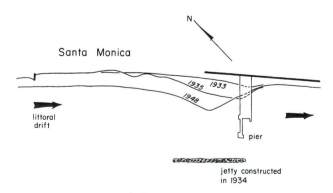

Santa Monica

N

littoral drift

1935 1933
1948

pier

jetty constructed in 1934

Figure 9-2 Sand deposition in the form of a tombolo in the sheltered lee of the Santa Mon- ica breakwater, California. [Adapted from The Littoral Drift Problem at Shoreline Harbors, J. W. Johnson, *Journal of Waterways and Harbor Division* 83, 1957. Reproduced with permission from the American Society of Civil Engineers.]

Figure 9-3 The Santa Barbara breakwater, California, with sand accretion on the updrift side and within the harbor as a spit of sand. [Courtesy of R. L. Wiegel]

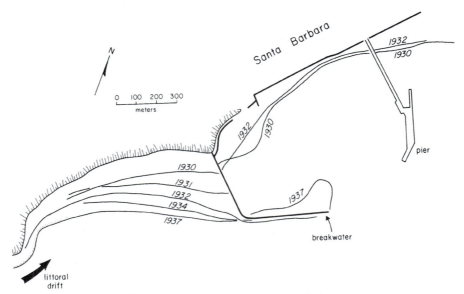

Figure 9-4 The pattern of accretion and erosion around the Santa Barbara break-water, California. [Adapted from The Littoral Drift Problem at Shoreline Harbors, J. W. Johnson, *Journal of Waterways and Harbors Division,* 1957. Reproduced with permission from American Society of Civil Engineers.]

Figure 9-5   Sandyland, 15 km downdrift from the Santa Barbara breakwater, showing the erosion that was experienced after breakwater construction. [Courtesy of F. P. Shepard]

the breakwater or jetties. In its simplest form, sand is scooped up from the updrift area and is carried past the jetties in trucks to the downdrift side. More elaborate schemes involve dredging sand on the updrift side and pumping it as a fluidized sand-water slurry to the downdrift beach where the sand can continue to move as part of the natural transport regime. Such bypassing schemes are diagrammed schematically in Figure 9-7. The oldest approach is illustrated in Figure 9-7(a) where a fixed dredging plant on the updrift side consists of a boom supporting an intake pipe that can be swung in an arc over the accretion zone. This intake pipe is connected to a pump and discharge line that crosses the inlet. Most floating dredges are sensitive to wave attack and must be protected in some way. The system used at Santa Barbara, California, is diagrammed in Figure 9-7(b), the harbor itself providing shelter for the dredge, which removes sand from the spit as it grows into the harbor. Figure 9-7(c) shows the system employed at the Channel Islands Harbor, California, where a detached breakwater shelters the dredge. Construction of such a breakwater for the sole use of a dredge is expensive, so a scheme such as that of Figure 9-7(d) is sometimes used where a low section of the updrift jetty, the weir, has an elevation at or near mean sea level. The design objective is to allow sand to wash over the weir into a depositional basin, where the dredge can operate in safety, the sand again being pumped to the downdrift side. An example of a weir system is found at Masonboro Inlet, North Carolina. It is apparent that considerable effort and expense is required to provide a sand-moving operation that replaces the natural sand transport processes that have been interrupted by jetty or breakwater construction.

Figure 9-6   Shoreline changes resulting from the construction of jetties at Ocean City, Maryland, with extreme erosion along Assateague Island in the downdrift direction.

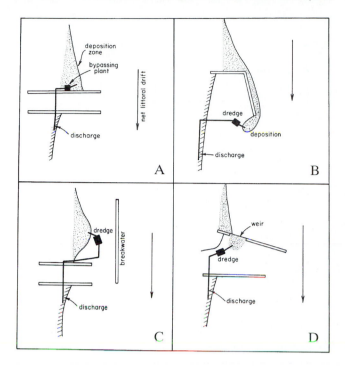

Figure 9-7 Different types of bypassing systems that have been used to transfer the blocked longshore sediment transport past jetties or breakwaters.

THE NET AND GROSS LONGSHORE SEDIMENT TRANSPORT

The *net longshore transport* of sediment is defined as the summation of the movement under all wave trains arriving at the shore from countless wave-generation areas, accounting for the different transport directions. For example, on a north-south trending beach, the sand may move northward for a time, due to waves arriving from the south, and later reverse to a southward movement under waves coming from the north. The net transport of sediment under these two wave trains would be the difference between the north and south movements. This net transport generally is much smaller than the total transport up and down the beach, a total transport that is termed the *gross littoral transport*. On some beaches the gross transport could be extremely large but with the net transport being close to zero.

The change in transport direction may be seasonal. For example, on the beaches of southern California the transport during the winter is predominately to the south due to storm waves arriving from the north Pacific. During the summer, swell waves arrive from the Southern Hemisphere, which is then in its storm-producing winter months. A northward transport of sand therefore prevails on California beaches during the summer. However, the southward transport in the winter is larger, so the net transport is to the south.

Multiple lines of evidence (Fig. 9-8) have been used to assess the direction of longshore sediment transport. Most reflect the net transport, the long-term result of many individual transport events. Blockage by structures such as jetties and groynes can provide the clearest indication of the transport direction. The impoundment adjacent to groynes is small and can respond to the shorter-term reversals in the transport direction (Chapter 12), while large jetties and breakwaters better reflect the long-term net transport. These examples of human-induced evidence for the transport direction are illustrated in Figure 9-8, together

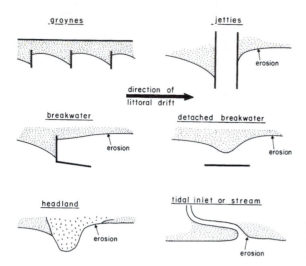

Figure 9-8 Indicators of the direction of the prevailing littoral sand transport along the beach.

with natural shoreline features such as the deflection of streams or tidal inlets by the longshore sand movement, and shoreline displacements at headlands that can act much like jetties.

Compositions and grain sizes of beach sediments also have been used to determine transport directions, as well as establishing the sources of the sediments. This was discussed in Chapter 3, particularly the use of unique heavy minerals contained within the sand to deduce transport paths away from the mineral sources. Trask (1952, 1955) used the heavy mineral augite as a tracer of longshore sand movement along the coast of southern California, demonstrating that the sand filling the harbor at Santa Barbara comes from a distance of more than 160 km up the coast, derived from volcanic rocks in the Morro Bay area. Noting the progressive dilution of the augite content within the beach by the addition of sand from other sources, Bowen and Inman (1966) were able to compute longshore transport rates along the California coast between Morro Bay and Santa Barbara, as well as establishing the direction of the net transport.

Longshore variations in grain sizes on the beach are often interpreted as evidence for a net longshore sediment transport, it generally being assumed that the mean grain size decreases in the direction of transport. Sunamura and Horikawa (1971) in particular have focused on interpretations of grain sizes to infer the longshore transport direction. However, as reviewed in Chapter 3, there can be multiple causes that produce longshore variations in grain sizes, including parallel variations in the wave energy, an asymmetry in the intensity of wave energy from opposite quadrants (as at Chesil Beach, England), or an exchange of sediment with the offshore as the sand otherwise moves alongshore. Exchange with the offshore generally involves a loss of finer sizes from the beach, resulting in progressively coarser beach sediments in the direction of the longshore transport [e.g., the study of McCave (1978) on the east coast of England]. Therefore, patterns of changing sediment grain sizes along the length of a beach cannot unequivocally be used to infer the net direction of transport.

Johnson (1956, 1957) compiled information derived from many sources in order to establish the directions and in some cases the quantities of the net longshore sediment transport for the world's shorelines. Table 9-1 lists the evaluated transport rates, while the pattern for a portion of the Atlantic coast of the United States is shown in Figure 9-9. Johnson found

TABLE 9-1 LONGSHORE SEDIMENT TRANSPORT RATES AT VARIOUS COASTAL SITES [Used with permission of American Society of Petroleum Geologists, from J. W. Johnson, Dynamics of Nearshore Sediment Movement, *Bulletin of the American Society of Petroleum Geologists.* Copyright © 1956 American Society of Petroleum Geologists.]

Location	Transport Rate (m³/yr)	Predominant Direction	Years of Record
U.S. Atlantic Coast			
Suffolk Co., N.Y.	255,000	W	1946–1955
Sandy Hook, N.J.	377,000	N	1885–1933
Sandy Hook, N.J.	334,000	N	1933–1951
Asbury Park, N.J.	153,000	N	1922–1925
Shark River, N.J.	255,000	N	1947–1953
Manasquan, N.J.	275,000	N	1930–1931
Barnegat Inlet, N.J.	191,000	S	1939–1941
Absecon Inlet, N.J.	306,000	S	1935–1946
Ocean City, N.J.	306,000	S	1935–1946
Cold Springs Inlet, N.J.	153,000	S	—
Ocean City, Md.	115,000	S	1934–1936
Atlantic Beach, N.J.	22,600	E	1850–1908
Hillsboro Inlet, Fla.	57,000	S	—
Palm Beach, Fla.	115,000–172,000	S	1925–1930
Gulf of Mexico			
Pinellas Co., Fla.	38,000	S	1922–1950
Perdido Pass, Ala.	153,000	W	1934–1953
Galveston, Texas	334,700	E	1919–1934
U.S. Pacific Coast			
Santa Barbara, Calif.	214,000	E	1932–1951
Oxnard Plain Shore, Calif.	756,000	S	1938–1948
Port Hueneme, Calif.	382,000	S	1938–1948
Santa Monica, Calif.	207,000	S	1936–1940
El Segundo, Calif.	124,000	S	1936–1940
Redondo Beach, Calif.	23,000	S	—
Anaheim Bay, Calif.	115,000	E	1937–1948
Camp Pendleton, Calif.	76,000	S	1950–1952
Great Lakes			
Milwaukee Co., Wis.	6,000	S	1894–1912
Racine Co., Wis.	31,000	S	1912–1949
Kenosha, Wis.	11,000	S	1872–1909
Ill. State line to Waukegan	69,000	S	—
Waukegan to Evanston, Ill.	44,000	S	—
South of Evanston, Ill.	31,000	S	—
Outside of the U.S.			
Monrovia, Liberia	383,000	N	1946–1954
Port Said, Egypt	696,000	E	—
Port Elizabeth, South Africa	459,000	N	—
Durban, South Africa	293,000	N	1897–1904
Madra, India	566,000	N	1886–1949
Mucuripe, Brazil	327,000	N	1946–1950

net transport rates on the order of a million cubic meters per year occurring along some coasts, demonstrating the considerable magnitudes involved in this process. The shoreline of the eastern United States illustrates the important influence of the coastal orientation on the net transport direction and rate. Beach sand on Long Island, which trends roughly east-west,

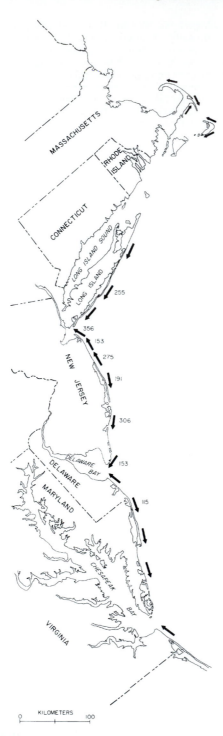

Figure 9-9 Directions and quantities (thousands of cubic meters per year) of littoral sediment transport rates along a portion of the east coast of the United States. [Adapted with permission of American Society of Petroleum Geologists, from J. W. Johnson, Dynamics of Nearshore Sediment Movement, *Bulletin of the American Society of Petroleum Geologists.* Copyright © 1956 American Society of Petroleum Geologists.]

is moving toward the west, while in northern New Jersey the sand is moving northward (Fig. 9-9). There is a node on the New Jersey coast such that in southern New Jersey the sand transport is toward the south. In general, along the northeast coast of the United States, the longshore transport of sediment on the beaches converges toward estuaries and bays, such as Chesapeake Bay, due to changes in coastal orientations.

FIELD-MEASUREMENT TECHNIQUES AND RESULTS

A long standing desire of coastal engineers and scientists has been to estimate longshore sediment transport rates from measurements of the waves and currents that cause the transport. With such a capability, estimates could be made of how much sediment would be blocked by the construction of jetties or a breakwater, and could assist in the design of bypassing systems to limit the adverse impacts of the structures. As will be seen in Chapter 10, equations relating sand transport rates to the waves and currents have an important application in the development of computer simulation models of shoreline change, resulting from the longshore sediment transport.

In view of the initial motivation for studying littoral transport of sediments, it is not surprising that the first evaluations were based on quantities of sediments blocked by the construction of jetties or other engineering structures. The first study that was successful in making quantitative correlations relating the sand transport rate to environmental factors of waves and currents was that of Watts (1953a), who assessed longshore sand transport rates at South Lake Worth Inlet, Florida, by measuring quantities of sand pumped past the jetties via a bypassing plant. The best correlation was obtained with month-long net quantities, relating the long-term sand movement to a balance of the wave parameters measured throughout the month.

The next significant contribution was that of Caldwell (1956), who also relied on the blockage of the littoral transport by jetties, those at Anaheim Bay, California. However, in his study the transport was evaluated from erosion rates of the beach downdrift from the jetties. Although the data gathered by Watts (1953a) and Caldwell are extremely scattered, their studies represent pioneering attempts to obtain quantitative measurements of longshore sediment transport rates in order to establish empirical correlations with the wave conditions.

Subsequent studies that have employed sediment blockage by engineering structures to measure transport rates include the investigations by Bruno and Gable (1976) and Bruno and co-workers (1980, 1981) at Channel Islands Harbor, California; the work by Dean et al. (1982), who measured sand accumulations in the spit growing across the breakwater opening at Santa Barbara, California, and that by Dean et al. (1987) at Rudee Inlet, Virginia. There are problems associated with the use of engineering structures to measure sediment transport rates, the foremost being the local effects of the structures on waves and currents and the long-term nature of the determinations. In some cases it takes a month or longer for sufficient quantities of sand to accumulate in order to make the volume determinations meaningful, an interval during which waves and currents are continuously changing.

Beginning in the 1960s, sand tracers came into use to determine shorter-term "instantaneous" littoral transport rates that could be related to relatively constant conditions of waves and currents. The development of a tracer generally involves tagging the natural beach sediment with a coating of fluorescent dye that does not alter the grain's hydrodynamic properties. Teleki (1966), Yasso (1966), and Ingle (1966) provide summaries of the techniques involved in tagging sand, together with examples of their use on beaches. Devices have been constructed to count the numbers of tagged grains in beach-sand samples (Teleki, 1966, 1967), and Farinato and Kraus (1981) have developed a spectrofluorometic method to determine tracer concentrations. Radioactive sand tracers also have been used but more often in the offshore than on beaches due to their potential hazards. Tracers also have been used to follow the movement of gravel-sized particles, in some cases by simply coating the clasts with paint. Wright, Cross, and Webber (1978) and Nicholls and Webber (1987) employed aluminum gravel particles, cast to have the shapes of natural gravel, the use of which has the advantage that they can later be found on the beach with a magnetometer, even when buried to considerable depths within the beach.

The study by Komar and Inman (1970) was the first to employ sand tracers to obtain quantitative measurements relating the longshore sand transport rate to waves and currents. That investigation, and a number of subsequent studies, employed what is now commonly referred to as the *spatial integration method*. Figure 9-10 shows an example of a tracer distribution obtained at El Moreno Beach on the Gulf of California, Mexico. This distribution of tracer concentrations was determined by injecting the tracer across the beach width, and then after a period of time (4 hours in this example), the beach was sampled on a grid to determine the distribution after transport. By counting the numbers of tracer grains in each grid sample, the local concentrations are determined, and they serve to establish the contours of tracer concentrations, as graphed in Figure 9-10. Those concentrations in turn permit an evaluation of the mean longshore distance of tracer movement during the period of transport, and hence

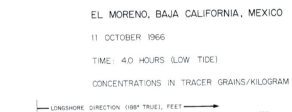

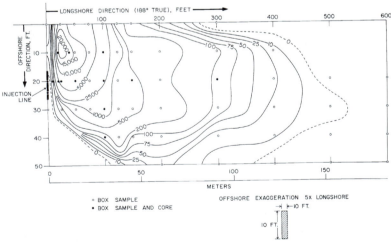

Figure 9-10   Distribution of sand tracer on El Moreno Beach, Baja California, Mexico. [Adapted with permission of American Geophysical Union, from P. D. Komar and D. L. Inman, Longshore Sand Transport on Beaches, *Journal of Geophysical Research* 85. Copyright © 1970 American Geophysical Union.]

yield a measure of the mean longshore advection velocity, $\overline{U}_\ell$, of the sand on the beach. As illustrated schematically in Figure 9-11, the volume transport rate, Q_ℓ, is obtained by multiplying $\overline{U}_\ell$ by the surf-zone width, X_b, and by the average thickness b of the moving layer of sand:

$$Q_\ell = \overline{U}_\ell X_b b \qquad (9.1)$$

The measurement of the thickness b is typically the most uncertain parameter in the use of sand tracers to evaluate Q_ℓ. Komar and Inman (1970) measured b as the depth to which the sand tracer is buried within the beach during its longshore transport, the rationale being that the tracer will rapidly mix downward to a depth governed by the thickness of the moving carpet of sand. They also used the technique devised by King (1951), which involves the burial of a column of dyed sand, later noting the thickness of its truncation by the moving layer of transported sediment. Gaughan (1978), Inman et al. (1980), Kraus et al. (1982), and Kraus (1985) have undertaken thorough analyses of this thickness of sand movement, examining how b depends on the wave energy and beach-sediment grain size, and how it varies across the beach width.

An alternative approach for using tracers to measure the sand advection $\overline{U}_\ell$ is by repeated sampling at a fixed longshore distance from the injection site to determine the time at which the maximum tracer concentration passes, a technique that is termed the *time integration method*. In this technique the mean time of tracer passage is calculated from the time-series measurements of tracer concentrations, and $\overline{U}_\ell$ is obtained by dividing the fixed longshore distance by that mean-time interval. The studies by Knoth and Nummedal (1977), Inman et al. (1980), and Kraus et al. (1982) attempted to use this approach, as well as applying the grid-sampling technique. It was found that the time series of tracer concentrations typically contain a number of peaks rather than being a uniform change through time. This makes the evaluation of $\overline{U}_\ell$ by the time-integration method uncertain and difficult to interpret.

Still another approach using tracers is the *dilution method,* first developed by Russell (1960). This method involves a continuous or quasi-continuous injection of tracer at a fixed rate, and then the measurement of its concentration at some distance down-transport. The equilibrium tracer concentration depends on the amount of dilution by the littoral transport and provides a measure of that transport. One advantage of this approach is that it does not require a determination of the thickness of movement b. Its obvious disadvantage is that the technique requires the continuous introduction of tracer over a prolonged period, several

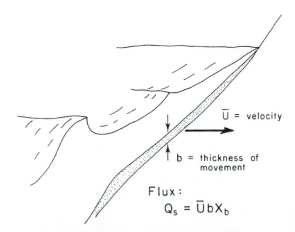

$\overline{U}$ = velocity

b = thickness of movement

Flux:

$Q_s = \overline{U}bX_b$

Figure 9-11 Schematic of the moving layer of sediment involved in the longshore transport, the flux or volume transport rate Q_ℓ being the product of the mean advection velocity $\overline{U}_\ell$, the thickness b of the moving layer, and its cross-shore dimension X_b.

hours and perhaps even days, long enough to achieve complete mixing and an equilibrium concentration of tracer. Russell used this approach to measure the transport of shingle along English beaches, and Duane and James (1980) undertook one experiment of this type at Point Mugu, California, in order to test its potential application on sand beaches.

Other techniques that have been used to measure sediment transport in the nearshore include various traps, pumps, and optical devices. Rather than yielding total quantities of littoral transport, those measurements are more relevant to questions concerning quantities of sediment transported in suspension. A discussion of such techniques and their results, therefore, will be deferred until later.

WAVE POWER EVALUATIONS OF SAND-TRANSPORT RATES

Table 9-2 lists the field studies that have yielded measurements of the total longshore transport of sand, together with the causative waves and currents. These data have been reviewed by Greer and Madsen (1978), Komar (1990), and Schoones and Theron (1993), with discussions of the quality of the measurements. The data listed in Table 9-2 are plotted in Figure 9-12 as the longshore-sediment transport rate versus

$$P_\ell = (ECn)_b \sin \alpha_b \cos \alpha_b \qquad (9.2)$$

where $(ECn)_b$ is the wave energy flux or power evaluated at the breaker zone, and α_b is the wave breaker angle. The rationale for using P_ℓ to represent the wave conditions generally involves a

TABLE 9-2 FIELD DATA FOR LONGSHORE SEDIMENT TRANSPORT RATES [Used with permission of Gulf Publishing Company, from P. D. Komar, Littoral Sediment Transport, *Handbook on Coastal and Ocean Engineering,* Copyright © 1990 Gulf Publishing Co.]

Source	Location	D_{50} (mm)	Number of Points	K	K'
Watts (1953a)[1]	Ft. Lake Worth Inlet, Fla.	0.40	4	0.89(0.73–1.03)	
Caldwell (1956)	Anaheim, Calif.	0.40	6	0.63(0.16–1.65)	
Moore and Cole (1960)	Cape Thompson, Alaska	1.00	1	0.18	0.18
Komar and Inman (1970)	El Moreno, Mexico	0.60	8	0.82(0.49–1.15)	
	Silver Strand, Calif.	0.18	4	0.77(0.52–0.92)	
Lee (1975)	Lake Michigan	?	8	0.42(0.24–0.72)	
Knoth and Nummedal (1977)	Bull Island, S.C.	?	5	0.62(0.23–1.0)	
Inman et al. (1980)	Torrey Pines, Calif.	0.20	2	0.69(0.26–1.34)	
Duane and James (1980)	Pt. Mugu, Calif.	0.15	1	0.81	
Bruno et al. (1981)[2]	Channel Islands Harbor, Calif.	0.20	7	0.87(0.42–1.5)	
Kraus et al. (1982)	Ajigaura	0.25	3		0.19 (0.16–0.22)
	Shimokita	0.18	2		0.32 (0.29–0.36)
	Hirono	0.59	2		0.091(0.09–0.10)
	Oarai	0.29	4		0.18 (0.16–0.19)
Dean et al. (1982)	Santa Barbara, Calif.	0.22	7	1.15(0.32–1.63)	
Dean et al. (1987)	Rudee Inlet, Va.	0.30	3	1.00(0.84–1.09)	

[1]Only the monthly averaged data of Watts (1953a) are used in the analysis.

[2]Includes only the data where the wave data are based on measurements by gauges, not those that are based on LEO visual observations.

"derivation" where $(ECn)_b$, the wave energy flux or power per unit wave-crest length, is converted to a unit-shoreline basis by inclusion of the $\cos\alpha_b$ factor to give $(ECn)_b \cos\alpha_b$, and then its multiplication by $\sin\alpha_b$ to yield the "longshore component," presumably only that portion of the wave power available for transporting sediment in the longshore direction. Based on this derivation, P_ℓ is often referred to as the "longshore component of wave power." Longuet-Higgins (1972) objected to this terminology, pointing out that since $(ECn)_b$ is a vector rather than a second-order tensor, the longshore component would be $(ECn)_b \sin\alpha_b$. In spite of such objections, P_ℓ has continued to be employed in longshore transport evaluations. Its significance as a process-controlled parameter will become more apparent later in this chapter.

The sand-transport rate can be expressed either as the volume transport rate, Q_ℓ, given by Equation (9.1), or as an immersed-weight transport rate, I_ℓ, defined as

$$I_\ell = (\rho_s - \rho)ga'Q_\ell \tag{9.3}$$

where ρ_s and ρ are, respectively, the sand and water densities, and a' is a pore-space factor such that $a'Q_\ell$ is the volume of solid sand alone, eliminating pore spaces included in the Q_ℓ volume transport rate (a' is usually taken as 0.6). The use of I_ℓ rather than Q_ℓ is based on the sediment-transport theories of Bagnold (1963, 1966) and applied specifically to sand transport on beaches by Inman and Bagnold (1963) and Inman and Frautschy (1966). One advantage of using I_ℓ is that this immersed-weight transport rate accounts for the density of the sediment grains. Also important is that I_ℓ and P_ℓ have the same units, so the relationship becomes

$$I_\ell = KP_\ell = K(ECn)_b \sin\alpha_b \cos\alpha_b \tag{9.4}$$

which is homogeneous, that is, the K proportionality coefficient is dimensionless.

The field data plotted in Figure 9-12 are limited to the longshore transport of sand-sized sediments, excluding the transport of coarser sediments such as gravel. The data are reasonably consistent with the direct proportionality predicted by Equation (9.4), with the solid line in the figure yielding the proportionality coefficient $K = 0.70$. This value is slightly less than the 0.77 coefficient obtained by Komar and Inman (1970), which is still often used in making

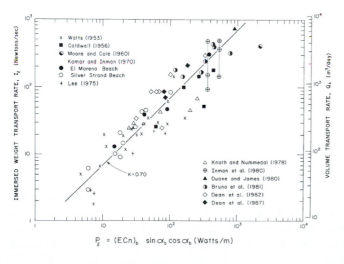

Figure 9-12 A compilation of field data for longshore sand-transport rates (I_ℓ and Q_ℓ) as a function of P_ℓ from Equation (9.2).

sand transport evaluations. It is possible to obtain a wide range of estimated average K coefficients, depending on which data sets are included and excluded. In a compilation of existing data, Schoones and Theron (1994) derived a coefficient that is equivalent to only $K = 0.40$, but this initial compilation included all data irrespective of quality, and included data where it is clear that only the suspension transport was measured, not the total transport (suspension plus bedload). When Schoones and Theron limited their analysis to only the high-quality data and where the total transport was measured, the proportionality coefficient doubled to $K = 0.82$. This exercise provides some indication of the uncertainty in the best choice value for K in Equation (9.4) and of resulting uncertainties in calculations of longshore sediment transport rates from any formulation.

Using $K = 0.70$, which is seen in Figure 7-12 to provide a good overall fit to the high-quality measurements, Equation (9.4) becomes

$$I_\ell = 0.70 P_\ell = 0.70 (ECn)_b \sin \alpha_b \cos \alpha_b \tag{9.5a}$$

or

$$I_\ell = 0.70 P_\ell = 0.70 \left(\frac{1}{8} \rho g H_{br}^2\right)(\sqrt{2gH_{br}}) \sin \alpha_b \cos \alpha_b \tag{9.5b}$$

$$I_\ell = 0.12 \rho g^{3/2} H_{br}^{5/2} \sin \alpha_b \cos \alpha_b \tag{9.5c}$$

having substituted $E_b = (1/8)\rho g H_b^2$ for the wave energy density, $C_b = \sqrt{g(h_b + H_b)}$ for the wave phase velocity, and $\gamma_b = H_b/h_b = 1$. By using Equation (9.3) and taking $\rho_s = 2650 \text{ kg/m}^3$ for quartz-density sand and $\rho = 1020 \text{ kg/m}^3$ for the density of seawater, one obtains the derivative relationships

$$Q_\ell = 6.2 P_\ell \tag{9.6a}$$

$$Q_\ell = 1.1 \rho g^{3/2} H_{br}^{5/2} \sin \alpha_b \cos \alpha_b \tag{9.6b}$$

where Q_ℓ has units m^3/day, and P_ℓ is Watts/meter (H_{br} is in meters, and $g = 9.8 \text{ m/sec}^2$). Equation (9.6) applies only to quartz-density sand, and the value of the proportionality coefficient depends on the units employed. In both Equations (9.5) and (9.6), the wave energies are based on the root-mean-square wave height, H_{br}, the height that corresponds to the correct assessment of the wave energy (Chapter 5). The proportionality coefficients must be changed if the calculations are in terms of the significant wave height, H_{bs}. Since $H_{bs}/H_{br} \approx 1.41$ (Longuet-Higgins, 1952), the calculated wave energies would differ by a factor $(1.41)^2 \approx 2$ and the phase velocities by $(1.41)^{1/2} \approx 1.2$, so the wave power differs by $2 \times 1.2 = 2.4$. The relationships then become

$$I_\ell = 0.30 P_\ell = 0.053 \rho g^{3/2} H_{bs}^{5/2} \sin \alpha_b \cos \alpha_b \tag{9.7a}$$

and

$$Q_\ell = 0.46 \rho g^{3/2} H_{bs}^{5/2} \sin \alpha_b \cos \alpha_b \tag{9.7b}$$

where Q_ℓ again has units m^3/day, and H_{bs} is in meters.

The above relationships are those that have seen most common use in applications. They provide simple correlations between the wave conditions and the resulting rates of longshore-sediment transport. However, although the correlations apparently have been successful, they are completely empirical with little thought having been given to the physical processes involved in the longshore transport of sediments. All of the data used to establish these equations come from sand beaches, so the resulting relationships do not contain the expected sediment grain-size dependence—this will be examined later in terms of possible variations in K with the grain size. Furthermore, the sediment transport is assumed to result solely from waves breaking at an angle to the shore, and indirectly to the longshore current generated in this situation (Chapter 8). The equations are not applicable to evaluations of longshore sediment transport that occur when the cell circulation, tides or the wind contribute to the magnitude of the longshore current, and thus to the sediment transport. Alternative models are needed that are more general, models that consider the processes of transport and are broader in their potential applications.

PROCESS MODELS OF SAND TRANSPORT ON BEACHES

Early workers such as Grant (1943) stressed that sand transport in the nearshore results from the combined effects of waves and currents, with the waves placing the sand in motion while the longshore current produces the net sand advection. Such a model was given a mathematical framework by Bagnold (1963) and applied specifically to the evaluation of the longshore sediment transport on beaches by Inman and Bagnold (1963). Their analysis yielded

$$I_\ell = K'(ECn)_b \frac{\overline{v}_\ell}{u_m} \tag{9.8}$$

where $\overline{v}_\ell$ is the longshore current velocity, in practice measured at the mid-surf position, and u_m is the maximum horizontal orbital velocity of the waves evaluated at the breaker zone. The ratio $(ECn)_b/u_m$ in effect is proportional to the mean stress exerted by the waves, which simply move the sand to-and-fro, while the longshore current gives the transport its direction as well as affecting the amount. K' is a dimensionless coefficient that must be calibrated using sediment-transport measurements.

Komar and Inman (1970) utilized their field data of longshore sand transport rates to make the first test of Equation (9.8), employing direct measurements of longshore currents as well as wave parameters. Their data are shown in Figure 9-13, yielding $K' = 0.28$ when considered alone. Kraus et al. (1982) subsequently obtained measurements from several beaches in Japan, and the addition of their data suggests a reduction to $K' = 0.25$, the coefficient for the line shown in Figure 9-13.

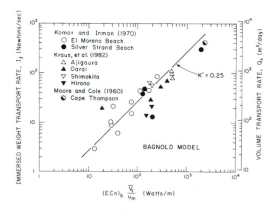

Figure 9-13 Field data for longshore sand-transport rates used to test the process-based Bagnold relationship, Equation (9.8).

Taking $E_b = (1/8)\rho g H_b^2$ for the wave energy density, and $u_m = \sqrt{2E_b/\rho h_b} = 0.5\sqrt{\gamma g H_b}$ for the orbital velocity according to linear-wave theory (Chapter 5), the relationship of equation (9.8) with $K' = 0.25$ reduces to

$$I_\ell = 0.088\rho g H_{br}^2 \bar{v}_\ell \qquad (9.9a)$$

$$I_\ell = 0.044\rho g H_{bs}^2 \bar{v}_\ell \qquad (9.9b)$$

where the breaker heights in the two relationships are, respectively, a root-mean-square wave height and the significant wave height. Using Equation (9.3) to convert I_ℓ to the volume transport rate, Q_ℓ, yields

$$Q_\ell = 0.088\rho g H_{br}^2 \bar{v}_\ell \qquad (9.10a)$$

$$Q_\ell = 0.044\rho g H_{bs}^2 \bar{v}_\ell \qquad (9.10b)$$

for quartz-density sand. The equivalence of the coefficients in Equations (9.9) versus (9.10) is fortuitous and results from $(\rho_s - \rho)a'/\rho_s \approx 1$ for quartz-density sand in seawater with $a' = 0.6$.

A relationship basically equivalent to Equation (9.10) was deduced by Kraus et al. (1982) from their measurements of sand-transport rates on beaches in Japan. However, they based its derivation on the $Q_\ell = \bar{U}_\ell X_b b$ relationship for tracer movement [Equation (9.1)], with empirical correlations between the sand advection velocity and the longshore current velocity ($\bar{U}_\ell = 0.014\bar{v}_\ell$) and between the thickness of sediment movement and the wave-breaker height ($b = 0.027 H_{bs}$). The surf-zone width was evaluated as $X_b = h_b/S = H_b/\gamma_b S$, where S is the beach slope, suggesting that the proportionality coefficients in Equations (9.9) and (9.10) might depend on the beach slope, however, such a dependence has not been established by the field data. A similar relationship is also derived by Kraus and Dean (1987) but based on a suspension transport model where the volume transport rate of sand is assumed to be proportional to the mean discharge of water associated with the longshore current.

Equations (9.9) and (9.10) have several advantages over the P_ℓ based sediment transport relationships obtained in the preceding section and should be used in more applications. From a practical standpoint, it is often easier and more accurate to measure the longshore current $\bar{v}_\ell$ than it is to determine the breaker angle α_b needed in the P_ℓ formulation. Equation (9.8) was derived by Bagnold (1963) and Inman and Bagnold (1963) on the basis of considerations of the processes of sand transport, and accordingly should be viewed as more fundamental than the empirical correlations with P_ℓ. In the derivation of Equations (9.8) through (9.10), the origin of the longshore current $\bar{v}_\ell$ was never specified, so its generation could be due to an oblique-wave approach, tidal currents, part of the cell circulation with rip currents, or driven by local winds. If the longshore current is generated solely by waves breaking obliquely to the shore, then from Chapter 8 we have the relationship

$$\bar{v}_\ell = 1.17\sqrt{gH_{br}} \sin \alpha_b \cos \alpha_b$$

for the magnitude of the current at the mid-surf position. Its substitution for $\bar{v}_\ell$ in Equation (9.9a) yields

$$I_\ell = 0.088\rho g H_{br}^2 \bar{v}_\ell$$

$$I_\ell = 0.088\rho g H_{br}^2 (1.17\sqrt{gH_{br}} \sin \alpha_b \cos \alpha_b)$$

$$I_\ell = 0.10\rho g^{3/2} H_{br}^{5/2} \sin \alpha_b \cos \alpha_b$$

which is essentially the same as the wave-power sand transport formula [Equation (9.5c)] obtained in the preceding section (the difference in the coefficients, 0.10 versus 0.12, results because the 0.10 value found here combines three empirical coefficients, K and K' for the sand transport relationships, and the 1.17 coefficient from the longshore current formula). The implication is that any of the P_ℓ relationships for sediment transport on beaches are suitable only for the limited case where the longshore current and resulting sediment transport are produced solely by waves breaking at an angle to the shoreline (Komar and Inman, 1970). This is specifically illustrated by the sand-transport measurements of Kraus et al. (1982) at Oarai Beach, Japan. At that site the data were obtained in the sheltered region of a breakwater where the longshore current results from the combined effects of obliquely incident waves and a longshore variation in wave heights (Chapter 8). Under such conditions an evaluation of the sand transport from P_ℓ alone would be erroneous. This is especially illustrated by one measurement series of Kraus et al. where the direction of the longshore current and sediment transport were opposite to that expected from the oblique-wave incidence.

LABORATORY STUDIES OF LONGSHORE SAND TRANSPORT

The relationships for the longshore sediment transport derived above are based entirely on field data where the sand transport was measured either by using tracers or from quantities of sand blocked by jetties and breakwaters. Studies of sediment transport processes on natural beaches are inherently difficult, and important variables such as the wave conditions cannot be controlled. This accounts for much of the data scatter seen in the graphs of Figures 9-12 and 9-13. This has induced investigators to turn to wave-basin experiments. The typical laboratory experiment involves the generation of waves with an oscillating paddle along one length of the basin, with the sandy beach oriented such that the waves break at an angle to the shore. The longshore currents accordingly are generated by obliquely breaking waves, which combine with the wave orbital motions to transport sand along the beach, the volume of sand then being trapped within a stilling basin at the downdrift end of the beach. Such a simple laboratory arrangement permits full control over the experimental conditions, including sediment grain sizes, the heights and periods of the waves, and breaker angles.

Several of the laboratory studies of longshore sand transport pre-date the field studies discussed above. The early works yielding data for sand transport rates versus wave conditions include those of Krumbein (1944), Saville (1950), Shay and Johnson (1951), Johnson (1953), Sauvage and Vincent (1954), and Savage (1959, 1962). The more recent experimental works include those of Özhan (1982) and Kamphuis and co-workers (Kamphuis and Readshaw, 1978; Kamphuis and Sayao, 1982; Kamphuis et al., 1986; Kamphuis, 1990). The research efforts by Kamphuis have gone the furthest to collect laboratory data and to use it in establishing predictive relationships for the longshore sediment transport. The accumulated data are analyzed empirically using dimensionless ratios of the various parameters, rather than through considerations of physical processes such as the generation of longshore currents and their role in transporting sediments. As summarized by Kamphuis (1990), the accumulated laboratory data support the empirical relationship

$$\frac{Q_\ell}{(\rho H_{bs}^3/T)} = 0.0013 S^{0.75}\left(\frac{H_{bs}}{L_\infty}\right)^{-1.25}\left(\frac{H_{bs}}{D_{50}}\right)^{0.25}\sin^{0.6}(2\alpha_b) \tag{9.11}$$

Figure 9-14 is a plot of the laboratory data, shown as the measured versus calculated sediment transport rates using Equation (9.11). When extrapolated, the relationship shows reasonable

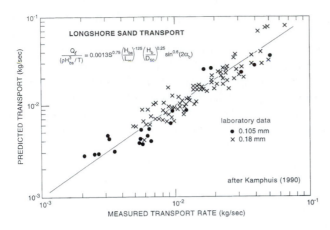

Figure 9-14 Measured longshore sand-transport rates derived from the laboratory wave-basin experiments of Kamphuis and co-workers versus rates computed with the empirical Equation (9.11). [Adapted from Littoral Sediment Transport Rate, J. W. Kampheris, *Proceedings of the 22nd Coastal Engineering Conference,* 1990. Reproduced with permission from the American Society of Civil Engineers.]

agreement with the field data, though not as good as the relationships discussed earlier for I_ℓ versus P_ℓ. Furthermore, a similar dimensional analysis by Kamphuis (1990) of the field data alone yields the relationship

$$\frac{Q_\ell}{(\rho H_{bs}^3/T)} = 0.0006 S^{1.0}\left(\frac{H_{bs}}{L_\infty}\right)^{0.5}\left(\frac{H_{bs}}{D_{50}}\right)^{1.0}\sin(2\alpha_b) \qquad (9.12)$$

that differs from Equation (9.11), particularly in finding a proportionality with $\sin(2\alpha_b) = 2\sin\alpha_b\cos\alpha_b$ rather than with $[\sin(2\alpha_b)]^{0.6}$ for the laboratory data, and there also are significant differences in the various empirical exponents. The principal contributions of these empirical equations are their suggested dependencies on the sediment grain diameters D_{50}, on the beach slope S, and on the wave steepness expressed as H_{bs}/L_∞. These dependencies will be examined in the next section where we will review a series of studies that have attempted to relate the rates of longshore sediment transport to grain sizes and other environmental factors.

The laboratory studies of longshore sediment transport have contributed to our understanding of the transport processes and have yielded carefully controlled measurements relating the transport rates to the wave conditions. However, there are uncertainties as to how these laboratory measurements relate to those collected on natural beaches. Because of the small scales of the laboratory experiments, most of the data represent P_ℓ factors and sediment transport rates that are roughly 2 orders-of-magnitude smaller than experienced under field conditions. Some experiments used granular coal or other low-density materials in attempts to scale the transport processes from prototype to laboratory conditions, but most work has been with quartz-density sand. There are difficulties in scaling sediment transport processes, and in the case of wave-basin tests of littoral sand transport, it is often observed that sand ripples extend across the entire beach and are totally out of scale with respect to natural prototype beaches. It perhaps is best to view these laboratory results as small-size prototypes, comparable to ponds or small lakes where the waves can be controlled. Products of the laboratory experiments such as the empirical equations derived by Kamphuis (1990) can provide guidance in making sediment transport calculations for grain sizes that are beyond the normal range of sand-size diameters, but they must be used with caution.

TRANSPORT DEPENDENCE ON ENVIRONMENTAL FACTORS

It can be expected that longshore sediment transport rates will depend on environmental factors such as the sediment grain size, the beach slope, and factors such as the wave steepness. Such dependencies have been suggested by Equations (9.11) and (9.12), derived from the dimensional analyses of Kamphuis and co-workers (1978, 1982, 1986, 1990). A number of investigations over the years have searched for such environmental controls, but the results often have been controversial, sometimes contradictory, and have generated debate among investigators.

Considering first the field measurements of sand transport rates on beaches, it can be seen in Figures 9-12 and 9-13 that there is considerable scatter within the data. This scatter is reflected in the K and K' proportionality coefficients compiled in Table 9-2 as averages and total ranges of values for individual data sets. The average K values are as high as 1.15 from Dean et al. (1982) and as low as 0.18 for the single measurement of Moore and Cole (1960). The question arises as to whether these variations in K and K' reflect environmental conditions such as beach-sediment grain sizes or whether they result from random errors and systematic differences in procedures between the studies.

The most obvious dependence that might be expected would be with the sediment grain size, intuition suggesting that beaches composed of coarser sediment experience smaller rates of longshore transport, other factors being equal. Accordingly, numerous investigators have attempted to explain the scatter in the K and K' coefficients as a result of differences in beach-sediment grain sizes or settling velocities (Castanho, 1970; Komar and Inman, 1970; Swart, 1976; Komar, 1978, 1988; Bruno, et al., 1980, 1981; Dean et al., 1982, 1987). The curve of Bruno et al. (1980) in Figure 9-15 (upper), based on field data from sand beaches, suggests that K decreases with an increase in the median diameter, D_{50}, of the beach sand. However, this specific empirical curve has been challenged by Komar (1988) on the basis that the trend is produced by the single measurement of Moore and Cole (1960) on a beach where $D_{50} = 1$ mm. Unfortunately, this K value from Moore and Cole is highly questionable. The sand transport rate was determined from accretion within a breached area of a spit, so there is uncertainty whether the total transport was measured. More important, the wave parameters were estimated visually. The velocity of the longshore current was measured, so it is possible to test the data of Moore and Cole against the Bagnold model of Equation (9.8) to determine a K' coefficient. This yields $K' = 0.18$, which is not significantly different from the 0.25 value based on the combined data of Komar and Inman (1970) and Kraus et al. (1982) and is well within the K' values for individual measurements (Table 9-2). The agreement of the Moore and Cole measurement with the other data is apparent in Figure 9-13, and it can be seen in Figure 9-15 (lower) that it does not support a decrease in the K' coefficient with increasing D_{50}. This implies that the mean breaker angle (25°) determined visually by Moore and Cole was too large, and this in turn made P_ℓ too large and hence K too small. As discussed earlier, it is easier to measure longshore currents than breaker angles, and this accounts for the data of Moore and Cole being more consistent with the Bagnold model of Equation (9.8) than with the wave-power relationship of Equation (9.4).

Without the low value of K from Moore and Cole (1960) at $D_{50} = 1$mm, little or no support remains within the field data from sand beaches for the curve of Bruno et al. (1981) in Figure 9-15 (upper), or for any other indication that K decreases with D_{50} within the range of sand-sized sediments. However, more support is obtained if the consideration is extended

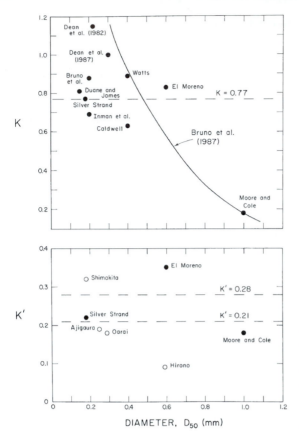

Figure 9-15 The average K and K' coefficients in the sand-transport relationships, [Equations (9.4) and (9.8)] for several data sets as functions of the median diameters, D_{50}, of the beach sediments. [From Environmental Controls on Littoral Sand Transport, P. D. Komar, *Proceedings of the 21st Coastal Engineering Conference*, 1988. Reproduced with permission from the American Society of Civil Engineers.]

to still coarser grained beaches, in the extreme to gravel and shingle. This is suggested by the analyses of del Valle, Medina, and Losada (1993), where K coefficients were estimated for beaches having sediment grain sizes up to 1.5 mm; the results in Figure 9-16 support a curve for K versus D_{50} which is similar to that originally proposed by Bruno et al. (1981). However, the data from del Valle and co-workers are uncertain in having been inferred from shoreline changes observed on aerial photographs covering a 30–year period and from a wave climate based primarily on visual observations. The area of study is the Adra River Deta on the Mediterranean coast of Spain, with different grain sizes being found along various portions of the delta shoreline, resulting from a complex history of accretion and erosion associated with a shift in the position of the river mouth during the last century. Therefore, the K values obtained by del Valle and co-workers are based on long-term integrated sediment transport rates inferred from the shoreline changes and on the integrated wave climate over those 30 years. Further questions regarding the validity of the K estimates come from the study of Guillen and Jimenez (1995) on the Ebro Delta, also on the Mediterranean coast of Spain. The focus of their investigation was on longshore variations in grain sizes along the length of the delta shoreline, and they found three instances where the beach sediments coarsened in the direction of longshore transport and two cases where the sediment becomes finer. They found that the coarseness of the beach sediment was correlated with the along-coast gradient of the longshore transport (dI_ℓ/dx or dQ_ℓ/dx), rather than with the transport rate itself. This does not necessarily negate the results of del Valle and co-workers (1993) and

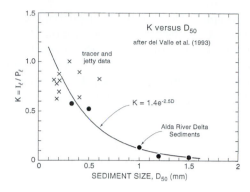

Figure 9-16 The variations in K versus the median sediment grain diameter D_{50}, inferred from the long-term shoreline changes along the Alda River Delta, Spain. [Adapted from Dependence of Coefficient K on Grain Size, R. del Valle, R. Medina, and M. A. Losada, *Journal of Waterway, Port, Coastal and Ocean Engineering*, 1993. Reproduced with permission from the American Society of Civil Engineers.]

their evaluations of K versus D_{50} in Figure 9-16 but does suggest that the longshore variations in grain sizes and transport dependencies on the grain sizes (or the inverse) are complex.

The dependence of K on D_{50} becomes more apparent if the consideration is extended to gravel and shingle beaches. This is indicated by the results of Nicholls and Wright (1991) using aluminum-pebble tracers to measure longshore transport rates on two shingle beaches in England. As expected, the K coefficients are substantially lower than those determined for sand beaches, ranging between 1 and 20 percent of the K values from sandy beaches. The results have a high degree of uncertainty, due to differential transport rates of the various size fractions within the beach gravel, but still clearly demonstrate that K (and K') decrease with increasing grain size when one considers the full range of beach sediments from sand through gravel. Unfortunately, the amount of field data derived from gravel beaches remains insufficient to establish quantitative relationships for the variations in K and K' as functions of the beach-sediment grain sizes. Furthermore, with such coarse beach materials, it becomes important to incorporate threshold evaluations, the stresses exerted by the waves and nearshore currents required to actually initiate gravel transport. This has been attempted by Morfrett (1989) in an application to the longshore transport of shingle, but his inclusion of grain threshold together with other modifications has significantly complicated evaluations of longshore transport rates.

When Equations (9.11) and (9.12) for the longshore sediment transport, based on the analyses of Kamphuis (1990), are rearranged algebraically into the form of Equation (9.4) or (9.7), the results imply that the K proportionality coefficient is given by

laboratory data:
$$K \propto S^{0.75}\left(\frac{H_b}{D_{50}}\right)^{0.25}\left(\frac{H_b}{L_\infty}\right)^{-0.75}[\sin(2\alpha_b)]^{-0.4} \qquad (9.13a)$$

field data:
$$K \propto S\left(\frac{H_b}{D_{50}}\right)\left(\frac{H_b}{L_\infty}\right) \qquad (9.13b)$$

The prediction is that K depends on the beach slope S and on the wave steepness expressed as H_b/L_∞, as well as on the sediment grain size. Of interest is the dependence on the beach slope, the results implying that an increase in beach slope will result in an increase in the transport rate. This is intuitively plausible, in that with a steeper beach the waves break close to shore by plunging, while on a fine-sand dissipative beach with a low slope, the waves break by spilling and their energy is dissipated over a wide surf zone so that more of the total energy flux is lost to turbulence than is applied toward the transport of sediments. Also of interest is that any dependence on S will act to offset the dependence on D_{50}, since on average coarser-sediment beaches tend to be steeper (Chapter 7). This may in part account for the difficulty in establishing a simple dependence between K and D_{50}, as seen above, since the accompanying increase in S could have been a mitigating factor. In Equations (9.13) from the laboratory experiments of Kamphuis, the dependence on S is stronger than on D_{50}, so it is possible that in some instances the longshore transport of sediment on a coarse-sand beach with a steep slope will be greater than on a fine-sand beach that has a very low slope.

Equation (9.13a) for the laboratory data and Equation (9-13b) for the field data also predict dependencies of K on the wave steepness, $H_b L_\infty$, although the dependencies are in opposite directions. In one of the earliest wave-basin studies of longshore sediment transport, Saville (1950) found a complex dependence on the wave steepness, with a maximum sand-transport rate occurring at a wave steepness $H_\infty/L_\infty = 0.025$, with a decrease for both larger and smaller steepness values. The thought was that the transport is enhanced when the beach is in transition from its steeper berm-type profile to its storm-induced bar-type profile having a lower slope (Chapter 7), but is reduced when those more stable configurations are achieved. The data of Saville supporting this trend were very scattered, and Özhan (1982) concluded that the transport coefficient K decreases with increasing wave steepness $[K = 0.015(H_\infty/L_\infty)^{-0.82}]$, much as found by Kamphuis (1990). Özhan explained this result in terms of the changing modes of wave breaking (spilling, plunging, collapsing, and surging), which depend on H_∞/L_∞ (Chapter 6), with the highest sand-transport rates occurring under collapsing breakers on steeper beaches.

This possible relationship to the breaker type suggests the potential for a dependence of K on the Iribarren number, $\xi_\infty = S/(H_\infty/L_\infty)^{1/2}$, which would combine the beach slope S with the wave steepness. Because of problems in selecting values for beach slopes in the complex profiles of the laboratory beaches, Özhan (1982) found that the dependence of K on ξ_∞ was more scattered than the dependence on the wave steepness alone. Bodge and Kraus (1991) further explored the dependence of K on the Iribarren number, both for field data and for various data sets derived from laboratory wave-basin experiments. Their results are shown in Figure 9-17, which support the relationship

$$K = 0.37\ln\xi_b + 0.59 \qquad (9.14a)$$

empirically based on the laboratory data alone, where the Iribarren number

$$\xi_b = \frac{S}{(H_b/L_\infty)^{1/2}} \qquad (9.14b)$$

is based on the wave-breaker height and deep-water wave length. It is apparent in Figure 9-17 that there is a strong trend established by the laboratory data, while the field data are randomly scattered. The laboratory data support the existence of a systematic increase in K with increasing ξ_b. This result implies that an increase in beach sediment grain size, resulting in an increase in beach slope S and thus in ξ_b for a fixed wave steepness, in turn produces a higher K coefficient and thus an increase in the longshore sediment transport rate for a given set of wave conditions. It should be recognized that the measurements establishing these trends are again limited to sand beaches, specifically to laboratory beaches, and it is probable that the trend will be reversed for gravel beaches where the sediment is

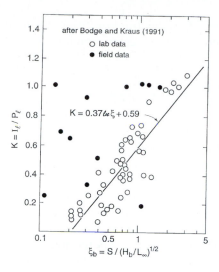

Figure 9-17 The K proportionality coefficient for use in the longshore sediment transport relationship of Equation (9.3), as a function of the Iribarren number of Equation (9.14b). The laboratory data support a definite trend, which could be used for guidance in field computations, even though the field data for K are highly scattered. [Adapted from Critical Examination of Longshore Transport Rate Magnitude, K. R. Bodge and N. C. Kraus, *Coastal Sediments '91*, 1991. Reproduced with permission from the American Society of Civil Engineers.]

sufficiently coarse to hinder its entrainment by nearshore waves and currents, reducing the overall rates of transport.

It is apparent from the sometimes conflicting evidence reviewed here that the dependence of the longshore sediment transport rate on environmental factors such as grain diameters, beach slopes, and wave steepnesses remains poorly established and that more research is needed.

THE CROSS-SHORE DISTRIBUTION OF LONGSHORE SEDIMENT TRANSPORT

Thus far our considerations have centered on the total transport of sediment along beaches. Many applications require evaluations of the distribution of the transport, that is, the cross-shore profile of the longshore sediment movement. For example, this knowledge is central to the effective design of groynes and jetties, especially jetties containing weirs as a sand-bypassing measure. In addition, equations for the prediction of the details of the transport within the nearshore are required in some computer simulation models that analyze the morphology of beaches and the effects of structures (Chapter 10).

Stresses exerted by waves vary in the cross-shore direction (Chapter 7), generally decreasing from initial wave breaking to the shoreline but not necessarily in a uniform manner due to the presence of bars and troughs. The longshore current also has a characteristic profile (Chapter 8). Since the sand transport results from the combined waves and currents, its distribution will be the product of their respective distributions.

Komar (1977a) utilized the Bagnold (1963) model to derive a theoretical distribution for the longshore sand transport as the product of the local wave stress times the local current velocity. In the analysis, x is the cross-shore coordinate with 0 at the shoreline, and a uniform bottom slope is assumed. Based on the Bagnold model, the local immersed-weight transport rate, $i(x)$, was taken to be proportional to the product of the stress distribution $\tau(x)$ produced by the wave-orbital motions, and the longshore-current distribution, $v(x)$. Taking the wave stress as $\tau = 0.5 f_w \rho u_m^2$, where f_w is the drag coefficient for oscillatory wave motions (Jonsson, 1966; Kamphuis, 1975), and u_m is the local maximum wave-orbital velocity, the derivation leads to the relationship

$$i(x) = \frac{\pi}{4} k_1 (0.5 f_w) \rho \gamma^2 h(x) v(x) \tag{9.15}$$

where $h(x)$ is the local water depth that enters the equation since $u_m \propto \sqrt{gh}$ from linear-wave theory. The dimensionless proportionality coefficient k_1 was calibrated by integrating $i(x)$ across the surf zone and placing the result equal to the total transport I_ℓ from Equation (9.4). The longshore current distribution, $v(x)$, was calculated from the solution of Longuet-Higgins (1970) but with its coefficients evaluated in such a way that the magnitude of the current at the mid-surf position agrees with Equation (8.16) (see discussion in Chapter 8). The resulting solution of $i(x)$ is an analytical expression having only one free parameter, P, that depends on the degree of horizontal eddy mixing and determines the longshore current profile in the Longuet-Higgins solution [equation (8.16e)]. The magnitudes of the $v(x)$ and $i(x)$ distributions are governed by the wave-breaker height and breaker angle. An example of a sand-transport distribution calculated by Komar (1977a) using this approach is shown in Figure 9-18 for $H_b = 100$ cm and $\alpha_b = 10°$ and with a beach slope $S = 0.100$. The sand transport has been converted from the $i(x)$ distribution to a $q_s(x)$ volume-transport distribution. The significance of these distributions is that $v(x)$ at the mid-surf position agrees with Equation (8.14) and thus with measurements of longshore currents (Chapter 8), and if the $q_s(x)$ distribution is integrated across the surf zone, it yields the correct total longshore sand transport rate for those wave conditions. Examining the resulting $q_s(x)$ distribution, it can be seen that it achieves a maximum at a cross-shore position that is intermediate between the maximum for the longshore current distribution and the breaker line where the wave stress is a maximum. This of course reflects the fact that $q_s(x)$ is the product of $v(x)$ and the distribution of wave stresses.

Bowen (1980) and Bailard and Inman (1981) applied the sediment transport theories of Bagnold (1963, 1966) to analyses of the instantaneous sand transport in the nearshore, including both cross-shore and longshore components. Bowen and Doering (1984) discuss the

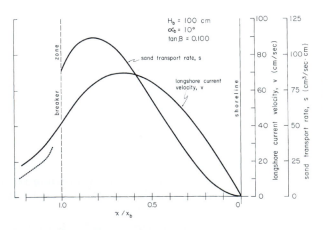

Figure 9-18 An example calculation of the longshore current and sand-transport distributions, the latter calculated with Equation (9.15) with the k_1 coefficient evaluated so that integration of the transport distribution yields the total transport according to Equation (9.5). [From Beach Sand Transport: Distribution and Total Drift, P. D. Komar, *Journal of Waterway, Port, Coastal and Ocean Division*, 1977. Reproduced with permission from the American Society of Civil Engineers.]

sand-transport processes, including the effects of wave asymmetries, undertow, and the down-slope component of gravity. That work has been used mainly in analyses of cross-shore sediment transport, and accordingly was reviewed in Chapter 7, but the analysis approach can also be applied in computations of the distribution of the longshore sediment transport. Holman and Bowen (1982) provide an example of an application of those equations in numerical models that simulate the growth of crescentic bars and oblique bars, respectively, beneath standing and progressive edge waves. Those models will be examined in Chapter 11, where we will consider the three-dimensional beach morphology.

Direct measurements of $q_s(x)$ distributions are difficult. Figure 9-19 is derived from the laboratory wave-basin study of Sawaragi and Deguchi (1978), which utilized circular bedload traps that enabled them to measure cross-shore sand movements as well as the longshore transport distribution. In this example the maximum sand transport occurs at approximately $X = x/X_b = 0.6$, while the maximum in the longshore current distribution is at $X = 0.4$. In a series of experiments, Sawaragi and Deguchi found that the results are relatively independent of the wave steepness but change with the sand-grain size. With finer sand the principal maximum in the sand-transport distribution shifts closer to the breaker zone ($X = 0.8$), with some tendency for a small secondary maximum at $X = 0.2$. The sand transport measurements were analyzed in a non-dimensional format, obtaining:

$$\frac{q_s(x)}{v(x)D_{50}} = 87\Phi^{3.7} \qquad \text{for } \Phi > 0.3 \qquad (9.16a)$$

$$= 23\Phi^{4.5} \qquad \text{for } \Phi < 0.3 \qquad (9.16b)$$

where

$$\Phi = \frac{\tau(x) - \tau_c}{(\rho_s - \rho)gD_{50}} \qquad (9.16c)$$

and D_{50} is the median sediment diameter, $\tau(x)$ is the bottom stress exerted by the waves, and τ_c is the critical threshold stress for the initiation of sand movement. In that both $v(x)$ and $\tau(x)$ vary across the surf zone, the $q_s(x)$ distribution is a function of both.

Equation (9.16) is a version of the Kalinske-Brown formula for sediment transport in rivers, adapted for use on beaches. The principal adaptation is that the stresses, $\tau(x)$, are evaluated so as to include both wave and current effects. Iwagaki and Sawaragi (1962)

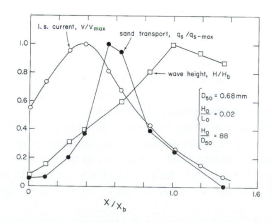

Figure 9-19 Laboratory measurements of the distributions of the longshore current and sand transport rates, based on the data of Sawaragi and Deguchi (1978).

had earlier attempted to apply a modified Kalinske-Brown formula for estimating sand-transport rates on beaches, the first to so modify a river-based equation. Madsen and Grant (1976) provide a later adaptation of the Kalinske-Brown equation to the general problem of sand transport under combined waves and currents, one application of which is to sand transport on beaches. The approach of Bijker (1968, 1971) for evaluating the longshore transport, used by many European engineers, also employs a modified river-based sediment transport relationship, in this case the bedload formula of Frijlink together with Einstein's method for evaluating the suspended-load transport. Bijker also undertook laboratory measurements of longshore sand-transport rates, and the results are similar to those of Sawaragi and Deguchi (1978). However, Bijker's measured distributions differ considerably from those calculated theoretically.

Tsuchiya (1982) provides a set of laboratory measurements of the cross-shore distribution of the longshore sediment transport but does not give information concerning how the data were collected. The transport rate within the distribution reaches a maximum at $X = x/X_b = 0.75$–0.8, so it does show the expected overall distribution as was already seen in Figures 9-18 and 9-19. Tsuchiya developed a theoretical model wherein the local longshore transport is given by

$$q_s(x) = C(x)h(x)v(x) \tag{9.17}$$

where $h(x)$ is the local water depth, $v(x)$ is again the longshore current as given by the Longuet-Higgins solution, and $C(x)$ is the depth-averaged sediment concentration. Based on earlier work, this sediment concentration was evaluated as

$$C(x) = \frac{0.2}{\sigma_s}\left[1 - \frac{\theta(x)}{\theta_c}\right] \tag{9.18}$$

where σ_s is the specific gravity of the sediment and $\theta(x) = \tau(x)/(\rho_s = \rho)gD_{50}$ is the Shields function, θ_c being the critical threshold value. Tsuchiya calculated three representative $q_s(x)$ profiles that agree reasonably well with the measured distribution.

Kamphuis (1991) presents a number of measured cross-shore distributions of the longshore sediment transport, derived from wave-basin experiments. The distributions generally were found to be bimodal, with one peak close to the breaker zone and the second in the swash zone. With smaller wave heights and smaller grain sizes, the two peaks come closer together and tend to blend into a single peak. None of the existing theoretical equations for the cross-shore distributions of longshore transport were successful in predicting the measured bimodal distributions.

The collection of field data for cross-shore distributions of longshore sand-transport rates is difficult and attempts have involved a range of techniques. The earliest approach was to use sand tracers. Zenkovitch (1960) determined the distribution shown in Figure 9-20, the result from averaging a large number of observations. In that example there are three maxima in the transport distribution, two over offshore bars, and a third in the swash zone. These correspond to zones of maximum concentrations of suspended sediments, although the proportions are different (the maximum concentration of suspended sediment is in the swash zone, but the maximum in the longshore transport rate is over the seaward-most bar). Kraus et al. (1982) provide a more recent example of the use of sand tracers to investigate the pro-

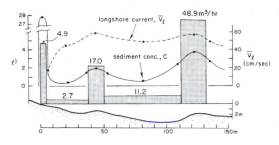

Figure 9-20 Distributions of the longshore current, the concentrations of suspended sediments, and of the longshore transport as determined with sand tracers. [Adapted with permission of C. Griffin, from V. P. Zenkovitch, Fluorescent Substances as Tracers for Studying the Movement of Sand on the Sea Bed, Experiments Conducted in the USSR, *Dock and Harbor Authority.* Copyright © 1960 C. Griffin.]

file of the longshore transport, in their case employing tracers with four distinct colors injected on a line crossing the surf zone.

Thornton (1972) employed sediment traps on a Florida beach to obtain measurements of the transport distribution. His data show a maximum transport on the seaward side of the offshore bar where the waves were breaking, with decreasing rates both shoreward and seaward of the breaker zone. However, only one sample location was situated shoreward of the breakers, so the overall distribution was not established. A theoretical analysis of the $q_s(x)$ distribution is presented by Thornton, one which again is based on the model of Bagnold (1963).

An interesting approach to measure the sediment transport distribution is that of Bodge and Dean (1987), which utilized a "short-term sediment impoundment" scheme consisting of the rapid deployment of a low-profile, shore-perpendicular barrier (a short groyne). Beach profile changes in the vicinity of the barrier were determined from repeated surveys over short intervals of time. A series of laboratory experiments were completed, and field measurements were undertaken at the Field Research Facility (FRF), Duck, North Carolina. One profile from the field experiments is given in Figure 9-21, showing a maximum in the outer surf zone just shoreward of the breaker zone and a second maximum in the swash zone ($X = 0$ is the still-water shoreline). The laboratory experiments were designed to examine changes in the transport profiles with varying breaker types. The results in general show bimodal distributions such as those in Figure 9-21, but with the relative significance of the peaks being a function of the breaker type. As the breakers vary from spilling to collapsing, the outer peak near the breaker zone decreases while that in the swash zone increases. Longshore transport in the swash zone was found to represent some 5 percent of the total transport with spilling breakers, increasing to over 60 percent for collapsing breakers. The longshore transport seaward of the breakpoint was found to represent about 10–20 percent of the total transport.

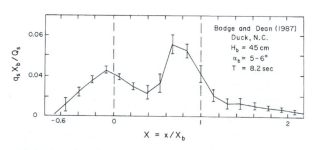

Figure 9-21 The cross-shore profile of the longshore sand transport measured at the Field Research Facility, Duck, N.C., using a deployable shore-perpendicular barrier (groyne). The error bars are due to uncertainties in the updrift limit of impoundment and the degree of groyne bypassing. [Adapted from Short-term Impoundment of Longshore Transport, K. R. Bodge and R. G. Dean, *Coastal Sediments '87*, 1987. Reproduced with permission from the American Society of Civil Engineers.]

THE MODES OF SEDIMENT TRANSPORT

The transport of sediment generally is categorized as consisting of a bedload and a suspension transport, the distinction being based on the primary mechanisms of grain movement. The bedload consists of grains that are transported by rolling or bouncing or where they are otherwise in frequent contact with the bed. In contrast, the suspended load is carried along within the water column, held above the bed by the turbulence of the flow. Uncertainty remains and differences of opinion exist as to the relative importance of bedload versus suspension transport on beaches.

The data reviewed thus far to establish the longshore transport of sediments on beaches, mainly derived using tracers or from the blockage of the transport by a jetty or breakwater, reflect the total transport, the sum of the bedload and suspended load. Of these two separate transport modes, it is much easier to measure the fraction in suspension, and this has been accomplished through a variety of techniques. The earliest approach was to pump water containing suspended sand from the surf zone (Beach Erosion Board, 1933; Watts, 1953b; Fairchild, 1972, 1977). The advantage of this technique is that large quantities can be processed, leading to some confidence that the samples are representative of concentrations found within the surf. However, comparisons between pumping and more sophisticated optical techniques of measuring suspension concentrations indicate that in the surf zone, pumping tends to over sample the suspended sediments (Black and Rosenberg, 1994). The main disadvantages of pumping are that one cannot investigate time variations in sediment concentrations at different phases of the wave motions, and the sampling programs of the early studies were limited to the collection from piers which likely influenced the results. Pumping did yield the first information on distributions of suspended sand across the beach profile, demonstrating that the highest concentrations are found within the breaker zone and at the base of the swash zone, areas of highest wave dissipation and turbulence. Attempts to relate the concentrations to factors such as wave heights yielded only rough trends due to the considerable scatter of the data (Beach Erosion Board, 1933; Watts, 1953b; Fairchild, 1972, 1977).

A more recent study based on pumped samples is that of Coakley and Skafel (1982), an investigation on a Lake Ontario beach. Their study utilized an instrumented sled that could be moved progressively across the width of the surf zone. Their results included the determination of a relationship between the mean suspension concentration for the nearshore as a whole and the energy density of the breaking waves.

Another method for measuring suspension concentrations is with "traps," usually consisting of a vertical array of three or four sample collectors, which can be used to examine the vertical distribution of suspended sediment. The samplers can be positioned at various locations across the surf zone and can be triggered when either a wave crest or trough passes. Fukushima and Kashiwamura (1959), Hom-ma and Horikawa (1962), and Hom-ma, Horikawa, and Kajima (1965) used samplers made of bamboo poles to examine vertical distributions of suspended sediments. Kana (1977, 1978) undertook a study on the coast of South Carolina, employing the suspension sampler shown in Figure 9-22 and described by Kana (1976). As shown in Figure 9-23, he found the expected upward decrease in concentrations with elevation above the bottom and found systematic differences depending on whether the waves are breaking by plunging or spilling, the plunging breakers throwing considerably more sand into suspension due to their intense interaction with the bottom. One unexpected result was that the concentration of suspended sediment decreased with increasing wave height, at least under the moderate wave energies prevailing during the experiments

Figure 9-22 The volume sampler apparatus of Kana (1976) designed to measure the vertical distribution of suspended sediment concentrations within the surf zone. The suspended sediment enters each trap from the side, which is then captured when the lids are activated. [Adapted with permission of the Society for Sedimentary Geology, from T. W. Kana, A New Apparatus for Collecting Simultaneous Water Samples in the Surf Zone, *Journal of Sedimentary Petrology* 46. Copyright © 1976 Society for Sedimentary Geology.]

($H_b < 1.5$ m). Under plunging waves the maximum concentrations were found approximately 3–5 m shoreward from the breakpoint, this corresponding to the plunge distance of the downward movement of the crest (Chapter 6), while under spilling waves the concentration was more uniform across the surf zone. Another study that used suspension traps was

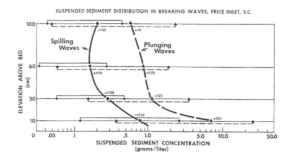

Figure 9-23   Vertical distributions of suspended-sediment concentrations showing approximately exponential decreases with distance above the bottom, with higher concentrations prevailing under plunging than under spilling breakers. [From Suspended Sediment Transport at Price Inlet, SC, T. W. Kana, *Coastal Sediments '77*, 1977. Reproduced with permission from the American Society of Civil Engineers.]

that of Inman et al. (1980), who reported on vertical distributions of sediment suspension across the surf zone at Torrey Pines Beach, California.

A number of studies have turned to the use of optical techniques for measuring suspension concentrations in the nearshore. Brenninkmeyer (1974, 1976) and Leonard and Brenninkmeyer (1978) employed opacity measuring devices, termed almometers, to determine instantaneous and continuous concentrations of suspended sediment across the surf zone. Their particular device was limited to measuring concentrations greater than 10 grams of sediment per liter of water and could only detect the occurrences of relatively high concentrations, which they termed "sand fountains." Brenninkmeyer found from the spectra of his measurements that during normal conditions, suspension is a low-frequency event with most of the movement centered at frequencies less than 0.25 Hz (periods > 4 sec), with the relatively high-frequency component of the waves contributing little to the total quantity of sediment being transported. In contrast, Leonard and Brenninkmeyer found that under storm conditions, higher-frequency sand movement is more common but is still not controlled by the prevailing wave and swell periods. It was found that sand is rarely thrown into suspension in the breaker zone at high concentrations detectable by the almometer and is highest in the transition zone at the base of the swash zone where the return backwash collides with incoming wave bores. Subsequent studies [e.g., Jaffe and Sallenger (1992)] have further established that the suspension of sand within the surf zone is intermittent, occurring as short periods of intense suspension events when concentrations are an order-of-magnitude greater than the mean.

Thornton and Morris (1977) measured suspension concentrations both optically and by pumping at Torrey Pines Beach, California. They were able to detect lower concentrations than measured with the almometer of Brenninkmeyer (1974, 1976), detecting mean concentrations that ranged between 0.05 and 0.32 grams of sand per liter. Spectral analyses were performed on the concentration measurements, as well as on simultaneous current measurements obtained with an electromagnetic current meter. It was found that the peaks of the suspension spectra occur at approximately twice the peak frequency of the velocity spectra, indicating that two or more suspension events occur per wave cycle. Suspension concentrations decreased exponentially with height above the bottom, and some correlation was found between the total concentration and the mean bed stress exerted by the waves.

A similar study was undertaken by Katoh, Tanaka, and Irie (1984), using an optical "densitometer" to measure suspended-sediment concentrations. Figure 9-24 shows typical concentration measurements obtained within the breaker zone, together with cross-shore

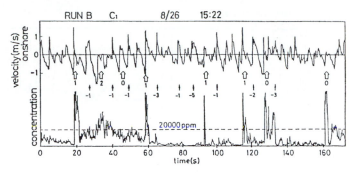

Figure 9-24 The occurrence of high-concentration suspension events in the breaker zone relative to whether individual breakers are seaward (open arrows) or landward (black arrows) of the instrumentation. [From Field Observation on Suspended-Load in the Surf Zone, K. Katoh, N. Tanaka and I. Irie, *Proceedings of the 19th Coastal Engineering Conference,* 1984. Reproduced with permission from the American Society of Civil Engineers.]

water velocities determined with an electromagnetic current meter. Of particular interest is the relationship of the suspension events to individual breaking waves whose positions were noted. The arrows in the graph show times of wave breaking, the open arrows signifying that the breaking took place seaward of the instruments while the black arrows indicate that they broke to the shoreward side (the numbers below the arrows give the distances in meters). All breakers were judged as plunging. It is apparent that the major suspension events resulted from wave breaking seaward of the instruments. Also of interest is the short duration of these high-concentration suspension events, the sediment rapidly settling and returning to the bottom, with little tendency to saturate the nearshore with suspended sand.

Downing (1984) and Sternberg, Shi, and Downing (1984) report on suspension measurements, respectively, at Twin Harbor Beach, Washington, and Santa Barbara, California. Both studies used optical backscatter sensors (OBS) to measure suspension concentrations, a small device (19 mm in diameter by 11 mm length) whose response is linear within the normal concentration range and can measure levels down to 0.1 parts per thousand (Downing, Sternberg, and Lister, 1981). The investigation of Sternberg and co-workers was part of the Nearshore Sediment Transport Study (NSTS) at Santa Barbara, so there was considerable supporting data establishing the wave and current conditions. The study employed several arrays of OBS sensors across the surf-zone, each array consisting of a vertical series of five sensors. Figure 9-25 is an example of the type of measurements and analysis. Moving vertically through this 7-minute time series, h is the instantaneous water depth at the array station, u and v are the cross-shore and alongshore components of the water motions measured with an electromagnetic current meter, BMO1 through BMO5 are outputs from the OBS sensors, G_s is the computed depth-integrated immersed weight suspended-sediment load, and i_s is the longshore transport of that load obtained by multiplying G_s by the instantaneous longshore current velocity v. The long-term net suspension transport is then determined by integrating i_s over the duration of a run, approximately 34 minutes. Apparent in Figure 9-25 is the vertical decrease in suspension concentrations, being highest at the BMO1 sensor at 3.5 cm elevation and lowest at the BMO5 sensor (19.5 cm). Concentrations near the seabed were found to vary over a wide range, and reached 180 kg/m^3 during individual

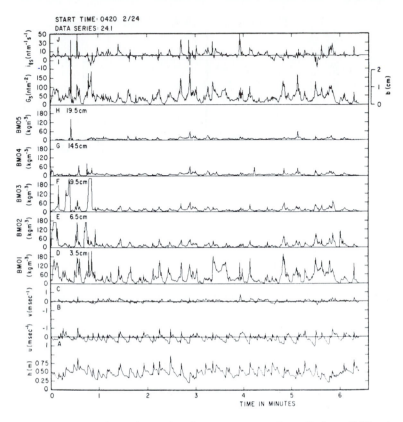

Figure 9-25 A 7–minute time series of measurements at Santa Barbara, California, of water motions and suspension concentrations, together with calculated longshore fluxes of the suspended sediments. [From Field Investigation of Suspended Sediment Transport in the Nearshore Zone, R. W. Sternberg, N. C. Shi, and J. P. Downing, *Proceedings of the 19th Coastal Engineering Conference,* 1984. Reproduced with permission from American Society of Civil Engineers.]

suspension events. The analysis indicates that the depth of erosion required to support the observed suspended load is approximately 1–2 cm; this is the *b* parameter in Figure 9-25 that is directly proportional to G_S. Individual suspension events were in phase with bores propagating across the surf zone. Low-frequency oscillations were observed in the measurements of water levels, the cross-shore component of the water velocity, and the resulting sediment suspension.

The Santa Barbara beach studied by Sternberg, Shi, and Downing (1984) has a steep slope and narrow surf zone, being closer to a reflective than a dissipative beach. In contrast, the beach studied by Downing (1984) on the Washington coast is characteristic of dissipative beaches, the average slope being low (0.02) with a very wide surf zone. On such beaches, low-frequency infragravity water motions ($f < 0.05$ Hz or $T > 20$ sec) dominate the inner surf zone where the energy of the incident bores is much reduced (Chapter 6). Downing found that sand suspension is correlated with strong offshore flows that recur at about one-fifth the incident wave frequency. Therefore, infragravity water motions are more important to sediment suspension than incident waves within the inner surf zone of dissipative beaches. Further evidence for the importance of infragravity motions to sediment transport on dissipative beaches is provided by the studies of Beach and Sternberg (1987, 1988, 1992) on the Oregon coast. Measurements again were obtained using arrays of OBS sensors to determine concentrations of suspended sediments, while currents were measured simultaneously with electromagnetic current meters. In their measurements on dissipative beaches, Beach and Sternberg found that infragravity water motions accounted for more than 85 percent of the spectral energy within the inner surf zone and suspended sediment loads were 3–4 times larger than those associated with incident waves. Much of their research focused on the response of the suspension transport within the surf zone to interactions between the waves and longshore current. In terms of the longshore transport of the suspended load, they found that the contribution by the longshore current is approximately 1.7 times the wave contribution (63 versus 37 percent), primarily because the longshore current carries suspended sediment higher within the water column where it is transported by stronger longshore currents. Their measurements also established that the longshore transport can be evaluated either as $\overline{C}\overline{v}$ or $\overline{Cv}$, where C and v are, respectively, the measured concentrations of suspended sediments and the longshore current velocity, that is, the transport can be calculated as the product of the mean concentration and mean current or as the mean of the product of the instantaneous values. Conformation of this detail is important to the use of measurements of suspension concentrations to evaluate the longshore suspension transport and in its comparisons with the bedload transport or total transport.

The OBS measurements of Sternberg, Shi, and Downing (1984) and of Beach and Sternberg (1992) further confirmed that there is a marked decrease in suspended-sediment concentrations upward from the bottom, already seen in Figure 9-23 from the sediment trap data of Kana (1977, 1978). The full profile of the vertical distributions of suspension concentrations can be obtained with acoustic concentrations meters (ACM), based on the backscatter of acoustic energy in a vertical beam (Young et al., 1982). The ACM yields the complete vertical distribution, while OBS sensors obtain data only at the specific elevations where they are mounted. Figure 9-26 shows gray-scale plots of suspended sand concentrations and

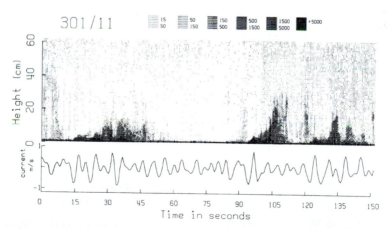

Figure 9-26 Gray-scale plots of concentrations (mg/liter) of suspended sediments vertically above the bed, and the current speeds, measured offshore from a beach in Nova Scotia. [From *Marine Geology* 96, C. E. Vincent, D. M. Hanes and A. J. Bowen, Acoustic Measurements of Suspended Sand on the Shoreface and the Control of Concentration by Bed Roughness, 1991 kind permission of Elsevier Science-NL, Sara Burgerhartstratt 25, 1055 KV Amsterdam, The Netherlands.]

current speeds measured by Vincent, Hanes, and Bowen (1991) in about 5 m water depth outside of the surf zone at Queensland Beach, Nova Scotia, Canada. The results provide a graphic representation of the concentrations within 60 cm of the bottom and their variations with the wave-induced oscillatory currents. Hanes et al. (1988) used the ACM device to obtain measurements of suspension concentrations within the surf zone of a beach on Prince Edward Island in the Gulf of St. Lawrence, Canada. Figure 9-27 shows vertical profiles of suspended sand concentrations averaged over 7 minutes of data collection, obtained under different wave conditions. All profiles show a remarkable increase in concentrations within about 5 cm of the bottom, with more gradual decreases farther above the bed. One advantage of acoustic concentration meters is that they provide an accurate determination of the elevation of the bottom and how it changes with time. One problem with the ACM sensors is that air bubbles in the water, common within the surf zone, can contaminate the measurements of suspended sediment concentrations.

In all of the techniques discussed thus far, the measurements were limited to determinations of suspended sediment concentrations, the quantities of sediment per unit volume of water. To use that data to evaluate longshore or cross-shore sediment transport rates, separate measurements of velocities are needed, generally obtained with electromagnetic current meters. A different approach has been employed by Kraus and Dean (1987), involving the use of vertical arrays of sediment traps called "streamers," shown in Figure 9-28. Each streamer is made of flexible polyester filter cloth of 0.105-mm mesh, which allows water to pass through but retains sand-size sediment. In practice the traps are aligned with the streamer mouths facing into the longshore current and trap a time-averaged sediment flux. This yields the suspension flux directly, not just the concentration, so it is not necessary to measure the local longshore current velocity. An example of their results is shown in Figure 9-29, the measured vertical distributions of longshore sediment fluxes across the width of the surf zone. The vertical distributions were found to follow exponential decreases with distance above the bed, the pattern typical of suspension transport.

These various measurements of suspension concentrations and transport serve as the primary basis for assessments of the importance of suspension versus bedload transport. Unfortunately, no reliable techniques have been developed to measure the bedload alone, so the comparisons generally involve an assessment of what fraction the suspension transport represents of the total transport. In most cases the longshore suspension transport is evaluated with the formula

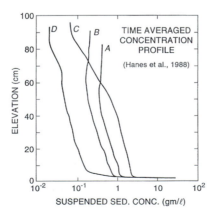

Figure 9-27 Vertical profiles of suspended sand concentrations measured with an acoustic concentration meter and averaged over 7 minutes. [Adapted from *Marine Geology* 81, D. M. Hanes et al., Acoustic Measurements of Suspended Sand Concentration in the C²S² Experiment at Stanhope Lane, Prince Edward Island, 1988 with kind permission of Elsevier Science-NL, Sara Burgerhartstratt 25, 1055 KV Amsterdam, The Netherlands.]

Figure 9-28 An array of "streamers" used by Kraus and Dean (1987) to measure the longshore flux of suspension transport within the surf zone. [Courtesy of R. Holman]

$$\text{suspension flux} = \overline{C}vA \tag{9.19}$$

where $\overline{C}$ is the average volume concentration of sand in suspension, $\overline{v}$ is the mean longshore current, and A is the total cross-sectional area of the nearshore region from the shoreline through the breaker zone (Dean, 1973; Galvin, 1972). This provides only a rough estimate, a more refined evaluation involving the integration of $C(x,z,t)v(x,z,t)$ across the surf-zone area. Employing such a relationship and measurements of suspension concentrations, Kana (1978) estimated daily longshore transport rates that were on the same order as the total transport evaluated with Equation (9.4). From this Kana concluded that the suspended load accounts for the major portion of longshore sand transport and that the bedload contribution must therefore be small. However, in an independent analysis undertaken at the same time, Komar (1978) arrived at just the opposite conclusion. Utilizing Equation (9.19) to evaluate the suspension transport, Equation (8.14) for the evaluation of the mean longshore current, and then dividing by I_ℓ of Equation (9.4) for the total transport, the ratio of suspension transport to the total transport was found to be

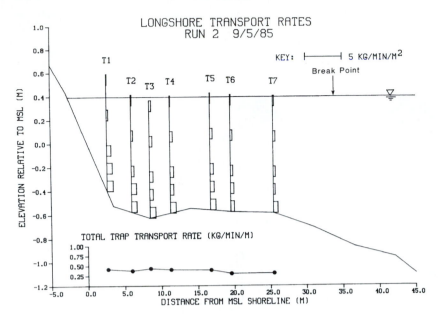

Figure 9-29 Vertical distributions of longshore sediment fluxes across the width of the surf zone at the Field Research Facility, Duck, N.C., measured with the streamer apparatus of Figure 9-28. [From Longshore Sediment Transport Rate Distributions Measured by Trap, N. C. Kraus and J. L. Dean, *Coastal Sediments '87,* 1987. Reproduced with permission from the American Society of Civil Engineers.]

$$\frac{I_{susp}}{I_\ell(total)} = 7.0 \frac{\rho_s - \rho}{\rho} \frac{\overline{C}}{\gamma S} \tag{9.20}$$

where S is the beach slope, and $\gamma \approx 1$ is the breaker height to depth ratio. Employing the measured values of sediment concentrations $\overline{C}$ determined by various investigators, it was found that this relationship yields ratios in the range 0.07–0.26, suggesting that suspension transport is comparatively low, less than 25 percent of the total transport rate, the bedload forming the remaining 75 percent or more.

Subsequent studies have continued the disagreement as to the relative proportions of suspension and bedload transport. These are summarized in Table 9-3 as $I_{susp}/I_\ell(total)$ ratios. Based on their simultaneous measurements of the total transport using sand tracers, and the amount of sand in suspension determined with traps, Inman et al. (1980) concluded that suspension transport accounts for approximately 15–20 percent of the total transport. Kana and Ward (1980) measured distributions of suspended sediments along the pier at the FRF, Duck, North Carolina, using the trap devices of Figure 9-22. Of interest were the contrasting conditions during a storm and then the post-storm period. The total load of sediment in suspension was approximately 10 times higher during the storm than during the post-storm period, and the longshore flux (concentration times the longshore current) was as much as 60 times more during the storm than in the post-storm period. According to their estimates, effectively all of the transport during the storm occurred in suspension, whereas during post-storm conditions less than one-third of the total transport was suspended load.

Sternberg et al. (1984) concluded that effectively all of the transport during the NSTS experiments in Santa Barbara occurred as suspended load. They noted that much of the dif-

TABLE 9–3 RATIO OF SUSPENSION TRANSPORT TO THE TOTAL LONGSHORE SEDIMENT TRANSPORT [Used with permission of Gulf Publishing Company, from P. D. Komar, Littoral Sediment Transport, *Handbook on Coastal and Ocean Engineering,* Copyright © 1990 Gulf Publishing Co.]

Study	Location	Lowest Measurement (cm)		I_{susp}/I_ℓ(Total)
Komar (1978)	various[1]	7.6–10		0.07–0.26
Inman et al. (1980)	Torrey Pines, CA	10.0		0.15–0.20
Kana and Ward (1980)	Duck, NC	5.0	storm	1.0
			post-storm	0.30
Cloakley and Skafel (1982)	Lake Ontario	10.0		0.5
Sternberg et al. (1984)	Santa Barbara, CA	3.5		1.0
Downing (1984)	Twin Harbor, WA	3.5		0.47
Kraus and Dean (1987)	Duck, NC	0.0		0.90

[1]Included the data of Watts (1953b) at Pacific Beach, CA, the data of Fairchild (1972) at Ventnor, NJ, and Nags Head, NC, and the data of Kana (1976) at Price Inlet, SC.

ference in results between the various studies is produced by sampling techniques, especially how close the lower most concentration measurement is to the bed. For example, their lowest sensor was at 3.5 cm above the bed, while in the study of Inman et al. (1980) suspension concentrations were determined only down to 10 cm from the bed. This lower level is critical, due to the rapid increase in concentrations as the bed is approached (e.g., Fig. 9-27). Semantic problems are introduced between definitions of suspension transport versus bedload within this lower region, but Sternberg et al. noted that the maximum volume concentration measured in their study at the 3.5–cm level was 0.07, which is below the 0.08 concentration which would produce significant intergranular collisions indicative of bedload transport. However, Downing (1984) used the same techniques as Sternberg et al., including measurements of concentrations down to the 3.5–cm level, but concluded that the suspension transport is about 47 percent of the total transport on the Washington beach of his study. Therefore, the closeness of the measurements to the bed cannot be the only factor causing differences in results between the various studies.

The final study listed in Table 9-3 is that of Kraus and Dean (1987), which also took place at the FRF and used streamers (Fig. 9-28). It is believed that the bottom most sampler collects bedload as well as some suspended load, and on that basis an approximate 90 percent proportion of suspension transport was estimated, the value given in Table 9-3.

Another indication of the relative importance of suspension versus bedload transport comes from the dispersal of sand tracers utilized in measurements of the total transport. Komar (1978) examined the farthest longshore displacements of tracer grains found in the grid sampling results from Komar and Inman (1970). That distance yields the maximum advection velocity, U_{max}, of tracer on the beach, which was compared with the measured longshore current velocity, $\bar{v}_\ell$. The reasoning was that if the ratio $U_{max}/\bar{v}_\ell$ is small, it must indicate that the sand lags well behind the transporting current and that there cannot have been a substantial suspension transport. At El Moreno Beach in the study of Komar and Inman, a coarse-grained reflective beach, $U_{max}/\bar{v}_\ell$ ranged 0.025–0.23 while $\bar{U}/\bar{v}_\ell$ ranged 0.0033–0.013, where $\bar{U}$ is the average tracer advection rate. These values indicate a considerable lag of sand movement behind the longshore current. Furthermore, it was found that the coarsest grains within the natural mixture of sizes present on the beach out distance the finer grains in their longshore movement (Komar, 1977b), a result that would be uncharacteristic of suspension transport. At Silver Strand Beach, a medium-grained dissipative beach, the $U_{max}/\bar{v}_\ell$ ratios are more uncertain, appearing to be on the order of 0.3–0.6, with $\overline{U}/\bar{v}_\ell$ in the range 0.0095–0.0125, still suggesting a considerable lag of sand movement relative to the longshore current. These results were interpreted to indicate that the suspended load could not be the primary mode of transport on beaches.

Considering the range of results and contrasting conclusions of the various studies, the question of the relative roles of suspension versus bedload transport on beaches must remain open and thus in need of additional research. Furthermore, nearly all of the data reflecting on this question have come from beaches composed of fine- to medium-grained sand; similar data are needed from beaches that consist of coarser sediments. In all probability, the results will depend greatly on the wave conditions as well as the sediment grain sizes. This is already evident in the results of Kana and Ward (1980), discussed above, where suspension transport prevailed during a storm, while bedload was more important during the post-storm interval. More consideration also needs to be given toward the basic definition of "bedload" versus "suspension," especially in light of the extreme variations in concentrations with distance above the bed, the uncertainty being as to where the high-concentration bedload gives way to the lower concentrations of the suspension transport.

SUMMARY

The longshore transport of sediment on beaches becomes a problem whenever this natural movement is prevented by the construction of jetties or breakwaters. Such structures act as dams to the sediment movement, causing a buildup of the beach on the structure's up-drift side and the simultaneous erosion in the downdrift direction. There are many examples of this structure-induced erosion along our coasts, in some cases the impacts to communities having been devastating. In order to anticipate such impacts, it is desirable to have the ability to predict quantities of the longshore sediment transport. The focus of this chapter has been on a review of the data that correlate the rate of sediment transport to the causative waves and nearshore currents. One equation based on that data relates the sand-transport rate to the "longshore component of the wave power," P_ℓ of Equation (9.2). The derivative relationships [Equations (9.5) though (9.7)] permit one to calculate the total transport from measurements of wave-breaker heights and angles. Such relationships have been those most commonly used in applications. However, it has been argued that the Bagnold relationship [Equation (9.8) and its derivative forms Equations (9.9) and (9.10)] are more fundamental in having accounted for the sediment transport in terms of the combined processes of waves and nearshore currents. Those relationships can be employed irrespective of the generation mechanism of the longshore current, whereas the wave power approach is applicable solely to the case where the current is driven by an oblique-wave approach to the shoreline. The Bagnold relationship should be used more frequently in applications.

Much remains to be learned about the processes of sediment transport on beaches, and improvements are needed in the accuracy to which we can estimate the transport. We still lack basic information regarding how the transport rates vary with sediment grain sizes; nearly all of our data have come from sand beaches and only recently have measurements been made of gravel transport rates. Although there have been a number of investigations of suspension transport on beaches, utilizing a variety of measurement techniques, the results are open to interpretation and have led to debate regarding the relative importance of bedload versus suspension transport. Fundamental questions therefore remain regarding the mechanisms of sediment transport on beaches.

PROBLEMS

1. Neglecting cross-shore sediment transport or other causative factors, how does the erosion, stability, or accretion of the beach at a specific location relate to the longshore sediment transport? It sometimes is assumed that the higher the longshore sediment transport rate, the greater the resulting beach erosion. Is this necessarily the case?

2. Work out the units for the parameters Q_ℓ, I_ℓ, and P_ℓ in the mks (meters-kilograms-seconds) metric system.

3. Why is the pore-space correction factor, a', included in Equation (9.3) for the immersed-weight sediment transport rate?

4. For a certain immersed-weight longshore sediment transport rate, governed by the prevailing wave conditions, how would the transport of quartz-density (2650 kg/m^3) sand compare with the transport on a beach composed of basalt-density sand (3500 kg/m^3)?

5. Waves breaking on a quartz-sand beach have a significant wave height of 4.0 m and a mean breaker angle of 5°. Evaluate the longshore immersed-weight and volume transport rates for these wave conditions. Later the angle increases to 10° while maintaining the same height. What is the transport under these new conditions?

6. According to Equations (9.5) and (9.6) for evaluating of the longshore sediment transport rate, by what factors will the transport rates increase if the wave-breaker heights are doubled or tripled, while maintaining the breaker angle constant?

7. A particular observer has an uncertainty of 2° in making visual estimates of wave breaker angles. If the significant wave breaker height is 2.5 m, what are the mean and range of calculated longshore volume transport rates corresponding to his angle estimate $\alpha_b = 7° \pm 2°$? If the mean angle increases to 15° while maintaining the same uncertainty, what now is the range of estimated transport rates?

8. Waves break on a quartz-sand beach with a significant wave height of 3.5 m and interact with a longshore current of 0.50 m/sec as measured at the mid-surf position. Evaluate the longshore immersed-weight and volume transport rates.

9. A longshore current flows toward the opening of a breakwater, even though the waves break parallel to shore. The magnitude of the current is 0.25 m/sec and the root-mean-square breaker height is 2.6 m. Calculate the daily and monthly volumes of quartz sand being carried into the harbor under these conditions.

10. Beginning with an initially straight shoreline and a cell circulation with rip currents and feeder longshore currents, what will be the sand transport pattern and how will the shoreline be modified? If you knew the magnitudes of the longshore current velocities and wave-breaker heights, how would you evaluate the spatially variable longshore sediment transport rates and the resulting shoreline changes?

11. A common practice in beach nourishment programs is to add sand to the beach that is somewhat coarser grained than the native sand (Chapter 12). How might this affect the morphology of the beach and longshore sediment transport rates, assuming that the wave conditions remain the same after the project?

REFERENCES

BAGNOLD, R. A. (1963). Mechanics of Marine Sedimentation. In *The Sea,* M.N. Hill (editor). Vol. 6, pp. 507–582, New York: Wiley-Interscience.

BAGNOLD, R. A. (1966). *An Approach to the Sediment Transport Problem from General Physics.* U.S. Geological Survey, Professional Paper 422–I.

BAILARD, J. A. and D. L. INMAN (1981). An Energetics Bedload Model for a Plane Sloping Beach: Local Transport. *Journal of Geophysical Research* 86: 2035–2043.

BEACH EROSION BOARD (1933). *Interim Report.* U.S. Army Beach Erosion Board, Office of Chief Engineer.

BEACH, R. A. and R. W. STERNBERG (1987). The Influence of Infragravity Motions on Suspended Sediment Transport in the Inner Surf Zone. *Coastal Sediments '87, Amer. Soc. Civil Engrs.,* pp. 913–928.

BEACH, R. A. and R. W. STERNBERG (1988). Suspended Sediment Transport in the Surf Zone: Response to Cross-Shore Infragravity Motions. *Marine Geology* 80: 671–679.

BEACH, R. A. and R. W. STERNBERG (1992). Suspended Sediment Transport in the Surf Zone: Response to Incident Wave and Longshore Current Interaction. *Marine Geology* 108: 275–294.

BIJKER, E. W. (1968). Littoral Drift as Function of Waves and Currents. *Proceedings of the 11th Coastal Engineering Conference, Amer. Soc. Civil Engrs.,* pp. 415–433.

BIJKER, E. W. (1971). Longshore Transport Computation. *Journal of Waterways, Harbors, and Coastal Engineering Div., Amer. Soc. Civil Engrs.* 97(WW4): 687–701.

BLACK, K. P. and M. A. ROSENBERG (1994). Suspended Sand Measurements in a Turbulent Environment: Field Comparison of Optical and Pump Sampling Techniques. *Coastal Engineering* 24: 137–150.

BODGE, K. R. and R. G. DEAN (1987). Short-Term Impoundment of Longshore Transport. *Coastal Sediments '87, Amer. Soc. Civil Engrs.,* pp. 469–483.

BODGE, K. R. and N. C. KRAUS (1991). Critical Examination of Longshore Transport Rate Magnitude. *Coastal Sediments '91, Amer. Soc. Civil Engrs.,* pp. 139–155.

BOWEN, A. J. (1980). Simple Models of Nearshore Sedimentation. Beach Profiles and Longshore Bars. In *The Coastline of Canada,* S. B. McCann (editor). Geological Survey of Canada, pp. 1–11.

BOWEN, A. J. and D. L. INMAN (1966). *Budget of Littoral Sands in the Vicinity of Point Arguello, California.* U.S. Army Coastal Engineering Research Center Technical Memo No. 19.

BOWEN, A. J. and J. C. DOERING (1984). Nearshore Sediment Transport: Estimates from Detailed Measurements of the Nearshore Velocity Field. *Proceedings of the 19th Coastal Engineering Conference, Amer. Soc. Civil Engrs.,* pp. 1703–1714.

BRENNINKMEYER, B. M. (1974). Mode and Period of Sand Transport in the Surf Zone. *Proceedings of the 14th Coastal Engineering Conference, Amer. Soc. Civil Engrs.,* pp. 812–827.

BRENNINKMEYER, B. M. (1976). In Situ Measurements of Rapidly Fluctuating, High Sediment Concentrations. *Marine Geology* 20: 117–128.

BRUNO, R. O. and C. G. GABLE (1976). Longshore Transport at a Total Littoral Barrier. *Proceedings of the 15th Coastal Engineering Conference, Amer. Soc. Civil Engrs.,* pp. 1203–1222.

BRUNO, R. O., R. G. DEAN, and C. G. GABLE (1980). Littoral Transport Evaluations at a Detached Breakwater. *Proceedings of the 17th Coastal Engineering Conference, Amer. Soc. Civil Engrs.,* pp. 1453–1475.

BRUNO, R. O., R. G. DEAN, C. G. GABLE, and T. L. WALTON (1981). *Longshore Sand Transport Study at Channel Island Harbor, California.* U.S. Army Corps of Engineers, Coastal Engineer Resource Center Technical Paper No. 81–2.

CALDWELL, J. (1956). *Wave Action and Sand Movement near Anaheim Bay, California.* U.S. Army Corps of Engrs., Beach Erosion Board, Technical Memo No. 68.

CAREY, A. E. (1903). The Sanding-Up of Tidal Harbours. *Minutes of the Proceedings of the Institute of Civil Engineers* 156: 215–302.

CASTANHO, J. (1970). Influence of Grain Size on Littoral Drift. *Proceedings of the 12th Coastal Engineering Conference, Amer. Soc. Civil Engrs.,* pp. 891–898.

COAKLEY, J. P. and M. G. SKAFEL (1982). Suspended Sediment Discharge on a Non-Tidal Coast. *Proceedings of the 18th Coastal Engineering Conference, Amer. Soc. Civil Engrs.*, pp. 1289–1304.

CORNICK, H. F. (1969). *Dock and Harbour Engineering,* 2nd edition. London, England: Charles Griffin. Vol. 2.

DEAN, R. G. (1973). Heuristic Models of Sand Transport in the Surf Zone. *Proceedings of the 1st Australian Conference on Coastal Engineering,* Engineering Dynamics in the Surf Zone, Sydney, pp. 209–214.

DEAN, R. G., E. P. BEREK, C. G. GABLE, and R. J. SEYMOUR (1982). Longshore Transport Determined by an Efficient Trap. *Proceedings of the 18th Coastal Engineering Conference, Amer. Soc. Civil Engrs.*, pp. 954–968.

DEAN, R. G., E. P. BEREK, K. R. BODGE, and C. G. GABLE (1987). NSTS Measurements of Total Longshore Transport. *Coastal Sediments '87, Amer. Soc. Civil Engrs.*, pp. 652–667.

DEL VALLE, R., R. MEDINA, and M. A. LOSADA (1993). Dependence of Coefficient K on Grain Size. *Journal of Waterway, Port, Coastal, and Ocean Engineering, Amer. Soc. Civil Engr.* 119: 568–574.

DOWNING, J. P. (1984). Suspended Sand Transport on a Dissipative Beach. *Proceedings of the 19th Coastal Engineering Conference, Amer. Soc. Civil Engrs.*, pp. 1765–1781.

DOWNING, J. P. R. W. STERNBERG, and C. R. B. LISTER (1981). New Instrumentation for the Investigation of Sediment Suspension Processes in the Shallow Marine Environment. *Marine Geology* 42: 19–34.

DUANE, D. B. and W. R. JAMES (1980). Littoral Transport in the Surf Zone Elucidated by an Eulerian Sediment Tracer Experiment. *Journal of Sedimentary Petrology* 50: 929–942.

FAIRCHILD, J. C. (1972). Longshore Transport of Suspended Sediment. *Proceedings of the 13th Coastal Engineeering Conference, Amer. Soc. Civil Engrs.*, pp. 1069–1088.

FAIRCHILD, J. C. (1977). *Suspended Sediment in the Littoral Zone at Vetnor, New Jersey, and Nags Head, North Carolina.* U.S. Army Corps of Engineers, Coastal Engineer Resource Center Technical Paper No. 77–5.

FARINATO, R. S. and N. C. KRAUS (1981). Spectrofluorometric Determination of Sand Tracer Concentrations. *Journal of Sedimentary Petrology* 51: 663–665.

FUKSHIMA, H. and M. KASHIWAMURA (1959). Field Investigation of Suspended Sediment by the Use of Bamboo Samplers. *Coastal Engineering in Japan* 2: 53–58.

GALVIN, C. J. (1972). A Gross Longshore Transport Formula. *Proceedings of the 13th Coastal Engineering Conference, Amer. Soc. Civil Engrs.*, pp. 953–970.

GAUGHAN, M. K. (1978). Depth of Disturbance of Sand in Surf Zones. *Proceedings of the 16th Coastal Engineering Conference, Amer. Soc. Civil Engrs.*, pp. 1513–1530.

GRANT, U. S. (1943). Waves as a Transporting Agent. *American Journal of Science* 241: 117–123.

GREER, M. N. and O. S. MADSEN (1978). Longshore Sediment Transport Data: A Review. *Proceedings of the 16th Coastal Engineering Conference, Amer. Soc. Civil Engrs.*, pp. 1563–1576.

GUILLEN, J. and J. A. JIMENEZ (1995). Processes Behind the Longshore Variation of the Sediment Grain Size in the Ebro Delta Coast. *Journal of Coastal Research* 11: 205–218.

HANDIN, J. W. and J. C. LUDWICK (1950). *Accretion of Beach Sand Behind a Detached Breakwater.* U.S. Army Corps of Engineers, Beach Erosion Board Technical Memo No. 16.

HANES, D. M., C. E. VINCENT, D. A. HUNTLEY, and T. L. CLARKE (1988). Acoustic Measurements of Suspended Sand Concentration in the C^2S^2 Experiment at Stanhope Lane, Prince Edward Island. *Marine Geology* 81: 185–196.

HOLMAN, R. A. and A. J. BOWEN (1982). Bars, Bumps and Holes: Models for the Generation of Complex Beach Topography. *Journal of Geophysical Research* 87(C1): 457–468.

HOM-MA, M. and K. HORIKAWA (1962). Suspended Sediment Due to Wave Action. *Proceedings of the 8th Coastal Engineering Conference, Amer. Soc. Civil Engrs.*, pp. 169–193.

HOM-MA, M., K. HORIKAWA, and R. KAJIMA (1965). A Study of Suspended Sediment Due to Wave Action. *Coastal Engineering in Japan* 3: 101–122.

INGLE, J. C. (1966). *The Movement of Beach Sand.* New York: Elsevier.

INMAN, D. L. and R. A. BAGNOLD (1963). Littoral Processes. In *The Sea,* M.N. Hill (editor). Vol. 6, pp. 529–553. New York: Wiley-Interscience.

INMAN, D. L. and J. D. FRAUTSCHY (1966). Littoral Processes and the Development of Shorelines. *Proceedings of the Coastal Engeering Speciality Conference, Amer. Soc. Civil Engrs.* (Santa Barbara, CA), pp. 511–536.

INMAN, D. L., J. A. ZAMPOL, T. E. WHITE, D. M. HANES, B. W. WALDORF, and K. A. KASTENS (1980). Field Measurements of Sand Motion in the Surf Zone. *Proceedings of the 17th Coastal Engineering Conference, Amer. Soc. Civil Engrs.,* pp. 1215–1234.

IWAGAKI, Y. and T. SAWARAGI (1962). A New Method for Estimation of the Rate of Littoral Sand Drift. *Coastal Engineering in Japan* 5: 67–79.

JAFFE, B. and A. SALLENGER (1992). The Contribution of Suspension Events to Sediment Transport in the Surf Zone. *Proceedings of the 23rd Coastal Engineering Conference, Amer. Soc. Civil Engrs.,* pp. 2680–2693.

JOHNSON, J. W. (1953). *Sand Transport by Littoral Currents. Proceedings of the 5th Hydrologic Conference,* Bulletin 34, State University of Iowa Studies in Engineering, pp. 89–109.

JOHNSON, J. W. (1956). Dynamics of Nearshore Sediment Movement. *Bulletin of the American Society of Petroleum Geologists* 40: 2211–2232.

JOHNSON, J. W. (1957). The Littoral Drift Problem at Shoreline Harbors. *Journal of Waterways and Harbors Division, Amer. Soc. Civil Engrs.* 83(WW1): 1–37.

JONSSON, I. G. (1966). Wave Boundary Layers and Friction Factors. *Proceedings of the 10th Coastal Engineering Conference, Amer. Soc. Civil Engrs.,* pp. 127–148.

KAMPHUIS, J. W. (1975). Friction Factors Under Oscillatory Waves. *Journal of Waterways, Harbors, and Coastal Engineering Division, Amer. Soc. Civil Engrs.* 101(WW2): 135–144.

KAMPHUIS, J. W. (1990). Littoral Sediment Transport Rate. *Proceedings of the 22nd Coastal Engineering Conference, Amer. Soc. Civil Engrs.,* pp. 2402–2415.

KAMPHUIS, J. W. (1991). Alongshore Sediment Transport Rate Distributions. *Coastal Sediments '91, Amer. Soc. Civil Engrs.,* pp. 170–183.

KAMPHUIS, J. W. and J. S. READSHAW (1978). A Model Study of Alongshore Sediment Transport Rate. *Proceedings of the 16th Coastal Engineering Conference, Amer. Soc. Civil Engrs.,* pp. 1656–1674.

KAMPHUIS, J. W. and O. J. SAYAO (1982). Model Tests on Littoral Sand Transport Rate. *Proceedings of the 18th Coastal Engineering Conference, Amer. Soc. Civil Engrs.,* pp. 1305–1325.

KAMPHUIS, J. W., M. H. DAVIES, R. B. NAIRN, and O. J. SAYAO (1986). Calculation of Littoral Sand Transport Rate. *Coastal Engineering* 10: 1–21.

KANA, T. W. (1976). A New Apparatus for Collecting Simultaneous Water Samples in the Surf Zone. *Journal of Sedimentary Petrology* 46: 1031–1034.

KANA, T. W. (1977). Suspended Sediment Transport at Price Inlet, S.C. *Coastal Sediments '77, Amer. Soc. Civil Engrs.,* pp. 366–382.

KANA, T. W. (1978). Surf Zone Measurements of Suspended Sediment. *Proceedings of the 16th Coastal Engineering Conference, Amer. Soc. Civil Engrs.,* pp. 1725–1743.

KANA, T. W. and L. G. WARD (1980). Nearshore Suspended Sediment Load During Storm and Post-Storm Conditions. *Proceedings of the 17th Coastal Engineering Conference, Amer. Soc. Civil Engrs.,* pp. 1159–1175.

KATOH, K., N. TANAKA, and I. IRIE (1984). Field Observation on Suspended-Load in the Surf Zone: *Proceedings of the 19th Coastal Engineering Conference, Amer. Soc. Civil Engrs.,* pp. 1846–1862.

KING, C. A. M. (1951). Depth of Disturbance of Sand on Sea Beaches by Waves. *Journal of Sedimentary Petrology* 21: 131–140.

KNOTH, J. S. and D. NUMMEDAL (1977). Longshore Sediment Transport Using Fluorescent Tracer. *Coastal Sediments '77, Amer. Soc. Civil Engrs.,* pp. 383–398.

KOMAR, P. D. (1977a). Beach Sand Transport: Distribution and Total Drift. *Journal of Waterway, Port, Coastal, and Ocean Division, Amer. Soc. Civil Engrs.* 103(WW2): 225–239.

KOMAR, P. D. (1977b). Selective Longshore Transport Rates of Different Grain-Size Fractions Within a Beach. *Journal of Sedimentary Petrology* 47: 1444–1453.

KOMAR, P. D. (1978). The Relative Significance of Suspension Versus Bed-Load on Beaches. *Journal of Sedimentary Petrology* 48: 921–932.

KOMAR, P. D. (1983). Coastal Erosion in Response to the Construction of Jetties and Breakwaters. In *Handbook of Coastal Processes and Erosion.* Boca Raton, Fla: CRC Press, pp. 191–204.

KOMAR, P. D. (1988). Environmental Controls on Littoral Sand Transport. *Proceedings of the 21st Coastal Engineering Conference, Amer. Soc. Civil Engrs.,* pp. 1239–1252.

KOMAR, P. D. (1990). Littoral Sediment Transport. In *Handbook on Coastal and Ocean Engineering,* J. B. Herbich (editor). Vol. 2, chapter 11, pp. 681–714. Houston, TX: Gulf Publishing Co.

KOMAR, P. D. and D. L. INMAN (1970). Longshore Sand Transport on Beaches. *Journal of Geophysical Research* 75(30): 5514–5527.

KRAUS, N. C. (1985). Field Experiments on Vertical Mixing of Sand in the Surf Zone. *Journal of Sedimentary Petrology* 55: 3–14.

KRAUS, N. C., M. ISOBE, H. IGARASHI, T. O. SASAKI, and K. HORIKAWA (1982). Field Experiments on Longshore Transport in the Surf Zone. *Proceedings of the 18th Coastal Engineering Conference, Amer. Soc. Civil Engrs.,* pp. 969–988.

KRAUS, N. C. and J. L. DEAN (1987). Longshore Sediment Transport Rate Distributions Measured by Trap. *Coastal Sediments '87, Amer. Soc. Civil Engrs.,* pp. 881–896.

KRUMBEIN, W. C. (1944). *Shore Currents and Sand Movment on a Model Beach.* U.S. Army Corps of Engineers, Beach Erosion Board, Technical Memo. No. 7.

LEE, K. K. (1975). Longshore Currents and Sediment Transport in West Shore of Lake Michigan. *Water Resources Research,* 11: 1029–1032.

LEONARD, J. E. and B. M. BRENNINKMEYER (1978). Storm Induced Periodicities of Suspended Sand Movement. *Proceedings of the 16th Coastal Engineering Conference, Amer. Soc. Civil Engrs.,* pp. 1744–1763.

LONGUET-HIGGINS, M. S. (1952). On the Statistical Distribution of the Height of Sea Waves. *Journal of Marine Research* 11: 245–266.

LONGUET-HIGGINS, M. S. (1970). Longshore Currents Generated by Obliquely Incident Waves, 2. *Journal of Geophysical Research* 75: 6790–6801.

LONGUET-HIGGINS, M. S. (1972). Recent Progress in the Study of Longshore Currents. In *Waves on Beaches,* R. E. Meyer (editor), pp. 203–248. New York: Academic Press.

MADSEN, O. S. and W. D. GRANT (1976). Quantitative Description of Sediment Transport by Waves. *Proceedings of the 15th Coastal Engineering Conference, Amer. Soc. Civil Engrs.,* pp. 1093–1112.

McCAVE, I. N. (1978). Grain-Size Trends and Transport Along Beaches: Example from Eastern England. *Marine Geology* 28: M43–M51.

MOORE, G. W. and J. Y. COLE (1960). Coastal Processes in the Vicinity of Cape Thompson, Alaska. In *Geologic Investigations in Support of Project Chariot in the Vicinity of Cape Thompson, Northwestern Alaska—Preliminary Report,* T. Kachadoorian et al. (editors). U.S. Geological Survey Trace Elements Investigation Report 753, pp. 41–55.

MORFRETT, J. C. (1989). The Development and Calibration of an Alongshore Shingle Transport Formula. *Journal of Hydraulic Research* 27: 17–30.

NICHOLLS, R. J. and N. B. WEBBER (1987). Aluminum Pebble Tracer Experiments on Hurst Castle Spit. *Coastal Sediments '87, Amer. Soc. Civil Engrs.,* pp. 1563–1577.

NICHOLLS, R. J. and P. WRIGHT (1991). Longshore Transport of Pebbles: Experimental Estimates of K. *Coastal Sediments '91, Amer. Soc. Civil Engrs.,* pp. 920–933.

ÖZHAN, E. (1982). Laboratory Study of Breaker Type Effect on Longshore Sand Transport. In *Mechanics of Sediment Transport,* B. M. Sumer and A. Müller (editors). Proceedings of Euromech 156, pp. 265–274, A.A. Balkema, Rotterdam.

RUSSELL, R. C. H. (1960). Use of Fluorescent Tracers for the Measurement of Littoral Drift. *Proceedings of the 7th Coastal Engineering Conference, Amer. Soc. Civil Engrs.,* pp. 419–444.

SAUVAGE DE SAINT MARC, G. and G. VINCENT (1954). Transport Littoral Formation de Fleches et de Tombolos. *Proceedings of the 5th Coastal Engineering Conference, Amer. Soc. Civil Engrs.*, pp. 296–328.

SAVAGE, R. P. (1959). *Laboratory Study of the Effect of Grains on the Rate of Littoral Transport.* U.S. Army Corps of Engineers, Beach Erosion Board Technical Memo No. 14.

SAVAGE, R. P. (1962). Laboratory Determination of Littoral Transport Rates. *Journal of Waterways, Harbors Division, Amer. Soc. Civil Engrs.* 88(WW2): 69–92.

SAVILLE, T. JR. (1950). Model Study of Sand Transport Along an Infinitely Long, Straight Beach. *Transactions American Geophysical Union* 31: 555–565.

SAWARAGI, T. and I. DEGUCHI (1978). Distribution of Sediment Transport Rate Across a Surf Zone. *Proceedings of the 16th Coastal Engineering Conference, Amer. Soc. Civil Engrs.*, pp. 1596–1613.

SCHOONES, J. S. and A. K. THERON (1993). Review of the Field-Data Base for Longshore Sediment Transport. *Coastal Engineering* 19: 1–25.

SCHOONES, J. S. and A. K. THERON (1994). Accuracy and Applicability of the SPM Longshore Transport Formula. *Proceedings of the 24th Coastal Engineering Conference, Amer. Soc. Civil Engrs.*, pp. 2595–2609.

SHAY, E. A. and J. W. JOHNSON (1951). *Model Studies on the Movement of Sand Transported by Wave Action Along a Straight Beach.* Institute of Engineering Research, University of California, Berkeley, Series 14.

SHEPARD, F. P. and H. R. WANLESS (1971). *Our Changing Coastlines.* New York: McGraw-Hill.

SPRING, F. J. E. (1919). Coastal Sand Travel near Madras Harbor. *Minutes of the Proceedings of the Institute of Civil Engineers* 210: 27–28.

STERNBERG, R. W., N. C. SHI, and J. P. DOWNING (1984). Field Investigation of Suspended Sediment Transport in the Nearshore Zone. *Proceedings of the 19th Coastal Engineering Conference, Amer. Soc. Civil Engrs.*, pp. 1782–1798.

SUNAMURA, T. and K. HORIKAWA (1971). Predominant Direction of Littoral Transport Along Kujyukuri Beach, Japan. *Coastal Engineering in Japan* 14: 107–117.

SWART, D. H. (1976). Predictive Equations Regarding Coastal Transport. *Proceedings of the 15th Coastal Engineering Conference, Amer. Soc. Civil Engrs.*, pp. 1113–1132.

TELEKI, P. G. (1966). Fluorescent Sand Tracers. *Journal of Sedimentary Petrology* 36: 469–485.

TELEKI, P. G. (1967). Automatic Analysis of Tracer Sand. *Journal of Sedimentary Petrology* 37: 749–759.

THORNTON, E. B. (1972). Distribution of Sediment Transport Across the Surf Zone. *Proceedings of the 13th Coastal Engineering Conference, Amer. Soc. Civil Engrs.*, pp. 1049–1068.

THORNTON, E. B. and W. D. MORRIS (1977). Suspended Sediments Measured Within the Surf Zone. *Coastal Sediments '77, Amer. Soc. Civil Engrs.*, pp. 655–668.

TRASK, P. D. (1952). *Sources of Beach Sand at Santa Barbara, California, as Indicated by Mineral Grain Studies.* Beach Erosion Board Technical Memo No. 28. U. S. Army Corps of Engineers.

TRASK, P. D. (1955). *Movement of Sand Around Southern California Promontories.* Beach Erosion Board Technical Memo No. 76. U.S. Army Corps of Engineers.

TSUCHIYA, Y. (1982). The Rate of Longshore Sediment Transport and Beach Erosion Control. *Proceedings of the 18th Coastal Engineering Conference*, pp. 1326–1334.

VERNON-HARCOART, L. F. (1881). Harbours and Estuaries on Sandy Coasts. *Minutes of the Proceedings of the Institute of Civil Engineers* 70: 1–32.

VINCENT, C. E., D. M. HANES, and A. J. BOWEN (1991). Acoustic Measurements of Suspended Sand on the Shoreface and the Control of Concentration by Bed Roughness. *Marine Geology* 96: 1–18.

WATTS, G. M. (1953a). *A Study of Sand Movement at South Lake Worth Inlet, Florida.* U.S. Army Corps of Engineers, Beach Erosion Board Technical Memo No. 42.

WATTS, G. M. (1953b). *Development and Field Test of a Sampler for Suspended Sediment in Wave Action.* U.S. Corps of Engineers, Beach Erosion Board Technical Memo No. 34.

WIEGEL, R. L. (1959). Sand Bypassing at Santa Barbara, California. *Journal of Waterways and Harbors Division, Amer. Soc. Civil Engrs.* 85(WW2): 1–30.

WIEGEL, R. L. (1964). *Oceanographical Engineering.* Englewood Cliffs, NJ: Prentice-Hall.

WRIGHT, P., J. S. CROSS, and N. B. WEBBER (1978). Aluminum Pebbles: A New Type of Tracer for Flint and Chert Pebble Beaches. *Marine Geology* 27: M9–M17.

YASSO, W. E. (1966). Formulation and Use of Fluorescent Tracer Coatings in Sediment Transport Studies. *Sedimentology* 6: 287–301.

YOUNG, R. A., J. MERRILL, J. R. PRONI, and T. L. CLARKE (1982). Acoustic Profiling of Suspended Sediments in the Marine Boundary Layer. *Geophysical Research Letters* 9: 175–178.

ZENKOVITCH, V. P. (1960). Fluorescent Substances as Tracers for Studying the Movement of Sand on the Sea Bed, Experiments Conducted in the USSR. *Dock and Harbor Authority* 40: 280–283.

10

Shoreline Planforms and Models to Simulate Their Evolution

It is astonishing and incredible to us, but not to Nature; for she performs with utmost ease and simplicity things which are even infinitely puzzling to our minds, and what is very difficult for us to comprehend is quite easy for her to perform.

Galileo
Dialogue Concerning the Two World Systems (1630)

Drawing an analogy with the graded stream, Tanner (1958) introduced the concept of an equilibrium beach as one whose curvature in planview and profile are adjusted in such a way that waves impinging on the shore provide precisely the energy required to transport the load of sediment supplied to the beach. The equilibrium profile was examined in Chapter 7, at least its two-dimensional aspects, and the three-dimensional beach morphology will be considered in Chapter 11. Here we examine the planform of the shoreline, the shape whose equilibrium curvature is in balance with the sediment supply and the capacity of the waves to redistribute that sediment alongshore. The time element for such an equilibrium shoreline is long term—years and decades.

The equilibrium shoreline and its evolution with time are best examined by the development of numerical models where computer routines incorporate equations that relate sediment movements to the nearshore waves and currents. Such analyses are not necessarily an esoteric examination of conceptual equilibrium shorelines. Numerical simulation models also have proved to be extremely powerful tools in analyzing the impacts of the construction of jetties and breakwaters on the shoreline configuration, duplicating or predicting the patterns of erosion and accretion.

WAVE PATTERNS AND THE EQUILIBRIUM SHORELINE

The most elementary shoreline is where there is no long-term net transport of sediment. This condition is closely approached by a pocket beach where there is little or no additional sediment being supplied to the beach and essentially no losses. Under this simplest of conditions, the shape of the beach depends on the wave conditions, in particular, the pattern of wave refraction and diffraction, yielding an arcuate shape typical of pocket beaches with bounding headlands (Fig. 10-1).

Consider the arrival of uniform swell waves to a pocket beach—these waves could be the prevailing or dominant waves in the area. Wherever wave crests strike the shoreline at an angle, they produce a longshore transport of beach sediment (Chapter 9). Sediment will then

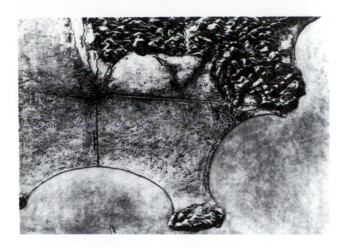

Figure 10-1 Leonardo da Vinci's drawing of his plan for draining the Pontine Marshes in Italy. The arcuate beaches between rocky promontories illustrate typical equilibrium shoreline configurations.

be shifted alongshore as long as the waves continue to break obliquely to the shore. The final equilibrium shoreline configuration within the pocket beach must necessarily be one where the breaker angles are zero everywhere along the shoreline, since the wave energy or power would not necessarily be reduced—only then would the longshore redistribution of sediment cease. The equilibrium condition, then, is one in which the shoreline is everywhere parallel to the crests of the incoming waves. Recall from Chapter 5 that as waves shoal they refract and diffract, bending and changing direction, the final shape depending on the nature of the offshore topography. These processes, which can occur well offshore, control the shape of the wave crests and therefore the exact form and orientation of the beach that is experiencing little or no net longshore sediment transport. This was realized in part by Lewis (1938), who examined beaches in England and demonstrated that they are oriented according to the prevailing waves. Lewis did not extensively develop the role played by refraction, though he appears to have recognized its importance. Davies (1958) more precisely demonstrated the significance of wave refraction and diffraction. Figure 10-2 from his paper shows a portion of the coast of Tasmania, which provides an excellent example of the way in which local beaches have become oriented to be parallel to the refracted waves, yielding a zero net longshore transport within each pocket. Although the long-term average shoreline configuration reflects the net-zero longshore transport, individual wave trains could produce an oblique-wave approach and a temporary longshore transport. The shoreline then wobbles about under the different wave trains arriving from slightly different directions, while the average beach configuration corresponds to the long-term condition of a net-zero longshore transport.

If there were a significant source of sediment to the pocket beach, then the equilibrium shoreline would have to make an angle with the refracted wave crests in order to redistribute that sediment alongshore away from the source. The clearest example of this is where a river supplies sediment and builds out a delta. In this case the delta takes on a long-term equilibrium shape whose orientation and curvature is exactly that required to produce the oblique-wave approach needed to redistribute the sediment supplied by the river to the full length of the shore, much as envisioned by Tanner (1958).

A number of studies have proposed specific geometric curves that approximate equilibrium beach shapes under certain conditions. Hoyle and King (1957) compared pocket beaches to arcs of great circles. Yasso (1965) utilized the logarithmic spiral to describe

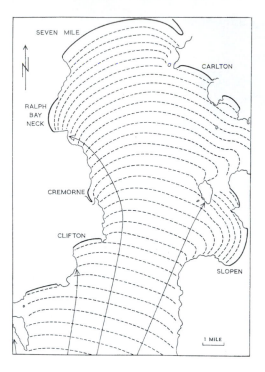

Figure 10-2 Wave refraction within Frederick Henry Bay, Tasmania, for a 14–sec southwesterly swell. Every fifth wave crest is shown as a dashed line, and the wave orthogonals are equidistant in deep water. Sand beaches, represented by thickened lines, are oriented and shaped to be congruent with the refracted wave crests, the equilibrium configuration of pocket beaches. [Used with permission of National Taiwan Normal University, from J. L. Davies, Wave Refraction and the Evolution of Shoreline Curves, *Geographical Studies 5,* p. 10. Copyright © 1958 National Taiwan Normal University.]

"headland-bay" beaches, such as found downdrift from an obstacle that blocks the longshore sediment transport. Figure 10-3 illustrates the near congruence of this geometric curve with the shoreline of Halfmoon Bay, California. The general form of the logarithmic spiral is shown in Figure 10-4, and its equation in polar coordinates is

$$r = r_0 e^{(\cot \alpha)\theta} \tag{10.1}$$

where r is the radius from the log-spiral center to a point on the curve, r_0 is the radius from the log-spiral center to the origin of the curve, θ is the variable angle between r and r_0 as shown in Figure 10-4, and α is the spiral angle defined as the angle between the radius vector and the tangent to the curve at any point. The angle α is a constant, so the logarithmic spiral is generated by the increase in length of the vector r as the angle θ swings around from its origin at the r_0 vector. In applications to curved shorelines, the center of the logarithmic spiral generally is close to the headland where the spiral begins; Halfmoon Bay in Figure 10-3 illustrates this. Yasso (1965) and other studies provide discussions of techniques for comparing measured shoreline curvatures to the mathematical form of Equation (10.1) in order to provide the best fit.

A series of investigations have made comparisons between curved headland-bay shorelines and the logarithmic spiral, variously considering its evolution with time and the processes of waves, currents, and sediment movement along such shorelines. Figure 10-5 documents the evolution of the shoreline in the downdrift direction from a seawall, measured by Terpstra and Chrzastowski (1992) on Lake Michigan. The erosion leading to this headland-bay beach was very rapid as the waves cut into loose fill that originally was meant to be

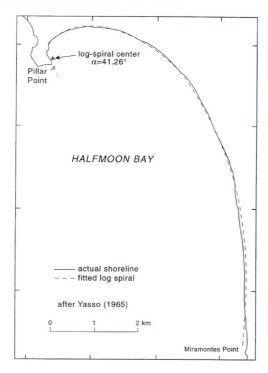

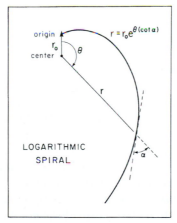

Figure 10-3 A comparison of the shoreline curvature of Halfmoon Bay, California, with the logarithmic spiral geometric curve given by Equation (10.1). [Adapted with permission of The University of Chicago Press, W. E. Yasso, Plan Geometry of Headland Bay Beaches, *Journal of Geology 73*, p. 707. Copyright © 1965 The University of Chicago Press.]

Figure 10-4 The logarithmic spiral defined with polar coordinates, the variable radius vector r and rotation angle θ from an initial radius r_o, while the angle α with the tangent line is held constant.

a feeder beach but subsequently was partly protected by the seawall. Their analyses established that the eroding shoreline quickly took the form of the logarithmic spiral where the value of r_0 in Equation (10.1) decreased with time as the center of the spiral approached the end of the seawall, the angle α decreased as the curvature of the overall shoreline lessened with time, and the total length of near congruence between the measured shoreline and the logarithmic spiral increased as the erosion progressed farther from the seawall. Phillips (1985) provides an example from Sandy Hook, New Jersey, where an eroding shoreline

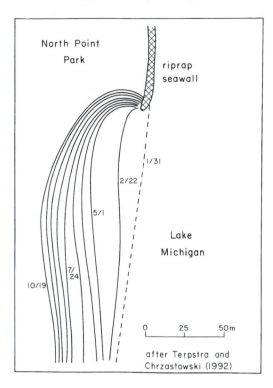

North Point
Park

riprap
seawall

1/31

2/22

5/1

Lake
Michigan

7/24

10/19

0   25   50m

after Terpstra and
Chrzastowski (1992)

Figure 10-5 The eroding logarithmic-spiral shoreline downdrift from a seawall constructed at the North Point Marina on the Illinois shore of Lake Michigan. [Adapted with permission of Journal of Coastal Research, from P. D. Terpstra and M. J. Chrzastowski, Geometric Trends in the Evolution of a Small Log-Spiral Embayment of the Illinois Shore of Lake Michigan, *Journal of Coastal Research 8,* p. 609. Copyright © 1992 Journal of Coastal Research.]

downdrift from a seawall developed the form of the logarithmic spiral but where the evolution spanned nearly 30 years.

Bremner and LeBlond (1974) compared the planimetric shape of the shoreline of Wreck Bay, Vancouver Island, Canada, to logarithmic spirals. As seen in Figure 10-6, this bay is highly complex, with offshore islands and shoals that affect the patterns of wave refraction and diffraction. The shoreline of the bay takes on the shapes of the wave crests, with the greatest curvature being adjacent to the bounding headlands. The lower half of Figure 10-6 shows the agreement between the northwestern shoreline and the logarithmic spiral of Equation (10.1), expressed in logarithmic form so that it yields a straight-line relationship between $\log(r)$ and θ. That study briefly considered the role of longshore currents and sediment transport in developing the shoreline forms in Wreck Bay, while LeBlond (1979) in more general terms considered the patterns of wave refraction and diffraction, the resulting wave-energy levels along the length of the spiral shoreline, and the parallel patterns of beach sediment grains sizes and beach slopes. Finkelstein (1982) provides a good example of analyses of the variations in sediment sizes and beach morphology along logarithmic-spiral, headland-bay shorelines found on Kodiak Island, Alaska.

In a series of papers, Silvester and co-workers analyzed what they term "crenulate-shaped bays" (Silvester, 1960, 1970; Silvester and Ho, 1972; Hsu, Silvester, and Xia, 1989)—the term comes from the scallops between ribs as found on seashells and leaves. Silvester (1960) provides examples from nature where pocket beaches are trapped between headlands, but much of his focus is on the shapes of shorelines between groynes or detached breakwaters that are constructed to stabilize the shore (Chapter 12). As illustrated in Figure 10-7, the crenulate

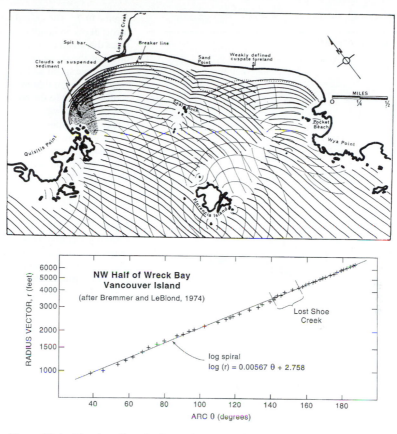

Figure 10-6   The shoreline planform of Wreck Bay, Vancouver Island, produced by the complex patterns of wave refraction and diffraction over offshore shoals and around islands. The lower graph shows the fit of the northwest shoreline of the bay to the logarithmic spiral of Equation (10.1). [Adapted with permission of Society for Sedimentary Geology, from J. M. Bremner and P. H. LeBlond, On the Planimetric Shape of Wreck Bay, Vancouver Is., *Journal of Sedimentary Petrology 44*, p. 1157. Copyright © 1974 Society for Sedimentary Geology.]

bay between a pair of breakwaters or seawalls consists of two components: (1) the shadow zone that is protected behind the upcoast structure, an area where the shoreline adopts the shape of a logarithmic-spiral curve; and (2) the tangential shoreline, the relatively straight shore near the downcoast structure. The shape of the crenulate shoreline again is governed by the pattern of refraction and diffraction as the waves generally approach the bay opening at an oblique angle in the offshore (Fig. 10-7). The downcoast tangential end of the crenulate shoreline is close to the original orientation of the offshore wave crests as they have undergone minimal refraction upon entering the bay. Silvester (1960) has used this correspondence to infer dominant wave directions from natural crenulate bays. The upcoast logarithmic-spiral segment of the shoreline is primarily produced by wave refraction, although there may be a small amount of diffraction around the tip of the structure. When the crenulate bay has reached equilibrium, the shoreline is everywhere parallel to the crests of the breaking waves. Silvester and Ho investigated this geometry in a series of wave-basin experiments and by comparisons with natural embayments. Example results are shown in Figure 10-8 where the ratio

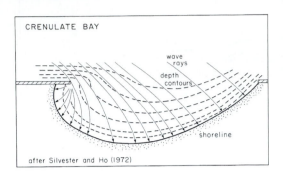

Figure 10-7 The crenulate-bay shoreline that forms between headlands or engineering structures due to an oblique-wave approach to the opening, the shoreline consisting of an upcoast logarithmic spiral and a downcoast tangential segment, a form that makes the shoreline everywhere parallel to the crests of the refracted and diffracted waves. [Adapted from Use of Crenulate-Shaped Bays to Stabilize Coasts, R. Silvester and S. K. Ho, *Proceedings of the 13th Coastal Engineering Conference,* 1972. Reproduced with permission from the American Society of Civil Engineers.]

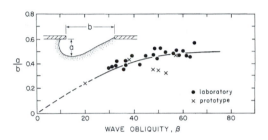

Figure 10-8 The geometry of the crenulate-bay shoreline defined by the *a/b* ratio shown in the diagram, determined by the obliquity of the offshore wave-approach angle β. [Adapted from Use of Crenulate-Shaped Bays to Stabilize Coasts, R. Silvester and S. K. Ho, *Proceedings of the 13th Coastal Engineering Conference,* 1972. Reproduced with permission from the American Society of Civil Engineers.]

of the degree of cut-back of the bay to the spacing between the "headlands" is dependent on the offshore angle of wave approach; an empirical curve is derived from the experiments, which establishes how the tangential angle α in Equation (10.1) for the logarithmic spiral decreases with an increase in the angle of wave obliquity. Recent analyses to further investigate the development of crenulate shorelines include the study of Tan and Chiew (1994) who undertook additional wave-basin studies to establish the equilibrium shape, and the study by Wind (1994) who developed analytical models that focus on variations with time of the developing shoreline geometry.

In spite of the findings of near congruence between shorelines and the logarithmic spiral mentioned above, one really cannot expect the shoreline shape to be that of some prescribed geometric form, since the exact curvature of the beach is governed by wave refraction and diffraction, which can depend on conditions well out to sea from the beach. However, the correspondence is sufficiently close to be useful in some applications, such as predicting the expected shapes of shorelines between designed groynes or offshore breakwaters. Such comparisons certainly are entertaining.

TECHNIQUES IN MODELING SHORELINES

The evolution of shorelines and how their shapes relate to wave conditions and sources of sediment can be analyzed using numerical simulation models developed to run on computers. The general approach to computer models of shoreline change involves the use of equations from Chapter 9 that relate the longshore sediment transport rate to the wave and nearshore current conditions, together with a continuity equation that keeps track of the total volume or mass of sand and ensures that none is unaccountably created or destroyed. The nearshore environment is divided into a large number of individual compartments or cells, and the shoreline model consists of following the shift of sand from one cell to another as governed by the sediment transport and continuity equations.

The models that have seen most widespread use are principally two dimensional, where a stretch of shoreline is divided into a series of cells as depicted in Figure 10-9(a). Such analyses often are termed "line models," in that they focus only on changes in the shoreline. Each individual cell has a sloping beach on its seaward end [Fig. 10-9(b)]. All of the cells diagrammed have the same width, Δx, though this need not be the case, and have individual lengths $y_1, y_2, \ldots y_i, \ldots y_n$ beyond some arbitrary baseline [Fig. 10-9(a)]. Changes in the position of the shoreline for any given cell are dependent on the quantity of sand entering that cell versus the amount leaving during a certain time interval Δt. As depicted in Figure 10-9(b), the respective rates of sediment input and exit from the cell are represented by Q_{in} and Q_{out}. The net volume of sand accumulation or erosion ΔV within the cell during the time interval Δt is then given by

$$\Delta V = (Q_{in} - Q_{out})\Delta t \qquad (10.2)$$

If the units of the sediment transport rate are m^3/day, and the time unit of Δt is also days, then the calculated ΔV has units m^3, representing either a net gain ($Q_{in} > Q_{out}$), that is shoreline accretion, or a net loss ($Q_{in} < Q_{out}$), representing erosion. This needs to be generalized, since there is the possibility that sand could exit a cell simultaneously in both directions, enter from

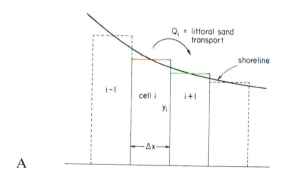

A

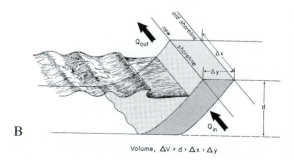

B

Volume, $\Delta V = d \cdot \Delta x \cdot \Delta y$

Figure 10-9 (a) The smooth shoreline divided into a series of cells of width Δx and variable y_i lengths beyond a base line. (b) One shoreline cell demonstrating how a change in sand volume ΔV is produced by the longshore sediment transport into and out of the cell and how this results in a change of shoreline position Δy. [From *Geological Society of America Bulletin 84*, P. D. Komar. Reproduced with permission of the publisher, the Geological Society of America, Bouder, Colorado USA. Copyright © 1973.]

both directions, or there could be various other combinations of sediment exchanges between adjacent cells. If we consider an arbitrary cell i, then

$$\Delta V_i = (Q_{i-1} - Q_i)\Delta t \tag{10.3}$$

where Q_i is the rate of longshore sediment exchange between cell i and cell $i + 1$, and Q_{i-1} is the sediment exchange between cell i and cell $i - 1$. The transport quantities can be positive or negative and therefore may represent sediment inputs or losses to cell i. To generalize still further, there could be other sand contributions to cell i, for example from an adjacent river or by onshore sand movement, or there could be losses due to sand being blown inland or transported offshore. These can be collectively represented by Q_r, so that Equation (10.3) becomes

$$\Delta V_i = (Q_{i-1} - Q_i \pm Q_r)\Delta t \tag{10.4}$$

It is apparent that the development of Equation (10.4) constitutes a mini-budget of sediment applied to cell i, but is conceptually the same as the larger-scale sediment budgets considered in Chapter 3. In view of this being a sediment budget, it is apparent that ΔV_i can be positive (net accretion), negative (erosion), or the various inputs and losses could balance to yield $\Delta V_i = 0$, in which case the shoreline position does not change in that particular cell.

In the models it is usually desirable to express the ΔV_i for the series of cells as patterns of shoreline change, that is, as a change in the length y_i of the cell [Fig. 10-9(a)]. If Δy_i is the corresponding change in y_i during the time increment Δt, then from the geometry of the cell depicted in Figure 10-9(b), we have

$$\Delta V_i = d\Delta y_i \Delta x \tag{10.5}$$

The accretional or erosional wedge has approximately the shape of a parallelepiped (a three-dimensional parallelogram) with length Δy_i, width Δx, and a "depth" d, which yields the correct volume of change. The value of the linear dimension d must be chosen to yield the correct correspondence between ΔV_i and Δy_i for a fixed Δx and will depend on the nature of the beach profile. The magnitude of d should be on the order of the closure depth of profile changes as reviewed in Chapter 7.

Combining Equations (10.4) and (10.5) and solving for Δy_i yields

$$\Delta y_i = (Q_{i-1} - Q_i \pm Q_r)\frac{\Delta t}{d\Delta x} \tag{10.6}$$

for the change in shoreline position of cell i as a function of sand inputs and losses. Since the parameters Δt, d, and Δx generally are fixed in most models, the Δy_i shoreline changes are produced simply by the balance of the Q-terms. Again, $\Delta y_i = -$ represents erosion and a shoreline retreat, while $\Delta y_i = +$ is a shoreline advance.

Equation (10.6) is more easily recognizable as a sediment continuity equation if the relationship is rearranged to be a function of $\Delta y/\Delta t$, the rate of shoreline change, and the finite elements are decreased to their limits to yield

$$\frac{dy}{dt} = -\frac{1}{d}\frac{dQ_\ell}{dx} \tag{10.7}$$

In deriving this form of the sediment continuity relationship, it has been assumed that $Q_r = 0$ and $\Delta Q = Q_i - Q_{i-1}$ has been taken as the change in the sediment transport rate in the longshore direction (the positive x-direction). This form of the continuity relationship is informative and useful in its own right in that it reveals that the time rate of shoreline change, dy/dt, is a function of the longshore gradient of the sediment transport rate, dQ_ℓ/dx. If $dQ_\ell/dx = +$, then Q_ℓ is increasing in the longshore direction, and this requires additional sand that comes from erosion of the beach (i.e., $dy/dt = -$, assuming there are no other sources). The greater the magnitude of dQ_ℓ/dx, the greater the corresponding rate of shoreline recession. In the opposite sense, a decrease in Q_ℓ in the longshore direction corresponds to $dQ_\ell/dx = -$ and yields $dy/dt = +$, an advance of the shore. It is important to recognize that shoreline changes, either erosion or accretion, are brought about by gradients in the longshore sediment transport rates and do not correspond to the absolute magnitude of the transport rate.

In computer models that simulate shoreline changes, the longshore transport rates between neighboring cells must be calculated, yielding the respective Q_i and Q_{i-1} values for each individual cell. From Chapter 9 we know that, in general, this requires specific values for wave heights, energies or powers, and either the angle of wave breaking at the shoreline or the magnitude of the longshore current that governs the sediment transport rate. For example, we could use the relationship

$$Q_\ell = 1.1\rho g^{3/2} H_{br}^{5/2} \sin \alpha_b \cos \alpha_b \tag{10.8}$$

for the longshore transport of quartz-density sand, where Q_ℓ has units m³/day, H_{br} is the root-mean-square wave-breaker height in meters, $g = 9.8$ m/sec², and α_b is the breaker angle. This longshore sediment transport rate has to be calculated for each cell or more precisely at the junctions between adjacent cells, so that in using Equation (10.8) we have to know the longshore variations of H_b and α_b. In some models this is directly specified, while in more sophisticated models it is determined by a wave refraction/diffraction analysis that is coupled with the shoreline-evolution model. Once the Q_ℓ magnitudes are evaluated between all adjacent cell pairs, then the corresponding Δy_i shoreline changes for the cells can be calculated for the specified Δt time interval, utilizing the sediment continuity equation. This procedure is repeated for the now altered shoreline through another Δt time interval, repeated over and over again for the total specified run time; for example, Δt could be 1 day, while the model duration could cover 25 years of shoreline evolution. It is important that the Δy_i changes in cell lengths remain small during any Δt time increment, so there are no sudden "jumps" in the shoreline position. This usually requires that the time increment be kept small, usually 1 day or less. From the continuity Equation (10.6), it is apparent that the smaller the cell width Δx selected for the model, the smaller the time increment required to avoid unwanted shoreline jumps.

ANALYSES BASED ON SEDIMENT CONTINUITY CONSIDERATIONS

The sediment continuity relationship in the form of Equation (10.7) provides a direct relationship between the rate of shoreline erosion or accretion, dy/dt, and the gradient of the longshore sediment transport, dQ_ℓ/dx. This correspondence can be useful in investigating the underlying causes of shoreline recession, and measurements of shoreline change can be used to infer the spatial and temporal variations in the sediment transport along a coast.

A simple example of the use of the continuity relationship is provided by the analysis of Frihy et al. (1991) of shoreline changes on the Nile Delta of Egypt. As related in Chapter 3, there has been extensive erosion along portions of the delta throughout this century, largely due to the construction of the Aswan Low Dam in 1907 and the High Dam in 1964, which has cut off essentially all sediment delivery to the coast. The resulting shoreline erosion has centered mainly near the mouths of the two river distributaries. Figure 10-10(a) shows the shorelines of the promontory adjacent to the Rosetta branch of the Nile as traced from 1955 and 1983 aerial photographs, as well as a ground survey made in 1988. This illustrates the remarkable erosion that has occurred on the tip of the promontory, along both the western and eastern sides of the distributary mouth. The shoreline comparisons also reveal that there has been significant accretion along the eastern most part of the promontory. It is apparent that the overall pattern of erosion and accretion is a response to an eastward transport of sediment, resulting from the prevailing wave arrival from the NW-NNW. The shift from erosion to accretion along the east flank reflects the longshore variations in the quantities of sand being transported. This is shown schematically by the lengths of the arrows in Figure 10-10(a). As one moves east from the river mouth, the erosion contributes more and more sand to the longshore transport, so the total quantity progressively increases. The maximum longshore sediment transport is positioned at the nodal point between the areas of erosion versus accretion. With the change to beach accretion, there is progressively less and less sand being carried alongshore, shown schematically by the decreasing lengths of the arrows toward the east. By this interpretation, the longshore sediment transport is everywhere to the east, but its varying quantities account for the shift from erosion to accretion.

The quantities of longshore sediment transport responsible for the observed shoreline changes have been assessed by Frihy et al. (1991) for the stretch of shoreline east of the river through application of the continuity relationship of Equation (10.7). The average rates of shoreline erosion or accretion, dy/dt, between the years 1955 and 1983 were determined from

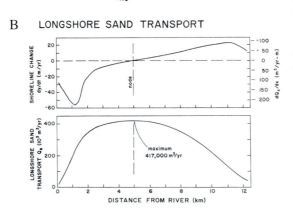

Figure 10-10 (a) The patterns of major erosion and accretion along the shoreline of the Rosetta Promontory of the Nile Delta. The direction of the arrows schematically depict the eastward longshore sand transport, while the length represents the variable magnitudes that account for the change from shoreline erosion to accretion with distance from the river. (b) Quantitative evaluations of the rates of shoreline change dy/dt between 1955 and 1983, with a scale for the inferred gradient of the longshore sand transport rate, dQ_ℓ/dx, and a separate graph for the transport Q_ℓ determined by the longshore integration of dQ_ℓ/dx. [Reprinted from *Coastal Engineering* 15, O. E. Frihy, A. M. Fanos, A. A. Khafagy, and P. D. Komar, Patterns of Nearshore Sediment Transport Along the Nile Delta, Egypt, 409–429, 1991 with kind permission of Elsevier Science-NL, Sara Burgerhartstratt 25, 1055 KV Amsterdam, The Netherlands.]

Figure 10-10(a) and are plotted in Figure 10-10(b). According to beach profiles from the area, the elevation-change parameter in the continuity relationship is approximately $d = 4$ m. Therefore, there is a simple proportionality between dy/dt and dQ_ℓ/dx, so that both scales can be given in Figure 10-10(b). Note again that shoreline erosion ($dy/dt = -$) corresponds to $dQ_\ell/dx = +$, that is, to an increasing transport rate in the longshore direction, just as discussed in the preceding section (shoreline accretion yields $dQ_\ell/dx = -$, a decreasing transport). The dQ_ℓ/dx local gradients have been integrated in the longshore direction from west to east to yield the graph of Q_ℓ in Figure 10-10(b). It was assumed that $Q_\ell = 0$ at the river mouth, that is, presently there is no sand being derived from the reduced river discharge and no exchange of sand with the area to the west, either with the river channel or with the western flank of the promontory. If there is in fact some net exchange, then the Q_ℓ curve of Figure 10-10(b) would simply shift up or down by that amount, while retaining its overall shape. It can be seen that Q_ℓ progressively increases with longshore distance from the river, reaches a maximum of approximately 417,000 m³/year at the nodal position, and then decreases. The quantities given in this graph correspond to the arrows in Figure 10-5(a) that schematically illustrate this variability of the longshore sand transport. According to the computations, the transport is reduced to 40,000 m³/yr at the eastern edge of the area analyzed, a volume that presumably is deposited farther to the east. The varying quantities of longshore sediment transport found in the analysis are due to systematic changes in breaker angles and wave energies along this stretch of shore, in turn caused by the refraction of the predominant NW-NNW waves around the shallow offshore of the Rosetta promontory.

Another example relating shoreline changes to the patterns of longshore sediment transport is provided by the analyses of Greenwood and McGillivary (1980), an application that focused on the central Toronto waterfront of Lake Ontario. The analysis is essentially the inverse of that above for the Nile Delta in that Greenwood and McGillivary first analyzed the wave conditions that would produce the longshore sediment transport and then considered the longshore gradients of that transport to infer the expected patterns of shoreline change. One example of their results is shown in Figure 10-11, which graphs the "longshore component" of wave power

$$P_\ell = (ECn)_b \sin \alpha_b \cos \alpha_b \qquad (10.9)$$

where $(ECn)_b$ is the wave-energy flux or power evaluated at the breaker zone, and α_b is the wave-breaker angle. We saw in Chapter 9 that there is a direct proportionality between P_ℓ and the resulting longshore sediment transport, so the spatial variations in P_ℓ along the coast, as graphed in Figure 10-11, can be used to infer the parallel variations in the longshore sediment transport (i.e., Q_ℓ and thus dQ_ℓ/dx). Along stretches of shoreline where P_ℓ and Q_ℓ are increasing, one would expect to have shoreline erosion; where P_ℓ decreases, there should be accretion. These trends are shown by the arrows along the P_ℓ graph, with the corresponding locations noted on the map of the Toronto waterfront. Although Greenwood and McGillivary did not employ the sediment continuity relation to make quantitative assessments of the resulting shoreline erosion or accretion, it can be seen that even in an area of complex shorelines such as the Toronto waterfront, considerations of spatial variations in dP_ℓ/dx and thus dQ_ℓ/dx can help explain observed patterns of shoreline recession or accretion. Davidson-Arnott and Pollard (1980) provide a similar analysis applied to the Ontario shoreline along Nottawasaga Bay, Lake Huron.

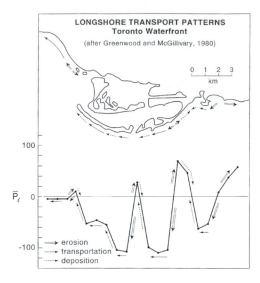

Figure 10-11   Variations in the wave power P_ℓ of Equation (10.8) along the Toronto waterfront, variations that produce parallel changes in the quantities of the longshore sediment transport that account for the patterns of shoreline erosion, transportation (no change), and deposition, as indicated by the arrows. [Used with permission of Gebruder Barntraeger, from B. Greenwood and D. G. McGillivary, Modelling the Impact of Large Structures upon Littoral Transport in the Central Toronto Waterfront, Lake Ontario, Canada, *Zeitschrift fur Geomorphologie N.F.,* p. 107. Copyright © 1980 Gebruder Barntraeger.]

ANALYTICAL SOLUTIONS OF SHORELINE PLANFORMS

In relatively simple applications it may be possible to obtain analytical solutions of the continuity and sand-transport equations to derived mathematical expressions for the shoreline planform. In order to make the problem tractable to mathematical treatment, however, the equations must be simplified. Pelnard-Considere (1954) examined variations in Q_ℓ resulting only from longshore variations in the breaker angle ($\partial \alpha_b/\partial x$), assuming that $(ECn)_b$ and the wave-breaker heights H_b are everywhere constant. This allowed the expansion of Equation (10.8) for Q_ℓ into a Taylor series as a function of α_b alone, which combined with the continuity equation yielded the relationship

$$\frac{\partial y}{\partial t} = \frac{1}{d}\left(\frac{\partial Q_\ell}{\partial \alpha_b}\right)\frac{\partial^2 y}{\partial x^2} \tag{10.10}$$

indicating that the rate of accretion or erosion ($\partial y/\partial t$) is dependent on the curvature of the shoreline ($\partial^2 y/\partial x^2$). In applying this approach, Pelnard-Considere investigated the shoreline configuration developed by a groyne blocking a longshore transport of sediment and obtained mathematical solutions for the updrift accretion, but not for the erosion within the shadow zone downdrift from the groyne. His solutions were partly confirmed by model tests. Bakker, Breteler, and Roos (1970) and Bakker (1968) further developed analytical solutions for shorelines within groyne systems.

   The principal analytical shoreline solution that is used in applications is that for the longshore spreading (diffusion) of a volume of sand placed on an otherwise straight beach, the application being to beach nourishment projects (Chapter 12). The volume of nourished sand can be considered as a perturbation to the shoreline that is out of equilibrium with the forces of waves and nearshore currents that have controlled the natural shoreline planform. Those processes act on the nourished sand, causing it to spread in the longshore direction until the natural curvature of the shoreline has been restored. This shoreline evolution has been analyzed by Le Méhauté and Soldate (1977), Dean (1983), and Dean and Yoo (1992) as a diffusion process, the spreading of the nourished sand along the shore, with the rate of diffusion in part governed by the wave energy or height. The controlling equation is then

$$\frac{\partial y}{\partial t} = G \frac{\partial^2 y}{\partial x^2}$$

(10.11)

where the rate of shoreline change again depends on its curvature, and the coefficient G is the so-called "alongshore diffusivity," which is given by

$$G = \frac{K H_b^{5/2} \sqrt{g/\gamma_b}}{8\left(\dfrac{\rho_s - \rho}{\rho}\right) a'(h_* + B)}$$

(10.12)

where $K = 0.70$ is the coefficient in the longshore sediment transport formula [equation (9.5)], $\gamma_b \approx 0.78$ is the ratio of the wave-breaker height to the water depth, ρ_s and ρ are the sand and water densities, $a' = (1 - \text{pore space}) = 0.6$ is the pore-space factor, and h_* and B are, respectively, the closure depth and dry elevation of the beach profile. The solution of Equation (10.11) for a rectangular beach fill of longshore length λ and seaward displacement Y is

$$y(x,t) = Y\left\{ erf\left[\frac{\lambda}{4\sqrt{Gt}}\left(\frac{2x}{\lambda} + 1\right)\right] - erf\left[\frac{\lambda}{4\sqrt{Gt}}\left(\frac{2x}{\lambda} - 1\right)\right]\right\}$$

(10.13a)

where "erf" represents the error function

$$erf[z] = \frac{2}{\pi}\int_0^z e^{-u^2} du$$

(10.13b)

Figure 10-12 illustrates how the solution represents the longshore spreading of an originally rectangular block of nourished sand of longshore length $\lambda = 1$ km and offshore extent $Y = 100$ m. In this example the wave height was set at $H_b = 0.5$ m, representative of average U.S. east-coast conditions where most beach nourishment projects have been undertaken. The greater the wave-breaker height, the larger the value of G in Equation (10.12) and the greater the rate of spreading. This becomes more apparent when comparing two nourishment projects having different alongshore lengths (λ_1 and λ_2) and experiencing different wave

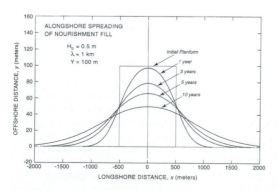

Figure 10-12 The alongcoast spreading of sand under the action of waves, where the sand originally was placed on the beach as in a nourishment project with a longshore length $\lambda = 1$ km and offshore projection $Y = 100$ m (note the extreme offshore distortion). The series of shorelines have been computed from the analytical solution of Equation (10.13), where G is given by Equation (10.12). In this example the wave-breaker height was set at 0.5 m, and the waves arrive from directly offshore.

climates. The solution of Equation (10.13) then yields an evaluation of the comparative times (t_{p1} and t_{p2}) required for a percentage p of the original volume to be transported out of the nourished area of beach fill, given by Dean (1983) as

$$t_{p2} = t_{p1} \frac{\lambda_2^2 \, H_1^{5/2}}{\lambda_1^2 \, H_2^{5/2}} \tag{10.14}$$

Important to the design of beach nourishment projects is the prediction that the fill life is proportional to the square of the fill length, so there are strong advantages in making the fill as long as possible. For example, if one-half of the volume of a fill is lost to longshore dispersion in 10 years, a fill having twice the length would lose half of its volume from the nourishment site in 40 years. The relationship also shows that doubling the wave height ($H_2 = 2H_1$) has the effect of reducing the life of the nourishment to 17.7 percent, illustrating the importance of individual storms to the longshore dispersion of the nourished sediment. The above analytical solution has seen widespread use in the preliminary design of beach nourishment projects to evaluate the longshore spreading of the sand (Chapter 12). However, more advanced analyses and predictions require the use of numerical shoreline models rather than this analytical solution.

The most complex problem that has been amenable to analytical solutions is the growth of a cuspate river delta, where sediment delivered to the coast by a river is redistributed alongshore by wave action. Pelnard-Considere (1954), Grijm (1960, 1964), Bakker and Edelman (1964), and Refaat, El-Din, and Tsuchiya (1992) have developed mathematical solutions for the growth of a delta. In order to make the problem solvable, they assumed (1) the river continuously supplies sediment at a fixed rate; (2) waves approach the shore at a constant deep-water angle and with a constant energy flux; and (3) wave refraction and diffraction are neglected. Even with these simplifying approximations, the resulting equations for the delta shoreline configuration are extremely complex. Figure 10-13 shows computed shorelines based on the solutions derived by Refaat et al. (1992), where the delta is slightly asymmetrical due to the oblique approach of the incident waves. The delta configurations are given in dimensionless form, where the dimensionless time is expressed as $t' = \varepsilon_2 t / h_k^2$; t is the actual time, h_k is the depth limit of the longshore sediment transport (i.e., approximately the closure depth), and ε_2 is a parameter that includes $\partial Q_\ell / \partial \alpha_b$ but is interpreted as a longshore diffusion coefficient. With ε_2 and h_k fixed, the series of delta planform solutions in Figure 10-13 can be viewed as the seaward growth of the delta with time.

Even with these relatively simple applications and making assumptions that further limit the reality of the results, the analytical solutions quickly become extremely complex.

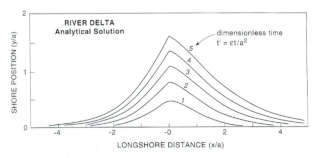

Figure 10-13 Delta shorelines computed using equations derived in an analytical solution, assuming constant wave conditions and sediment delivery by the river. The slight asymmetry of the delta is due to the general obliquity of the incident waves. The solutions are given in dimensionless form, including the time t'. [Adapted from Formation and Reduction Processes of River Deltas: Theory and Experiments, H. Refaat, A. A. El-din, and Y. Tsuchiya, *Proceedings 23rd Coastal Engineering Conference*, 1992. Reproduced with permission from the American Society of Civil Engineers.]

Furthermore, the analytical approach is incapable of dealing with the reality of such natural processes as time variations in wave conditions, the refraction of waves in the offshore, and fluctuations in the delivery of sediments to the coast by a river. The treatment of the full complexities requires the development of numerical models where the processes are simulated on a computer.

NUMERICAL MODELS OF SHORELINE CHANGE

Numerical simulation models of shoreline change have the capacity to more fully account for the variability of natural processes, both spatially and with time. The simple "line models" reduce the focus to the shoreline alone but are able to examine how its configuration changes with variations in deep-water wave conditions, offshore refraction and diffraction, changes in sediment supply, or due to the placement of a structure within the nearshore. Such line models have received the most attention in their development and in applications, so much of this section will be devoted to their review. In their most advanced development, numerical shoreline models are fully three dimensional, necessitating evaluations of cross-shore sediment transport and profile changes (Chapter 7), as well as calculations of the longshore sediment transport.

The numerical line models of shoreline planforms employ the sediment continuity relationship, usually in the finite-difference form of Equation (10.6), in conjunction with an evaluation of the longshore sediment transport that is responsible for the shoreline change. As discussed above, this transport generally is calculated with a formula like Equation (10.8), with its dependence on wave-breaker heights and angles, but a Bagnold-type transport formula with a dependence on the longshore current could be employed. The simulation models usually involve the following sequence of operations:

1. Define an initial shoreline configuration;
2. Divide the shoreline into a series of cells of width Δx;
3. Establish the various sources and losses of sand to the beach, the Q_r terms of Equation (10.6);
4. Provide the model with the offshore wave conditions (heights, periods, and approach angles);
5. Specify how the longshore sediment transport will be governed by the wave parameters, usually employing a relationship such as Equation (10.8); and
6. Run the model through increments of time, Δt, for some total span of time to determine how the shoreline evolves from its initial configuration, applying the sediment continuity equation to evaluate the shoreline changes and to ensure conservation of sediment volume.

A simple numerical shoreline model that illustrates this approach is given in Figure 10-14, an example where a jetty blocks a longshore transport of sediment. Figure 10-14(a) shows the individual cells that approximate the shoreline, each cell being $\Delta x = 25$ m wide. In total the calculations involved 100 such cells but only the few closest to the jetty are shown in the diagram. The jetty itself forms the right-hand limit to the model, since it is assumed to represent a total barrier to the transported sediment. Sediment enters the model at its far left, into cell $i = 1$, the quantity of the transport being governed by the wave conditions, which in this example are

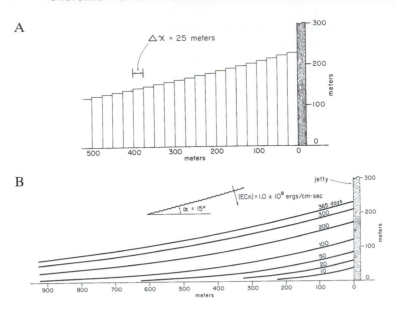

Figure 10-14   (a) Numerical model of a jetty blocking the longshore sediment transport from left to right in the diagram. Only the last 20 cells are shown ($i = 80$ through 100), the model actually extending updrift of the jetty for 2.5 km. (b) The progressive advance of the shore obtained in the model, smooth shorelines having been drawn through the 25–m wide cells. [Used with permission from John Wiley & Sons, Inc., from P. D. Komar, Modeling of Sand Transport on Beaches and the Resulting Shoreline Evolution, *The Sea*, p. 505. Copyright © 1977 John Wiley & Sons, Inc.]

set at a fixed power $(ECn)_b = 10^8$ ergs/cm·sec $= 10^3$ Watts/m with a breaker angle $\alpha_b = 15°$. At the beginning of the run, the initial shoreline is straight, corresponding to the x-axis ($y_i = 0$), and the model is run through time increments $\Delta t = 0.1$ day. Figure 10-14(a) shows cells 80 through 100 after 1 year of the littoral drift having been blocked by the jetty and the shoreline having built out due to the accumulated sand. As expected, cell 100 adjacent to the jetty has accumulated the most sand and has built out the farthest, with systematically smaller quantities having accumulated in the cells more distant from the jetty. Figure 10-14(b) shows the progressive accretion throughout the year, smooth curves having been fitted to the finite-width cells to more realistically represent the shorelines. The shorelines calculated with the simulation model are seen to be very similar to those actually observed when a new jetty or groyne blocks a net longshore sediment transport. Even with this simple model one could predict rates of shoreline advance at various positions updrift from a proposed jetty. In a real application rather than this hypothetical example, one would need to account for daily or at least seasonal fluctuations in wave conditions that might produce reversals in the sediment transport directions. The use of a constant set of wave conditions, as in the model of Figure 10-14, would at best represent the prevailing waves or the effective wave conditions that yield the correct net longshore sediment transport. In addition, of potentially greater interest in an actual application would be a model for the shoreline erosion in the downdrift direction from the jetty, a model that would predict rates of shoreline recession induced by the construction.

     The first study that involved the development of computer models of shoreline change was that of Price, Tomlinson, and Willis (1972), a study that examined changes in a beach brought about by construction of a long groyne or jetty that blocked the longshore sediment transport. Their investigation is of particular interest in that the model is on the scale of a wave basin rather than on a prototype scale, and the results from the computer model were compared with actual tests undertaken in a wave basin, a physical model. Figure 10-15 shows an example of the results obtained by Price and co-workers, with the waves approaching the initially straight beach at a 4° angle with a 4–cm offshore height. The shorelines predicted with the computer model compare favorably with the physical model within the wave basin. Being only a short stretch of beach, the shoreline in effect rotates so as to become parallel to the wave crests such that $\alpha_b \approx 0$ and $Q_\ell \approx 0$ when equilibrium is achieved, at which time little subsequent change is experienced. This is shown by both the computer model and physical experiment.

   A prototype-scale comparison between the observed effects of jetty construction and that shown in a computer model is provided by the study of Komar, Lizarraga-Arciniega, and Terich (1976). Their analysis focused on the shoreline changes associated with the construction of a pair of jetties at the mouth of the Siuslaw River on the central Oregon coast. The observed shoreline changes are shown in Figure 10-16(a). A condition of net-zero longshore transport occurs on the Oregon coast, since it in effect consists of a series of large pocket beaches or littoral cells separated by rocky headlands. With a net-zero transport, when jetties have been constructed the sand has accumulated immediately adjacent to the jetties and the shorelines have advanced to both the north and south as seen in Figure 10-16(a) for the Siuslaw River jetties. The jetty construction in effect produced embayments between the structures and the pre-jetty shoreline, the 1889 shoreline of Figure 10-16(a), which curved inward toward the river mouth. Prior to jetty construction this curved shoreline was apparently in equilibrium with the ocean waves and river and tidal flow through the inlet. Jetty construction removed the contribution from the river and tidal flows, leaving only the waves that broke obliquely to the inward-curving shoreline and therefore tended to carry sand into the embayments and toward the jetties. As seen in Figure 10-16(a), the embayments progressively filled both to the north and south of the jetties, the process taking some 20 years for the complete filling of the larger embayment to the north. The shoreline advance continued only until it became parallel to the prevailing waves, at which time a new equilibrium had been achieved, again one with a net-zero longshore sediment transport where no further rearrangements occurred. This was achieved by the mid-1930s, the present day shoreline being basically the same.

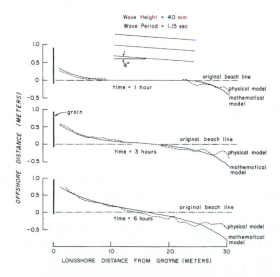

Figure 10-15   Numerical simulation model of the shoreline reorientation between two groynes, compared with the observed changes of a physical model within a laboratory wave basin. [From Predicting Changes in the Plan Shape of Beaches, W. A. Price, K. W. Tomlinson and D. H. Willis, *Proceedings of the 13th Coastal Engineering Conference*, 1972. Reproduced with permission from the American Society of Civil Engineers.]

A

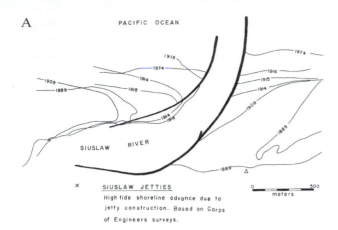

× SIUSLAW JETTIES
High tide shoreline advance due to
jetty construction. Based on Corps
of Engineers surveys.

B

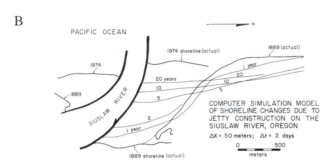

Figure 10-16 (a) Shoreline changes brought about jetty construction at the mouth of the Siuslaw River, Oregon, showing accretion to both sides. (b) A computer simulation analysis of the shoreline changes to the north of the jetties, the actual 1889 shoreline having been used as the starting configuration in the calculations. [From Oregon Coast Shoreline Changes Due to Jetties, P. D. Komar, J. R. Lizarraga-Arciniega and T. A. Terich, *Journal of Waterways, Harbors and Coastal Engineering*, 1976. Reproduced with permission from the American Society of Civil Engineers.]

The computer model of Komar, Lizarraga-Arciniega, and Terich (1976) was developed to simulate this filling of the large embayment to the north of the jetties, with the results shown in Figure 10-16(b). The surveyed 1889 shoreline was used in the computer model as the initial configuration prior to jetty construction. The shoreline was divided into a series of cells of uniform width $\Delta x = 50$ m, and the model was run with a time increment $\Delta t = 2$ days for a total period of 20 years. The waves were held constant with an energy flux that is representative of the average wave height and period on the Oregon coast and were made to approach parallel to the shoreline to the far right of the jetties beyond the effects of the river mouth, this yielding the observed net-zero longshore transport along that portion of the coast. The severest shortcoming of the model is the absence of considerations of wave refraction within the embayment, so that computed breaker angles would be larger than actually had occurred, resulting in higher sand-transport rates. In spite of this shortcoming, it can be seen in Figure 10-16 that the shoreline changes calculated with the computer model agree reasonably well with those that had occurred. The main disagreement results from the actual jetty construction not having kept pace with the shoreline advance, so that some sand bypassed the jetty and entered the estuary. Note that the model not only simulated deposition within the embayment adjacent to the jetty but also evaluated the beach erosion farther to the north from the jetty, which was required to supply sand to the accreting area. There is even an intermediate zone that first eroded and then later accreted. Such predictions would have been useful at the time the jetties were actually constructed, especially if homes had been located within the zone of shoreline erosion. This type of model makes it possible to predict actual magnitudes of shoreline erosion and accretion at various positions along the coast resulting from jetty construction. This particular example developed for the Siuslaw River jetties is a "hindcast" model, making possible an immediate comparison between model results and observed shoreline changes. However, it is apparent that such a modeling approach could be used to predict shoreline changes and magnitudes of erosion for proposed jetties.

LeBlond (1973) and Rea and Komar (1975) developed numerical line models to analyze the downdrift erosion from a headland or jetty, primarily with the objective of understanding the development of headland-bay, crenulate beaches that were examined earlier. Figure 10-17 from Rea and Komar, shows the formation of the erosive beach in the lee of a headland or breakwater as determined with a model that calculates the longshore sediment transport at all shoreline positions and then evaluates the patterns of erosion and accretion as accounted for by the sediment-continuity relationship. A distinct shadow zone of erosion develops, which shows good agreement with a logarithmic spiral that transforms downcoast into a straighter shoreline, much as found in the crenulate-bay shoreline as defined by Silvester (1960, 1970) and shown in Figure 10-7. Rea and Komar examined how the shape of the erosional shoreline depends on the offshore obliquity of wave approach and how it evolves with time. Similar models have been developed by Perlin (1979) and Perlin and Dean (1978) for shorelines affected by jetties and groynes and by Dean (1978) for the roughly circular shorelines that form landward from small gaps in segmented breakwaters that are built in the offshore approximately parallel to the original shoreline (Chapter 12).

Early in the chapter we examined the equilibrium beach within an embayment and how the shoreline shape conforms or becomes congruent with the curvature of the refracted and diffracted waves such that $\alpha_b = 0$ everywhere along the shore. This can be demonstrated with numerical models; Fig. 10-18(a) shows the model results for a pocket beach of 1-km length that reorients itself from an originally straight shoreline until it takes on the shape of the refracted wave crests. In this example the wave curvature is arbitrarily taken as parabolic, but any shape could have been used. The time increment is $\Delta t = 0.002$ day, and the longshore width of each cell is $\Delta x = 10$ m. The effects of different wave trains arriving at the pocket beach are shown in Figure 10-18(b). When the waves change, the beach responds by altering its curvature and orientation to correspond to the new wave conditions. The shoreline then wobbles about within the embayment because of the changing wave directions. In this example with a 1-km long bay, it takes several days to achieve a new equilibrium orientation where $\alpha_b \approx 0$ along the entire shore. As would be expected, additional models reveal that a larger beach takes still longer to reach equilibrium for the same wave conditions. In nature, waves change on a daily basis, so in general the larger the scale of the beach embayment, the smaller the degree of wobble under varying waves since the beach cannot fully respond.

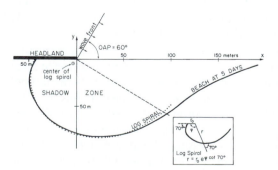

Figure 10-17 The results of a computer line model of the curved shoreline that develops downdrift from a headland or engineering structure, in this example showing good agreement with the logarithmic-spiral geometric curve of Equation (10.1) in the shadow zone where wave refraction and diffraction are greatest. [Used with permission of Society for Sedimentary Geology, from C. C. Rea and P. D. Komar, Computer Simulation Models of a Hooked Beach Shoreline Configuration, *Journal of Sedimentary Petrology 45*, p. 870. Copyright © 1975 Society for Sedimentary Geology.]

A

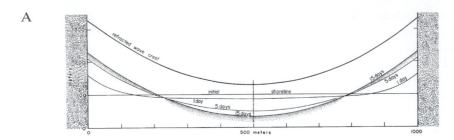

B

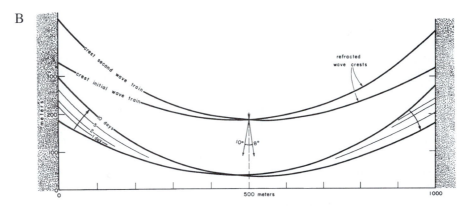

Figure 10-18   (a) The results of a numerical model that demonstrates that an initially straight shoreline between two rocky headlands takes on the same curvature as the refracted wave crests, such that after 15 days the sediment transport is reduced nearly to zero since $\alpha_b \approx 0$, this representing the equilibrium configuration. (b) A continuation of the computer-simulation analysis, showing how the pocket beach is reoriented by a change in wave direction, with accretion occurring at one end of the beach while erosion takes place simultaneously at the opposite end.

With a source of sand to the beach, as from a river, the equilibrium shoreline has to make an angle with the refracted wave crests in order for the sand to move away from the river mouth and become distributed along the shoreline. This is illustrated in Figure 10-19 where a stream supplying sand at the rate 1000 m³/day has been introduced into the beach embayment; the conditions of the model are otherwise the same as in Figure 10-18. The shoreline shown in the diagram is the equilibrium form achieved after several years of simulated shoreline change. The refracted wave crest breaks at an angle to the flanks of the delta, and the angle of wave breaking is greatest adjacent to the river mouth. The breaker angle is slightly larger on the left flank than on the right, and this results because more of the river-supplied sand must move to the left where there is a longer stretch of shore to build out than exists to the right of the river. The breaker angles systematically decrease with longshore distance away from the river mouth, as there is progressively less sand to be transported, some being left behind to account for the accretion. Although greatly simplified, the shoreline of Figure 10-19 provides an excellent example of how an equilibrium configuration is achieved, where sediments supplied to the beach are evenly distributed along the shore by the prevailing wave conditions.

The geometry of delta shorelines has been explored in detail by Komar (1973) through the use of numerical simulation models. Figure 10-20(a) shows the 1-year growth of a delta

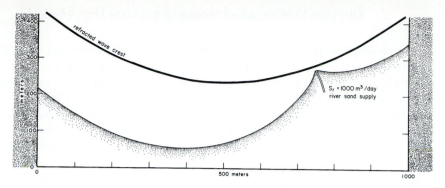

Figure 10-19 Computer simulation model of the equilibrium shoreline configuration within a pocket beach of 1 km length, where a river supplies sand to form a delta. In this example, $ECn = 1.5 \times 10^7$ ergs/cm·sec.

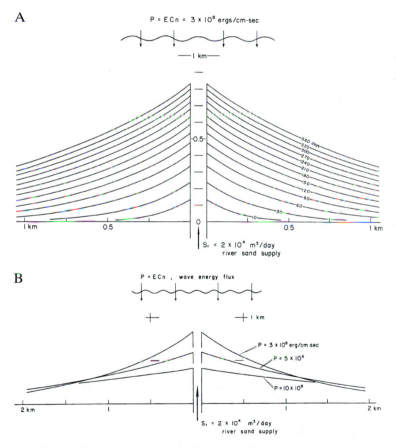

Figure 10-20 (a) Computer line model of the first 360 days growth of a river delta where waves arrive straight onshore to an initially straight shoreline. Values of wave energy flux and river sediment input are shown. (b) Effects on the 360-day delta configuration resulting from different wave energy fluxes, while maintaining a constant river sediment supply. [From *Geological Society of America Bulletin 84*, P. D. Komar. Reproduced with permission of the publisher, the Geological Society of America, Boulder, Colorado USA. Copyright © 1973.]

where the river supplies sand at the rate 2×10^4 m³/day, tending to build out a delta, while waves of energy flux 3×10^8 ergs/cm·sec $= 3 \times 10^3$ Watts/m work to flatten the delta form. After some 150 days there is a balance such that the delta continues to grow outward at a steady rate, maintaining its overall shape in the process. An equilibrium has been achieved in which there is a shoreline curvature where the waves are just able to redistribute the sand along the shore at the rate it is supplied by the river. In Figure 10-20(b) the effect of varying the wave-energy flux is illustrated; the greater the energy flux for a constant river sand supply, the smaller the breaker angles required to redistribute the sand and the flatter the resulting equilibrium delta. Decreasing the river sand supply while maintaining the wave energy flux would have had the same effect.

Figure 10-21 illustrates one possible type of application of computer models of shoreline change, an example where a river first enters the ocean with its mouth at the 0 km position, but after 10 years of delta growth the mouth suddenly shifts to the flank of the original delta. The resulting growth lines are indicated, and these can be imagined as representing series of old beach ridges (Chapter 2). A complex relationship of beach ridges results from this simple shift of the river mouth. The shoreline erodes at the first position of the river after it has shifted because the waves continue to transport sand alongshore, but the river source has moved. Ridges are truncated, while at the same time the delta builds out rapidly at the new mouth. Much more sand moves alongshore to the left of the new mouth than to the right, owing to the mass of sand around the old river mouth causing reduced breaker angles in that direction. The ridges on the left are therefore more widely spaced than on the right. It is apparent that such computer models could be used to test hypotheses regarding the sequence of events in the growth of systems of beach ridges. Not only might the sediment source shift position as in this example, but the wave directions and energy fluxes could change and alter the patterns of ridges. One intriguing possibility is that a model such as that in Figure 10-21, could be run backward through time rather than forward (Komar, 1977). One could start with the configuration of ridges, as shown in Figure 10-21, and compute the sand transport along the beach but have it move backward, that is, the reverse of the actual transport directions. Running time in reverse, the sand would systematically be

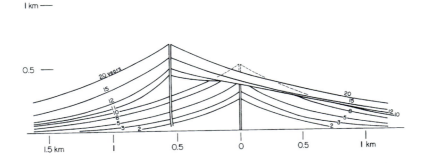

Figure 10-21 Numerical line model of a delta that consists of a complex pattern of shore-growth lines that could be represented in nature as beach ridges, resulting from the shifted position of the river's mouth after 10 years of delta growth. The shift results in the truncation of part of the original delta, followed by the overlap of the shorelines from the newly growing delta. [Used with permission from John Wiley & Sons, Inc., from P. D. Komar, Modeling of Sand Transport on Beaches and the Resulting Shoreline Evolution, *The Sea*, p. 510. Copyright © 1977 John Wiley & Sons, Inc.]

removed from the delta and beach ridges and be "pushed" back into the river. This approach would provide a still more powerful technique for unraveling the history of development of beach-ridge patterns.

The shoreline simulation models developed by Motyka and Willis (1974) provide a particularly good example of their use in applications and also illustrate a problem where it was necessary to couple the shoreline model with wave refraction analyses. The necessity for computing wave refraction considerably increases the complexity of the models in that an offshore array is required to account for the bottom topography. Both the offshore topography and shoreline configuration could evolve through time. The problem analyzed by Motyka and Willis involved an examination of whether dredging of sand and gravel from the continental shelf could alter the wave refraction to a sufficient degree that it would induce shoreline erosion. This is shown schematically in Figure 10-22(a), where a circular dredge hole in the offshore affects the wave refraction, which causes the waves to break at larger angles along the shore, resulting in localized beach erosion in the lee of the dredged area. One example of the model calculations of Motyka and Willis is shown in Figure 10-22(b). This specific example considers the effects of dredging a 4-m deep hole in 7-m water depth, 500 m offshore. The model used realistic profiles of the beach and offshore, with a dredged hole inserted. The root-mean-square wave height was 0.4 m and periods of 5 and 8 sec were used. Wave directions were selected to yield a net longshore sediment transport of 30,000 m^3/year. The model performed the following sequence of operations:

1. Calculate breaking wave conditions by refracting the deep-water waves over the sea bed,
2. Calculate the longshore sand transport rates at each cell position along the shoreline,

A

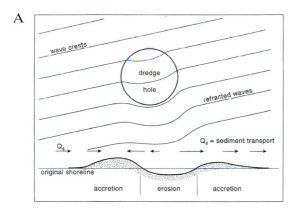

B

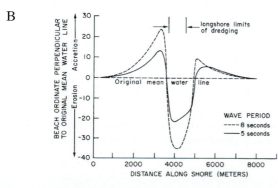

Figure 10-22 (a) Schematic diagram of the effects of offshore dredging that creates a deeper hole that alters the wave refraction and induces shoreline changes. (b) Patterns of shoreline erosion and accretion obtained with a computer model designed to simulate the effects of offshore dredging. [From The Effect of Refraction Over Dredged Holes, J. M. Motyka and D. H. Willis, *Proceedings of the 14th Coastal Engineering Conference*, 1974. Reproduced with permission from the American Society of Civil Engineers.]

3. Calculate the resulting change in beach plan shape,
4. Distribute the accretion and erosion over the inshore bed, and
5. Recalculate the wave refraction and return to step 2.

The results in Figure 10-22(b) predict that the dredged hole would cause erosion of the shoreline in its lee and an advance to either side. The pattern is asymmetrical owing to the superimposed longshore sediment transport resulting from the overall oblique-wave approach. The shoreline alterations are greater under the 8-sec waves than with the 5-sec waves in that the longer-period waves undergo more refraction. The models predict major erosion resulting from offshore dredging, a shoreline retreat of 20–35 m existing over a kilometer of shoreline length. The model also predicts that there are zones of shoreline accretion marginal to the area of erosion directly in the lee of the dredged hole. This results from the decreasing breaker angles with distance from the erosion area, which results in decreasing quantities of transported sediment ($dQ_\ell/dx = -$), yielding the shoreline accretion ($dy/dt = +$) according to Equation (10.7). The predictions of the Motyka and Willis model appear to correspond to an example of shoreline erosion that occurred on Grand Isle, Louisiana, induced by offshore dredging (Combe and Soileau, 1987). The dredging occurred over a wide area about 500 m offshore and lowered the bottom by 3–6 m. The photograph of Figure 10-23 shows the resulting development of an erosional embayment between accretional cusps, appearing to correspond to that obtained in the numerical models of Motyka and Willis depicted in Figure 10-22(b). However, research underway suggests that in this specific example, the shoreline response may have been more complex than analyzed by Motyka and Willis, with wave reflection from the landward wall of the dredged hole reducing the wave energy in the lee of the hole.

The most recent advances in numerical line models to simulate shoreline change have been incorporated into GENESIS, an acronym for Generalized Model for Simulating Shoreline Change. One of its chief contributions is in providing a flexible model that can be applied to an arbitrary prototype situation, one that calculates wave transformations as they shoal and undergo refraction and diffraction, calculates the patterns of longshore sediment transport, and then determines the resulting shoreline changes. The full technical development of GENESIS is provided in the report by Hanson and Kraus (1989), while the report by Gravens, Kraus, and Hanson (1991) is a workbook and user's manual.

One of the principal modifications of the GENESIS model is in the calculation of longshore sediment transport rates, an approach that includes both the transport due to waves breaking obliquely to the shoreline and from longshore variations in wave-breaker heights. From the Bagnold model in Chapter 8 [Equation (8.10)], we have

$$Q_\ell = 0.088\rho g(H_{br})^2 v_\ell \tag{10.15}$$

for the volumetric longshore sediment transport rate, where H_{br} is the root-mean-square wave-breaker height, and v_ℓ is the longshore current. In this case the longshore current is given by Equation (7.17) derived in Chapter 7,

$$v_\ell = 1.17\sqrt{gH_b}\,\sin\alpha_b\cos\alpha_b - a\sqrt{gH_b}\,\frac{dH_b}{dx} \tag{10.16}$$

which shows the dependence on the longshore gradient of the wave height, dH_b/dx, where x is the longshore coordinate, as well as the dependence on the breaker angle α_b. In SBEACH this is expressed as

Figure 10-23 The erosional embayment and adjacent accretional cusps that developed on the shoreline of Grand Isle, Louisiana, in response to offshore dredging. [From Behavior of Man-Made Beach and Dune, Grand Isle, Louisiana, A. J. Combe and C. W. Soileau, *Coastal Sediments '87*, 1987. Reproduced with permission from the American Society of Civil Engineers.]

$$Q_\ell = (H^2 C_g)_b \left(a_1 \sin \alpha_b \cos \alpha_b - a_2 \cos \alpha_b \frac{dH_b}{dx} \right) \qquad (10.17)$$

where $C_g \approx \sqrt{gh_b} \approx \sqrt{gH_b/\lambda}$ is the wave group velocity. The coefficients a_1 and a_2 contain the sediment and water densities and beach-sand porosity but also include dimensionless coefficients that are empirically varied to calibrate SBEACH in applications. The inclusion of the dH_b/dx factor accounts for longshore currents and sand transport produced by longshore variations in wave-breaker heights, a process that can be important within the partially sheltered regions of breakwaters, jetties, and groynes (Chapters 12). The importance of this inclusion in numerical shoreline models was first demonstrated by Kraus and Harikai (1983) in their analyses of Oarai Beach, Japan, where wave diffraction behind a long breakwater is a dominant process. The importance has been shown further in some applications of GENESIS where the processes are affected by shoreline structures.

Hanson, Gravens, and Kraus (1988) applied GENESIS to several example analyses, including a simulation of the shoreline development on Homer Spit, Alaska, and the erosion downdrift from a seawall on Sandy Hook, New Jersey. The most complex analysis focused on the shoreline changes at Lakeview Park, Lorain, Ohio, on Lake Erie, an example that has been examined in greater detail by Hanson and Kraus (1991a). This application involved simulating with GENESIS the shoreline changes observed at Lakeview Park following the construction of three detached breakwaters in the offshore, two bounding groynes, and the placement of sand to create a new beach (Fig. 10-24). The simulation involved an analysis of wave diffraction through the gaps between the breakwaters and the resulting readjustment of the shoreline from its original smooth curvature after sand placement. Figure 10-24 shows a measured shoreline compared with one calculated by GENESIS after several months of sand movement by the waves. Cusps or salients have developed in the sheltered region behind each breakwater, with intervening bays between and opposite the gaps between the breakwaters. The agreement between the measured shoreline and that computed by GENESIS is seen to be excellent, but it needs to be recognized that there were five empirical coefficients that could be adjusted in the analysis to maximize the fit, a_1 and a_2 in Equation (10.17) for the evaluation of the sediment transport, and three coefficients that represent the amount of wave-energy transmitted through the three breakwaters. However, Hansen and Kraus found that the agreement is still very good even if it is accepted that all three breakwaters had the same energy transmission, thereby decreasing the number of adjustable coefficients in the analysis. After the model had been calibrated and tested versus observed shoreline changes, Hanson and Kraus explored alternative project designs for maintaining the beach fill in place. This included analyses of the beach retention for various lengths of the bounding groynes and for the absence of groynes. Such analyses again demonstrate the usefulness of numerical shoreline models, and specifically of GENESIS, as many design alternatives can be examined with minimal expense and probably with better results than could be obtained with physical models undertaken in scaled laboratory experiments. Hanson and Kraus (1991b) compared predictions from GENESIS to the results obtained in physical models, again for a series of detached breakwaters such as those at Lakeview Park, but also for the trapment of sand by a series of groynes built across the beach. In all cases the numerical models closely reproduced the shoreline changes that occurred in the physical models.

The numerical models discussed thus far have been "line models," limited to a depiction of the shoreline and how it changes in response to varying wave conditions, structures, etc. A multi-line model has been developed by Perlin and Dean (1978) and Perlin (1979) that analyzes changes in a series of depth contours as well as at the shoreline and thereby provides some three dimensionality to the results. Only a few attempts have been made to develop fully three-dimensional analyses of the nearshore topography, models that include evaluations of the cross-shore sediment transport as well as the longshore transport. To accomplish this, the models generally have employed the Bagnold approach to evaluate the sediment transport, as formulated by Bowen (1980) and Bailard and Inman (1981). A few of the three-dimensional numerical models were developed to explore the processes that affect the morphology of a beach. For example, Holman and Bowen (1982) examined the role of standing and progressive edge waves in the formation of crescentic and transverse bars; their study will

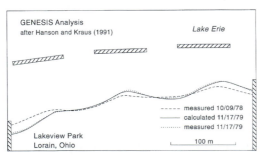

Figure 10-24 Application of GENESIS to simulate shoreline changes at Lakeview Park, Lorain, Ohio, resulting from the construction of three offshore breakwaters, two groynes, and the placement of sand on the beach. The computed shoreline corresponds closely to the measured configuration. [Adapted from Numerical Simulation of Shoreline Change at Lorain, Ohio, H. Hanson and N. C. Kraus, *Journal of Waterway, Port Coastal and Ocean Engineering,* 1991. Reproduced with permission from the American Society of Civil Engineers.]

be reviewed in Chapter 11. Yamashita and Tsuchiya (1992) developed a three-dimensional numerical model to simulate the formation of a pocket beach between shore-parallel break-waters. With the presence of structures and a non-uniform bottom, the analyses had to include evaluations of the complete refraction/diffraction patterns of the waves, the circulation induced by longshore variations in wave-breaker heights, and the cross-shore and alongshore components of the sediment transport. Figure 10-25 illustrates their results, with the upper diagram depicting the computed patterns of the nearshore circulation, while the lower diagram shows the topography of the beach that eroded behind a single breakwater. The configuration is seen to be similar to that in Figure 10-15 obtained by Rea and Komar (1975) for the shoreline derived from a line model and to the crenulate-bay shoreline of Figure 10-7 as defined by Silvester (1960, 1970), but the analysis has been extended by Yamashita and Tsuchiya to the full three dimensions.

SUMMARY

The large-scale planform configuration of the shoreline tends to approach an equilibrium in which the wave climate provides precisely the energy and mean wave approach angles required to transport and redistribute sediments supplied to the beach. The beach configuration is controlled mainly by the curvature and orientation of the refracted wave crests and by the locations and relative importance of the beach-sediment sources and losses. In the simplest case, the pocket beach between headlands or structures such as groynes takes on a curvature that is congruent to the refracted and diffracted wave crests. It has been found that this curvature can be approximated by the logarithmic-spiral geometric form of Equation (10.1) or by the crenulate beach, which consists of a log-spiral shore and a tangential straight shoreline in the downdrift direction. However, more comprehensive analyses of shoreline shapes involve the development of numerical computer models that evaluate patterns of shoreline erosion and accretion due to the longshore sediment transport, accounting for the time-evolution of the shoreline shape and position. Numerical shoreline models have been used in geomorphic analyses, for example, to examine the shapes of deltas controlled by the rate of delivery of sediment to the coast by the river versus the rate of sediment redistribution alongshore by waves. The greatest potential for such models in practical applications comes in examinations of designs and impacts of engineering structures such as breakwaters, jetties, and groynes. In the past the expected shoreline changes and patterns of erosion and accretion have been examined with physical models in laboratory wave basins, scaled down from the prototype. Such models are expensive and the results somewhat uncertain due to the scaling of sediment transport processes. Most analyses of engineering structures and shoreline impacts can now be undertaken with computer simulation models. A number of examples have been illustrated in this chapter, ranging from sand blockage by jetties and groynes, to sand redistribution behind offshore breakwaters and to the impact on shoreline erosion by offshore dredging which alters the patterns of wave refraction. During the past 20 years, the use of computer simulations to examine and predict shoreline changes has become standard practice in engineering analyses. Most of these models have been simple line models where the coast is represented solely by the shoreline. The more advanced numerical models include a consideration of the full three dimensions of the beach and nearshore processes and include calculations of wave refraction and diffraction patterns, nearshore currents, and cross-shore as well as alongshore sediment

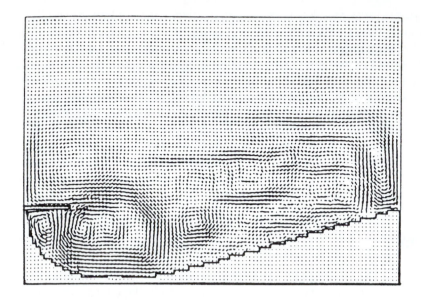

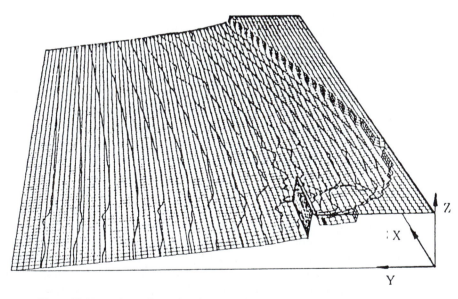

Figure 10-25 A three-dimensional numerical model of the erosion of a crenulate bay behind a shore-parallel breakwater or seawall. The upper diagram shows the computed pattern of wave-induced nearshore currents, while the lower diagram gives the computed beach topography. [From Numeric Simulation of Pocket Beach Formation, T. Yamashita and Y. Tsuchiya, *Proceedings of the 23rd Coastal Engineering Conference*, 1992. Reproduced with permission from the American Society of Civil Engineers.]

transport rates. The development of three-dimensional models can be expected to see continued advances but will have to be accompanied by an increased understanding of nearshore processes such as the generation of nearshore currents and the processes responsible for the formation of complex nearshore-bar systems.

PROBLEMS

1. Calculate and graph the log-spiral shoreline form with Equation (10.1) that corresponds to the shape of Halfmoon Bay in Figure 10-3, using the values $r_0 = 500$ m and $\alpha = 41.25°$ determined by Yasso (1965) for closest congruence. Change the α coefficients to 30° and 50° to establish how the curvature of the shoreline varies with this parameter.

2. Along a 200–m stretch of beach, the wave-breaker angle increases from 5° to 15° while the root-mean-square breaker height remains constant at 2.5 m. Estimate the rate of shoreline change (dy/dt) that would occur due to the resulting gradient in the longshore sediment transport rate. Evidence suggests that nearshore sediments extend out to a depth of about 5 m.

3. A river supplies sediment at a rate 2×10^4 m³/day to a growing delta, with half of the sediment moving toward each of the delta flanks, producing a fairly uniform shoreline advance of 0.75 m/year for the first 2 km from the river mouth, with the sediment extending out to a depth of 6 m. What gradient in the longshore transport accounts for this shoreline advance rate, and what is the longshore sediment transport rate Q_ℓ at 2 km distance from the river? If the root-mean-square wave breaker height is 1.5 m everywhere along the length of the delta shoreline, evaluate the longshore variation in the breaker angles that account for the shoreline accretion.

4. Program Equations (10.13) for the calculation of the longshore dispersion of a beach-nourishment sand fill. Calculate the evolution for a quartz-density sand fill having a vertical dimension $h_* + B = 6$ m, longshore length of 1 km, and offshore extent of 100 m; perform calculations for breaking-wave heights 0.5 m and 1.5 m in order to examine the effect of the wave-energy level on the rate of dispersion. Repeat the analysis for a fill having a longshore length of 3 km to examine the impact of having a longer fill.

5. Discuss conceptually how the application of Problem 4 (above) for a beach-nourishment fill could be analyzed using numerical computer-simulation techniques. What processes and variables could be included in this simulation analysis that were neglected in the analytical solution leading to Equations (10.13)?

REFERENCES

BAKKER, W. T. (1968). The Dynamics of a Coast with a Groyne System. *Proceedings of the 11th Coastal Engineering Conference, Amer. Soc. Civil Engrs.,* pp. 492–517.

BAKKER, W. T., E. H. BRETELER, and A. ROOS (1970). The Dynamics of a Coast with a Groyne System. *Proceedings of the 12th Coastal Engineering Conference, Amer. Soc. Civil Engrs.,* pp. 1001–1020.

BAKKER, W. T. and T. EDELMAN (1964). The Coastline of River Deltas. *Proceedings of the 9th Coastal Engineering Conference, Amer. Soc. Civil Engrs.,* pp. 199–218.

BAILARD, J. A. and D. L. INMAN (1981). An Energetics Bedload Model for a Plane Sloping Beach: Local Transport. *Journal of Geophysical Research* 86: 2035–2043.

BOWEN, A. J. (1980). Simple Models of Nearshore Sedimentation. Beach Profiles and Longshore Bars. In *The Coastline of Canada,* S. B. McCann (editor). Pp. 1–11. Geological Survey of Canada.

BREMMER, J. M. and P. H. LeBLOND (1974). On the Planimetric Shape of Wreck Bay, Vancouver Island. *Journal of Sedimentary Petrology* 44: 1155–1165.

COMBE, A. J. and C. W. SOILEAU (1987). Behavior of Man-Made Beach and Dune, Grand Isle, Louisiana. *Coastal Sediments '87, Amer. Soc. Civil Engrs.,* pp. 1232–1242.

DAVIDSON-ARNOTT, R. G. D. and W. H. POLLARD (1980). Wave Climate and Potential Longshore Sediment Transport Patterns, Nottawasga Bay, Ontario. *Journal of Great Lakes Research* 6: 54–67.

DAVIES, J. L. (1958). Wave Refraction and the Evolution of Shoreline Curves. *Geographical Studies* 5: 1–14.

DEAN, R. G. (1978). Diffraction Calculation of Shoreline Planforms. *Proceedings of the 16th Coastal Engineering Conference, Amer. Soc. Civil Engrs.,* pp. 1903–1917.

DEAN, R. G. (1983). Principles of Beach Nourishment. In *CRC Handbook of Coastal Processes and Erosion,* P. D. Komar, (editor). Boca Raton, FL: CRC Press, pp. 217–232.

DEAN, R. G. and C. YOO (1992). Beach-Nourishment Performance Predictions. *Journal of Waterway, Port, Coastal, and Ocean Engineering, Amer. Soc. Civil Engrs.* 118(6): 567–586.

FINKELSTEIN, K. (1982). Morphological Variations and Sediment Transport in Crenulate-Bay Beaches, Kodiak Island, Alaska. *Marine Geology* 47: 261–281.

FRIHY, O. E., A. M. FANOS, A. A. KHAFAGY, and P. D. KOMAR (1991). Patterns of Nearshore Sediment Transport Along the Nile Delta, Egypt. *Coastal Engineering* 15: 409–429.

GRAVENS, M. B., N. C. KRAUS, and H. HANSON (1991). *GENESIS: Generalized Model for Simulating Shoreline Change, Report 2: Workbook and System User's Manual.* Technical Report CERC-89-19, U.S. Army Engr. Waterways Exp. Station, Coastal Engineering Research Center, Vicksburg, Mississippi.

GRIJM, W. (1960). Theoretical Forms of Shorelines. *Proceedings of the 7th Coastal Engineering Conference, Amer. Soc. Civil Engrs.,* pp. 197–202.

GRIJM, W. (1964). Theoretical Forms of Shorelines. *Proceedings of the 9th Coastal Engineering Conference, Amer. Soc. Civil Engrs.,* pp. 219–235.

GREENWOOD, B. and D. G. McGILLIVARY (1980). Modelling the Impact of Large Structures upon Littoral Transport in the Central Toronto Waterfront, Lake Ontario, Canada. *Zeitschrift für Geomorphplogie N.F.* 34: 97–110.

HANSON, H., M. B. GRAVENS, and N. C. KRAUS (1988). Prototype Applications of a Generalized Shoreline Change Numerical Model. *Proceedings of the 21st Coastal Engineering Conference, Amer. Soc. Civil Engr.,* pp. 1265–1279.

HANSON, H. and N. C. KRAUS (1989). *GENESIS: Generalized Model for Simulating Shoreline Change, Report 1: Reference Manual and Users Guide.* Technical Report CERC-89–19, U.S. Army Engineer Waterways Experiment Station, Coastal Engineer Research Center, Vicksburg, Mississippi.

HANSON, H. and N. C. KRAUS (1991a). Numerical Simulation of Shoreline Change at Lorain, Ohio. *Journal of Waterway, Port, Coastal, and Ocean Engineering, Amer. Soc. Civil Engrs.,* 117: 1–18.

HANSON, H. and N. C. KRAUS (1991b). Comparison of Shoreline Change Obtained with Physical and Numerical Models. *Coastal Sediments '91, Amer. Soc. Civil Engrs.,* pp. 1785–1799.

HOLMAN, R. A. and A. J. BOWEN (1982). Bars, Bumps, and Holes: Models for the Generation of Complex Beach Topography. *Journal of Geophysical Research* 87(C1): 457–468.

HOYLE, J. W. and G. T. KING (1957). The Origin and Stability of Beaches. *Proceedings of the 6th Conference on Coastal Engineering, Amer. Soc. Civil Engrs.,* pp. 281–301.

HSU, J. R., R. SILVESTER, and Y. M. XIA (1989). Static Equilibrium Bays: New Relationships. *Journal of Waterways, Port, Coastal, and Ocean Engineering, Amer. Soc. Civil Engrs.* 115(3): 285–298.

KOMAR, P. D. (1973). Computer Models of Delta Growth Due to Sediment Input from Rivers and Longshore Transport. *Geological Society of America Bulletin* 84: 2217–2226.

KOMAR, P. D. (1977). Modeling of Sand Transport on Beaches and the Resulting Shoreline Evolution. In *The Sea,* E. Goldberg et al. (editors). Vol. 6, pp. 499–513. New York: Wiley-Interscience.

KOMAR, P. D., J. R. LIZARRAGA-ARCINIEGA, and T. A. TERICH (1976). Oregon Coast Shoreline Changes Due to Jetties. *Journal of Waterways, Harbors, and Coastal Engineering, Amer. Soc. Civil Engr.,* 102(WW1): 13–30.

KRAUS, N. C. and S. HARIKAI (1983). Numerical Model of the Shoreline at Oarai Beach. *Coastal Engineering* 7: 1–28.

LEBLOND, P. H. (1973). On the Formation of Spiral Beaches. *Proceedings of the 13th Coastal Engineering Conference, Amer. Soc. Civil Engr.*, pp. 1331–1345.

LEBLOND, P. H. (1979). An Explanation of the Logarithmic Spiral Plan Shape of Headland-Bay Beaches. *Journal of Sedimentary Petrology* 49: 1093–1111.

LEMEHAUTE, E. M. and M. SOLDATE (1977). *Mathematical Modeling of Shoreline Evolution.* U.S. Army Corps of Engineers, Coastal Engineering Research Center Misc. Report No, 77-10.

LEWIS, W. V. (1938). The Evolution of Shoreline Curves. *Proceedings of the Geological Association of England* 49: 107–127.

MOTYKA, J. M. and D. H. WILLIS (1974). The Effect of Refraction over Dredged Holes. *Proceedings of the 14th Coastal Engineering Conference, Amer. Soc. Civil Engrs.*, pp. 615–625.

PELNARD-CONSIDERE, R. (1954). *Essai de Théorie de l'Évolution des Formes de Revage Plages de Sable et de Galets.* Société Hydrotechnique de France, IVes Journées de l'Hydraulique, Les Energies de la Mer, Paris, Question 3.

PERLIN, M. (1979). Predicting Beach Planforms in the Lee of a Breakwater. *Proceedings of the Coastal Structures '79, Amer. Soc. Civil Engrs.*, pp. 792–808.

PERLIN, M. and R. G. DEAN (1978). Prediction of Beach Planforms with Littoral Controls. *Proceedings of the 16th Coastal Engineering Conference, Amer. Soc. Civil Engrs.*, pp. 1818–1838.

PHILLIPS, J. D. (1985). Headland-Bay Beaches Revisited: An Example from Sandy Hook, New Jersey. *Marine Geology* 65: 21–31.

PRICE, W. A., K. W. TOMLINSON, and D. H. WILLIS (1972). Predicting Changes in the Plan Shape of Beaches. *Proceedings of the 13th Coastal Engineering Conference, Amer. Soc. Civil Engrs.*, pp. 1321–1329.

REA, C. C. and P. D. KOMAR (1975). Computer Simulation Models of a Hooked Beach Shoreline Configuration. *Journal of Sedimentary Petrology* 45: 866–872.

REFAAT, H., A. A. EL-DIN, and Y. TSUCHIYA (1992). Formation and Reduction Processes of River Deltas: Theory and Experiments. *Proceedings of the 23rd Coastal Engineering Conference, Amer. Soc. Civil Engrs.*, pp. 2772–2785.

SILVESTER, R. (1960). Stabilization of Sedimentary Coastlines. *Nature* 188(4749): 467–469.

SILVESTER, R. (1970). Growth of Crenulated Shaped Bays to Equilibrium. *Journal of Waterways and Harbors Division, Amer. Soc. Civil Engrs.* 96(WW2): 275–287.

SILVESTER, R. and S. K. HO (1972). Use of Crenulate-Shaped Bays to Stabilize Coasts. *Proceedings of the 13th Coastal Engineering Conference, Amer. Soc. Civil Engrs.*, pp. 1347–1365.

TAN, S.-K. and Y.-M. CHIEW (1994). Analysis of Bayed Beaches in Static Equilibrium. *Journal of Waterway, Port, Coastal, and Ocean Engineering, Amer. Soc. Civil Engrs.* 120: 145–153.

TANNER, W. F. (1958). The Equilibrium Beach. *Transactions American Geophysical Union* 39: 889–891.

TERPSTRA, P. D. and M. J. CHRZASTOWSKI (1992). Geometric Trends in the Evolution of a Small Log-Spiral Embayment on the Illinois Shore of Lake Michigan. *Journal of Coastal Research* 8: 603–617.

WIND, H. G. (1994). An Analytical Model of Crenulate Shaped Beaches. *Coastal Engineering* 23: 243–253.

YAMASHITA, T. and Y. TSUCHIYA (1992). Numerical Simulation of Pocket Beach Formation. *Proceedings of the 23rd Coastal Engineering Conference, Amer. Soc. Civil Engrs.*, pp. 2556–2566.

YASSO, W. E. (1965). Plan Geometry of Headland Bay Beaches. *Journal of Geology* 73: 702–714.

11

Nearshore Morphodynamics

The world thus appears as a complicated tissue of events, in which connections of different kinds alternate or overlap or combine, and thereby determine the texture of the whole.

Werner von Heisenberg
1901–1976

A two-dimensional analysis of beach profiles was presented in Chapter 7, including an examination of the overall slope of the beach, the development of bars and troughs, and the processes of cross-shore sediment transport that are responsible for changes in the profile morphology. However, beaches commonly have morphologies that are more complex (Fig. 11-1), being three dimensional with systems of offshore bars that are crescentic or trend obliquely toward the offshore, systems that can have a distinct longshore rhythmicity. This three dimensionality extends to the exposed part of the beach and to the beach face and berm, which often contains alternating cusps and embayments having scales of a few meters (beach cusps) to hundreds of meters (giant cusps), and up to tens of kilometers if capes are included. This chapter focuses on the full complexity of three-dimensional beaches and, in particular, on the variety of rhythmic shoreline forms: beach cusps, crescentic bars, transverse bars, and the embayments and cusps associated with rip currents. These diverse morphological systems can develop at the same time, and this contributes to the overall complexity of the beach. However, we now have a reasonable understanding of the origins for many of these systems of bars and cuspate shorelines, and this knowledge can assist us in deciphering the elements within the complexity of the nearshore morphology. This growing understanding has permitted the development of genetic classifications that incorporate many aspects of the beach morphology and associated processes of waves and currents—the classification where beaches range from dissipative to reflective. This classification was introduced in Chapter 3, limited at that time to a two-dimensional view; now the classification can be examined in its full three dimensions in order to establish how the various patterns of nearshore bars and rhythmic shorelines fit into the scheme.

RHYTHMIC SHORELINE FORMS

Beaches are seldom straight or smooth in curvature for any significant longshore distance (Fig. 11-1). Instead, they commonly contain seaward projections of sediment that trend at right angles to the shoreline and are known variously as beach cusps, sand waves, shoreline rhythms, or giant cusps (Shepard, 1952; Bruun, 1954; Bakker, 1968; Hom-ma and Sonu, 1963; Zenkovitch, 1967; Dolan, 1971). Although such features can be isolated, they more commonly occur in series with a fairly uniform spacing, the horizontal distance between successive cusps.

Figure 11-1 Beaches can be extremely irregular and three dimensional. In this example at Nestucca Spit, Oregon, the irregularity is due mainly to the patterns of nearshore currents and rip currents, which erode troughs and embayments, leaving a series of cusps extending out from the berm.

A wide range of cusp spacings can be found on beaches. Along the shores of ponds and small lakes, the spacings may vary from less than 10 cm to 1 m (Johnson, 1919, p. 467; Evans, 1938; Komar, 1973). Small-scale cusps can also be generated in laboratory wave basins (Longuet-Higgins and Parkin, 1962; Flemming, 1964; Guza and Inman, 1975; Kaneko, 1985). On ocean beaches with small waves, the cusp spacings may be less than 2 m, while those built by large storm waves may be 50 m or more in their spacings. Russell and McIntire (1965) provide numerous cusp observations and statistics from ocean beaches, with spacings ranging from 6 to 57 m. Dolan's (1971) measurements of "shoreline rhythm" from the North Carolina coast yielded spacings between successive cusps ranging from 150 to 1,000 m, with most between 500 and 600 m. The cusps projected on average some 15–25 m seaward from the embayments. Shepard's (1952) "giant cusps" ranged up to 1,500 m in spacings.

Attempts to classify rhythmic shoreline forms generally have stressed the magnitudes of their spacings. Beach cusps were considered to have smaller spacings, less than 25 m (Dolan and Ferm, 1968; Dolan, Vincent, and Hayden, 1974), while sand waves, rhythmic topography, and giant cusps have larger spacings. These latter terms can be considered to be nearly synonymous, different names for the same or very similar features. However, since it is now apparent that cuspate shorelines with a wide range of spacings can be produced by a single mechanism, and more than one mechanism may be capable of producing cuspate shorelines having the same spacings, only a genetic classification will be satisfactory, one that clearly reflects the processes of cusp formation. An attempt at such a classification is given in Figure 11-2, derived from the review of Komar (1983a). The discussion here considers this series of cuspate features and associated bars in the order shown in the figure, which generally also

RHYTHMIC SHORELINE FORMS

reflective beach cusps

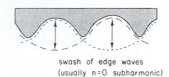

swash of edge waves
(usually n=0 subharmonic)

rip current embayment-cusp system

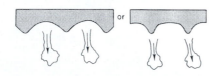

crescentic bar-cusp system

transverse and oblique bars

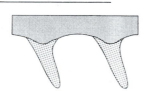

Figure 11-2   A genetic classification of rhythmic shoreline forms, where the origin of the series of cusps and embayments can be attributed to different processes of waves and currents within the nearshore. [Used with permission of Claredon Press, from P. D. Komar, Rhythmic Shoreline Features and Their Origins, *Mega-Geomorphology*, p. 107. Copyright © Claredon Press.]

parallels their progressively increasing sizes, beach cusps tending to be the smallest and crescentic bars and transverse bars the largest rhythmic shoreline forms. For completeness, this review will include a brief examination of the origin of capes such as those on the southeast coast of the United States, a shoreline form where the spacings between successive cusps extend to hundreds of kilometers.

Beach Cusps

The most intriguing morphological structures observed on beaches are the cuspate deposits of sand and gravel built by wave action and known as *beach cusps* (Fig. 11-3). Because of their marked regularity, beach cusps have attracted many observers and much speculation as to their origin. So much has been written about beach cusps that practically every observation and idea concerning their origin advanced by one author seems to have been directly contradicted by another. Because of such contradictions, arguments persist regarding the origin of beach cusps and which processes of wave motion and sediment transport control their rhythmic spacings.

A

B

C

Figure 11-3 Regularly spaced beach cusps in a variety of settings: (a) cusps on the shingle beach in Alum Bay, England; (b) on a sandy beach in Mexico; (c) on the shore of Mono Lake, California. [(a) Used with permission of The University of Chicago Press, from P. H. Kuenen, The Formation of Beach Cusps, *Journal of Geology 56,* p. 35. Copyright © 1948 The University of Chicago Press.] (b) Courtesy of *Sunset Magazine.* (c) [Used with permission of Geological Society of America, from P. D. Komar, Observations of Beach Cusps at Mono Lake, California, *Geological Society of America Bulletin 84.* Copyright © 1973 Geological Society of America.]

The morphology of beach cusps depends on the grain size of the sediment, the beach slope (which also depends on grain size), and on the tidal range. Cusps show their best development on gravel or shingle beaches in areas of small tidal range (Fig. 11-4). At such locations the cusp horns stand high above the arcuate embayments, and offshore from each embayment is an underwater delta. A similar morphology can occur on sandy beaches, but the overall relief is smaller and the deltas tend to merge into a uniform step at the base of the swash zone. As the tidal range increases, the cusps become stretched out down the beach face, forming a series of ridges. These are most obvious when the ridges are composed of gravel resting on an otherwise sandy beach.

Beach cusps can form in any type of sediment. Russell and McIntire (1965) observed cusps in materials ranging from basaltic boulders and cobbles, through normal quartz-feldspar sand, to fine-grained calcareous sand. One of the common features of beach cusps is the sorting of the sediment by grain size, with the cusp ridges generally containing coarser sediment than the embayments between the cusps. This contrast in grain sizes between cusps and embayments results in differences in their permeability, and this can influence the processes of formation and the preservation of the cusps. Longuet-Higgins and Parkin (1962) stressed the importance to cusp formation of a vertical stratification of material and corresponding differences in permeability. This difference allows the wave swash to maintain its energy so that the coarse material can be readily moved about. Once the cusps have developed, the coarse-material horns are less subject to erosion relative to the bays, due to their higher permeability, which dissipates the swash energy.

Beach-cusp formation is clearly most favorable when the waves approach normal to the beach, that is, with their crests parallel to the shoreline (Johnson, 1919; Timmermans, 1935; Longuet-Higgins and Parkin, 1962). This may explain why the embayments of pocket beaches are particularly favorable sites for cusp formation, since oblique waves are less likely to occur. Even this has been disputed, however, as Evans (1938) and Otvos (1964) maintain that

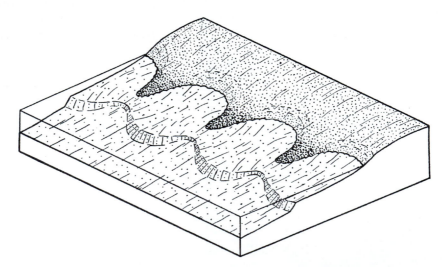

Figure 11-4   Sketch of beach cusps formed on the shingle beach in Alum Bay on the Isle of Wight, England. Offshore from each embayment is an underwater delta, such that the step at the base of the beach is a sinuous mirror reflection of the cuspate shoreline. [Used with permission of The University of Chicago Press, from P. H. Kuenen, The Formation of Beach Cusps, *Journal of Geology 56,* p. 37. Copyright © 1948 The University of Chicago Press.]

the wave direction is irrelevant, and at least one theory of cusp formation (Dalrymple and Lanan, 1976) requires an oblique-wave approach and intersecting waves. However, a substantial longshore transport of sediment under oblique waves clearly does destroy beach cusps by first making them highly asymmetric and then washing them away entirely. If the transport is not too great, the cusps may persist in an asymmetrical form. Krumbein (1944) produced cusps in a laboratory basin with waves approaching the beach at a 15° angle; the cusps migrated slowly down the beach, the maximum rate being approximately 30 cm/hr. It is possible, however, that the cusps observed by Krumbein were another form of rhythmic topography, that associated with the cell circulation and rip currents (Fig. 11-2), a pattern that commonly migrates alongshore, whereas the classical form of beach cusps is usually more stable in position.

In addition to confirming that beach cusps form best when the wave crests are parallel to the shore, Longuet-Higgins and Parkin (1962) noted that regular waves with long crest lengths are particularly conducive to cusp formation. Beach cusps generally are not formed by irregular, confused seas.

Various investigators have described contrasting patterns of water circulation induced by the wave swash around cusp horns and within the embayments. In his investigation of the classical form of beach cusps at Alum Bay, the Isle of Wight (Fig. 11-4), Kuenen (1948) observed an alternating surge in and out of the embayments. Water flows from one embayment where the run-up has been a maximum, around the nearest horn and into a neighboring bay where it rushes up the beach face with the next incident wave. Thus, the maximum run-up switches in its timing between adjacent embayments. This pattern described by Kuenen contrasts with the circulation observed by Bagnold (1940), which is diagrammed in Figure 11-5. Bagnold noted that the incident waves break evenly over a straight step, but the wave surge then piles up against the steep horns of the cusps and is divided into two divergent streams by the promontories, each stream flowing into one of the adjoining embayments. These streams head off the wave surge that had flowed directly up the bay over the shallow underwater deltas. The two side streams from the cusps on either side meet in the center of the bay and together form a seaward flow of considerably greater intensity than the previous upward

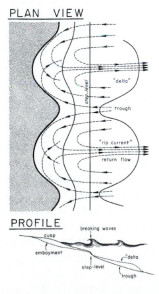

Figure 11-5 The wave swash and water movement around beach cusps and within embayments. [Adapted with permission of The Institute of Civil Engineers, from R. A. Bagnold, Beach Formation by Waves: Some Model Experiments in a Wave Tank, *Journal of the Institute of Civil Engineers,* p. 46. Copyright © 1940 Institute of Civil Engineers.]

surge. This return flow can resemble a rip current, but unlike true rip currents as described in Chapter 8, the flow is discontinuous through time and the processes of formation are quite different. Because of the sideways flow of water within the cusp system, there is no dead period when the water comes to rest. As will be discussed later, these contrasting patterns of water circulation observed, respectively, by Kuenen and Bagnold within beach-cusp systems may suggest that more than one mechanism can give rise to the formation of such systems, or alternately that there is a shift in the circulation pattern as the cusps form and their developing morphologies are sufficient to have a feedback effect in modifying the water circulation.

Numerous hypotheses have been proposed to account for the formation of beach cusps. For a theory to be acceptable, it must account for the uniformity of spacings within an observed series of cusps and the way in which this spacing is related to the wave parameters. Johnson (1910, 1919), Kuenen (1948), and Russell and McIntire (1965) proposed models for cusp formation that envisioned that a regular succession of swash flows on a smooth beach will begin to erode any slight depression they encounter. The backwash carries the eroded sediment out of the hollows to build deltas opposite them. By continued erosion, the series of hollows are enlarged into embayments. As the depth in the outer part of an embayment is increased, the rate of erosion gradually slackens. At this stage, the model envisions that refraction of the swash within the bays causes material to be transported out of the bays and toward adjacent cusp horns. The coarser material is deposited on the cusp horns, while the finer material is washed back into the embayments. The growth of the bays and progradation of the cusp horns gradually decreases as the depth of the central area of the bay approaches a maximum and wave swash prevents greater deposition on the horns. According to the model, it can be expected that initially there would be a great deal of variation in the cusp spacings, but the larger cusps grow at the expense of the smaller ones, and small cusps encroach on larger ones that already have reached their maximum depth. In this way, the theory proposes that the initially irregular pattern of cusps eventually evolves into a regular rhythmic pattern having a nearly uniform spacing and development of embayments.

The processes of beach cusp formation generally are not so random and time consuming as depicted by this theory. Experiments in wave tanks and observations of cusp formation on the shores of lakes and ocean beaches show that beach cusps can develop rapidly and achieve remarkably uniform spacings immediately upon formation. Evans (1945) observed the formation of a perfect set of cusps on a Lake Michigan beach, where the generation occurred in less than 2 min from an initially plane beach. Komar (1973) similarly reports on the formation by surging waves of small cusps along the shores of Mono Lake, California, developing within minutes after the beach had been artificially smoothed. Huntley and Bowen (1978) and Guza and Bowen (1981) describe instances when nearly uniform cusps formed within a short time on ocean beaches during experiments that involved measurements of surf processes.

Another shortcoming of the cusp-formation models proposed by Johnson (1910, 1919), Kuenen (1948), and Russell and McIntire (1965) is that they provide no prediction of the conditions under which beach cusps might or might not form and yield no testable predictions of how the final, equilibrium cusp spacing or other dimensions might be related to wave parameters and beach conditions. In the absence of theoretical predictions, these early studies directed their efforts toward finding empirical correlations between cusp spacings and wave parameters, particularly the wave height. Johnson (1910) observed that doubling the heights of the waves roughly doubles the spacings of the beach cusps. Longuet-Higgins and Parkin (1962) made measurements of beach cusps on Chesil Beach, England, demonstrating a trend of increasing cusp spacings with increasing wave heights but with a considerable scatter within the data. The best trend in their measurements was between cusp spacings and swash distances, a correlation that mainly reflects the consistent geometry of cusps, independent of their overall sizes. Takeda and Sunamura (1983) compiled field and laboratory wave-tank data of beach-cusp spacings and did find a reasonably good correlation with the wave-breaker heights. However, a better correlation was achieved with calculated wave lengths of edge waves, which depend on wave periods and beach slopes and, like Longuet-Higgins and Parkin (1962), a good correlation also was found with the swash length of the waves. Such correlations relate directly to the recent hypotheses of beach-cusp formation, that beach cusps are formed by subharmonic edge waves (Guza and Inman, 1975) or through a "self organiz-ing" process under the action of the swash (Werner and Fink, 1993).

The longshore rhythmicity of edge waves, particularly of standing edge waves, is strongly suggestive of their playing a role in the formation of beach cusps. The pattern of swash on a beach due to standing edge waves is diagrammed in Figure 11-6, consisting of zones of maximum swash run-up at the antinodes versus positions of nodes where the edge waves do not contribute to the run-up. It is easy to envision that beach cusps would quickly develop under this regular pattern of swash, with the cusps forming at the nodes as suggested in the diagram. This schematic illustration corresponds closely to results from the wave-tank experiments of Guza and Inman (1975), where a 2-cm layer of sand initially covered an other-wise uniformly sloping solid beach but was then quickly rearranged into the series of cusps shown in Figure 11-7. The experiments confirmed that the embayments develop at the an-tinode positions of maximum run-up, while the horns of the cusps form at the nodes of the edge waves. The horns were observed to grow seaward and to become thicker, indicating that accretion of the horns is important to cusp development, not merely an absence of erosion by the swash. A particularly significant finding by Guza and Inman is that in most instances it is the mode $n = 0$, subharmonic edge wave that is important to beach cusp generation, that is, edge waves that have a period that is twice the period of the incident waves ($T_e = 2T_i$). Based on the work of Guza and Davis (1974), it is the subharmonic edge wave that is most likely to develop on reflective beaches (Chapter 6). Therefore, beach cusps generated by sub-harmonic edge waves should be most common on reflective beaches, but Guza and Inman noted that the swash zone of otherwise dissipative beaches can be steep and locally reflec-tive, so that subharmonic edge waves could generate beach cusps there as well. In their wave-tank experiments Guza and Inman found that cusp generation can sometimes be associated with synchronous edge waves ($T_e = T_i$), particularly when conditions for the formation of subharmonic edge waves are less favorable.

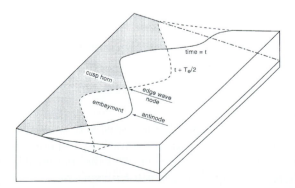

Figure 11-6 The regular pattern of swash on the beach face induced by standing edge waves. The swash could be expected to rearrange the beach sediment into a series of cusps where the horns develop at the nodes of the edge waves, while embayments are eroded by the strongest swash at the antinode positions.

Figure 11-7 Wave-tank experiments where a system of beach cusps developed in response to the presence of subharmonic edge waves as diagrammed in Figure 11-6. Water has been drained from the tank—the former still-water shoreline is marked by the arrow (A). Offshore deltas have formed seaward from the embayments, much as observed in natural beach cusps. A meter stick is shown for scale. [Used with permission of American Geophysical Union, from R. T. Guza and D. L. Inman, Edge Waves and Beach Cusps, *Journal of Geophysical Research 80*, p. 3007. Copyright © 1975 American Geophysical Union.]

From Figures 11-6 and 11-7, it is apparent that when formed by subharmonic edge waves, two cusps develop within the range of one edge-wave length—the spacings of cusps must be one-half the wave length of the edge waves (when formed by synchronous edge waves, the cusp spacing is equal to the edge-wave length). The relationship between cusp spacings and edge-wave lengths was first proposed by Bowen (1972) in terms of cusp formation by synchronous edge waves, was then clearly demonstrated in the wave-tank experiments of Guza and Inman (1975), and was independently found by the field measurements of Darbyshire (1977) on the ocean beach in Hell's Mouth Bay, Wales. Thus, there is substantial confirmation that

$$\lambda_c = L_e/2 \qquad \text{(subharmonic edge waves)} \qquad (11.1a)$$

$$\lambda_c = L_e \qquad \text{(synchronous edge waves)} \qquad (11.1b)$$

where λ_c is the cusp spacing and L_e is the edge-wave length. The longshore wave length of edge waves, according to Eckart (1951), is (Chapter 6)

$$L_e = \frac{g}{2\pi} T_e^2 (2n + 1)\tan\beta \qquad (11.2)$$

where T_e is the period of the edge waves, n is the modal number ($n = 0, 1, 2, \ldots$), and β is the angle of the beach slope. Assuming $T_e = 2T_i$ and $n = 0$ for cusp generation by subharmonic edge waves as generally found by Guza and Inman (1975), the combined Equations (11.1a) and (11.2) yield a predictive relationship for the beach-cusp spacing:

$$\lambda_c = \frac{g}{\pi} T_i^2 \tan\beta \qquad \text{(subharmonic edge waves)} \qquad (11.3a)$$

The corresponding relationship for cusps formed by synchronous edge waves is

$$\lambda_c = \frac{g}{2\pi} T_i^2 \tan\beta \qquad \text{(synchronous edge waves)} \qquad (11.3b)$$

The cusp spacing becomes a function of the period of the incident waves and of the beach slope. One sometimes sees Equation (11.3) written in terms of $\sin\beta$ rather than $\tan\beta$, depending on whether Equation (11.2) from Eckart (1951) has been used for the edge-wave length or the theoretical form derived by Ursell (1952), which yields the $\sin\beta$ dependence. For beaches of low to intermediate slopes, the difference between $\sin\beta$ and $\tan\beta$ is small, and the corresponding difference in the predictions of beach-cusp spacings is negligible. Inman and Guza (1982) discuss the limitations of the above-mentioned relationships in predicting cusp spacings, particularly the problems associated with the choice of the wave period when there is a spectrum of waves present, and the use of Equation (11.2) for evaluating the edge-wave length, a relationship that was derived theoretically, based on an assumption of there being a uniform beach slope. Because of these limitations, one cannot expect Equations (11.3) to provide a precise prediction of beach-cusp spacings.

Equation (11.3a), based on the hypothesis of beach-cusp formation by subharmonic edge waves (Guza and Inman, 1975), provides a prediction of cusp spacings as functions of wave periods and beach slopes, one that is amenable to testing with laboratory and field data. Inman and Guza (1982), Takeda and Sunamura (1983), and Kaneko (1985) compiled field and laboratory data to test these predictions—the graph from the latter study is given in Figure 11-8. The results generally are consistent with the formation of beach cusps by subharmonic edge waves; although in a few cases it appears that generation may have been by synchronous edge waves. Therefore, the available laboratory and field measurements of beach-cusp spacings are consistent with their formation by edge waves.

Several investigations provide direct observational evidence for edge waves being the mechanism in beach-cusp formation (Huntley and Bowen, 1978; Sallenger, 1979; Guza and Bowen, 1981; Inman and Guza, 1982; Seymour and Aubrey, 1985; Sherman, Orford, and

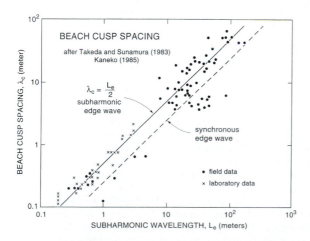

Figure 11-8 A compilation of laboratory and field measurements of beach-cusp spacings versus predicted spacings from Equation (11.3), based on the hypothesis of Guza and Inman (1975) that cusp generation is in response to the presence of subharmonic edge waves. For the most part the data support this hypothesis, while in a few instances the beach cusps appear to have been formed by synchronous edge waves. [Adapted from *Coastal Engineering* 9, A. Kaneko, Formation of Beach Cusps in a Wave Tank, 1985 with kind permission of Elsevier Science-NL, Sara Burgerhartstratt 25, 1055 KV Amsterdam, The Netherlands.]

Carter, 1993). Sallenger (1979) monitored the development of beach cusps on Parramore Island, Virginia, using a grid of narrow rods driven into the beach to measure the associated
sand-elevation changes. Figure 11-9 shows block diagrams derived from his grid measurements, including (a) the initial formation of a ridge with equally spaced channels distributed
along its length, followed by (b) the full development of the beach cusps as the ridge migrated
shoreward. A similar development of a ridge later modified into cusps was found in a few of
the wave-tank experiments of Guza and Inman (1975). In both studies it was observed that
the ridge acted to pond water on its landward side, which then returned seaward through the
equally spaced series of channels. The spacings of these channels were found to conform with
Equation (11.3a) and therefore presumably were controlled by the maximum run-up (antinode) positions of subharmonic edge waves. This spacing was retained as the ridge and channel system evolved into the classical form of beach cusps, as shown in Figure 11-9(b) from the
field observations of Sallenger.

The most direct field confirmation of the role of edge waves in the formation of beach
cusps has been provided by the study of Huntley and Bowen (1978). The primary objective
of their research was to obtain measurements of edge waves, utilizing systems of electromagnetic current meters for their detection, as reviewed in Chapter 6. However, beach cusps
were observed to form in the midst of their experiment, cusps whose mean spacing of 12.7 m
corresponded to those expected if formed by the measured edge waves—the incident waves
had a period of 6.9 sec, and the measured edge waves were the $n = 0$ subharmonic ($T_e = 13.8$
sec) which from Equation (11.3a) yields 12.0 m for the predicted cusp spacing, very close to
that observed. Guza and Bowen (1981) report on additional instances when cusps formed
during field experiments that documented the presence of subharmonic edge waves, providing further confirmation of their role in beach-cusp development.

Although the evidence for an edge-wave origin of beach cusps is fairly conclusive, questions remain concerning cusp interactions with the incident waves and the possibility of the
cusps continuing to grow and achieving new equilibrium configurations after the edge waves

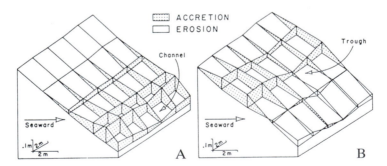

Figure 11-9 Block diagrams showing the development of beach cusps on Parramore Island, Virginia. (a) During flood tide, a ridge was deposited on the foreshore with equally spaced channels along its length. (b) During flood tide, the ridge
migrated shoreward, while during ebb tide the mouths of channels were progressively widened by swash erosion until adjacent mouths met, effecting a cuspate
shape. The spacings of the initial channels and subsequent beach cusps conformed
with Equation (11.3) and thus with the hypothesis that they developed in response
to the swash of subharmonic edge waves. [Reprinted from *Marine Geology* 29,
A. H. Sallenger, Beach-cusp Formation, p. 25, 1983 with kind permission of Elsevier Science-NL, Sara Burgerhartstratt 25, 1055 KV Amsterdam, The Netherlands.]

have disappeared. In their wave-tank experiments of cusp formation due to subharmonic edge waves, Guza and Inman (1975) observed that as the cusps developed and modified the morphology of the beach, the morphology exerted a negative feedback that detuned the conditions necessary to maintain the energy transfer from incident waves to edge waves. It was surmised that the outward growth of the cusp horns at the nodal positions of the edge waves interfered with the longshore-directed water motions of the standing edge waves, which are a maximum at the nodes—a similar detuning was brought about by the placement of a 5-cm high metal bar across the surf zone at a nodal position. Guza and Bowen (1981) further explored the morphological feedback of the growing cusps in detuning the edge waves responsible for cusp formation. They showed theoretically that as the cusps grow, the edge-wave dispersion relationship changes, and that at some value of cusp steepness, the edge waves can no longer extract energy from the incoming waves. This was substantiated by wave-tank tests, which showed that the edge-wave amplitudes decrease as the cusps grow, being reduced to about half of the original amplitude by the presence of well-formed cusps. Guza and Bowen hypothesized that as the cusps build out and there is a reduced level of edge-wave energy, an equilibrium balance is achieved between the tendency of the weakened edge waves to maintain the cusps and the tendency of the incident waves to erase them. For this approximate balance, they derived the relationship

$$\frac{\eta_{c,\max}}{\lambda_c} = 0.13 \tan\beta \tag{11.4}$$

for the cusp steepness, the ratio of the maximum possible vertical height $\eta_{c,\max}$ of the cusp horns above the beach as a whole, and λ_c, the cusp spacing. Based on a rather different model, one that attributes the development of cusp heights to the maximum run-up level reached by the wave swash, Inman and Guza (1982) derived a similar relationship for the cusp steepness,

$$\frac{\eta_{c,\max}}{\lambda_c} = 0.24 K_3 \tan\beta \tag{11.5}$$

where $K_3 \approx 1$ is an empirical coefficient that is related to the dependence of run-up elevations on wave heights (Chapter 6). Figure 11-10 includes field data provided by Inman and Guza on cusp heights that support Equation (11.5) as an approximate upper limit to the

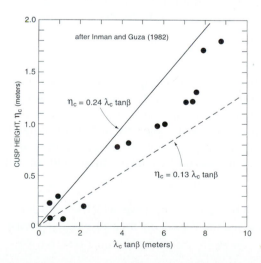

Figure 11-10 Field measurements of cusp heights versus $\lambda_c \tan \beta$, where λ_c is the observed spacing. The lines correspond to Equations (11.4) and (11.5), which predict the limits of maximum cusp steepness. [Adapted from *Marine Geology* 51, D. L. Inman and R. T. Guza, The Origin of Swash Cusps on Beaches, p. 144, 1982 with kind permission of Elsevier Science-NL, Sara Burgerhartstratt 25, 1055 KV Amsterdam, The Netherlands.]

growth of beach cusps. Utilizing Equation (11.3a) for the predicted spacings of cusps generated by subharmonic edge waves, Equation (11.5) becomes

$$\eta_{c,\max} = 0.076g[T_i \tan\beta]^2 \tag{11.6}$$

for the maximum possible vertical relief of the cusps. According to these relationships, cusp
heights should be greatest on steep beaches exposed to long swell, and observations indicate
that this is certainly true (Guza and Bowen, 1981).

Inman and Guza (1982) suggested that although edge waves may be necessary for the
initial cusp formation, particularly in establishing their spacings, the edge waves need not persist for the development of the mature cusp topography that may be controlled more by the
incident waves. They point to the observed water circulation within cusp systems as evidence
for the importance of the incident waves to the mature equilibrium morphology. The surging
motion, as described by Kuenen (1948) and Guza and Inman (1975), where the maximum
run-up alternately occurs in adjacent embayments, conforms with the expected water motions for cusp generation by subharmonic edge waves. However, as the edge waves are detuned by the growing cusp morphology, a circulation pattern could develop that is controlled
more by the incident waves, a circulation like that described by Bagnold (1940) where the
cusp horns divide the wave run-up, while the return flow occurs as "rip currents" within the
embayments (Fig. 11-5). Inman and Guza suggest that the mature morphology of the cusp
system will be in equilibrium with this latter circulation pattern and that the necessity for
maintenance of strong rip currents will limit the sizes of stable beach cusps. Thus, although
edge waves of long periods on steep beaches are capable of forming beach cusps with spacings up to about 150 m, spacings greater than about 75 m are seldom observed, since the
longer spacings will not be stable under the incident waves that dominate after the edge
waves are detuned and effectively disappear.

Dean and Maurmeyer (1981) analyzed the equilibrium condition of beach cusps when
the circulation pattern is like that described by Bagnold (1940). With this circulation there is
a dominance of the wave uprush on the cusp horns. Important is that the period of the incident waves and the time interval of the swash within the cusps are nearly equal—the rip-
current backrush within the embayments will then interfere with the uprush of the next wave
into the embayments. According to the model of Dean and Maurmeyer, at equilibrium the
lateral slope from the horns into the embayments and the associated water movement are
just sufficient to transport the sediment carried up the horns by the local dominance of uprush. In their analysis they idealized the cusp topography by the relationship

$$h(x,y) = Ax(1 + \varepsilon \sin ky) \tag{11.7}$$

in which h is the beach elevation at the shore-normal (x) and shore-parallel (y) positions,
$k = 2\pi/\lambda_c$ is the longshore wavenumber of the beach cusps, A is the average beach slope, and
ε is the parameter that accounts for the cusp elevation. The kinematics of water-particle motions within this cusp system were analyzed by Dean and Maurmeyer, neglecting friction and
percolation. The maximum displacement of a water particle in the shore-parallel direction
should be more or less equal to one-half the cusp spacing and should occur at one wave period
in order to interfere with the next incoming wave. The analysis then yields the relationship

$$\lambda_c = 3.9\sqrt{\varepsilon\zeta_x} \tag{11.8}$$

where ζ_x is the maximum displacement of the water up-rush, which should correspond to the
cross-shore extent of the cusp embayments. Based on measurements of cusps along the

beaches of Point Reyes and Drakes Bay, California, Dean and Maurmeyer took $\varepsilon = 0.15$ as a representative value, so that Equation (11.8) becomes

$$\lambda_c = 1.5\zeta_x \tag{11.9}$$

defining the equilibrium cusps in terms of their spacings versus the cross-shore lengths of their embayments. However, the λ_c/ζ_x ratios differed systematically for Point Reyes (1.9) versus Drakes Bay (3.2), a difference that might reflect their contrasting sediment sizes and hence the amount of friction and percolation (Point Reyes beach is coarser and steeper). Furthermore, as cusps develop along a given beach, according to the geometric relationship of Equation (11.7), the amplitude parameter ε would be expected to progressively increase, affecting the proportionality in Equation (11.8). It should be noted that Dean and Maurmeyer did not actually observe the progressive formation of the cusps at Point Reyes and Drakes Bay, the cusps already having achieved a mature equilibrium form with a circulation like that described by Bagnold (1940).

A "self-organization" model for the formation of beach cusps has been advanced by Werner and Fink (1993), supported by the development of computer simulation analyses of cusp development. In many respects their simulations are comparable to the early descriptions of Johnson (1910, 1919), Kuenen (1948), and Russell and McIntire (1965), reviewed above, where incipient topographic depressions in a beach are amplified by attracting and enhancing the water flow, thereby increasing erosion to produce the cusp embayments. The simulations of Werner and Fink are important in testing these concepts through analyses that incorporate modeling of the wave swash, the sediment transport under the swash, and the resulting morphology changes. In the present stage of development the simulations are simple, with only the swash front being followed where the water particles move along parabolic paths governed by Newton's kinematic equations. The simulated beach is composed of thin, square slabs of sediment stacked on a rectangular grid of square cells. The water particles erode the sediment slabs, transport them in the direction of swash advance, with the sediment-transport rate being proportional to the cube of the flow velocity (in agreement with many bedload transport theories). Erosion of topographic lows into cusp embayments occurs mainly during the run-out of the swash edge. Swash particles that move over topographic highs decelerate and deposit sediment, and this leads to the formation of the cusp horns. Deltas form offshore from cusp bays, and depressions develop offshore from cusp horns, yielding simulated cusps that are similar to those observed on steep, coarse-sediment beaches (Werner and Fink, 1993). According to the analyses, the development of uniformly spaced simulated beach cusps from a plane beach requires 50–1000 swash cycles, which would correspond to time intervals of 6 min to 3 hrs for the specific case of 10-sec waves.

In the simulations by Werner and Fink (1993), the cusps have cross-shore embayment lengths that are equal to the swash run-up excursion distance. The steady-state beach-cusp spacing is also observed to be proportional to the run-up distance, so the results are comparable to Equation (11.9) obtained by Dean and Maurmeyer (1981) for equilibrium cusps. In their simulations, Werner and Fink obtained proportionality coefficients between 1 and 3, depending on assumptions in the model regarding water-particle interactions and smoothing. The similarity in the findings of Werner and Fink and Dean and Maurmeyer is not surprising in that the analyses are similar with respect to kinematic calculations of the swash run-up, with both studies neglecting friction and percolation that could affect the flow. Dean and Maurmeyer focused on the final equilibrium of the swash around the cusps, while Werner and Fink simulated the development of the cusps toward that equilibrium.

Dean and Maurmeyer (1981) and Werner and Fink (1993) presented their models as alternatives to the edge-wave hypothesis of Guza and Inman (1975). Dean and Maurmeyer argued against any role of subharmonic edge waves [see the response of Inman and Guza (1981)]. Werner and Fink noted that predictions of cusp spacings by their self-organization model are on the same order as those predicted by the edge-wave hypothesis and concluded that the currently available observational data are insufficient to discriminate between the two models.

In summary, two hypothesis have emerged as contenders in offering explanations for the formation of beach cusps. There is strong supporting evidence for the role of edge waves, particularly of standing subharmonic edge waves as proposed by Guza and Inman (1975). That hypothesis provides testable predictions of cusp spacings as functions of incident wave periods and beach slopes [Equation (11.3)], predictions that agree with laboratory and field data (Fig. 11-8). Particularly strong confirmation comes from observations like those of Huntley and Bowen (1978), where beach cusps rapidly formed with nearly uniform spacings in the midst of field measurements that documented the presence of subharmonic edge waves that account for the cusp development and spacings. The second hypothesis is the self organization model of Werner and Fink (1993), with support from the general observations of Johnson (1910, 1919), Kuenen (1948), and Russell and McIntire (1965) and from the cusp-equilibrium analysis of Dean and Maurmeyer (1981). It is premature to judge this hypothesis, as additional simulations need to be undertaken with more realistic analyses of swash dynamics and sediment transport, together with attempts at verification of the model in the laboratory and field. At this stage, the shortcoming of the self-organizing model appears to be the length of time required for the swash to develop beach cusps, intervals that are on the order of 6 min to 3 hrs, according to the simulations completed by Werner and Fink. This interval undoubtedly will be significantly longer when more natural irregular waves are used in the analyses. It should be recognized that this may not be an either/or choice of hypothesis, in that beach cusps could be formed independently by the different mechanisms and, in some instances, the two mechanisms could dominate at different stages in cusp development. In particular, there is a distinct possibility that subharmonic edge waves are responsible for the initial formation of cusps having nearly uniform spacings but that the developing morphology has a feedback effect on the edge waves, causing them to decrease in energy. At that stage, the incident waves could become dominant, modifying the cusp morphology by a self-organizing process that might alter the spacings somewhat but would mainly affect the embayments so they are compatible with the proportionality of Equation (1.8). It is apparent that more research is needed to document the respective roles of edge waves and incident waves in the formation of beach cusps.

Rip-Current Embayments and Cuspate Shorelines

Within the series of cuspate shorelines diagrammed in Figure 11-2, generally the next larger form beyond classical beach cusps is the system of erosional embayments and intervening cusps formed by circulation cells consisting of feeder longshore currents and seaward-flowing rip currents (Chapter 8). In most instances the rip currents erode sand from the beach and transport it offshore, forming embayments at the rip-current positions and cusps midway between (Riviere, Arbey, and Vernhet, 1961; Bowen and Inman, 1969; Komar, 1971). Rip currents, and hence the embayments and cusps, typically have spacings that range from tens to hundreds of meters. As such, the resulting cuspate shoreline has been referred to variously as "rhythmic topography," "sand waves," or "giant cusps."

This rhythmic shoreline form, together with the associated cell circulation, is diagrammed schematically in Figure 11-11. The cusps correspond to positions of zero longshore sediment transport produced by waves combining with the feeder currents that flow alongshore toward the rip currents. According to the Bagnold relationships reviewed in Chapter 9, the rate of sand transport depends on the magnitude of the longshore current, and since that velocity is zero at divergence points where the feeder currents begin to flow in opposite directions, the sand transport must also be effectively zero at those points. The longshore currents accelerate toward the rip currents, and this results in progressively higher velocities and increasing quantities of transported sediment. As discussed in Chapter 10, this positive gradient in the transport ($dQ_\ell/dx = +$) will result in beach erosion ($dy/dt = -$). This suggests that the embayments are purely erosional, with the cusps being remnants. In reality, the cusps may be partly accretional, due to sand that is carried shoreward from the bars and shoals that generally exist offshore from the cusp positions.

The shoreline graphed in Figure 11-11, consisting of embayments and cusps, is only the surface expression of the underwater topography that is molded by the circulation currents and waves. Bowen and Inman (1969) suggested that a seaward-flowing rip current would tend to erode a channel, the result being a segmented offshore bar and a system of cusps along the shoreline midway between rip currents that occupy the bays. Accordingly, each cusp may be part of a shallow zone that extends out to the remaining segment of the offshore bar. Such a topography and relationship to the cell circulation was found by Sonu (1972a) in his investigations of beaches in the Gulf of Mexico. With waves arriving normal to the shore, the rip currents occupied troughs as described above, with shoals present between the rip currents, but in this case cusps did not develop on the shoreline in response to the offshore topography. However, under similar wave conditions in Lake Michigan, Davis and Fox (1972) did find cusps associated with the nearshore shoals positioned between rip-current troughs, so the morphology there consisted of alternating embayments and cusps as hypothesized by Bowen and Inman (1969).

This form of rhythmic topography with embayments cut by rip currents may be important to property erosion along the coast, in that the embayments can remove most of the buffer protection offered by the beach. Investigations of erosion along the Oregon coast have found that the development of rip-current embayments is the primary factor in the erosional losses of property on foredunes of sand spits (Komar and Rea, 1976; Komar, 1983b) and in cutting away sea cliffs backing beaches (Komar and McDougal, 1988; Shih and Komar, 1994). Figure 11-12 shows an example of beach erosion and property losses on Siletz

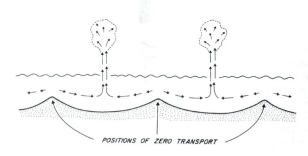

POSITIONS OF ZERO TRANSPORT

Figure 11-11 Schematic diagram of the cell circulation consisting of rip currents and feeder longshore currents and the embayments they erode from an initially uniform beach, leaving cusps midway between the rips. [From *Geological Society of America Bulletin 82*, P. D. Komar. Reproduced with permission of the publisher, the Geological Society of America, Boulder, Colorado USA. Copyright © 1971.]

Spit, where the presence of an unusually large rip embayment eliminated the beach berm. Although the rip embayments themselves do not cause much foredune erosion and loss of property, they provide an area of deeper water where storm waves can approach close to shore before breaking, so the maximum episodes of erosion have resulted from a combination of rip embayments, a storm that generates unusually high waves, and spring tides that produce high-water levels.

The above discussion has focused on conditions where the rip currents erode embayments, this being the most common situation, especially on sandy ocean beaches. To confuse matters, in some instances beach sediment may be deposited in the lee of a rip current so the cusps correspond to the rip locations. This appears to occur most commonly on steep beaches under relatively low wave conditions (Komar, 1971), in which case it produces a system of cusp ridges on an otherwise flat beach. In many instances these cusps consist of accumulations of gravel on otherwise sandy beaches and have all the attributes of classical beach cusps described in the preceding section. Cusps also have been observed to form in the lee of rip currents in laboratory wave-basin experiments. In experiments described by Komar (1971), the rip currents that produced the cusps ultimately disappeared, but the cusps did not erode under the continued action of the incident waves. As discussed in Chapter 9, this development of an equilibrium cuspate shoreline involved waves breaking obliquely to the cusps, which normally would generate a longshore current and transport sand alongshore, but this force was opposed and balanced by the driving mechanism that originally produced the rip currents, namely a longshore variation in breaker heights and wave set-up. Such an occurrence suggests that a cuspate shoreline could be formed by rip currents that subsequently disappear and may not be present when the cusps are observed. In some instances the resulting features could mistakenly be interpreted as classical beach cusps.

Crescentic Bars

Crescentic bars can be remarkable in the regularity of their lunate shapes and uniform spacings (Fig. 11-13) but unfortunately this regularity generally cannot be appreciated by observers on the dry beach since most of the feature is under water. There may be an associated series of cusps on the beach if the crescentic bars in effect attach to the shore, in which case the topography is like that diagrammed in Figure 11-14. Crescentic bars are found throughout the world, commonly on long straight beaches (Shepard, 1952; Hom-ma and Sonu, 1963), but are particularly well developed within bays bounded by headlands, as can be seen in Figure 11-13 (Clos-Arceduc, 1962, 1964). Their development appears to be confined to regions of small to medium tidal range (King and Williams, 1949; Shepard, 1963; Bowen and Inman, 1971). Crescentic bars also appear to form best where the beach slope is low, that is, on dissipative beaches.

The shapes of the arcuate crescentic bars are generally symmetrical, but there are cases where they become skewed, apparently in response to oblique waves and a small net longshore transport of sediment (Hom-ma and Sonu, 1963, p. 255). With a strong longshore transport, the rhythmic bars smooth into straight bars parallel to the shoreline. The presence of crescentic bars, therefore, implies the prevalence of waves breaking nearly parallel to the shore with a near-zero longshore sediment transport.

Crescentic bars are much larger features than beach cusps and are somewhat larger than the rhythmic topography due to rip currents. In some instances, large crescentic bars form the outer-bar system of a beach (Fig. 11-15) while the inner bar is segmented by the

Figure 11-12 A rip-current embayment cut into the beach on Siletz Spit, Oregon, which allowed storm waves during December 1972 to erode properties on the fore-dunes. One home was lost, while others were saved by the rapid placement of riprap revetments, which formed promontories extending onto the beach as the erosion of unprotected lots continued. [From P. D. Komar and C. C. Rea, Erosion of Siletz Spit, Oregon, *Shore & Beach 44,* 1976.]

Figure 11-13 Well-developed crescentic bars within a pocket beach near Cape Kalaa, Algeria. [Used with permission of Bulletin de la Société Française Photogramme, from A. Cordons Littoraux, Etude Sur les Vues Aeriennes, des Alluvions Littorales d'Allure Periodique, Cordons Littoraux et Fostons. Copyright © 1962 Bulletin de la Société Française Photogramme.]

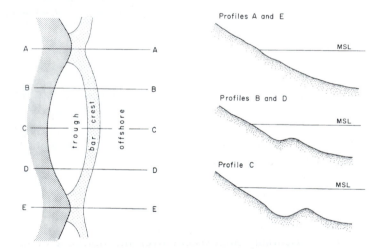

Figure 11-14 Crescentic bars with associated cusps formed along the shore. Also shown is the pattern of beach profiles, which vary systematically depending on their position relative to a crescentic bar.

more closely spaced rip currents. The range in lengths of crescentic bars is difficult to establish, since for many reported occurrences, it is not possible to determine conclusively whether crescentic bars or some other form of rhythmic topography is being described. Crescentic bars appear to range from about 100 to 2,000 m in length, with a predominance of 200 to 500 m. Hom-ma and Sonu (1963) describe a system of multiple crescentic bars,

A   Rhythmic Topography on Inner Bar

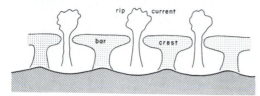

B   Crescentic Bars

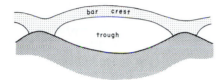

C   Combination

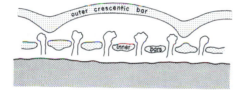

Figure 11-15 Two types of rhythmic morphologies that can produce cuspate shorelines, with the crescentic bars generally being of larger scale than those associated with rip-current embayments.

the maximum spacing being 1,000 m; the farther offshore the bars, the larger their spacings. Figure 11-16 shows a system of two crescentic bars measured from aerial photographs on the Mediterranean coast of Israel by Goldsmith, Bowman, and Kiley (1982); the outer set is characterized by a considerable regularity in spacing just under 100 m, while the inner set is more irregular and gives rise to a corresponding set of irregularly spaced cusps along the shore. Goldsmith and co-workers established that, at this location, the two sets of crescentic bars form independently and even at different times. The inner bar is just as likely to be roughly parallel to shore rather than being crescentic, or to be skewed due the presence of a longshore transport during low wave conditions when sediment movement is essentially nonexistent over the dormant offshore crescentic bars, which remain symmetrical.

Like beach cusps, the regularity and even spacings of crescentic bars have inspired a number of suggestions as to their origin. Sonu (1969, 1973) attributed them to bed perturbations by the longshore current to form dunes much like those found in rivers, but this is unlikely since crescentic bars tend to be obliterated under strong longshore currents. The mechanism first proposed by Bowen and Inman (1971) for crescentic-bar generation provides the most reasonable explanation, a hypothesis that is supported by their wave-basin experiments and field comparisons. The hypothesis involves the velocity field associated with edge waves on a sloping beach, particularly with the net drift velocities. According to this

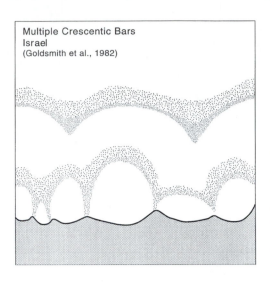

Multiple Crescentic Bars
Israel
(Goldsmith et al., 1982)

Figure 11-16 Sets of two crescentic bars measured from aerial photographs on the coast of Israel, with the outer set consisting of symmetrical, evenly spaced crescentic bars, while the inner set is irregular. [Adapted with permission of Society for Sedimentary Geology from V. Goldsmith, D. Bowman and K. Kiley, Sequential Stage Development of Crescentic Bars, *Journal of Sedimentary Petrology* 52, p. 241. Copyright © 1982 Society for Sedimentary Geology.]

mode of formation, beach sediment will drift about under the currents of the edge waves, until the sand reaches zones where the velocity is below the threshold of sediment motion, where it then accumulates. A schematic diagram of the expected pattern of deposition and crescentic-bar formation for edge-wave modes $n = 1$ and 2 is shown in Figure 11-17. Holman and Bowen (1982) generated comparable simulations of crescentic-bar formation using Bagnold-type sediment transport equations. The similarity of the theoretical forms of crescentic bars depicted in Figure 11-17 to the observed forms is remarkable and provides strong evidence in support of the Bowen and Inman hypothesis. According to their model, the bars should have a longshore wave length that is one-half that of the edge waves. In the case of

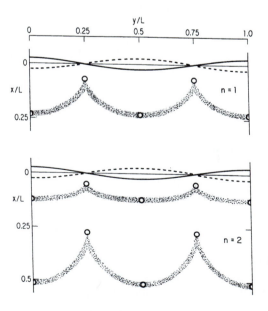

Figure 11-17 Theoretical forms of crescentic bars based on the hypothesis that their formation is by long-period edge waves. The solid and dashed lines represent the edge-wave motions, consisting of alternating nodes and antinodes in the longshore direction, while the stippled patterns represent the areas of expected bar formation across and along the beach. A mode $n = 1$ edge wave generates one series of crescentic bars, while the $n = 2$ edge wave has the capacity to generate two series, although experiments indicate that the inner bar may not be stable. [Used with permission of American Geophysical Union, from A. J. Bowen and D. L. Inman, Edge Waves and Crescentic Bars, *Journal of Geophysical Research 76,* p. 8666. Copyright © 1971 American Geophysical Union.]

$n = 2$, a system of two crescentic bars could form. Bowen and Inman conducted a series of laboratory wave-basin experiments that confirmed this basic depositional pattern. As shown in Figure 11-18, they were successful in generating perfectly shaped crescentic bars in their experiments. They did find that the inner bar of the $n = 2$ case tends to become unstable and may eventually erode away, leaving only the outer crescentic bar.

The field examination by Bowen and Inman (1971) of their hypothesis for the generation of crescentic bars centered on the example shown in Figure 11-13 of bars on the Mediterranean Sea coast of Algeria. This example was analyzed earlier by Clos-Arceduc (1962, 1964), who attributed their formation to standing longshore oscillations between the headlands. As pointed out by Bowen and Inman, this explanation is a good attempt by someone unfamiliar with edge waves, but his explanation is not an alternative since the standing waves of

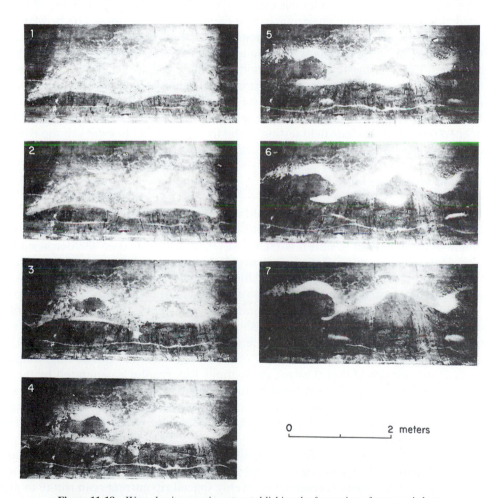

Figure 11-18 Wave-basin experiments establishing the formation of crescentic bars by a mode $n = 2$ edge wave. Both inner and outer crescentic bars initially are formed, as diagrammed in Fig. 11-17, but the inner series proved to be unstable and eventually disappeared. [Used with permission of American Geophysical Union, from A. J. Bowen and D. L. Inman, Edge Waves and Crescentic Bars, *Journal of Geophysical Research 76*, p. 8668. Copyright © 1971 American Geophysical Union.]

Clos-Arceduc (1962, 1964) become edge waves when they exist on a sloping beach. The wave length of the crescentic bars in Figure 11-13 is approximately 500 m, so the edge-wave length needed to account for their formation would have to be about 1,000 m. The morphology of the bars suggest a mode $n = 2$, and on this basis and using reasonable estimates for the beach slope, Bowen and Inman utilized Equation (11.2) to evaluate the required edge-wave period for bar formation. They obtained $T_e = 40–50$ sec, which is much longer than periods of incident wind-generated waves. At the time of their study in the early 1970s, the existence of such long-period (infragravity) edge waves had not been documented conclusively in the nearshore. Accordingly, Sonu (1972b) argued against the edge-wave hypothesis for the generation of crescentic bars. However, as reviewed in Chapter 6, there have been many field experiments in recent years that have documented the existence of long-period infragravity edge waves containing considerable energy at periods equal to and in excess of the 40–50 sec needed to account for the crescentic bars in Figure 11-13, as proposed by Bowen and Inman.

The hypothesis of Bowen and Inman (1971) that crescentic bars are formed by edge waves with periods generally within the infragravity range has been fairly conclusively established. The hypothesis accounts for both their lunate shapes and spacings associated with infragravity periods known to be important within the spectra of nearshore water motions. At this time there are no viable alternative hypothesis for crescentic-bar formation that similarly accounts for their regularity in shapes and spacings. However, additional research is needed to provide further field testing of the hypothesis and to establish the conditions under which crescentic bars are likely to form.

Welded and Transverse Bars

A variety of sand bars have been observed within the nearshore that run obliquely to the longshore trend of the beach. These have been characterized as "welded" or "transverse." An example of a welded bar is shown in Figure 11-19, photographed on the ocean shore of Cape Cod, Massachusetts, consisting of a system of bars and a cuspate shoreline where there is a distinctive longshore rhythmicity of several hundred meters. A number of suggestions have been made for the origin of such bars and for other types of bars that trend obliquely to the shore.

When the skewed rhythmic topography, such as that in Figure 11-19, occurs under an oblique-wave approach and strong longshore currents, investigators have suggested an analogy between the bars and bed forms generated by the flow in a river. Bruun (1954) indicated that there may be a correspondence between the ridges and the similar migrating river bars. Sonu (1969) proposed that this type of rhythmic topography is generated by the instability of the loose surf zone bed perturbed by the longshore current; that is, the oblique ridges of sediment across the surf zone are analogous to dunes produced in a river or by winds. Dolan (1971) suggested that the origin may somehow be similar to the development of meanders in a river.

Sonu (1972a) provides another possible explanation for the origin of welded bars like those in Figure 11-19 by having observed that when waves arrive at an oblique angle to the beach, the shoals or bars segmented by evenly spaced rip currents can rotate to align themselves with the incoming wave crests, as do the rip currents and their troughs; this rotation is depicted in Figure 11-20. Where the shoals join with the shore, a system of cusps develop, being the subaerial expression of the shoals, and the rip current troughs occupy the embayments between the cusps. Horikawa and Sasaki (1968) [see discussion in Sonu (1973, p. 60)] similarly produced a system of cusps along the shoreline in a wave basin by the offshore bar ro-

Figure 11-19 Welded bars and intervening channels on Cape Cod, Massachusetts. The long-shore spacing is several hundred meters. [Courtesy of David S. Aubrey, Woods Hole Oceanographic Institution]

tating and attaching to the shore. The resulting system of alternating bars and obliquely trending rip-current troughs appears similar to that pictured in Figure 11-19 from Cape Cod.

Holman and Bowen (1982) provide a distinctly different explanation for the origin of welded bars like those on Cape Cod, one that involves the superimposed effects of two edge waves of different modes but having the same periods. As discussed above, when two edge waves are of the same mode and period, traveling in opposite directions along the length of the beach, they give rise to a pattern of standing edge waves that can form symmetrical crescentic bars. Holman and Bowen illustrated the formation of crescentic bars through a full simulation that included calculations of water motions beneath the edge waves and quantitative evaluations of the resulting sediment transport rates and directions. When they next undertook a similar analysis but with edge waves having different modes, they obtained obliquely trending sand bars that have a strong resemblance to the Cape Cod welded bars. An example of their model simulations is given in Figure 11-21 for edge waves of mode $n = 1$ and $n = 2$ acting together to modify the beach topography, with the diagram on the left

A   Normal   Waves

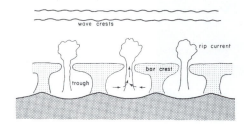

B   Oblique   Waves

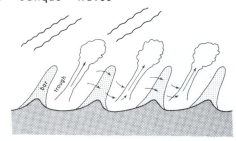

Figure 11-20   Realignment of bars as observed by Sonu (1972a), under an oblique-wave approach to give a cuspate shoreline of alternating cusps and embayments occupied by rip currents.

showing the resulting pattern of drift velocities while the diagram on the right gives the resulting welded sand-bar morphology. In addition to the bar itself, the analysis yields a transverse channel that has the classical form of a rip channel, even though a cell circulation with rip currents did not exist in the model. Holman and Bowen point out that with the breaking of incident waves as would be present on natural beaches, a rip current would quickly develop within this channel, even though it did not originally form the channel. Such circumstances would make it difficult to conclusively establish the origin of such oblique bars and channels when found in the field.

Another type of transverse bar had been investigated by Niedoroda and Tanner (1970). This form is limited to ocean coasts of very low wave energy, and also has been found in lakes (Evans, 1938; Carter, 1978). The transverse bars specifically studied by Niedoroda and Tanner occurred on St. James Island, Florida, where the average annual breaker height is only 6 cm. The transverse bars tend to occur in families that run parallel to one another, forming a large angle with the beach but not necessarily normal to the shoreline trend. Two families with different orientations may exist on the same beach. For a given family, the longshore spacings are crudely regular but range widely from as little as 50 m to several kilometers for different sets. Their lengths are quite variable when one compares different families; in general, the ratio of their lengths to spacings is on the order of 2–4. The average relief of the bar above the adjacent trough is generally less than 50 cm, decreasing in the offshore direction. At each point where a transverse bar joins the beach, a large cusp-like feature is developed on the beach face. Transverse bars can be a fairly permanent feature—those investigated by Niedoroda and Tanner on St. James Island were observed in aerial photographs spanning 25 years and showed no tendency to migrate alongshore during that time.

Niedoroda and Tanner (1970) investigated the effects of transverse bars on the nearshore water circulation, making field observations and undertaking wave-basin experiments of wave

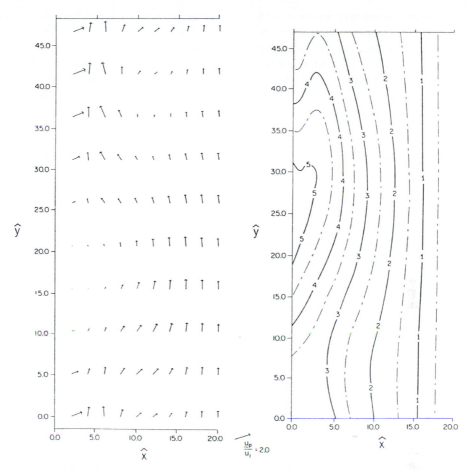

Figure 11-21 A simulated welded bar (right) formed by the interaction of two progressive edge waves having modes $n = 1$ and $n = 2$. The diagram on the left shows the normalized drift velocities of the combined edge waves, which are important in the generation of the bar. [Used with permission of American Geophysical Union, from R. A. Holman and A. J. Bowen, Bars, Bumps, and Holes: Models for the Generation of Complex Beach Topography, *Journal of Geophysical Research 87*, p. 463. Copyright © 1982 American Geophysical Union.]

refraction and current generation over a bottom preformed into the shape of a bar. The bar caused significant refraction, the waves developing V-shaped patterns as seen from above, with the bar crest beneath the apex of each V. Eventually the converging wave crest segments cross at the apex of the V, and this produces a zone of spilling breakers that progress along the axis of the bar. Niedoroda and Tanner found that if the offshore extension of the bar is small, wave energy is focused on the bar position by wave refraction, producing higher waves than over the troughs between the bars. This drives a current that flows shoreward along the bar and returns seaward within the troughs. Although this pattern is similar to the circulation over welded bars, the currents are much weaker and the return flow does not qualify as a rip current. With a long transverse bar the current is in the opposite direction, flowing seaward along the bar. This is accounted for by the wave energy dissipation through frictional drag over the long bar, causing lower waves over the bar than in the troughs. Experiments by Niedoroda and Tanner with fluorescent sand tracers indicated that sand movement is active along the length of the bar, produced by the spilling breakers combining with the currents over the bar axis.

Niedoroda and Tanner (1970) concluded that transverse bars can remain in a state of dynamic equilibrium for long periods of time, maintained by waves moving sand onto the bars, which is then distributed along the bar length by the currents that flow along its axis. The ultimate origin of transverse bars is uncertain and is not specifically addressed by Niedoroda and Tanner. Barcilon and Lau (1973) modeled their formation as caused by longshore currents, being analogous to dunes in rivers. Such an origin is questionable, and in a study of transverse bar formation in Lough Neagh, Northern Ireland, Carter (1978) found no support for such a mode of formation. Instead, he found that the initial bar nucleus appeared to arise from the interaction of opposing longshore currents, but once initiated, the bar extended lakeward and generated its own diagnostic current system as found by Niedoroda and Tanner. Carter further concluded that after one transverse bar is formed, it acts to divert the longshore movement of sediment to form additional bars, so that in time a hierarchy of evenly spaced bars of irregular lengths develops.

Large-Scale Capes and Erosional Embayments

The presence of a hierarchically arranged series of cuspate shorelines has been documented by Dolan along the barrier islands of the east coast of the United States (Dolan and Ferm, 1968; Dolan, 1971). These range in size from very small "cusplets," up through beach cusps having spacings on the order of 10 m, to other cusp systems and rhythmic shorelines in the range 100–1,000 m that probably are associated with rip-current embayments and crescentic bars, as discussed above. Dolan continues this series up through secondary capes having spacings on the order of 10 km, and finally the full Carolina Capes that occur roughly every 1,000 km along the southeast coast of the United States. Although one can argue with Dolan's partitioning of these cuspate shoreline features into approximate factor-of-ten intervals, more important is his observation that their simultaneous presence gives rise to a highly variable shoreline configuration that has differing degrees of susceptibility to erosion. Their importance to erosion is illustrated in Figure 11-22, which shows the North Carolina coast in the vicinity of the Cape Hatteras lighthouse during 1970. The shoreline positions and beach widths are highly irregular, there being a series of cusps and embayments of varying scales, with the deepest embayment posing a threat to the lighthouse. Dolan and Hayden (1981) have further shown that the occurrence of overwash on the barrier islands has longshore periodicities that correspond to the spacings of the embayments between the cuspate shorelines, their analysis being specific to the major Ash Wednesday storm that occurred during 1962.

Much of the shoreline erosion pictured in Figure 11-22 appears to have been associated with the scale of cusps and embayments produced by rip currents and crescentic bars, but the orientation of the coast as controlled by the larger capes likely was also important. Dolan et al. (1977) analyzed the variability in the erosion along the Atlantic coast barrier islands in terms of the coastal orientation, which is highly changeable due to the superposition of the shoreline cusps and capes. They established that there is a strong correlation between the long-term rate of erosion and the orientation of the shoreline, a dependence that results mainly from the dominant direction of storm-generated waves that approach the coast.

When the hierarchically arranged series of cuspate forms affects the shoreline configuration, their summation can result in a highly irregular shoreline configuration and accom-

Figure 11-22   An irregular longshore pattern of alternating cusps and embayments on the Cape Hatteras coast of North Carolina, photographed during 1970, with the largest embayment threatening the historic lighthouse. [From *Geological Society of America Bulletin 82,* R. Dolan. Reproduced with permission of the publisher, the Geological Society of America, Boulder, Colorado USA. Copyright © 1971.]

panying variations in total beach widths. This is analogous to the summation of ocean waves having a range of heights and periods, which are analyzed by spectral analysis to separate the different wave trains (Chapter 5). Comparable analyses can be made of irregular shorelines to separate the series of longshore wavelengths of the cuspate forms. This has been done by Dolan, Hayden, and Felder (1979a), based on measurements along the coast between Cape Hatteras and Cape Lookout. Their measurements included the shoreline configuration, the rates of shoreline change (erosion or accretion), and the longshore variations in storm-penetration distances due to overwash. The resulting spectra obtained by Dolan and co-workers shows series of "energy" peaks (the energy is related to the amplitude of the shoreline variation, just as in analyses of ocean waves). The peaks correspond to a wide range of longshore wave lengths, from about 1.6 km to more than 25 km. The analysis was not sufficiently detailed to identify spacings that might result from cuspate shorelines associated with rip currents and crescentic bars. Instead, the spacings are more on the order of those associated with capes or with the longshore variability of wave energy produced by wave refraction over a complex offshore bathymetry. Dolan and co-workers suggested that the series of energy peaks and longshore spacings might be associated with standing edge waves trapped between major capes such as Cape Hatteras and Cape Lookout, but this is questionable as very long periods would be required to account for these large spacings, periods that would be associated with shelf waves that are trapped to the slope of the continental shelf, rather than of edge waves that are trapped to the sloping beach.

Dolan, Hayden, and Felder (1979b) investigated the possible role of linear shoals off the Delmarva coast (Delaware, Maryland, and Virginia) as a control in the large-scale variability in shoreline erosion and barrier-island washovers. These shoals are found on the continental shelf and generally do not have a direct morphological effect within the nearshore. On the Delmarva coast the linear shoals consist of ridges that trend obliquely to the coast, forming northward-opening angles, have heights up to 10 m, and extend for tens of kilometers in length (Swift, 1980); they occur elsewhere on the Atlantic coast (Swift et al., 1972; Duane et al., 1972) and on other continental shelves (Swift et al., 1978). Dolan and co-workers undertook spectral analyses of both the shoreline patterns and the offshore bathymetry affected by the linear shoals, but there was little correspondence between the spectra, and on that basis they concluded that none of the shoreline rhythmicity could clearly be attributed to the effects of the offshore shoals.

If offshore shoals are sufficiently shallow to affect wave refraction, their regular spacings could then give rise to a correspondingly regular pattern of wave refraction that consists of alternating zones of energy convergence and divergence (Chapter 5). This longshore variation in wave energy could in turn affect the variability of shoreline erosion versus accretion, yielding a rhythmic shoreline having a wave length that roughly corresponds to the spacings of the offshore shoals. This has not been specifically examined for series of offshore shoals, giving rise to a rhythmic shoreline, but there have been numerous analyses that document the effects of individual shoals in altering wave refraction in such a way that they cause enhanced shoreline erosion during storms. Goldsmith (1976) has undertaken such analyses of the Delmarva coast, establishing that the larger-scale patterns of wave refraction over the continental shelf give rise to a high variability in wave heights along the coast, which roughly corresponds to historical shoreline changes during intervals of 48–105 years. Another example is the study of Healy (1987) on the coast of New Zealand where offshore wave refraction focuses the wave energy on the shore and has resulted in enhanced wave heights and shoreline erosion causing embayments to be cut into the dune line, embayments that were on the order of 200–500 m long and 10–20 m in amplitude. In one case at Mangawhai Harbor, the erosion led to the breaching of a barrier spit, creating a dual inlet system. Robinson (1980) and MacDonald and O'Connor (1994) investigated the effects of offshore shoals in the North Sea, which alter the wave refraction and have caused shoreline erosion, studies that were reviewed in Chapters 2 and 5.

The largest cuspate forms identified by Dolan and Ferm (1968) in their heirarchy for the southeast coast of the United States are the individual capes that together form the Carolina Capes (Fig. 11-23): Capes Hatteras, Lookout, Fear, and Romain. Also part of the series is Cape Canaveral in Florida and the less prominent but similar capes developed along the Georgia coast at Tybee Island and Little St. Simons Island. Together they have an average cusp spacing of roughly 100 km. Most of the capes are part of the barrier island system that rims the southeast coast. Shoals are present seaward from the capes, extending for tens of kilometers offshore. For example, depths as shallow as 1 m exist 15 km offshore from Cape Fear on the Frying Pan Shoals.

The earliest suggestions regarding the formation of the Carolina Capes attributed their development to a series of secondary rotational cells or "eddy currents," which develop along

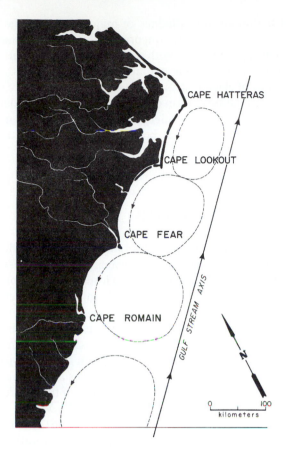

Figure 11-23 The Carolina Capes and suggested eddies from the Gulf Stream, once thought to be the cause of the capes.

the western margin of the Gulf Stream as diagrammed in Figure 11-23 (Tuomey, 1848; Abbe, 1895). Bumpus (1955) investigated the circulation on the shelf in the Cape Hatteras region but found no evidence for such eddies. He indicated that southwesterly winds may pile up water on the south sides of the capes, creating a current in the offshore direction over the projecting shoals, and suggested that such currents might help to maintain the capes. The role of waves in local beach erosion and deposition in cape formation has been stressed by Cooke (1936) and Zenkovitch (1967). According to Cooke, the waves scoop out the arcs between the capes, transporting the sediment alongshore and depositing it on the capes to either side, maintaining a smoothly curved shoreline. Zenkovitch explained the capes as "a result of diagonal wave incidence from the north and south." A direct causative role played by waves is difficult to accept, since the waves should tend to concentrate their energy on the projecting capes because of wave refraction. This should act to erode the capes and straighten the overall shoreline. Indeed, the Carolina Capes presently appear to be retreating under wave attack. The barrier islands are narrowest in the vicinity of the capes and are being driven landward over the onshore lagoonal deposits (Hoyt and Henry, 1971). In general, the barrier islands at each of the capes consist of a single beach ridge, which attests to their lack of accretion and progradation. Historical records plus recent active erosion also demonstrate cape retreat.

White (1966) discussed the importance of the seaward shoals in maintaining the capes. Because of the shoals, the capes are self-maintaining, and relict capes have been able to localize new capes. White stressed the important influence of major rivers on the Carolina Cape positions. Hoyt and Henry (1971) noted that the concurrence of capes and major rivers is too prevalent to be fortuitous. Each cape coincides with one or more river mouth; some of these rivers, however, presently empty into lagoons behind the barrier islands. Hoyt and Henry hypothesized that during periods of glacial advance, the river gradients were increased due to the lowered sea level. This caused the formation of deltas, which prograded across the shelf as the sea level lowered. Thus, a series of delta ridges were deposited perpendicular to the coast, which during the subsequent rise in sea level, became the loci for cape formation, including their offshore shoals.

A MORPHODYNAMIC CLASSIFICATION OF BEACH CYCLES

In Chapter 3 beaches were categorized as dissipative versus reflective, and those terms have been used frequently throughout this text. Dissipative beaches are characteristically low in slope such that the waves break well offshore and continue to lose energy as bores as they cross the wide surf zone—the beach is dissipative in the sense of being highly effective in dissipating the deep-water energy of the incident waves. It also was noted that dissipative beaches can be dominated by long-period infragravity water motions, some of which may be in the form of edge waves. In the opposite extreme, reflective beaches are steep and the incident waves break close to shore and immediately wash up the beach face. Accordingly, reflective beaches are dominated by the incident waves, and if edge waves are present they are most probably subharmonic edge waves where the period is twice that of the incident waves.

This classification of beaches has been elaborated upon greatly by Wright and Short (1983), with the addition of several intermediate beach morphologies. Furthermore, the resulting morphodynamic classification is now fully three dimensional, incorporating the various cuspate shoreline forms that have been reviewed in this chapter. The models originally were developed independently by Wright et al. (1978, 1979) and Short (1978, 1979), but the taxonomic differences between their schemes led to confusion. They therefore developed a unified and updated model (Wright and Short, 1983), and it is this morphodynamics classification that is in common use. This classification is diagrammed in Figure 11-24, where it can be seen that the most extreme limits are represented by the dissipative and reflective beach states, but four intermediate states have been added. The dissipative and reflective extremes are the most two dimensional, whereas the intermediate states show varying degrees of three dimensionality. Starting at the top, the dissipative state (a) is the condition already described, while the longshore bar-trough state (b) differs in there being more relief in the bar-trough elevations and the beach face is steeper than on the dissipative beach. In response to this morphology, most of the wave dissipation occurs over the bar. The waves reform within the deeper trough and break for a second time on the beach face. The steep beach face can be locally reflective, so beach cusps often develop. The rhythmic bar and beach state (c) is similar to state (b), but the distinctive feature is the development of crescentic bars and the accompanying large-scale cusps along the shore. The transverse bar and rip topography state (d) consists of welded bars and intervening oblique troughs occupied by strong rip currents. The ridge-and-runnel/low-tide terrace state (e) is characterized by the presence of a swash bar that migrates shoreward, confining a narrow and deep longshore trough at the base of the beach face. Small, weak, and irregularly spaced rip currents may be present, in places cutting through the bar.

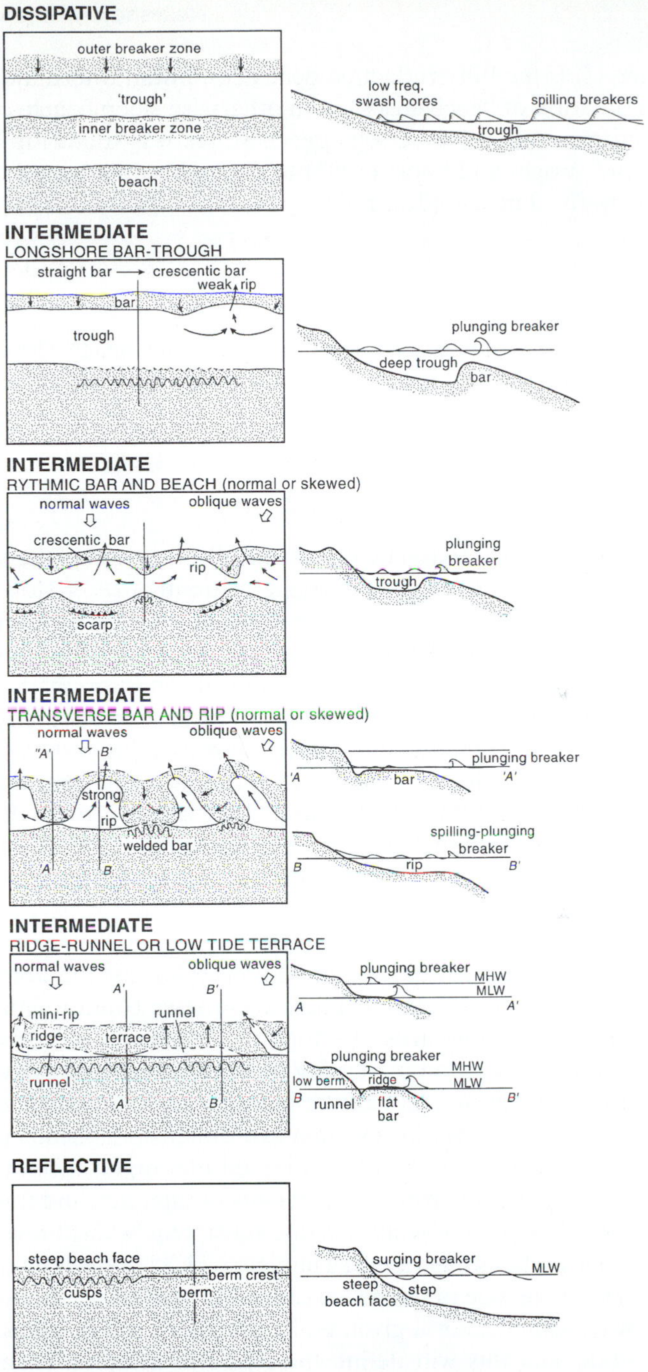

DISSIPATIVE

INTERMEDIATE
LONGSHORE BAR-TROUGH

INTERMEDIATE
RYTHMIC BAR AND BEACH (normal or skewed)

INTERMEDIATE
TRANSVERSE BAR AND RIP (normal or skewed)

INTERMEDIATE
RIDGE-RUNNEL OR LOW TIDE TERRACE

REFLECTIVE

Figure 11-24 The morphodynamic classification of beaches developed by Wright and Short (1983), based on observations in Australia. The plan and profile configurations of six beach states are recognized, depending on the overall beach slope and wave conditions. A beach can range through several of the states, depending on the occurrence of storms versus low-wave conditions. [Adapted with permission of CRC Press, from L. D. Wright and A. D. Short, Morphodynamics of Beaches and Surf Zones in Australia, Handbook of Coastal Processes and Erosion, p. 39. Copyright © 1983 CRC Press.]

The final stage (f) is the fully reflective beach condition, where the sole three-dimensional element is in the form of beach cusps on an otherwise steep beach.

The occurrence of these various beach states was found by Wright and co-workers (1978, 1979) and Wright and Short (1983) to depend on the beach slope and wave parameters according to the dimensionless ratio:

$$\varepsilon = \frac{H_b \omega^2}{2gS^2} = \frac{2\pi^2 H_b}{gT^2 S^2} \tag{11.10}$$

where $\omega = 2\pi/T$ is the radian frequency, and S is the beach slope. This ratio can be rearranged into the form of an Iribarren number, using $L_\infty = gT^2/2\pi$ for the deep-water wave length, yielding

$$\xi_b = \frac{S}{(H_b/L_\infty)^{1/2}} = \left(\frac{\pi}{\varepsilon}\right)^{1/2} \tag{11.11}$$

The dissipative extreme was found by Wright and Short to occur when ε ranges from 30 to 100, corresponding to very low values of the Iribarren number ($\xi_b = 0.2$–0.3). At the opposite extreme, the reflective-beach condition occurred when $\varepsilon < 1$ or approximately $\xi_b > 2$. The four intermediate stages are more difficult to assess in terms of these dimensionless ratios, since their values can range widely depending on the local conditions of the bottom slope and wave parameters, permitting for example the outer surf zone to be dissipative in character while the steep beach face of the inner surf zone is more reflective [as in state (b) of Fig. 11-24].

When the wave conditions change, as during a storm, the beach state within the sequence, shown in Figure 11-24, will change. With a storm, ε generally increases while ξ_b decreases, and this acts to force the beach morphodynamics from reflective towards dissipative. When viewed as a change in the beach profile, this is seen to effectively be a shift from a berm-type profile to a bar-type profile as categorized in Chapter 7, where our attention was limited to the two-dimensional profile. The full sequence as outlined in Figure 11-24 is more complex, especially with the return of lower wave conditions following the storm. During the storm, the beach may shift quickly to the state (a) fully dissipative condition having a bar-type profile, but with lower wave conditions after the storm the beach may successively pass through states (b), (c), etc., characterized by an increasing three dimensionality having the various cuspate or rhythmic shoreline features (crescentic bars, rip-current embayments, and welded bars). Characterizing the morphology by a simple two-dimensional beach profile is obviously then an oversimplification. Short (1978, 1979) stressed the importance of these successive beach states with increasing or decreasing wave heights or energies, and this is further discussed by Wright and Short (1983). They point out that most beaches will not typically range through all the morphological states diagrammed in Figure 11-24 but will be constrained within limits that depend on the grain size of the beach sediment (which largely controls its average slope) and the total wave climate. For a given grain size, there will be a variability in wave climate about some mode, and this will define the range in morphodynamics about some corresponding mode. Thus, a fine-grained beach that experiences a high-energy wave climate will be dominated by states (a) through (c), that is, toward the dissipative end within the spectrum of beach types. In contrast, a coarse-grained ocean beach or a sandy beach in a lake will be observed to cycle through states (d), (e) and (f), depending on the wave conditions.

The modality of the beach morphodynamics will control the overall variability of beach changes. Beaches that are at the extreme, either dissipative or reflective, tend to show the least variability in their three-dimensional morphology or in a simple set of beach profiles. This is illustrated in Figure 11-25 for various beaches in Australia. Goolwa Beach and North Seven Mile Beach, which are dissipative, show the least amount of profile variation over a year. This is largely because such beaches of low slope are effective in dissipating the energy of storm waves, so that much of the inner surf zone is "saturated" by depth limiting the wave-bore energies. Dissipative beaches are also not affected significantly by rip-current embayments or other three-dimensional features that can alter an individual profile line. In contrast, beaches that have a central modality within the intermediate beach states show a greater range in morphologies; this is well illustrated by Narrabeen Beach in Figure 11-25, where the beach elevation at a particular cross-shore position can change by up to 5 m, governed by the respective elevations of the rip-current troughs versus the welded bars as they shift positions in the longshore direction.

The morphodynamic classification of Wright and Short (1983) was based on observations along the coast of Australia but has been shown to be applicable to other coasts, in some cases with minor modifications. An interesting test of the classification was that undertaken by Lippmann and Holman (1990), based on 2 years of daily observations of bar patterns and wave conditions at the Field Research Facility (FRF) in Duck, North Carolina. The bar observations were derived from time-exposure images created from video recordings of the surf zone by directly averaging individual frames over a 10-min period using an image-processing

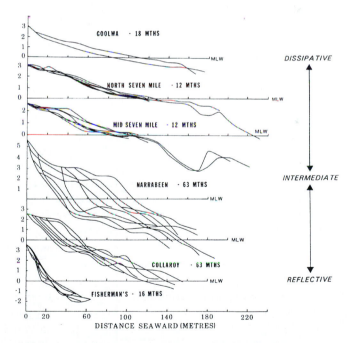

Figure 11-25 A series of profiles obtained on beaches in Australia that generally range from dissipative to reflective. The mobility is greatest for the beaches that center toward the intermediate states as defined in Figure 11-24, due principally to their high degree of three dimensionality, whereas the more two-dimensional dissipative and reflective beaches at opposite ends in the morphodynamic states show less variability. [Used with permission of CRC Press, from L. D. Wright and A. D. Short, Morphodynamics of Beaches and Surf Zones in Australia, Handbook of Coastal Processes and Erosion, p. 59. Copyright © 1983 CRC Press.]

system. This remote sensing technique is ideal for monitoring bar changes, since it is not constrained by high-energy storms when the bars are undergoing their most rapid changes. The technique is based on the preferential breaking of incident waves over the shallow bars, and records the contrasts in light intensity between breaking and nonbreaking regions (see Chapter 7 for a review of these techniques).

The classification of Lippmann and Holman (1990) for the beach at the FRF is shown in Figure 11-26, which includes a direct comparison with the morphodynamics classification of Wright and Short (1983). Examples of the corresponding time-exposure video images are shown in Figure 11-27. The sequence is largely the same as developed by Wright and Short, but there are two revisions offered by Lippmann and Holman. The longshore bar and trough state of Wright and Short is divided into two bar types, one representing linear bars with no longshore variability (bar-type G) and the other representing bars with nonrhythmic longshore variability (bar-type F). The transverse bar and rip state is divided into two bar types, each attached to the shore and with longshore variability distinguished by the presence (bar-type D) or absence (bar-type C) of dominant periodicity. The other morphodynamic forms

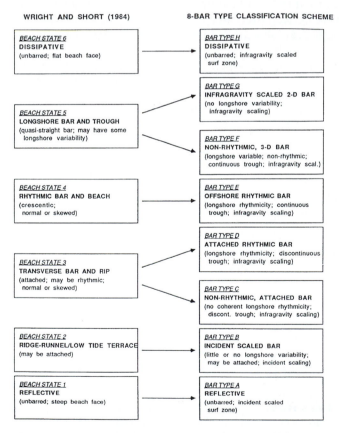

Figure 11-26 An eight-stage morphodynamics classification scheme based on daily observations of bar morphologies at the Field Research Facility, Duck, North Carolina, compared with the classification of Wright and Short (1983). [Used with permission of American Geophysical Union, from T. C. Lippmann and R. A. Holman, The Spatial and Temporal Variability of Sand Bar Morphology, *Journal of Geophysical Research 95*, p. 11,577. Copyright © 1990 American Geophysical Union.]

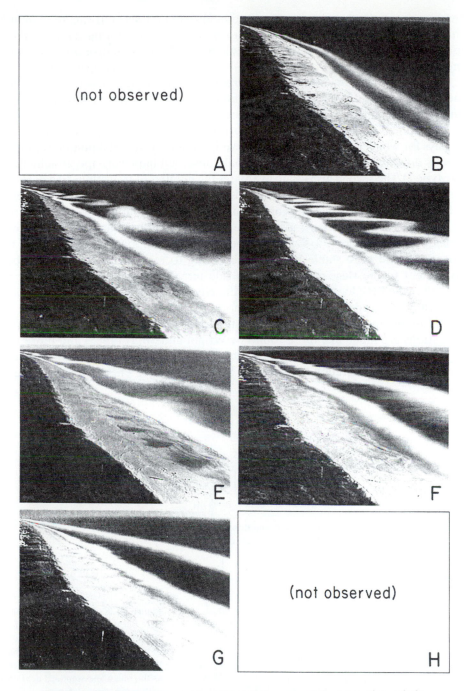

Figure 11-27 Time-exposure video images that represent the range of morphologies in the classification of bar types listed in Figure 11-26, based on observations at the Field Research Facility, Duck, North Carolina. [Used with permission of American Geophysical Union, from T. C. Lippmann and R. A. Holman, The Spatial and Temporal Variability of Sand Bar Morphology, *Journal of Geophysical Research 95,* p. 11,578. Copyright © 1990 American Geophysical Union.]

remain essentially the same as proposed by Wright and Short. It can be seen in Figure 11-27 that two categories were actually never observed at FRF during the 2-year study of Lippmann and Holman, the extreme end members, the fully dissipative, and fully reflective beach morphologies.

The daily documentation of the bar conditions using video techniques allowed Lippmann and Holman (1990) to evaluate the frequency of occurrence of the various bar forms and their residence times. The results are given in the histograms of Figure 11-28. The top most graph gives the probability of occurrence of each bar type, P_i, evaluated as the number of days for which the state "i" was observed, divided by the total number of sample days. The results document that the most commonly observed bars were the attached rhythmic bars, bar-type D. The second histogram evaluates the number of transitions to a particular bar state, N_i, while the third histogram is the residence time, τ_i, a measure of bar-type persistence once it has formed (shown as means and standard deviations). It can be seen that the transitions to bar-types E and F occur most frequently, but that these bar forms have lower residence times, persisting for only 4–5 days; this accounts for their lower percentages of total observations (P_i). The most stable bar form according to residence times is bar-type B, which is equivalent to the ridge-and-runnel/low tide terrace category of Wright and Short (1983), which for the FRF location is the most reflective condition achieved by the beach. This is the condition achieved during prolonged periods of low-wave conditions.

Lippmann and Holman (1990) established that during a storm with high waves, the transitions are much like those found by Wright and Short (1983) and more fully discussed by Short (1978, 1979). The order of the bar types yields a good first-order approximation of erosional versus accretional responses. With rising wave energy, the bar-type shifts higher

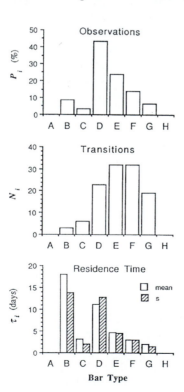

Figure 11-28 Histograms of the probabilities P_i of bar occurrences at the Field Research Facility, Duck, North Carolina, the total numbers of transitions N_i to the various bar types, and the residence time τ_i for each bar type (graphed as the mean and standard deviation, s). [Used with permission of American Geophysical Union, from T. C. Lippmann and R. A. Holman, The Spatial and Temporal Variability of Sand Bar Morphology, *Journal of Geophysical Research 95*, p. 11,582. Copyright © 1990 American Geophysical Union.]

within the listings shown in Figures 11-26 and 11-27, that is, toward the more dissipative forms. The higher the wave energy the greater the jump and the more dissipative the beach becomes. Under declining wave conditions, the bars progressively shift through the sequence, toward a more reflective beach condition. According to Figure 11-28, the residence times of the different bar types are quite variable, with bar-type D (attached rhythmic bars) lasting the longest (high residence time), so this form is observed most commonly (highest P_i value). Lippmann and Holman also have documented the wave conditions associated with these various bar-type forms, evaluating the waves alternately in terms of the deep-water significant wave height, the wave steepness, and the wave power. These various evaluations of the wave conditions work equally well in their associations with bar types, all showing that an increase in wave intensity during storms shifts the bar types toward the more dissipative end of the series, while the other bar types are associated with progressively lower wave-intensity conditions.

SUMMARY

The focus of this chapter has been on the complexity of the three-dimensional beach, where the morphology is affected by a variety of rhythmic shoreline forms. These forms range in scale from beach cusps, to embayments eroded by rip currents, crescentic bars and associated shoreline cusps, welded and transverse bars, and on up to the large-scale capes that have developed on some coasts. Many of these shoreline forms can occur simultaneously, and this accounts particularly for the morphological complexity of beaches. The origins of some of these shoreline forms are fairly well understood, as in the case of rip-current embayments and crescentic bars; others, like beach cusps, are still subject to considerable debate regarding their origin. This subject can be expected to remain the focus of future research. Interest in these various rhythmic shoreline forms is not simply academic. As we have seen, they are an extremely important element in the overall beach morphology, and some forms like rip-current embayments are significant to the occurrence of erosion and property losses directly landward from the beach.

REFERENCES

ABBE, C. J. (1895). Remarks on the Cuspate Capes of the Carolina Coast. *Proceedings of the Boston Society of Natural History* 26: 489–497.

BAGNOLD, R. A. (1940). Beach Formation by Waves: Some Model Experiments in a Wave Tank. *Journal of the Institute of Civil Engineers* 15: 27–52.

BAKKER, W. T. (1968). A Mathematical Theory About Sand Waves and Its Application on the Dutch Waden Isle of Vlieland. *Shore & Beach* 36: 4–14.

BARCILON, A. I. and J. P. LAU (1973). A Model for Formation of Transverse Bars. *Journal of Geophysical Research* 78: 2656–2664.

BOWEN, A. J. (1972). Edge Waves and the Littoral Environment. *Proceedings of the 13th Coastal Engineering Conference, Amer. Soc. Civil Engrs.,* pp. 1313–1320.

BOWEN, A. J. and D. L. INMAN (1969). Rip Currents, 2: Laboratory and Field Observations. *Journal of Geophysical Research* 74: 5479–5490.

BOWEN, A. J. and D. L. INMAN (1971). Edge Waves and Crescentic Bars. *Journal of Geophysical Research* 76(36): 8662–8671.

BRUUN, P. (1954). Migrating Sand Waves and Sand Humps, With Special Reference to Investigations Carried Out on the Danish North Sea Coast. *Proceedings of the 5th Coastal Engineering Conference, Amer. Soc. Civil Engrs.*, pp. 269–295.

BUMPUS, D. F. (1955). The Circulation over the Continental Shelf South of Cape Hatteras. *Transactions American Geophysical Union* 36: 601–611.

CARTER, R. W. G. (1978). Small-Scale Transverse Bars in Lough Neagh, Northern Ireland. *Journal of Earth Sciences, Royal Dublin Society* 1: 205–209.

CLOS-ARCEDUC, A. (1962). Etude Sur les Vues Aeriennes, des Alluvions Littorales d'Allure Periodique, Cordons Littoraux et Fostons. *Bull. Soc. Fr. Photogramm.* 4: 13–21.

CLOS-ARCEDUC, A. (1964). La Photographie Aerienne et l'Étude des Depots Prelittoraux. Etude de Photointerpretation No. 1, Institut Geographique National, Paris.

COOKE, C. W. (1936). *Geology of the Coastal Plains of South Carolina.* U.S. Geological Survey, Bulleting No. 867.

DALRYMPLE, R. A. and G. A. LANAN (1976). Beach Cusps Formed by Intersecting Waves. *Geological Society of America Bulletin* 82: 57–60.

DARBYSHIRE, J. (1977). An Investigation of Beach Cusps in Hell's Mouth Bay. In *Voyage of Discovery: George Deacon 70th Anniversary Volume*, M. Augel (editor), pp. 405–427. New York: Pergamon Press.

DAVIS, R. A. and W. T. FOX (1972). Coastal Processes and Nearshore Sand Bars. *Journal of Sedimentary Petrology* 42: 401–412.

DEAN, R. G. and E. M. MAURMEYER (1981). Beach Cusps at Point Reyes and Drakes Bay Beaches, California. *Proceedings of the 17th Coastal Engineering Conference, Amer. Soc. Civil Engrs.*, pp. 863–884.

DOLAN, R. (1971). Coastal Landforms: Crescentic and Rhythmic. *Geological Society of America Bulletin* 82: 177–180.

DOLAN, R. and J. C. FERM (1968). Crescentic Landforms Along the Atlantic Coast of the United States. *Science* 159: 627–629.

DOLAN, R., L. VINCENT, and B. HAYDEN (1974). Crescentic Coastal Landforms. *Zeitschrift für Geomorphologie* 18: 1–12.

DOLAN, R., B. HAYDEN, J. HEYWOOD, and L. VINCENT (1977). Shoreline Forms and Shoreline Dynamics. *Science* 197: 49–51.

DOLAN, R., B. HAYDEN, and W. FELDER (1979a). Shoreline Periodicities and Edge Waves. *Journal of Geology* 87: 175–185.

DOLAN, R., B. HAYDEN, and W. FELDER (1979b). Shoreline Periodicities and Linear Offshore Shoals. *Journal of Geology* 87: 393–402.

DOLAN, R., and B. HAYDEN (1981). Storms and Shoreline Configuration. *Journal of Sedimentary Petrology* 51: 737–744.

DUANE, D. B., M. E. FIELD, E. P. MEISBURGER, D. J. P. SWIFT, and S. J. WILLIAMS (1972). Linear Shoals on the Atlantic Continental Shelf, Florida to Long Island. In *Shelf Sediment Transport, Process and Pattern*, D. J. P. Swift et al., (editors), pp. 447–498. Stoudsburg, PA: Dowden, Hutchinson and Ross.

ECKART, C. (1951). *Surface Waves on Water of Variable Depth.* Wave Report 100, University of California, Scripps Institution of Oceanography, La Jolla, CA.

EVANS, O. F. (1938). The Classification and Origin of Beach Cusps. *Journal of Geology* 46: 615–627.

EVANS, O. F. (1945). Further Observations on the Origins of Beach Cusps. *Journal of Geology* 53: 403–404.

FLEMMING, N. C. (1964). Tank Experiments on the Sorting of Beach Material During Cusp Formation. *Journal of Sedimentary Petrology* 34: 112–122.

GOLDSMITH, V. (1976). Wave Climate Models for the Continental Shelf: Critical Links Between the Shelf Hydraulics and Shoreline Processes. In *Beach and Nearshore Sedimentation,* Special Publication No. 24, pp. 24–47, Society of Econ. Paleon. and Mineralogists, Tulsa, Oklahoma.

GOLDSMITH, V., D. BOWMAN, and K. KILEY (1982). Sequential Stage Development of Crescentic Bars: Hahoterim Beach, Southeastern Mediterranean: *Journal of Sedimentary Petrology* 52: 233–249.

GUZA, R. T. and R. DAVIS (1974). Excitation of Edge Waves by Waves Incident on a Beach. *Journal of Geophysical Research* 79: 1285–1291.

GUZA, R. T. and D. L. INMAN (1975). Edge Waves and Beach Cusps. *Journal of Geophysical Research* 80(21): 2997–3012.

GUZA, R. T. and A. J. BOWEN (1981). On the Amplitude of Beach Cusps. *Journal of Geophysical Research* 86: 4125–4132.

HEALY, T. R. (1987). The Importance of Wave Focusing in the Coastal Erosion and Sedimentation Process. *Coastal Sediments '87, Amer. Soc. Civil Engrs.*, pp. 1472–1485.

HOLMAN, R. A. and A. J. BOWEN (1982). Bars, Bumps, and Holes: Models for the Generation of Complex Beach Topography. *Journal of Geophysical Research* 87(C1): 457–468.

HOM-MA, M. and C. J. SONU (1963). Rhythmic Patterns of Longshore Bars Related to Sediment Characteristics. *Proceedings of the 8th Coastal Engineering Conference, Amer. Soc. Civil Engrs.*, pp. 248–278.

HORIKAWA, K. and T. SASAKI (1968). Some Considerations on Longshore Current Velocities. *Proceedings of the 15th Coastal Engineering Conference, Japan Soc. Civil Eng.*, pp. 126–135 [in Japanese].

HOYT, J. H. and V. J. HENRY (1971). Origin of Capes and Shoals Along the Southeastern Coast of the United States. *Geological Society of America Bulletin* 82: 59–66.

HUNTLEY, D. A. and A. J. BOWEN (1978). Beach Cusps and Edge Waves. *Proceedings of the 16th Coastal Engineering Conference, Amer. Soc. Civil Engrs.*, pp. 1378–1393.

INMAN, D. L. and R. T. GUZA (1982). The Origin of Swash Cusps on Beaches. *Marine Geology* 49: 133–148.

JOHNSON, D. W. (1910). Beach Cusps. *Geological Society of America Bulletin* 21: 604–624.

JOHNSON, D. W. (1919). *Shore Processes and Shoreline Development*. New York: John Wiley & Sons. [facsimile edition: Hafner, New York (1965)]

KANEKO, A. (1985). Formation of Beach Cusps in a Wave Tank. *Coastal Engineering* 9: 81–98.

KING, C. A. M. and W. W. WILLIAMS (1949). The Formation and Movement of Sand Bars by Wave Action. *Geographical Journal* 107: 70–84.

KOMAR, P. D. (1971). Nearshore Cell Circulation and the Formation of Giant Cusps. *Geological Society of America Bulletin* 82: 2643–2650.

KOMAR, P. D. (1973). Observations of Beach Cusps at Mono Lake, California. *Geological Society of America Bulletin* 84: 3593–3600.

KOMAR, P. D. (1983a). Rhythmic Shoreline Features and Their Origins. In *Mega-Geomorphology*, R. Gardner and H. Scoging (editors). Pp. 92–112. Oxford, England: Claredon Press.

KOMAR, P. D. (1983b). The Erosion of Siletz Spit, Oregon. In *Handbook of Coastal Processes and Erosion*, P. D. Komar (editor). Pp. 65–76. Boca Raton, FL: CRC Press.

KOMAR, P. D. and C. C. REA (1976). Erosion of Siletz Spit, Oregon. *Shore & Beach* 44: 9–15.

KOMAR, P. D. and W. G. MCDOUGAL (1988). Coastal Erosion and Engineering Structures. The Oregon experience. *Journal of Coastal Research* 4: 77–92.

KRUMBEIN, W. C. (1944). *Shore Currents and Sand Movement on a Model Beach*. U.S. Army Corps of Engrs., Beach Erosion Board Technical Memo No. 7.

KUENEN, P. H. (1948). The Formation of Beach Cusps. *Journal of Geology* 56: 34–40.

LIPPMANN, T. C. and R. A. HOLMAN (1990). The Spatial and Temporal Variability of Sand Bar Morphology. *Journal of Geophysical Research* 95(C7): 11,575–11,590.

LONGUET-HIGGINS, M. S. and D. W. PARKIN (1962). Sea Waves and Beach Cusps. *Geographical Journal* 128: 194–201.

MACDONALD, N. J. and B. A. O'CONNOR (1994). Influence of Offshore Banks on the Adjacent Coast. *Proceedings of the 24th Coastal Engineering Conference, Amer. Soc. Civil Engrs.*, pp. 2311–2324.

NIEDORODA, A. W. and W. F. TANNER (1970). Preliminary Study of Transverse Bars. *Marine Geology* 9: 41–62.

OTVOS, E. G. (1964). Observations of Beach Cusps and Pebble Ridge Formation on the Long Island Sound. *Journal of Sedimentary Petrology* 34: 554–560.

RIVIERE, A., F. ARBEY, and S. VERNHET (1961). Remarque sur l'Evolution et l'Origine des Structures de Plage a Caractere Periodique. *C. R. Acad. Sci.* 252: 767–769.

ROBINSON, A. H. W. (1980). Erosion and Accretion Along Part of the Suffolk Coast of East Anglia, England. *Marine Geology* 37: 133–146.

RUSSELL, R. J. and W. G. MCINTIRE (1965). Beach Cusps. *Geological Society of America Bulletin* 76: 307–320.

SALLENGER, A. H. (1979). Beach-Cusp Formation. *Marine Geology* 29: 23–37.

SEYMOUR, R. J. and D. G. AUBREY (1985). Rhythmic Beach Cusp Formation: A Conceptual Synthesis. *Marine Geology* 65: 289–304.

SHEPARD, F. P. (1952). Revised Nomenclature for Depositional Coastal Features. *Bulletin American Association of Petroleum Geologists* 36: 1902–1912.

SHEPARD, F. P. (1963). *Submarine Geology,* 2nd edition. New York: Harper & Row.

SHERMAN, D. J., J. D. ORFORD, and R. W. G. CARTER (1993). Development of Cusp-Related, Gravel Size and Shape Facies at Malin Head, Ireland. *Sedimentology* 40: 1139–1152.

SHIH, S. -M. and P. D. KOMAR (1994). Sediments, Beach Morphology and Sea Cliff Erosion Within an Oregon Coast Littoral Cell. *Journal of Coastal Research* 10: 144–157.

SHORT, A. (1978). Wave Power and Beach Stages: A Global Model. *Proceedings of the 16th Coastal Engineering Conference, Amer. Soc. Civil Engrs.,* pp. 1145–1162.

SHORT, A. (1979). Three Dimensional Beach Stage Model. *Journal of Geology* 87: 553–571.

SONU, C. J. (1969). Collective Movement of Sediment in Littoral Environment. *Proceedings of the 11th Coastal Engineering Conference, Amer. Soc. Civil Engrs.,* pp. 373–400.

SONU, C. J. (1972a). Field Observation of Nearshore Circulation and Meandering Currents. *Journal of Geophysical Research* 77(18): 3232–3247.

SONU, C. J. (1972b). Comments on Paper by A. J. Bowen and D. L. Inman, "Edge wave and crescentic bars." *Journal of Geophysical Research* 77(33): 6629–6631.

SONU, C. J. (1973). Three-Dimensional Beach Changes. *Journal of Geology* 81: 42–64.

SWIFT, D. J. P. (1980). Shoreline Periodicities and Linear Offshore Shoals: A Discussion. *Journal of Geology* 88: 365–368.

SWIFT, D. J. P., B. HOLLIDAY, N. AVIGNONE, and G. SHIDELER (1972). Anatomy of a Shoreface Ridge System, False Cape, Virginia. *Marine Geology* 12: 59–84.

SWIFT, D. J. P., G. PARKER, N. W. LANFREDI, G. PERILLO, and K. FIGGE (1978). Shoreface-Connected Sand Ridges on American and European Shelves: A Comparison. *Estuarine and Coastal Marine Research* 17: 257–273.

TAKEDA, I. and T. SUNAMURA (1983). Formation and Spacing of Beach Cusps. *Coastal Engineering in Japan* 26: 121–135.

TIMMERMANS, P. D. (1935). Proeven over den invloed van golven op een strand. *Leidsche Geol. Meded.* 6: 231–386.

TUOMEY, M. (1848). *Report on the Geology of South Carolina.* Bulletin Geological Survey of South Carolina.

URSELL, F. (1952). Edge Waves on a Sloping Beach. *Proceedings of the Royal Society of London,* Series A, 214: 79–97.

WERNER, B. T. and T. M. FINK (1993). Beach Cusps as Self-Organized Patterns. *Science* 260: 968–971.

WHITE, W. A. (1966). Drainage Asymmetry and the Carolina Capes. *Geological Society of America Bulletin* 77: 223–240.

WRIGHT, L. D., B. G. THOM, and J. CHAPPELL (1978). Morphodynamic Variability of High-Energy Beaches. *Proceedings of the 16th Coastal Engineering Conference, Amer. Soc. Civil Engrs.,* pp. 1180–1194.

WRIGHT, L. D., J. CHAPPELL, B. G. THOM, M. P. BRADSHAW, and P. COWELL (1979). Morphodynamics of Reflective and Dissipative Beach and Inshore Systems: Southeastern Australia. *Marine Geology* 32: 105–140.

WRIGHT, L. D. and A. D. SHORT (1983). Morphodynamics of Beaches and Surf Zones in Australia. In *Handbook of Coastal Processes and Erosion,* P. D. Komar (editor), pp. 35–64. Boca Raton, FL: CRC Press.

ZENKOVITCH, V. P. (1967). *Processes of Coastal Development.* Translated by D. G. Fry and J. A. Steers. Edinburgh, Scotland: Oliver and Boyd.

12

The Protection of Our Coasts

The shattered water made a misty din. Great waves looked over others coming in,
And thought of doing something to the shore. That water never did to land before.

Robert Frost, 1874–1963
"Once by the Pacific"

In the opening chapter we observed that there are growing developmental pressures in the coastal zone, with about two-thirds of the world's population now living within this narrow belt directly landward from the ocean's edge. This concentration of population has resulted in increasing demands for the recreational use of beaches, and problems with erosion as homes, hotels, roads, and parks often lie in the path of shoreline recession and the destructive impacts of storms. Along the barrier islands of the Atlantic coast of the United States, many of them heavily developed, shoreline recession rates average approximately 1 m/year, while the mean rate for the Gulf coast barriers is nearly 2 m/year (Dolan, Hayden, and May, 1983; May, Dolan, and Hayden, 1983). On the west coast, expensive homes have been built along the edges of receding bluffs. As a result of this growth in coastal development and increasing demand for shore-front properties, what formerly had been natural occurrences of shoreline retreat and storm inundation have become major human and economic "problems."

The objective of this concluding chapter is to examine the response options to eroding coasts. One response is establishing coastal management programs designed to limit or to prevent construction within hazardous zones. Those management efforts were reviewed in Chapter 1, so our considerations here will focus more on what can be done regarding erosion problems within already developed areas. This will include protection of threatened properties, and where possible the restoration of the beach itself. There are four options available in reacting to receding shorelines:

1. no action,
2. retreat and relocation,
3. beach nourishment (the "soft" solution), and
4. stabilization structures (the "hard" solution).

These responses are ordered from the most passive to the most active in terms of hardening the coast with structures and are examined here in that order.

NO ACTION—RETREAT AND RELOCATION

Taking no action in the face of coastal erosion is an option that unfortunately is not often followed. It simply involves letting nature take its course, allowing the erosion to continue unabated. Being a land-based organism, *Homo sapiens* have an instinctive reaction that begrudges any loss of land to the sea. Of course, if that bit of land happens to be developed and contains a hotel worth millions of dollars, or your own more modest home, that inner instinct becomes stronger. Yet, in some instances "no action" or "retreat and relocation" are clearly the wisest choices. An example of this policy is the response to the erosion that has continued for decades in the Cape Shoalwater area of Washington, an erosion problem that represents the greatest loss of land along the west coast of the United States. One small community located there is aptly called Washaway Beach.

Cape Shoalwater is located on the north shore of the inlet to Willapa Bay, Washington (Fig. 12-1). The erosion has continued for approximately a century, with a shoreline retreat rate that has averaged between 30 and 40 m/year (Terich and Levenseller, 1986). The progress of the erosion is apparent in charts of the inlet to Willapa Bay, reproduced in Figure 12-1. The chart of Figure 12-1(a) was published in 1911, but the topographic and hydrographic surveys upon which it is based date back to 1871 to 1891. Cape Shoalwater is shown as a south-trending sand spit, projecting well into the inlet connecting the ocean to Willapa Bay. At that early date, the width of the channel opening was about 5 km. Figure 12-1(b) shows the area with a survey dating to 1911. The channel has increased in width to 6.5 km, mainly due to the northward retreat of the Cape Shoalwater shoreline. Figure 12-2 includes a compilation of shoreline positions at various dates between 1891 and 1967, derived from aerial photographs as well as from survey charts (Terich and Levenseller, 1986). During that period, some 75 years, the shoreline retreated a total of 3,750 m. This yields a maximum shoreline retreat rate of about 50 m/year; the average for the entire Cape was about 40 m/year. The erosion has slowed somewhat since 1967, averaging about 30 m/year.

The inlet survey shown in Figure 12-1(a) reveals that a century ago there were two channels with small shoals midway across the inlet. By the time of the 1911 survey [Fig. 12-1(b)], there was a single deep channel at the north and a large shoal had developed adjacent to

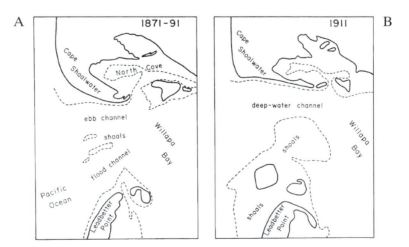

Figure 12-1   (a) The entrance to Willapa Bay, Washington, based on a Coast and Geodetic Survey chart containing survey data from 1871–1991, showing Cape Shoalwater at the north and Leadbetter Point (the Long Beach Peninsula) to the south. (b) The 1911 chart of the same area shows the extensive erosion that had occurred along Cape Shoalwater.

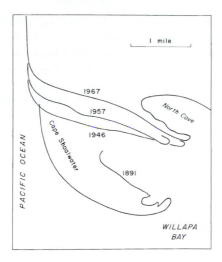

Figure 12-2 A compilation of shorelines along Cape Shoalwater derived from survey charts and aerial photographs. [Adapted with permission of Journal of Coastal Research, from T. Terich and T. Levenseller, The Severe Erosion of Cape Shoalwater, Washington, *Journal of Coastal Research 2,* p. 471. Copyright © 1986 Journal of Coastal Research.]

Leadbetter Point, the tip of the Long Beach Peninsula. However, Leadbetter Point itself had not grown to the north, so the total width of the inlet had increased due to the erosion of Cape Shoalwater. The trends noted in these two charts have been the main factors in the erosion of the cape—a northward migration and progressive deepening of the channel within the inlet. This is documented more fully in the series of inlet cross sections compiled in Figure 12-3. The migrating channel has impinged on the cape, resulting in its erosion. Although the shoreline retreat along the cape has been the primary concern since it has resulted in the loss of thousands of acres of land, it is apparent that the shoreline erosion is but a small part of the much larger changes that have occurred within the inlet.

Although the erosion of Cape Shoalwater has been faster and has lasted for a longer time than at any other site along the Pacific coast of the United States, the problem has received little attention beyond the immediate region. This is because the area is rural, and the erosion has endangered or destroyed relatively few man-made structures; a lighthouse was lost some years ago, and a few homes have succumbed to the erosion. The main highway along the coast was threatened during the 1970s and had to be moved inland. At the same time, a historic pioneer cemetery was relocated. This retreat inland has been the main response to the erosion. During the 1960s and 1970s, the Corps of Engineers undertook lengthy studies to determine the cause of the erosion and what might be done to halt further land loss. It was concluded that no short-term measures are feasible, and long-term solutions such as jetty construction would be extremely costly. It was recommended that any government funds that might be dedicated to alleviating the erosion would be better spent in purchasing the threatened land.

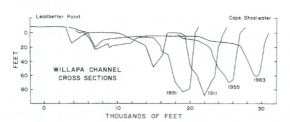

Figure 12-3 Cross sections of the inlet to Willapa Bay, showing the progressive northward shift of the main channel where it has cut into Cape Shoalwater. [Adapted with permission of Journal of Coastal Research, from T. Terich and T. Levenseller, The Severe Erosion of Cape Shoalwater, Washington, *Journal of Coastal Research 2,* p. 475. Copyright © 1986 Journal of Coastal Research.]

This response to the erosion at Cape Shoalwater is clearly a wise one, representing no action with some effort at retreat and relocation in the form of moving the highway and a cemetery. The decision not to build jetties to control the shifting inlet saved taxpayers millions of dollars. Furthermore, in that there is a net northward transport of sediment along this stretch of coast, the construction of jetties might have resulted in a new set of problems.

Retreat and relocation is often the best response to shoreline erosion and can be the most cost-effective solution. Relocation generally involves the movement of individual structures, shifting a house away from an eroding beach or back from the edge of a sea cliff that is being undercut by waves. At times, very large structures have been moved (Fig. 12-4).

In deciding whether to retreat and relocate versus "holding the line" with a seawall, one has to consider the relative expenses and whether a hard structure such as a seawall would actually represent a solution. Seawalls and revetments can be very expensive, so it is often less costly to move a home back on its lot or to relocate it entirely. More important, the construction of the seawall or revetment may not represent a long-term solution, as such structures can fail under the intense onslaught of storm waves and surf.

BEACH NOURISHMENT—THE "SOFT" SOLUTION

Beach nourishment involves the placement of large quantities of sand or gravel in the littoral zone to advance the shoreline seaward (Hall, 1952; Stauble, 1990; Seymour et al., 1995). Such nourishment can be used to create or to maintain a recreational beach or to build out the shore in order to improve the capacity of the beach to protect coastal properties from wave attack and storm inundation. Beach nourishment represents a "soft" solution that contrasts with "hard" solutions, which involve the construction of seawalls or other fixed structures. Beach nourishment is the only form of shore protection that attempts to maintain a naturally appearing beach. Costs are considerable, commonly ranging up to $5 million per kilometer of beach nourished, but the economic benefits to a community or even to the government in the form of tax revenues can exceed the costs (Houston, 1996). Some projects have been successful in providing a wide beach for a number of years, while in other cases sand added to the beach has rapidly dispersed.

The largest-scale beach-nourishment project that has been undertaken in the United States is that at Miami Beach, Florida (Fig. 12-5). This restoration project was completed during the 5-year period between 1976 and 1981, and involved the placement of 13 million cubic meters of sand dredged from offshore at a cost of more than $60 million (CERC, 1984; Egense and Sonu, 1987; Wiegel, 1992). The main objective of the project was to provide an expanded recreational beach to replace the one that had progressively decreased in size over the years. The widened beach also has served to protect the highly developed beach-front area of Miami Beach from hurricane waves and storm surge. The original project included about 16 km of shoreline length, and the placement of the 13 million cubic meters of sand produced a dry beach 55 m wide at an elevation of 3 m above mean low water. The project provided for periodic renourishment to compensate for erosional losses during the first 10 years following the initial placement. The sand for the fill came from dredging in the offshore, the source area being trenches that ran parallel to the shore 1.8–3.7 km offshore at water depths between 12 and 18 m. The nourishment sand from this source generally had a high carbonate content, as it consisted of shells and coral fragments. The project is viewed as having been highly

Figure 12-4   Etchings showing the relocation in 1888 of the Brighton Beach Hotel on Coney Island, New York. The upper illustration shows waves undermining the hotel and the partial destruction of the music stand in the offshore. The hotel was placed on freight cars on 24 lines of track and pulled inland 150 m by six locomotives. [From *Scientific American*, Vol. VIII, no. 15, April 14, 1888, p. 230]

Figure 12-5   The restoration of Miami Beach, Florida, by a nourishment project that imported sand from the offshore. The upper photo was taken in February 1978 just prior to the project, while the lower photo from October 1979 shows the expanded beach resulting from the nourishment. [From CERC, *Shore Protection Manual*, 1984, U.S. Army Corps of Engineers.]

successful in that the nourished beach has functioned for more than 20 years and has survived two major hurricanes, Hurricane David in 1979 and Hurricane Andrew in 1992 (Wiegel, 1992).

A more typical beach-nourishment project is that at Ocean City, Maryland, where about 8.5 million cubic meters of sand was placed along 11 km of beach, with the project completed in 1991 (Fig. 12-6). Nourishment was undertaken in two phases, with the State of Maryland initially building a recreational beach during 1988, using 3.5 million cubic meters of sand, and then during 1990–1991 the Corps of Engineers added 5 million cubic meters for storm protection, placing the sand in the form of dunes and an elevated berm. This project is of particular interest in that a thorough monitoring program was underway during and following the beach fill, including the collection of cross-shore profiles from the dune base to beyond the offshore closure depth of profile change. As a result, this is the best documented beach-nourishment project undertaken thus far, with a number of publications derived from the implementation and monitoring programs (Hansen and Byrnes, 1991; Grosskopf and Stauble, 1993; Grosskopf and Behnke, 1993; Stauble and Grosskopf, 1993; Kraus and Wise, 1993). This

Figure 12-6   The placement of sand nourishment on the beach at Ocean City, Maryland. [Courtesy of Great Lakes Dredge & Dock Company]

project is also of interest in that it has been subjected to a series of major storms including the Halloween storm of 1991, a northeaster that lasted 4 days, to some extent by Hurricane Bob during August 1991, and finally by a major storm on January 4, 1992. Although each storm removed sand from the berm and finally cut into dunes backing the beach, the eroded material from the exposed beach has remained within the littoral system, that is, it has not moved offshore to beyond the closure depth of the envelope of profile changes (Chapter 7); much of the sand subsequently has been transported onshore, back onto the subaerial beach. The project must be viewed as having been a success, since the nourished beach and dunes have acted to protect properties from major storms, and the sand has remained to enhance the continued recreational uses of the beach.

Other beach-nourishment projects have met with less success. An example is that at Hunting Island, South Carolina, a 7-km long barrier island surrounded by a complex series of tidal entrances (Walton and Purpura, 1977). During December 1968 some 650,000 cubic meters of sand were placed on the beach, having been derived from the landward side of the island. During 1969 and 1970, storm activity caused higher than expected erosional losses, and the initial beach and a feeder beach had all but disappeared by the end of 1970. At the end of the first 18 months, erosion had caused losses in excess of the original fill. During May–August 1971, a second nourishment of the beach was carried out, involving the placement of another million cubic meters of sand. Within six months, approximately 97 percent of that sand had been lost as well.

It can be seen in the above examples that sand for beach nourishment can potentially come from a variety of sources. In the past a primary source has been the dredging of bays and lagoons, but many of those environments are biologically rich and sensitive to disturbance, and now are generally off limits as sediment-source areas. Dredging within harbors and inlets can still be a sediment source for nourishing nearby beaches. In many cases that dredged sand originally came from the beach, and accordingly should be returned, rather than deposited in the deep-water offshore where it may be permanently lost from the littoral zone. Dean (1988) documented that in the previous 50 years, more than 60 million cubic meters of good quality sand had been dredged from Florida's east coast inlets and discarded offshore. His calculations indicate that this volume would have been sufficient to advance the shoreline by more than 7 m over the entire 600-km sandy shoreline of the east coast of Florida. In recent years a primary source of sand for beach nourishment has been offshore deposits. This material is dredged from the seabed, transported to the beach, and either dumped or pumped into the littoral zone (Fig. 12-7). In some instances a sand source can be found on land, including dune sand or another beach that has an over abundance of sand. An interesting example is the nourishment project on Sandy Hook, New Jersey, Figure 12-8 (Nordstrom et al., 1979). There is a significant northward longshore transport of sand along this spit. The construction of groynes and other structures to the south has interrupted that transport and induced erosion, particularly at the South Recreational Beach. Sand eroded from South Beach moves north as littoral transport, and has been deposited at the north end of the spit and within Sandy Hook Channel beyond the end of the spit. The beach-nourishment project simply involves using the North Recreational Beach and Channel as source areas (Fig. 12-8) and trucking the sand back to the South Recreational Beach where it is recycled through the system.

During recent years there has been increased study of the science and engineering involved in beach-nourishment projects. This research has focused on the development of basic criteria for the selection of the sediment, and the design of the geometry of the fill to

Figure 12-7 A beach-nourishment project on the Mediterranean coast of Spain, where the sand has been derived from the offshore and pumped onto the new beach. [From Projects, Works and Monitoring at Barcelona Coast, C. Peña, V. Carrion and A. Castañeda, *Proceedings of the 23rd Coastal Engineering Conference,* 1992. Reproduced with permission from the American Society of Civil Engineers.]

increase its retention on the beach and to maximize its effectiveness in providing a recreational beach and buffer for coastal defense. In the past there had been insufficient monitoring of completed nourishment projects, but as illustrated by the Ocean City nourishment, there now are increased efforts to monitor all aspects of a project so the factors that control their success or failure can be better determined.

Particularly critical is the selection of the sediment to be used as nourishment material. If the sediment is too fine, the turbulence of wave breaking will suspend the grains, allowing them to drift into the deep water of the offshore, that is, to beyond the closure depth of nearshore profile changes, and thus be lost from the littoral zone. In the case of nourishment with a wide range of sediment grain sizes, the finer fractions may be lost offshore while the coarser portions remain on the beach. This is commonly the case when sediment for nourishment is obtained from the offshore or from dredging within bays and lagoons, as sediments from those environments commonly contain silt and clay, even when predominantly composed of sand that would be suitable for nourishment use. One criterion that can be employed

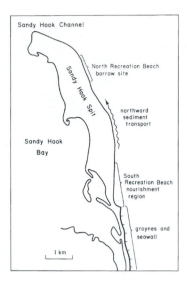

Figure 12-8   Sandy Hook, New Jersey, where sand for nourishing the South Recreational Beach is obtained from the accreted North Recreation Beach near the north end of the spit, in effect recycling the sand that has moved northward as littoral drift.

to evaluate grain sizes that would be retained on the beach is that developed by Dean (1973, 1983). If the fall time of suspended material is greater than the wave period T, then that material will probably remain permanently in suspension and be carried seaward into deep water. Considering that the material becomes suspended to some fraction, δ, of the total water depth h within the nearshore, this criterion developed by Dean becomes

$$w_s < \frac{\delta h}{T} \qquad (12.1)$$

for the settling velocities of sediments that would be lost offshore. For a suspension height δh of 10 cm above the bottom and a wave period of 10 sec, Equation (12.1) yields $w_s < 1$ cm/sec for the settling velocities of sediments that would remain in suspension throughout the wave period and so could be lost from the beach. This limit corresponds to a sediment diameter of 0.1 mm for quartz-density grains, which is usually the approximate lower limit of material found in the surf zone of natural beaches (Dean, 1983). Dette (1977) has shown that losses from beach nourishment fills occur mainly in the finest grain-size fractions, mostly the sizes less than 0.2 mm, suggesting the appropriateness of Equation (12.1) as a criterion for the limiting grain sizes that would be stable on a beach.

In practice, the choice of material for a nourishment project is controlled by availability and cost, with only an application of the "rule of thumb" that the median diameter of the borrowed material be equal to or somewhat coarser than the median diameter of the native sediment on the beach. A number of beach-nourishment projects have established that the use of a coarser-sediment fill does lead to reduced losses (e.g., Chew, Wong, and Chin, 1974; Simoen, Verslype, and Vandenbossche, 1988). More elaborate theoretical analyses of the stability of beach-fill sediments have been developed by Krumbein and James (1965), James (1974, 1975), and Dean (1974); a detailed review of their approaches is provided by Swart (1991). They compare the grain-size distribution of the natural sediment on the beach with the distribution of the source sediment in an attempt to predict the compatibility of the source material. Often the analysis involves a prediction of what fraction of the nourished material will be quickly lost into deep water and therefore how much overfill is required to supply the desired amount of "usable" sediment on the nourished beach. Unfortunately, there have been only limited attempts to examine these analysis procedures during subsequent monitoring programs. In his 1991 review, Swart concluded that information on past nourishment projects is inadequate to permit a quantitative test of the theoretical approaches. Furthermore, in a recent investigation of the stability of nourishment fills on barrier islands along the North Sea coast of Germany, Eitner (1996) found that sediment grain size had only a small effect on beach-fill longevity, while being strongly influenced by grain density. He accounted for this by analyses of the hydrodynamic properties of the sediment particles, their settling velocities and threshold stresses for entrainment, finding that the heavy minerals contained within the beach-fill borrow material formed the most stable sand fractions. This demonstrates the need for more refined analyses of sediment stability in the design of beach-nourishment projects, analyses that consider the dynamic behavior of the sediment particles under the action of waves and currents.

Various design schemes have been used for the placement of nourished sand on the beach, and there is more controversy regarding this decision than in any other aspect of beach-nourishment projects. Several placement possibilities are illustrated in Figure 12-9 and include (a) placing all of the sand as a dune backing the beach, (b) using the nourished sand to build a wider and higher berm above the mean water level, (c) distributing the added sand over the entire beach profile, or (d) placing the sand in the shallow offshore as an artificial

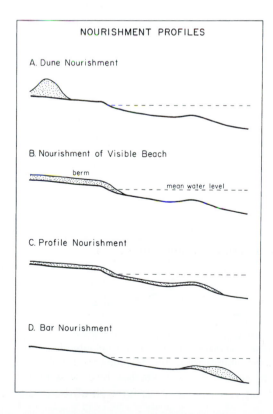

Figure 12-9 Schematic representation of a series of beach-fill profile designs used in nourishment projects, ranging from placing the sand in dunes backing the beach to its placement in the offshore as a mound or bar.

bar. The selected design depends in large part on the location of the source material and the method of delivery to the beach. If the source area is on land and the sand is trucked to the beach, then placement on the berm or in a dune is obviously the most economical solution; whereas, if the material is derived from offshore dredging, it may be more practical to dump the sand to form an artificial bar. Another factor in the selected placement design is the overall objective of the nourishment project, in particular, whether the primary goal is the construction of a recreational beach or the protection of properties from erosion and inundation.

Whichever design is chosen to place the nourished sediment on the profile (Fig. 12-9), the sediment is usually quickly redistributed in the cross-shore direction into the form of a more natural profile that is determined by the sediment size and the prevailing wave conditions (Chapter 7). This is illustrated schematically in Figure 12-10 for nourished sediment placed on the beach as a "construction profile" to form a wide, elevated berm. It is apparent that this placement artificially steepens the beach and should result in the offshore movement of the sand as it is reworked by waves and redistributed over the profile out to its closure depth. The "adjusted profile" of Figure 12-10 is the profile predicted by the design analysis for the fate of the fill after its initial reworking, an adjustment that may involve only a few weeks. Figure 12-11 provides an example of the profile changes, derived from the monitoring program at Ocean City, Maryland (Houston, 1991a). The nourishment fill was confined to the berm and inner-surf zone, placed in the form of an even sand slope over the pre-fill profile. After 4 months (01/17/89), the profile shows the expected adjustment into a more natural profile, the sand having been eroded from the nourishment wedge to form a longshore trough, with the eroded sand having moved offshore to form a bar. The profile of 04/20/89 in Figure 12-11(b) shows the effect of the first major storm that enhanced the formation of the longshore trough and bar, with significant erosion of the nourished sand placed on the subaerial part of the beach. That erosion of the berm did not represent a permanent loss of sand from the beach, however, as it was deposited on the offshore bar and therefore was still landward

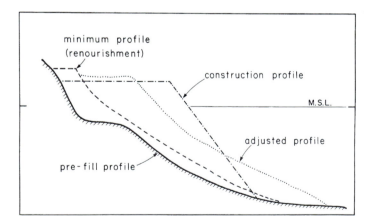

Figure 12-10   Schematic illustration of the placement of the nourished sand on the berm as the "construction profile" and as the "adjusted profile" after the waves have redistributed the sand across the entire beach profile. Also shown is the minimum profile required for protecting backshore properties, the stage when renourishment is required.

**BEACH NOURISHMENT
Ocean City, Maryland**
(after Houston, 1991)

A   Initial Profile Adjustment

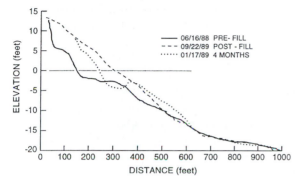

B   Storm Impact

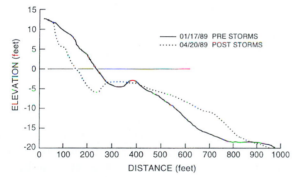

C   Recovery Phase

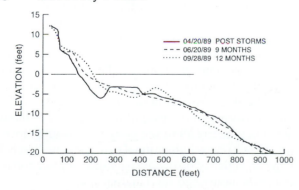

Figure 12-11 Profile changes at Ocean City, Maryland, following the placement of nourishment sand. (a) The initial adjustment of the sand into a more natural profile after its placement as an evenly sloping sand wedge. (b) The offshore shift of the nourished sand in response to a major storm. (c) The return of the nourished sand to the subaerial beach as the profile recovered following the storm. [Adapted from J. R. Houston, Beachfill Performance, *Shore & Beach* 59, 1991.]

of the closure depth of profile changes. The profiles of Figure 12-11(c) show the subsequent onshore return of a major portion of that sand during the lower energy conditions following the storm, with the width of the berm expanding.

Larson and Kraus (1991) analyzed the readjustment of constructed profiles in beach nourishment as the fill is first eroded by normal waves and subsequently attacked by a hurricane or northeaster. They utilized the SBEACH model (Chapter 7) for evaluating the profile changes in response to "synthetic" storms and obtained computed results that are very similar to those seen in Figure 12-11 from Ocean City.

In cases such as is illustrated schematically in Figure 12-10 and in Figure 12-11 for Ocean City, the general public may perceive the loss of nourished sand from the dry part of the beach as a sign of failure of the project. This aspect of public involvement has been addressed by Smith and Jackson (1990), who point out that for the layperson, the health of the beach is defined solely by the width of sand that can be seen. The public expects that any beach-nourishment project should instantly widen the visible beach and that the sand will remain there. In that most nourishment projects involve some sand placement on the berm, with its expected redistribution over the profile within a few weeks, most projects are predestined to "failure" in the view of the public. It is apparent that the public needs to be better informed about the expected changes in a nourishment fill, so they can understand that some initial offshore sediment movement will occur, resulting in a decreased width of the dry part of the beach. It should also be explained to them that as long as the sand remains within the littoral zone, it has not actually been "lost." Furthermore, the presence of the sand in the immediate offshore acts to break storm waves and helps dissipate their energy before they reach the shore, and accordingly, the nourished sand within the offshore is still meeting the objective of protecting the coast from property erosion.

Hansen and Byrnes (1991) have investigated the design of the beach-fill cross section that provides the optimum protection for the backshore against storm impacts. The beach profile response was modeled using SBEACH, and the analysis was based on the nourishment project undertaken at Ocean City, Maryland, involving the initial placement of the 3.5 million cubic meters of sand on the beach in June 1988. Six months after the project was completed, a strong northeaster produced erosion of the beach with significant profile changes. Approximately one-third of the fill material was transported from the subaerial beach and deposited as a bar between the 3 and 5.5-m water depths, but the total quantity of sand was conserved landward of the profile closure depth. Hansen and Byrnes used this measured change to calibrate SBEACH. They then used the calibrated model to examine the responses of different beach-fill designs (Fig. 12-9) that would have occurred in response to that northeaster, including both measured wave heights and water levels in the model calculations. According to the analyses, all designs withstood the impact of one northeaster or hurricane. In simulations of back-to-back northeasters, the design involving placement of all of the sand into a dune [Fig. 12-9(a)] provided the maximum protection of backshore properties, with some dune remaining even after two major storms. In the case of placing most of the nourished sand on the berm, SBEACH predicted that nearly all of the berm sand would move into the offshore bar during the two storms. Therefore, excluding the desire to immediately have a wide berm for recreational purposes, the objective of providing backshore protection is best met by a profile design that places most of the nourished sand in dunes.

The use of large dunes as a coastal protection measure has been long recognized in the Netherlands (Watson and Finkl, 1990; Verhagen, 1990; Roelse, 1990; Louisse and van der Meulen, 1991). The coastal dunes found along the Dutch coast are for the most part man-made and are designed to withstand the one in 10,000-years condition of wave intensity and storm surge. This extreme level of protection is justified since entire cities lie behind the coastal defense, and the failure of the dune system would have catastrophic consequences. The maintenance of these dunes in part involves artificial nourishment, with some sand also placed on the fronting beach.

Bruun (1988, 1990a) has been the primary proponent for nourishing the entire beach profile [Fig. 12-9(c)] which he terms "profile nourishment." The main advantage of this approach is that the sand is placed in approximately the configuration of the existing profile so the initial readjustments are minimized, in particular the rapid erosion of a nourished berm. This would in part avoid problems with the adverse perception of laymen, but the public might still wonder why so much sand is being dumped offshore rather than on the subaerial beach where it is "useful."

Beach nourishment also has involved the placement of dredged sand in the offshore [Fig. 12-9(d)] (McLellan, 1990; McLellan and Kraus, 1991). Dredged material is deposited in shallow water, typically using split-hull barges, either as a mound or in the form of a long linear ridge that simulates a naturally occurring longshore sand bar. It is expected that the sand will progressively move onto the beach, but even before that occurs, there may be benefits since the created bar could cause storm waves to break farther offshore or might reduce their energy through bottom friction. Such a reduction in the wave energy has been shown in numerical models that calculate the theoretical wave attenuation due to the presence of an offshore mound (Allison and Pollock, 1993) and also by field measurements of waves seaward and landward from a mound constructed of dredged sediments placed offshore from the entrance to Mobile Bay, Alabama (Burke and Williams, 1992).

Initially, the use of offshore mounds for beach nourishment yielded disappointing results. For example, in 1935 the Corps of Engineers built a sand bar along the 6 to 7-m water-depth contours offshore from the updrift end of the eroding beach south of the Santa Barbara breakwater (Chapter 9). The objective was to supply sand to the eroding downdrift beach, but after 21 months of monitoring, there had been no appreciable movement of the bar and no reduction in the shoreline erosion (Hands and Allison, 1991). After several such disappointments using offshore mounds, there finally were reported successes at Durban, South Africa (Zwamborn, Fromme, and Fitzpatrick, 1970), at Copacabana Beach in Brazil (Vera-Cruz, 1972), and in Denmark (Mikkelsin, 1977). These successes rekindled interest in beach nourishment by offshore disposal, and in recent years this approach has been used at increasing numbers of sites.

The question remains as to why in some instances sand from the offshore nourishment mound moves onshore to the beach, while in other cases the sediment remains as a stable deposit and does not move shoreward. Hands and Allison (1991) reviewed a number of projects in an attempt to answer this question. They compared the disposal depth with the closure depth of beach-profile changes as predicted by the analysis approach of Hallermeier (1981), and found that if the disposal depth is less than the closure depth, the disposal sediment would be active and move quickly onto the subaerial beach. An example of the associated profile changes is shown in Figure 12-12, derived from a project on the Gold Coast of Australia (Smith and Jackson, 1990). Such activity of the nourishment mound or bar placed at a depth that is shallower than the closure depth is not surprising, since this placement immediately introduces the sand into the nearshore zone of active profile change where the material can be readily distributed across the nearshore. More uncertain is the fate of disposal sand when it is placed at water depths greater than Hallermeier's closure depth. Hands and

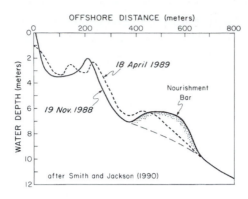

Figure 12-12 Nourishment of a beach on the Gold Coast of Australia, by placement of sand as an offshore bar. Within a few months the waves transported most of the sand onshore to form an elevated profile. [Adapted from W. S. Smith and L. A. Jackson, The Siting of Beach Nourishment Placements, *Shore & Beach* 58, 1990.]

Allison (1991) found that, in half the cases, the material was still active and moved onto the beach, while in the remaining half the disposal sediment was stable and did not nourish the adjacent beach. They compared "stable" versus "active" disposal mounds and bars with the local wave climate and had reasonable success in characterizing the sediment movement on the basis of the annual distribution of near-bottom wave-orbital velocities over the mound. As expected, if the orbital velocities are sufficiently high due to large waves and/or shallow water depths, then the disposal sand remained active and tended to move onshore. Stable mounds, like that placed offshore from Santa Barbara during the 1930s, were explainable in terms of the low wave-orbital velocities experienced over the mound. De Lange and Healy (1994) further analyzed the stability of offshore sediment mounds through calculations of grain threshold and sediment transport rates, and developed spreadsheet analyses for use on personal computers to make these evaluations.

It can be expected that in the near future we will have a better understanding of the movement of offshore disposal sediments and can better establish criteria for their onshore movement so they successfully nourish adjacent beaches. Much of this understanding will be derived from projects where the disposal mounds or bars are carefully monitored. Recent examples of monitoring programs are provided by Andrassey (1991) and Healy, Harms, and de Lange (1991), respectively, near San Diego, California, and off Tauranga Harbor, New Zealand. Healy and co-workers found that sediment dispersion from the offshore mound was rapid in the first 2 years, with some sand moving onshore to nourish the beach, but progressively slowed and became stable after 7 years, as the depth over the mound increased and a lag of coarse-grained material restricted further sediment movement.

Much of the sand lost from the immediate area of a nourishment project can result from its longshore transport to neighboring beaches. For example, Dean (1988) reported that all of the 3 million cubic meters of sand used in 1974 to nourish the beach at Port Canaveral, Florida, could be accounted for in downdrift beaches to the south. Everts, DeWalls, and Czeriak (1974) documented a systematic longshore movement of nourishment sand by monitoring the beach fill at Atlantic City, New Jersey. Sand volumes on beaches southwest of the nourished areas increased in a time-ordered sequence, while during the same period the nourished beaches experienced erosion. This is apparent in Figure 12-13, where profiles 1, 2, and 3 at the north, within the initial nourishment area, reveal a cumulative erosion spanning several years, while profiles 4 through 7 in the downdrift direction experienced an extended period of accretion. Although this longshore movement of sand can be viewed as a loss from the immediate project area, the sand continues to benefit adjacent beaches.

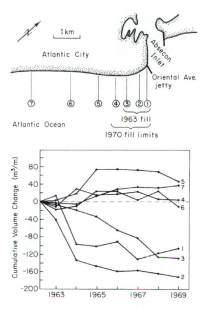

Figure 12-13 The southward movement of nourishment sand at Atlantic City, New Jersey, following its initial placement at profile lines 1 through 4. The lower graph shows that there was progressive erosion and loss of the nourished beach at the original placement site, but there was a simultaneous increase in the amount of sand on profiles 5 through 7, as the sand drifted toward the south under the prevailing waves. [Adapted from Behavior of Beach Fill at Atlantic City, NJ, C. Everts, A. E. Dewall, and M. T. Czeriak, *Proceedings 14th Coastal Engineering Conference,* 1974. Reproduced with permission from the American Society of Civil Engineers.]

As shown schematically in Figure 12-14, the volume of nourished sand can be considered as a perturbation to the shoreline, which is out of equilibrium with the forces of waves and nearshore currents that have controlled the natural shoreline planform. Those forces will act on the nourished sand, causing it to spread in the longshore direction until the natural curvature of the shore has been restored. This shoreline evolution has been analyzed as a diffusion process, the spreading of the nourished sand, with the rate of diffusion in part governed

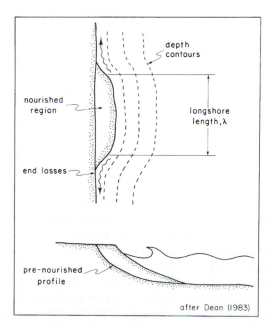

Figure 12-14 A schematic representation of sand nourishment as a perturbation of the natural shoreline, leading to the longshore movement of the sand and "end losses" to the fill. [Adapted with permission of CRC Press, from R. G. Dean, Principles of Beach Nourishment, *Handbook of Coastal Processes and Erosion,* p. 219. Copyright © 1983 CRC Press.]

by the wave energy or wave heights. Analytical models have been developed to predict the evolution of the shoreline planform (Le Méhauté and Soldate, 1977; Dean, 1983; Walton, 1994); these models were reviewed in Chapter 10, with Figure 10-12 depicting an example calculation of shorelines that result from the spreading of a fill that initially extended for 1 km alongshore with a 100-m offshore projection. One product of the analytical solutions is a relationship for the comparative times t_p required for a percentage p of the original volume to be transported out of the nourished area:

$$t_{p2} = t_{p1} \frac{\lambda_2^2}{\lambda_1^2} \frac{H_1^{5/2}}{H_2^{5/2}} \tag{12.2}$$

where λ_1 and λ_2 are the longshore lengths of two fills, and H_1 and H_2 are the respective wave heights experienced at the two sites. The important result is the prediction that the fill life is proportional to the square of the fill length λ. Doubling the length of the fill will increase the retention time in the project area by a factor of $2^2 = 4$, so it clearly is desirable for nourishment projects to have as extensive alongcoast lengths as possible, since short projects will experience rapid loses due to longshore spreading. Equation (12.2) also shows the importance of the wave-energy level and in part accounts for the rapid dispersal of nourished sediments during major storms. Doubling the wave height ($H_2 = 2H_1$) has the effect of reducing the life of the nourishment in the project area to 17.7 percent. Dean and Yoo (1992) expanded the analyses of the alongshore dispersion of nourishment fills to account for an oblique-wave approach acting on the fill, the presence of a background erosion that affects the entire stretch of coast, and the incorporation of structures such as groynes or jetties that may reduce the spreading.

A more detailed analysis of the longshore dispersion of nourished sand placed on the beach can be undertaken with computer models such as GENESIS, which simulate shoreline changes, the type of models discussed in detail in Chapter 10. Such models analyze shoreline changes resulting from longshore sediment transport, with the net erosion versus accretion at a site depending on the longshore gradient of the transport (dQ_ℓ/dx). As applied to the prediction of the fate of a beach fill, such models can examine the response to the typical wave climate of the site but more interesting would be the potential response to "abnormal" storm events or the cause of "hot spots" in the project area where the erosion is unexpectedly large. The modeling efforts should continue as part of the monitoring program, where the measured wave conditions are used as input to the model, and there are comparisons between the model-predicted beach responses and those observed in measured profile series.

The existence of a longshore transport at the site of a proposed beach-nourishment project can be an important factor in its design. In addition to the direct placement of sediment on the beach, a stockpile or feeder beach is often placed at the updrift end to serve as a source of littoral-drift material (Hall and Watts, 1957; Everts, DeWalls, and Czeriak, 1974). This situation is illustrated by Ediz Hook, on the Strait of Juan de Fuca coast of Washington (Fig. 12-15). Ediz Hook is a spit formed by the eastward longshore transport of gravel and cobbles derived from the Elwha River and cliff erosion into glacial outwash sediments (Galster and Schwartz, 1990). Erosion of the spit began early in the century as the river was dammed, cutting off its estimated supply of 38,000 m³/year of sediment to the littoral zone and then by the construction of a bulkhead along the eroding sea cliff, eliminating its 200,000 m³/year sediment contribution to the beach. Not unexpectedly, Ediz Hook began to erode, with maximum erosion at its western end, while the terminal end of the spit continued to grow toward the east. The maintenance of Ediz Hook is important as it forms the natural protection for Port

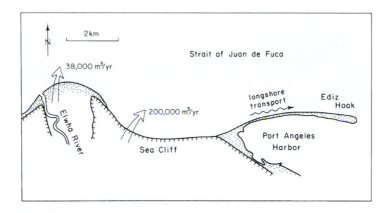

Figure 12-15 The erosion of Ediz Hook, Washington, resulting from the loss of its natural sediment sources due to dam construction on the Elwha River and construction of a seawall along the eroding sea cliffs (Galster and Schwartz, 1990). The nourishment of the beach on Ediz Hook represents a replacement of those losses within the budget of sediments.

Angeles Harbor. The response to the growing erosion was the construction of a revetment along the length of the spit, together with nourishment of the fronting beach. The initial nourishment involved gravel and cobbles, derived from inland sources, placed along the length of the spit. It is apparent, however, that nourishment would be more effective if placed at the western end of the spit as a feeder beach, basically replacing the sand and gravel that formerly was contributed by the natural sources. This provides an excellent example that beach nourishment often represents human intervention into the overall budget of littoral sediments (Chapter 3). At Ediz Hook, the sediment budget was affected first by cutting off the two major sources, sediments derived from the Elwha River and from sea cliff erosion. Beach nourishment represents a further human manipulation of the budget in an attempt to restore the lost sources. Accordingly, all beach-nourishment projects should take a broad view in attempting to develop a budget of sediments for the site, in order to better understand the basic causes of the erosion and the reason for needing a nourishment project.

In some cases the longshore loss of sand from a nourished beach can be eliminated or much reduced by the placement of terminal groynes that act like jetties in blocking longshore sand movements and particularly by exceptionally long groynes that in effect become artificial headlands. Dunham (1965) advocated their use and provided examples from California where they have been successfully employed. Prestedge, Strickland, and Watson (1991) and Prestedge (1992) documented their use in Port Elizabeth, South Africa, where a long submerged groyne with an overhead pier improved sand retention within a pocket beach. In the extreme, groynes and other structures can be part of a major project to restore a coast or to establish new artificial beaches by nourishment. Figure 12-16 is an example from the Mediterranean coast of Spain where the nourished beaches consist of a series of pockets between groynes. It is apparent that essentially no sand would be lost from such a system, except possibly under very extreme storm conditions.

In recent decades, nourishment has become common in the restoration of beaches for recreation, and in most locations has become the preferred method for shore protection from erosion and storm-surge inundation of coastal properties. The use of beach nourishment is

Figure 12-16   The creation of an artificial beach on the Mediterranean coast of
Spain by nourishment placed between groynes that act to retain the sand. [Courtesy
of Dragados Y Construciones, Spain]

now worldwide: in England (May, 1990), the Netherlands (Roelse, 1990), Belgium (Simoen,
Verslype, and Vandenbossche 1988), France (Hallegout and Guilcher, 1990), Spain (Peña,
Carrion, and Castañeda, 1992), Portugal (Psuty and Moreira, 1990), the Black Sea coast of
Russia (Kiknadze et al., 1990), Chile (Paskoff and Petiot, 1990), New Zealand (Healy, Kirk,
and de Lange, 1990), Australia (Chapman, 1980; Bird, 1990; Bourman, 1990), and Japan
(Koike, 1990). On the other hand, Pilkey and co-workers have criticized the use of beach
nourishment, suggesting that the failure of projects has been common and costly (Pilkey and
Clayton, 1989, 1990; Leonard, Clayton, Dixon, and Pilkey, 1989; Pilkey, 1990; Leonard, Clayton and Pilkey, 1990; Pilkey and Leonard, 1991). Others have come to the defense of beach
nourishment as a successful approach (Houston, 1990, 1991a, 1991b; Smith, 1990; Bruun,

1990b). This difference in opinion as to failure versus success is partly one of perception. The analyses and conclusions of Pilkey and co-workers are based on the visible beach, and following the opinions of laymen, any decrease in the width of the dry berm represents a sand loss and failure of the project. In opposition to such an interpretation, Houston, Smith, and Bruun all point out that sand movement from the berm to the shallow-water offshore is to be expected and does not represent a loss from the littoral zone unless the sand moves into deep water beyond the closure depth of profile changes. This debate again emphasizes the need for better monitoring programs to follow beach fills, and also the need for more effort at public education as to what is expected from beach-nourishment projects.

STABILIZATION STRUCTURES—THE "HARD" SOLUTION

Relocation as a response to erosion is not always possible and a "soft" solution through beach nourishment may not be viable. A "hard" solution in the form of structures that armor the coastline may then be required. Alternative structures designed to resist the onslaught of waves and to fix or advance the position of the shoreline include seawalls built parallel to the shore, groynes that act much like small jetties to trap some of the littoral drift to build out a protective beach, and offshore breakwaters that locally shelter the shore from wave attack. These and related structures are described in this section, with some discussion of their effectiveness in halting coastal erosion.

Seawalls and Revetments

The most commonly used structures for shore protection are seawalls and revetments, constructed essentially parallel to the shoreline. Such structures are built mainly for the purpose of protecting properties from the erosive action of nearshore waves and currents—a secondary objective may be to diminish the slumping of sea cliffs backing the beach. In general, seawalls are rather massive structures designed to resist the full force of the waves and may be constructed of solid or block concrete, steel sheets, or timber. Revetments may be built from natural stone known as riprap, or if large stones are unavailable, they may be constructed of concrete armor units such as dolos.

A well-known seawall is that constructed in Galveston, Texas (Fig. 12-17). The seawall was built in response to the destruction of the September 8, 1900 hurricane that killed an estimated 6,000 Galveston citizens and caused $25,000,000 in property losses. At that time, Galveston was one of the most important seaports in the United States, so that no action or retreat and relocate were not reasonable options. A decision was made to build a seawall and to raise the elevation of the city some 3 m by dredging and pumping sand fill from the bay. The initial 5-km segment of the wall was completed in July 1904. Since that time the wall has been progressively lengthened and now is about 16 km long. The face of the wall is curved so as to deflect the wave swash upward and then seaward, helping to prevent water from running over the embankment behind the wall. The Galveston seawall is founded on wooden piles and is protected from being undermined by sheet piling and a layer of loose rock 8 m wide and 1 m thick extending out from the toe of the wall (Davis, 1961).

The Galveston seawall has proved its value over the years by saving the city from a number of potentially destructive hurricanes. In 1915, soon after completion of its first segment, a hurricane having a greater storm surge and longer duration than the one in 1900, struck the

Figure 12-17   The Galveston seawall that was built following the September 8, 1900 hurricane which destroyed much of the city. The upper photograph shows the concave seawall within its wooden mold, and the lower diagram is an artist's rendition of the completed project with a riprap support for the toe of the concrete wall. [Courtesy of Rosenberg Library, Galveston]

Texas coast. This time twelve lives were lost in Galveston, and the main physical destruction (estimated at $4,500,000) was limited to the embankment behind the seawall, caused by water washing over its crest. The only damage to the wall itself consisted of two chips caused by a four-masted schooner having been washed over the wall during the height of the storm (Davis, 1961).

Most shore-protection structures are not built with the immense scale of the Galveston structure. Examples of other types of structures are shown in Figure 12-18, including vertical seawalls of concrete and timber and a rubble-mound riprap revetment. The selection of the

A

B

Figure 12-18   (a) Seawalls at Avalon, New Jersey, one constructed of bolted timbers, the second being composed of concrete and formed with a concave face. (b) A riprap revetment at St. Simon's Sound, Georgia. [Lower photo courtesy of J. R. Weggel.]

type of structure used depends on: (1) foundation conditions, (2) exposure to wave action, (3) availability of materials, (4) both initial costs and expected repair costs, and (5) past performance records of structures used in the area (CERC, 1984). In all cases it is important to have a design that provides a deep foundation so the structure will not be undermined by scour at its toe, resulting from the intense surge as the waves collide with the wall. If the foundation is inadequate, scour at the base of the structure could result in its failure. This is most dramatic for vertically faced seawalls that tend to fall forward onto the beach, but riprap structures can also fail as the waves displace the armor units and move them out of the structure.

Concrete or wooden seawalls are rigid structures whose vertical or concave faces reflect wave energy upward and back out to sea. This reflection of the wave energy may assist in the protection of coastal properties, but it may lead to greater scour of the fronting beach. As seen at the Galveston seawall, riprap is sometimes placed at the base of vertical walls to reduce wave reflection and to limit the occurrence of scour along the immediate front of the wall that might lead to its demise. Such problems with wave reflection, as well as the high costs of constructing solid seawalls, have led many to favor revetments built entirely of riprap. The following advantages are cited for riprap structures:

1. They generally cost less than vertically faced seawalls;
2. They dissipate more of the wave energy, producing less scour of the fronting beach than do vertical seawalls;
3. Because of their roughness, they allow less overtopping by run-up than do seawalls of the same height;
4. Because of their flexibility, they may settle under wave attack, but do not succumb to rapid and complete failure; and
5. They are easily maintained by the placement of additional rock.

A well-designed and constructed riprap revetment (Fig. 12-19) incorporates several components. An outer layer of large armor rock is backed by smaller rocks and a finer matrix; this grading of the material acts to prevent a strong surge of water through the structure, which could sap the sea cliff or other ground materials behind the structure. At times a geotextile or filter fabric is placed beneath the riprap, which helps prevent the armor stones from sinking into the sand. If a fabric is used, it should have a pore space that is small enough to prevent passage of the underlying sediment, while at the same time allowing the passage of water so there is no buildup of pore-water pressure on the landward side of the structure. It is important that the toe of the structure is trenched well down under the beach sediment (Fig. 12-19) and is armored by large stones so that it will not move under the forces of the most extreme waves that are expected at the site. The *Shore Protection Manual* (CERC, 1984) recommends that riprap subject to breaking waves be placed at slopes of 1.5h:1v or less (1.5 horizontal to 1 vertical), that is at 35° or less. The design elevation is based on expected storm wave heights and run-up levels, and expected mean water elevations due to tides and storm surge.

The stones of a riprap revetment must be durable and free from cracks and composed of a material that will not readily abrade or dissolve. Rounded stones such as river boulders, flattish stones with one very short axis, and stones with one very long axis should be avoided as they are less stable, being easily rolled or slid out of position by the wave surge. Estimates of the weights of armor units required to resist the forces of waves generally are based on Hudson's formula (Hudson, 1953, 1959, 1961; CERC, 1984). The derivation of this formula in-

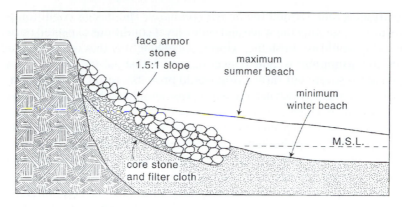

Figure 12-19 The components of a typical riprap revetment, incorporating layers of different sizes of rocks, with the base founded within an excavated trench that extends well below expected levels of scour by attacking waves.

volves an analysis of the forces acting on an individual rock placed on the sloping surface of the riprap—the forces exerted by the waves, the component of gravity acting down the riprap slope, and the frictional resistance or mechanical constraint of the underlying units that act to hold the exposed rock in position. Bruun (1976) provides a detailed analysis of many of these factors. In its most general form, as given in the *Shore Protection Manual* (CERC, 1984, p. 7–205), Hudson's formula is

$$W = \frac{\rho_s g H^3}{K_D [S_s - 1]^3 \cot \theta} \tag{12.3}$$

where

W = weight of the individual armor unit;

ρ_s = density of the armor-unit material;

S_s = specific gravity of the armor material relative to water;

H = the height of the largest wave expected to impact the structure within its designed lifetime (i.e., 20 years, etc.);

θ = angle of the structure slope measured from the horizontal; and

K_D = a stability coefficient that depends on the shapes of the armor units, the roughness of the structure's surface, and the degree of interlocking obtained in placing the units [see Table 7-8 in CERC (1984, p. 7–206)]

The formula indicates that there is a strong dependence of the armor-unit weight on the design-wave height, such that even a moderate increase in H results in a large increase in the weight W; for example, a 10 percent increase in H produces a 30 percent increase in W, while doubling H requires an 800 percent increase in armor weight to have a stable structure.

It is important to recognize that all shore-protection structures have an expected lifetime inherent in their design, even if maintained by the periodic addition of new stone. Coastal structures are built to survive the ocean environment of waves and currents up to some extreme event, termed the *design condition*. The design condition is typically the 50-year storm, representing the wave and water-level conditions expected once on average every 50 years. This would establish the extreme wave height H that is used in Hudson's formula to calculate the armor-unit weight W necessary to withstand the forces of the storm, recognizing that a more severe but rarer storm would probably result in the failure of the revetment. One could design and construct a structure that should withstand the 100-year storm, but this structure would be much larger and substantially more expensive. Judgment in the design, therefore, involves the expected lifetime of the properties being protected and their value in relationship to the cost of building the revetment. But in all cases, irrespective of the need for maintenance, there are design limitations as to what wave forces and water elevations the structure can withstand and be effective in offering protection.

The failure of a riprap revetment usually comes from a combination of settlement and stone dispersal. Such modes of failure have been described by Griggs and Fulton-Bennet (1988) for structures built along the California coast. Unless constructed immediately on top of a bedrock platform, riprap tends to settle with time, requiring the addition of more rock on top of the degraded structure. Much of this settlement can occur during storm-wave attack, so the structure loses much of its effectiveness just when needed the most. It is best to anticipate that some settlement will occur, possibly as much as 1–2 m if built over sand, requiring that the structure be substantially higher than expected to halt wave run-up during extreme storms. The other common mode of failure for riprap has been described variously as sliding or rolling of the individual stones out of position, allowing them to move downward and out onto the fronting beach. Griggs and Fulton-Bennett indicated that repairs are commonly required in California, every 5–10 years following the occurrence of major storms of even moderate wave energy.

A variation on hard structures, which are designed to be immobile in the face of wave attack, are "rubble beaches" or *dynamic revetments* that may provide protection to coastal developments while remaining more flexible. This involves the construction of a gravel or cobble beach, often at the back of a sandy beach that is offering inadequate protection for the properties. Such a morphology is naturally found on many coasts, so placement of a rubble beach as a shore-protection measure constitutes an attempt to develop a more natural solution than offered by a fixed revetment. Indeed, as first presented by Muir Wood (1970), the objective was to construct a rubble beach to be as close as possible in form to natural gravel beaches in order to ensure its stability. In this respect, the construction of a dynamic revetment is a blend in the design of a beach nourishment project and the standard riprap revetment. Downie and Saaltink (1983) and Johnson (1987) provide example applications that have been successful in being stable and providing property protection, respectively, in Vancouver, Canada, and at various sites in the Great Lakes. An interesting extension of this approach for shore protection is the artificial gravel beach at the Port of Timaru, on the east coast of the South Island of New Zealand (Kirk, 1992). The breakwater of the port had suffered degradation due to direct attack by high-energy waves, so a protective beach was established along the length of the breakwater by constructing a short groyne at its end, which partially blocked the longshore gravel transport that previously had bypassed the breakwater. The artificial gravel beach has been so successful in dissipating the wave energy that large rocks of the breakwater have been "mined" for use in structures elsewhere.

Questions have been raised as to whether the presence of a seawall or revetment can have deleterious effects on the fronting beach, and whether the structures might accelerate erosion of adjacent properties that are not protected. Answers to such questions are still debated by coastal scientists and engineers, and cannot be fully resolved based on our present understanding. Kraus (1988) reviewed the literature up to mid-1988 regarding such questions, and Kraus and McDougal (1996) updated the review.

It is important to recognize, as discussed by Weggel (1988), that the purpose of a seawall is to protect the land and developments behind it from wave attack and erosion, and its success or failure must be based on that criterion. In protecting the property, the seawall necessarily alters the patterns of wave swash, nearshore currents, and sediment movements. Weggel further discusses these potential modifications of the processes, and notes that they will depend in large part on the position of the seawall relative to the active beach face. A seawall located landward of the mean shoreline will not influence the nearshore processes except during exceptionally high water as might occur during a storm surge, while seawalls located within the surf will actively modify the processes of waves and currents and therefore the beach morphology. On this basis, Weggel developed a classification of seawall impacts into six types, depending on the wall's location on the beach and on the water depth at its toe, factors that determine how the presence of the seawall modifies the nearshore processes.

In examining the potentially harmful effects of seawalls and revetments on the beach and adjacent properties, it is necessary to distinguish between the longer-term background erosion that is not induced by the structure and erosion that is directly due to interactions of the wall with local coastal processes. Pilkey and Wright (1988) have used the terms "passive erosion" and "active erosion" for these two components. In field studies it is not always a simple matter to separate the two, especially if the study is initiated after construction of the seawall and there is no local control section without a seawall for comparison. Laboratory experiments permit greater control over the factors related to the structure-induced active erosion but are limited by their small scales.

A number of studies, both field and laboratory, have found that scour tends to develop on the fronting beach at the toe of a seawall or revetment. This effect is illustrated by the laboratory wave-channel experiments of Barnett and Wang (1988), with example results presented here in Figures 12-20 and 12-21. Each experiment began with a smooth concave profile, with the upper series in the figures showing the profile development without a seawall, while the lower series shows the profiles where a seawall has been placed at the still-water shoreline. In Figure 12-20 there is some excess scour at the base of the seawall compared with the natural profile series, but the development of offshore bars and the overall configurations of the profiles are not significantly different between the experiments without and with a seawall. The differences are much greater in Figure 12-21 where the mean water depth has been increased. There has been substantial scour at the toe of the seawall and a system of well-defined multiple bars has developed, which is largely absent in the parallel experiment without a seawall. The development of multiple bars is strong evidence that the presence of the seawall has increased wave reflection and the formation of standing waves (Chapter 7), and further suggests that much of the scour at the base of the seawall has occurred in response to that reflection.

Common wisdom has been that wave reflection from seawalls can be important and accounts for the lowering of the beach that is often observed in front of structures. Furthermore, it generally is assumed that non-reflective structures such as sloping riprap revetments are preferred over vertical seawalls in that the decrease in reflected wave energy will result in less toe scour. Recently, however, McDougal, Kraus, and Ajiwibowo (1996) argued against the role of wave reflection in producing toe scour, based on the lack of such scour during the 1991 Supertank wave-channel experiments and their development of a modified-SBEACH profile

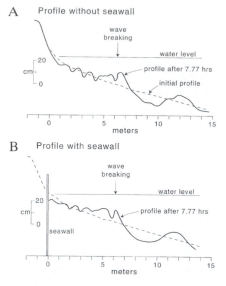

Figure 12-20 (a) Beach profile changes in laboratory wave-channel experiments, under waves of period 1.81 sec and offshore heights of 11.75 cm. (b) The similar change of the profile where a vertical seawall was placed at the still-water shoreline. Some scour has occurred at the base of the wall, but the profile otherwise has been modified very little due to the wall's presence. [Adapted from Effects of a Vertical Seawall on Profile Response, M. R. Barnett and H. Wang, *Proceedings of the 21st Coastal Engineering Conference,* 1988. Reproduced with permission from the American Society of Civil Engineers.]

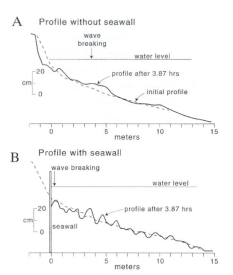

Figure 12-21 Beach profile changes without (a) and with (b) a seawall, under the same wave conditions as the experiments in Figure 12-20, but with the water depth increased. In this instance, more scour has occurred at the base of the wall, and well-developed multiple bars indicate that wave reflection from the wall has been significant. [Adapted from Effects of a Vertical Seawall on Profile Response, M. R. Barnett and H. Wang, *Proceedings of the 21st Coastal Engineering Conference,* 1988. Reproduced with permission from the American Society of Civil Engineers.]

model that includes wave reflection in the analysis. It may be that much of the toe scour that is observed is associated more with local turbulence generated by wave interactions with the structure, rather than having been produced simply by wave reflection. It is also likely that differences will be found where the fronting beach is dissipative versus reflective, and another important factor will be the position of the structure within the nearshore as suggested by Weggel (1988). In general, the farther the structure extends into the surf, the greater the resulting scour; some experiments have indicated that the scour is maximum if the seawall is located between about the middle of the surf zone to a point two-thirds out from the shoreline (Kraus, 1988).

   Most observations regarding the possible effects of a seawall or revetment on the fronting beach are based on the controlled conditions of laboratory wave channels, such as those undertaken by Barnett and Wang (1988). There has been relatively little verification from field studies, mainly due to the difficulty in gathering data during extreme storms. One interesting field investigation was that of Kriebel (1987) at Clearwater, Florida, at a time when Hurricane Elena (August 28–September 1, 1985) produced erosion along the coast. Beach profiles were obtained in front of a seawall, as well as in a control section lacking a wall. Significant toe scour (−1.6 m) occurred on the profiles backed by the seawall, erosion that exceeded the lowering of the beach profile where there is no seawall. Kriebel further noted that the volume of sand lost due to scour at the seawall was approximately equal to the volume eroded on the adjacent beach without a seawall. Apparently, the beaches with and without seawalls had similar magnitudes of offshore sediment transport under the erosive hurricane waves, but the seawall limited the extent of the landward erosion so most of the sand for the offshore transport was derived from scour at the base of the wall. Subsequent monitoring of the profiles following the storm demonstrated that both areas responded in the same way to the return of mild waves, with the beach building out to the same degree in front of the seawall as compared with the stretch of beach lacking a seawall.

In that the presence of a seawall tends to withhold sediment from the littoral system when erosion occurs, it has been argued that this results in enhanced erosion of adjacent properties, with more sand being eroded from those unprotected properties to make up for the lack of contribution from the land protected by the seawall. The greater the length of the seawall, the greater the volume of sand impounded and the greater the expected impact on adjacent properties. This explanation was used by Walton and Sensabaugh (1978) [discussed by Chiu (1977)] to account for the observed erosion during Hurricane Eloise in 1975 on the Gulf Coast. Such an interpretation is supported by wave-basin experiments undertaken by Mc-Dougal, Sturtevant, and Komar (1987), where a vertical seawall was inserted into an equilibrium beach with its toe at or above the still-water level. The return of waves resulted in erosion that was a maximum immediately adjacent to the ends of the seawall and progressively decreased with distance from the structure. The measurements were used to determine the erosion parameters r and s depicted in Figure 12-22, respectively, the inland and longshore extent of the wall-induced erosion. The laboratory data alone yielded the regression $r = 0.10L_s$, where L_s is the length of the seawall. As seen in Figure 12-22, this laboratory-based relationship agrees with the field measurements of Walton and Sensabaugh (1978), even though the scales of these data sets differ by 2 orders of magnitude. The longshore extent of the erosion, s, was found to be about 70 percent of the seawall length ($s = 0.7L_s$). These dependencies on the length of the seawall are best explained in terms of the structure denying sand to the littoral system during the erosion, resulting in excess erosion of unprotected adjacent properties.

Excess erosion of adjacent properties can also be induced by the presence of a seawall when the structure extends sufficiently into the surf that the structure plus the sheltered property act together like a groyne or jetty by intercepting a portion of the longshore sediment transport. This situation is diagrammed in Figure 12-23, illustrating that sand accumulates on the updrift side of the seawall so the erosion there is reduced, but at the same time enhanced

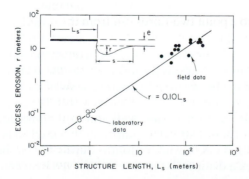

Figure 12-22 Erosion of the unprotected shoreline adjacent to a seawall, based on the wave-basin experiments of McDougal and co-workers (1987) and the field data of Walton and Sensabaugh (1978). [From Laboratory and Field Investigations of Shoreline Stabilization Structures on Adjacent Properties, W. G. McDougal, M. A. Sturtevant, and P. D. Komar, *Coastal Sediments '87*, 1987. Reproduced with permission from the American Society of Civil Engineers.]

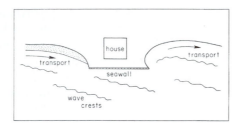

Figure 12-23 A seawall and protected property may extend across the width of the active beach to such an extent that they act like a groyne in blocking the longshore sediment transport, so the beach and adjacent properties in the down-drift direction experience erosion.

erosion occurs in the downdrift direction. This effect has been modeled in three-dimensional wave-basin experiments by Toue and Wang (1990). The similar laboratory experiments of Kamphuis, Rachet, and Jui (1992) found that the rate of longshore sediment transport is decreased in front of the seawall as a result of structure-induced increased water depths. They further established that the peaks in the cross-shore distributions of the bedload and suspension transport associated with the zone of wave-breaking shift offshore, and the transport within the swash zone disappears entirely as the beach in front of the seawall is eroded.

Only during the last decade have there been extensive field programs to monitor the potential impacts of seawalls on adjacent beaches and properties. Basco, Bellomo, and Pollock (1992) investigated the effects of seawalls along the coast of Virginia and in particular at Sandbridge where roughly 60 percent of the 7.7-km length of the study area is protected by a variety of timber, steel, or concrete seawalls. Basco (1990) analyzed the rates of shoreline change over the past 120 years and concluded that during the last 50 years since seawall construction the rate of shoreline recession has not been statistically different from rates that existed prior to wall construction. Although the highest rates of shoreline recession (3 m/year) have occurred in Sandbridge where the concentration of seawalls is greatest, Basco concludes that the higher recession is due to the steeper offshore bathymetry and higher wave energy in that region and is not caused by the presence of the seawalls. Basco and co-workers also focused their investigation in the Sandbridge area, collecting monthly profiles at twenty-eight locations—twelve contain seawalls, ten are across dunes, and six are located near the ends of seawalls. They found that while the landward loss of sediment was much lower at the seawall sections, demonstrating their effectiveness in offering protection to the properties, there was a slightly higher loss of beach seaward of the walls compared with areas lacking walls. This resulted in a greater reduction in the level of the berm fronting a seawall. However, as found in the longer-term analyses by Basco (1990), statistical evaluations of shoreline recession rates showed no differences between the sections having seawalls versus the natural beach/dune sections.

An extensive monitoring program focusing on the potential effects of seawalls on the adjacent beach and properties has been conducted by Griggs and his students in the vicinity of Santa Cruz, California (Griggs and Tait, 1988; Griggs, Tait, and Scott, 1991; Plant and Griggs, 1992; Griggs, Tait, and Corona, 1994). Their monitoring programs included four sites having different types of protective structures at various positions across the beach profiles. Figure 12-24 is site no. 4, South Aptos Seascape, a curved concrete seawall with a riprap toe, extending 75 m seaward on the profile from the base of the sea cliff. Biweekly surveys were made at each site, including profiles in front of the wall and in the immediately adjacent areas, and where possible, a distant control stretch of unprotected beach was also surveyed. It was found that with the onset of the fall to winter months during the cut back of the profiles in their general seasonal cycles, a distinct difference is observed between the profiles in

Figure 12-24 A curved concrete seawall with a riprap toe near Santa Cruz, California, protecting houses built on the berm of the active beach. This is one of several sites included in studies by Griggs and co-workers of the effects of seawalls on the adjacent beach and properties. [Courtesy of G. Griggs]

front of the seawalls and the profiles on the adjacent unmodified beaches, with an attainment of winter profiles occurring much sooner in front of the seawalls. This enhanced erosion of the berm profile in front of the seawalls produced a 0.6–1.2 m vertical difference in beach elevation compared with beaches without seawalls, and the berm on the unprotected beach may be as far as 12 m seaward of the seawall position. Similar rates of berm erosion took place in front of dissipative riprap revetments and vertical seawalls, suggesting that differences in their degrees of wave reflection and porosity were not important. Furthermore, although there was enhanced erosion of the berm, excess toe scour was never observed at the immediate foot of the structure, but this may have been due to the relatively mild wave conditions that prevailed throughout the duration of their study. Once the berm crest passed the seawall during the erosion phase, no significant differences between the two sets of profiles could be discerned that might be attributed to wave reflection from the structure. However, at two sites the structures extended out sufficiently to act like groynes, with excessive erosion occurring in the downdrift directions, lowering the profiles even below those fronting the seawalls. Rebuilding of the berm on the profiles during the late spring to summer proceeded independently of the structures, resulting in a continuous, uniform berm along the length of the beach.

Plant and Griggs (1992) investigated the same series of structures in the Santa Cruz area but focused on the interactions between the structures and the nearshore processes in order to better understand the observed beach responses. They used video techniques and found that swash reflected from the seawall is directed seaward several seconds earlier than the swash on an adjacent natural beach, increasing the duration and velocity of the backwash. Six piezometers were installed in two shore-parallel rows positioned along the upper and lower beach slopes to monitor groundwater levels. It was found that rock revetments inhibit groundwater flow and increase water-table elevations at the boundary of the wall and beach. This increased water-table elevation feeds the backwash flow and increases the offshore sediment transport in front of the wall. Waves that reflect from the structure increase the water depth and accentuate swash run-up on the downcoast beach immediately adjacent to the wall. On the adjacent section of beach, Plant and Griggs observed that the backwash begins later than it does both in front of the wall and farther downcoast, and lasts longer since the water drains off the flank of the seawall and down the beach slope.

In summary, both laboratory and field studies of the physical effects of seawalls and revetments demonstrate that they can have adverse impacts to both the fronting beach and adjacent properties. In most cases the reduction in the level of the fronting beach is small, so the impact there is relatively insignificant. It may be more important when the structure extends out into the surf, resulting in a permanent deepening of the water. In this case a greater potential impact comes from the structure acting like a groyne, inducing erosion of downdrift adjacent properties. It is important, therefore, to avoid the situation where structures originally built landward of the shore are sufficiently massive to withstand the most extreme storms, so that with the general retreat of the coast they eventually become positioned within the surf where they have a more negative impact. The studies also have shown that the presence of a structure can impound sand, especially when they protect foredunes that would normally erode during a storm and supply sand to the fronting beach. The result is greater erosion of adjacent, unprotected foredunes and properties, in effect acquiring an equivalent volume of sand denied by the area protected by the seawall. Proponents of seawalls point to the recovery of the beach following the storm, and that the beach fronting the seawall recovers to the same degree as the unprotected beach, concluding that there is no long-term impact. This misses the point, since the subsequent recovery of the beach following a storm is of little consolation to individuals who have lost their homes due to the wall-induced erosion during the height of the storm. Seawalls and revetments clearly can serve to protect developments in the face of coastal erosion and accordingly their use should not be arbitrarily prohibited. On the other hand, their potentially adverse impacts need to be recognized and evaluated whenever installation of a structure is under consideration.

Not addressed above is the visual impact of seawalls and revetments, which when they achieve a massive scale can seriously degrade the natural aesthetics of the coast. Fortunately, smaller-scale structures sometimes become buried beneath beach and dune sand, and reappear only when there is an episode of extreme erosion, at which time they play their intended role. This is illustrated in Figure 12-25 for a massive riprap revetment that was built on the Oregon coast to protect houses following a major storm in 1978 (Komar, 1978). In subsequent years, sand has returned to the beach to such an extent that the revetment is completely covered, and the problem now is with sand accumulating to such an extent that houses have become partially buried and their windows "frosted" by sand blasting. Sand burial also occurred on the Taraval seawall, built on the ocean shore of San Francisco in 1941 in response to erosion problems (Berrigan, 1985a, 1985b). Shortly after construction, the wall was covered by high dunes, and its existence was virtually forgotten until the severe storms of 1977–1978 and 1982–1983 uncovered the wall, at which time it again served to protect coastal properties. In order to decrease the visual impact of seawalls and revetments, they sometimes are designed to have a sand cover. For example, Headland (1992) describes the design of a combined dune and seawall system to protect a military facility at Dam Neck, Virginia. Immediately

A

B

Figure 12-25　(a) A large riprap revetment constructed in 1978 to protect houses in Pacific City, Oregon, from erosion induced by a major storm. (b) In the intervening years, sand has returned to the beach and foredunes, covering the revetment and beginning to bury the houses.

following its construction, the seawall was buried within a sand dune placed as part of a beach nourishment operation, the expectation being that the wall will serve as backup protection only during the most extreme storms and would be visible only during those rare events.

Groynes and Breakwaters

A conceptually different approach to harden the coast is the use of groynes and detached breakwaters, both being designed to trap sand and build out a buffering beach. A *groyne* (American spelling, groin) (Fig. 12-26) is a rib built approximately perpendicular to the shore, designed to trap a portion of the longshore sediment transport in order to build out the beach. It was seen earlier that groynes may also be constructed to create a localized pocket beach, generally to contain nourished sand. In either case, the widened beach serves as a buffer between the surf and land, so the construction of groynes can diminish erosion of coastal properties. Groynes are relatively narrow in width and may vary in length from less than 10 m to over 200 m. In this regard they appear similar to jetties, although their function and purpose are very different. Groynes and jetties have the same effect in damming the longshore sediment transport, so the shoreline builds out along the updrift side and erodes in the downdrift direction. To protect a large area from erosion, a series of groynes, a groyne field, may be constructed to act together. This builds out an extended stretch of beach and shifts the zone of erosion out of the immediate area, but in doing so generally transfers the erosion problem to the downcoast neighbors.

Once a groyne is filled, it allows any additional longshore sediment transport to pass around its seaward end, so it traps only a finite quantity of sand. While sand is accumulating between the groynes, that sand is prevented from reaching the beaches in the downdrift direction, so erosion is enhanced there, analogous to the erosion that occurs downdrift from jetties (Chapter 9). To prevent damage to the adjacent areas, groynes may be filled artificially by beach nourishment. If this is done, the groynes will not derive their sand fill from the natural littoral drift, which should in theory continue to reach the downdrift beaches just as it did prior to groyne construction.

The segment of beach between two adjacent groynes acts like a small pocket beach. Comparable to the computer-simulated beach oscillations developed in Chapter 10, the beach between two groynes aligns itself with the crests of the incoming waves and therefore generally will depart from the shore alignment that existed prior to groyne construction. Similarly, the beach between groynes will oscillate due to waves arriving from various offshore directions. Because the pockets between groynes are generally small, this rotation of the beach in response to changing wave conditions can be quite rapid.

The construction of groynes is similar to that of seawalls and revetments. They commonly are built of stone, again with a surface armor of large stones and a core of smaller material. Groynes also are constructed of wood, either planks or logs. The main design elements involve the selection of the lengths of the groynes and their longshore spacings. The conventional practice is that the groyne length should be approximately 40–60 percent of the average surf zone width, and the spacing between groynes should be about four times their length. This length will allow the structure to trap some of the littoral drift, but once the compartment is filled, the subsequent longshore transport will be able to pass around the end of the structure which is still within the surf zone. The spacing between groynes is governed principally by the expected orientations of the fill in response to the changing wave directions. The spacing together with groyne lengths establishes the shoreline orientation that would allow

Figure 12-26 Groynes at Seabright, New Jersey, which are designed to trap sand in order to build out a beach that provides protection to the coastal developments. [Courtesy of J. R. Weggel]

the beach to extend out to the end of the structure and spill sand around its tip. It follows that the longer the groyne and the greater the spacing, the greater the extent of the accreted beach trapped by the structure.

The failure of designed groynes to meet their objective of trapping sand to create a protective beach often results from the presence of nearshore currents. As illustrated in Figure 12-27(a), the groyne may deflect longshore currents generated by waves breaking obliquely to the shore, in effect creating a rip current adjacent to the groyne that can scour away the impounded sand and transport it offshore and to the downdrift side of the groyne. This effect is potentially greatest with larger groyne spacings, since there is an increased probability for obliquely breaking waves along the shore and the large spacing permits the acceleration of the longshore current to its value governed by the heights of the waves and their breaker angles (Chapter 8). An interesting possibility is that the designed groyne spacing might also trap standing edge waves. The worst case would be if the selected spacing results in resonate edge waves at the incident wave frequency, as this would give rise to a strong cell circulation as depicted in Figure 12-27(b), with rip currents positioned at both limiting groynes (Gaughan and Komar, 1977). This situation is comparable to the wave-basin experiments of Bowen and Inman (1969), where such resonant conditions were created purposefully to investigate the generation of the cell circulation with rip currents, experiments that were discussed in Chapter 8.

Detached breakwaters are "detached" in the sense that they are not connected to the shore like groynes but instead are aligned approximately parallel to the local shoreline (Fig. 12-28). When used as a shore-protection structure, detached breakwaters typically are 25–100 m long and are placed just offshore from the average width of the surf zone. Detached breakwaters often are built as a series, analogous to the groyne field, in order to protect a long

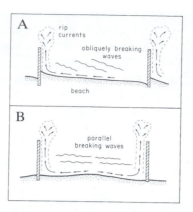

Figure 12-27 (a) The deflection of a wave-generated longshore current by a groyne, forming a rip current that may carry sand seaward from the beach. (b) The formation of a cell circulation with rip currents adjacent to the groynes, resulting from the selection of a groyne spacing that produces resonate conditions for trapped standing edge waves as in the wave-basin experiments of Bowen and Inman (1969).

Figure 12-28 Detached breakwaters on the Mediterranean coast of Spain, the trapped sand having built out to the point where the beach is now attached to the breakwaters, forming tombolos. [Courtesy of Dragados Y Construciones, Spain]

stretch of coast; in this form they are often referred to as *segmented breakwaters*. The gap distance between breakwater segments becomes an important parameter in their design, together with the lengths of the breakwaters and offshore distances. Detached breakwaters are designed to provide a sheltered beach area, with calmer water than found on the open coast. In principle, the sediment can pass alongshore between the structure and the shoreline. However, the beach in the lee of the structure is generally built out because of the structure's shadowing effect that reduces the wave energy and controls the pattern of wave refraction and diffraction. In general, there are three potential shoreline responses to a detached breakwater: (a) the development of a tombolo with an attachment to the structure (Fig. 12-28); (b) the formation of a salient or cusp in the lee of each breakwater, but without attachment to the structure; or (c) only limited modification of the shoreline configuration. In this way, detached breakwaters help to build out the beach and reduce property erosion, while unlike groynes, they do not tend to deflect beach sediment into deep-water offshore. The main disadvantages of breakwaters include their higher cost of construction compared with land-connected groynes, and design criteria are still not well established.

Although the first detached breakwaters were constructed as early as the 1930s (Magoon, 1976), it has been only within the last decade that they have been commonly employed as shore-protection structures. Monitoring of completed projects is expected to provide new information and data as to their design and the shoreline responses. Dally and Pope (1986), Pope and Dean (1986), Suh and Dalrymple (1987), and Chasten et al. (1993) provide reviews of the functional design of detached breakwaters and the shoreline responses, based on the results of numerical and physical models and on completed projects. Particularly important to the design of detached breakwaters is the prediction of whether the beach response will be in the form of tombolos, salients, or whether there will be minimal effects on the shoreline configuration. It is clear that this depends in part on the offshore distance of the breakwaters from the average shoreline, their lengths compared with gap distances, and on the wave-energy level. The dependence on these factors is supported by the analysis of Pope and Dean (1986) based on completed projects, with the results shown in Figure 12-29. The shoreline form is a function of the ratio of the structure length to gap length (L_s/L_g) and to the ratio of the offshore distance to water depth at the structure position (X/d_s). The water depth limits the heights of waves that can enter the gaps and diffract in the sheltered lee of the structure. Although this analysis does not include other relevant factors such as beach sediment parameters or the wave transmission due to "permeability" of the breakwaters, the results of Figure 12-29 can still provide guidance as to the expected shoreline response. These results have been further verified with the GENESIS numerical simulation of shoreline planforms behind a detached breakwater, undertaken by Hanson, Kraus, and Nakashima (1989) and Hanson and Kraus (1990).

The detached breakwaters installed on the Lake Erie shoreline at Lorain, Ohio, provide an example of their use and effectiveness (Pope and Rowen, 1983). Three segmented breakwaters were constructed in 1977 to protect the eroding shore and to build a recreational beach. The project consists of three 75-m long rubble-mound breakwaters separated by 50-m gaps (Fig. 12-30) and two end groynes, a short west groyne and a 100-m long east groyne. The orientations of the detached breakwaters and respective lengths of groynes were designed in response to the prevailing wave approach from the west and the dominant longshore sediment transport toward the east. The compartment created by the construction was filled with 84,000 m^3 of sand having a 0.5 mm median diameter, which is significantly coarser than the native beach sand. The response of the shore is documented in Figure 12-30; the 10/24/77 shoreline is representative of only 1 month after completion of the project but distinct bulges already are apparent behind each breakwater. The shoreline of 10/18/82 shows the subsequent equilibrium configuration with three salients, demonstrating that the beach has rotated counterclockwise with maximum erosion adjacent to the west groyne. A physical model of the project (1:50 scale), using crushed coal as sediment, indicated that the material eroded next to the west groyne tends to migrate eastward and offshore between the breakwaters and beachfill, with no appreciable amount moving alongshore east of the east groyne. The model tests also indicated that the beach would be wider if the west groyne were not present, due to the eastward longshore sediment transport. This is also suggested by the field observations which reveal that the fine-grained natural sand is progressively replacing the coarser nourished sand in the western portion of the protected beach. This project can be considered as a success since the detached breakwaters have maintained a beach that protects the coast from wave erosion and also has provided a recreational beach.

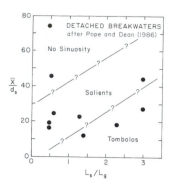

Figure 12-29 The three forms of shorelines protected by detached breakwaters, the form depending on the offshore distance of the structure (X), the water depth at the structure's position (d_s), and on the ratio of the structure's length to its gap distance (L_s/L_g). [Adapted from Development of Design Criteria for Segmented Breakwaters, J. Pope and J. L. Dean, *Proceedings of the 20th Coastal Engineering Conference,* 1986. Reproduced with permission from the American Society of Civil Engineers.]

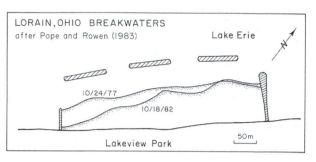

Figure 12-30 The detached breakwaters and associated groynes designed to protect the coast at Lorain, Ohio, on Lake Erie, and to establish a recreational beach. The shoreline of 10/24/77 is the extent of the nourished sand fill soon after completion of the project, while the shoreline of 10/18/82 shows its subsequent reorientation and the development of three minor salients in the protected zones behind the breakwaters. [Adapted from Breakwaters for Beach Protection at Lorain, OH, J. Pope and D. D. Rowen, *Proceedings of Coastal Structures '83,* 1983. Reproduced with permission from the American Society of Civil Engineers.]

SUMMARY

There are a number of options in protecting shorelines from the threat of erosion and the possible destruction of coastal properties. These range from the most passive response of taking no action or retreating out of the zone of danger, to attempts to build a beach buffer through nourishment projects, to the hardening of the coast through the construction of seawalls, groynes, and breakwaters. Different situations require different responses, and this has been illustrated by various real-case examples presented throughout this chapter. Any response needs to be based on analyses of the physical processes of waves, nearshore currents, and sediment transport, as well as by sound engineering analyses and designs that consider the various options. The choice needs to be made within the overall framework of coastal-zone management issues that maintain a broad perspective on the long-term fate of the coast and its natural hazards. An interesting example is provided by Florida, where it has been decided that beach nourishment is the favored response to provide protection to coastal properties and to restore recreational beaches. Where hard structures are required, they are designed to withstand only minor storms but to fail for a larger storm such as a hurricane. The idea is that the natural force of a major storm should be allowed to reshape the coast in a uniform and more natural manner. For example, if a property is protected by a massive seawall that survives the 100-year event, while the adjacent shorelines erode and retreat, that wall will end up within the surf zone with no fronting beach and could act as a barrier to the longshore movement of beach sediments. Such well-reasoned approaches to shoreline protection are required if we are to maintain the coastal zone in an attractive form for recreational use, while at the same time provide an acceptable degree of safety for development.

QUESTIONS AND PROBLEMS

1. General practice in beach-nourishment projects is to use a sand source that is coarser than the native sand on the beach. How would this increase in grain size potentially change the morphology of the beach and its dynamic response to storms? What might be the negative versus positive consequences in choosing a coarser sediment size for the nourished material?

2. In comparing two beach-nourishment projects, other factors being equal, one project is expected to experience wave heights that are 2 times as large as at the second project. According to Equation (12.2), how much greater should the length of the placement be so that both projects have the same rate of alongshore spreading?

3. Discuss the relative merits of the different nourishment schemes in placing the sand across various portions of the beach profile, as diagrammed in Figure 12-9.

4. You are responsible for the design of a beach-nourishment project and have decided to place part of the sand on the berm and part in dunes backing the beach. Write a letter to the mayor of the city justifying this choice and explain why so much of the sand has disappeared during the first storms of the winter.

5. You need to design a riprap revetment to withstand storms having a design wave height of 4 m. The rock material available is granite with a specific gravity of 2.8. Table 7-8 in the *Shore Protection Manual* (CERC, 1984, p. 7–206) yields an approximate value of $K_D = 2.0$ for the stability coefficient. Calculate the rock weights required for a range of reasonable structure slopes. How will the run-up of waves on the structure depend on its slope and armor-rock sizes? Discuss how the cost of the structure will depend on the choice of its slope, on the corresponding rock sizes required for that slope, and on the total elevation of the structure needed to prevent overtopping by extreme wave run-up.

6. A beach has an average slope of $S = 0.05$, and the waves typically have periods of 10 sec. Construction of a groyne field has been proposed as a shore-protection structure. What groyne spacings should be avoided so that resonant edge waves do not develop between adjacent groynes, edge waves having the same period as the incident waves such that rip currents might be generated according to the hypothesis of Bowen and Inman (1969)?

7. Sketch the expected patterns of refraction and diffraction of waves as they pass through the gaps in a series of segmented breakwaters, and the patterns of nearshore currents and sediment transport that would be generated along an initially straight shoreline. How might these patterns evolve into new equilibrium shorelines represented by the formation of salients or tombolos in the protected areas behind the breakwaters?

REFERENCES

ALLISON, M. C. and C. B. POLLOCK (1993). Nearshore Berms: An Evaluation of Prototype Designs. *Coastal Zone '93, Amer. Soc. Civil Engrs.*, pp. 2938–2950.

ANDRASSY, C. J. (1991). Monitoring of a Nearshore Disposal Mound at Silver Strand State Park. *Coastal Sediment '91, Amer. Soc. Civil Engrs,* pp. 1970–1984.

BARNETT, M. R. and H. WANG (1988). Effects of a Vertical Seawall on Profile Response. *Proceedings of the 21st Coastal Engineering Conference, Amer. Soc. Civil Engrs.*, pp. 1493–1507.

BASCO, D. R. (1990). The Effect of Seawalls on Long-Term Shoreline Change Rates for the Southern Virginia Ocean Coastline. *Proceedings of the 22nd Coastal Engineering Conference, Amer. Soc. Civil Engrs.*, pp. 1292–1305.

BASCO, D. R., D. A. BELLOMO, and C. POLLOCK (1992). Statistically Significant Beach Profile Change With and Without the Presence of Seawalls. *Proceedings of the 23rd Coastal Engineering Conference, Amer. Soc. Civil Engrs.*, pp. 1924–1937.

BERRIGAN, P. D. (1985a). The Taraval Vertical Seawall. *Shore & Beach* 53(1): 2–7.

BERRIGAN, P. D. (1985b). Seasonal Beach Changes at Taraval Seawall. *Shore & Beach* 53: 9–15.

BIRD, E. C. F. (1990). Artifical Beach Nourishment on the Shores of Port Phillip Bay, Australia. *Journal of Coastal Research, Artificial Beaches,* M. L. Schwartz and E. C. F. Bird (editors). Special Issue 6, pp. 55–68.

BOURMAN, R. P. (1990). Artificial Beach Progradation by Quarry Waste Disposal at Rapid Bay, South Australia. *Journal of Coastal Research, Artificial Beaches,* M. L. Schwartz and E. C. F. Bird (editors). Special Issue 6, pp. 69–76.

BOWEN, A. J. and D. L. INMAN (1969). Rip Currents, 2: Laboratory and Field Observations. *Journal of Geophysical Research* 74: 5479–5490.

BRUUN, P. (1976). Parameters Affecting Stability of Rubble Mounds. *Journal of the Waterways, Harbors, and Coastal Engineering Division, Amer. Soc. Civil Engrs.* 102(WW2): 141–164.

BRUUN, P. (1988). Profile Nourishment: Its Background and Economic Advantages. *Journal of Coastal Research* 4: 219–228.

BRUUN, P. (1990a). Beach Nourishment—Improved Economy Through Better Profiling and Backpassing from Offshore Sources. *Journal of Coastal Research* 6: 265–327.

BRUUN, P. (1990b). Discussion of: Leonard, L., Clayton, T., and Pilkey, O., 1990. An Analysis of Replenished Beach Design Parameters on U.S. East Coast Barrier Islands. *Journal of Coastal Research* 6: 1037–1039.

BURKE, C. E. and G. L. WILLIAMS (1992). Nearshore Berms—Wave Breaking and Beach Building. *Ports '92 Conference, Amer. Soc. Civil Engrs.,* pp. 96–110.

CERC (1984). *Shore Protection Manual.* U.S. Army Corps of Engineers, Waterways Experiment Station, Coastal Engineering Research Center, Volumes I and II, Superintendent of Documents, Washington, D.C.

CHAPMAN, D. M. (1980). Beach Nourishment as a Management Tool. *Proceedings of the 17th Coastal Engineering Conference, Amer. Soc. Civil Engrs.,* pp. 1636–1649.

CHASTEN, M. A., J. D. ROSATI, J. W. MCCORMICK, and R. E. RANDALL (1993). *Engineering Design Guidance for Detached Breakwaters as Shoreline Stabilization Structures.* Technical Report CERC-93-19, U.S. Army Corps of Engineers.

CHEW, S. Y., P. P. WONG, and K. K. CHIN (1974). Beach Development Between Headland Break Waters. *Proceedings of the 14th Coastal Engineering Conference, Amer. Soc. Civil Engrs.,* pp. 1399–1418.

CHIU, T. Y. (1977). Beach and Dune Response to Hurricane Eloise of September 1975. *Coastal Sediments '77, Amer. Soc. Civil Engrs.,* pp. 116–134.

DALLY, W. R. and J. POPE (1986). *Detached Breakwaters for Shore Protection.* Technical Report CERC-86-1, Coastal Engineering Research Center, U.S. Army Engineer Waterways Experiment Station, Vicksburg, Mississippi.

DAVIS, A. B. (1961). The Galveston Sea Wall. *Shore & Beach* 29: 6–13.

DEAN, R. G. (1973). Heuristic Models of Sand Transport in the Surf Zone. *Proceedings of the 1st Australian Conference on Coastal Engineering,* Engineering Dynamics in the Surf Zone, Sydney, pp. 209–214.

DEAN, R. G. (1974). Compatibility of Borrow Material for Beach Fills. *Proceedings of the 14th Coastal Engineering Conference, Amer. Soc. Civil Engrs.,* pp.1319–1330.

DEAN, R. G. (1983). Principles of Beach Nourishment. In *CRC Handbook of Coastal Processes and Erosion,* P. D. Komar (editor). Pp. 217–232. Boca Raton, FL: CRC Press.

DEAN, R. G. (1988). Managing Sand and Preserving Shorelines. *Oceanus* 31: 49–55.

DEAN, R. G. and C. YOO (1992). Beach-Nourishment Performance Predictions. *Journal of Waterway, Port, Coastal, and Ocean Engineering, Amer. Soc. Civil Engrs.,* 118(6): 567–586.

DE LANGE, W. and T. HEALY (1994). Assessing the Stability of Inner Shelf Dredge Spoil Mounds Using Spreadsheet Applications on Personal Computers. *Journal of Coastal Research* 10: 946–958.

DETTE, H. H. (1977). Effectiveness of Beach Deposit Nourishment. *Coastal Sediments '77, Amer. Soc. Civil Engrs.,* pp. 211–227.

DOLAN, R., B. HAYDEN, and S. MAY (1983). Erosion of U.S. Shorelines. In *Handbook of Coastal Processes and Erosion,* P. D. Komar (editor). Pp. 285–299. Boca Raton, FL: CRC Press.

DOWNIE, K. A. and H. SAALTINK (1983). An Artificial Cobble Beach for Erosion Control. *Coastal Structures '83, Amer. Soc. Civil Engrs.,* pp. 846–859.

DUNHAM, J. W. (1965). Use of Long Groins as Artificial Headlands. *Coastal Engineering, Santa Barbara Speciality Conference, Amer. Society of Civil Engrs.,* pp. 755–762.

EGENSE, A. K. and C. J. SONU (1987). Assessment of Beach Nourishment Methodologies. *Coastal Zone '87, Amer. Soc. Civil Engr.,* pp. 4421–4433.

EITNER, V. (1996). The Effect of Sedimentary Texture on Beach Fill Longevity. *Journal of Coastal Research* 12: 447–461.

EVERTS, C., A. E. DEWALLS, and M. T. CZERIAK (1974). Behavior of Beach Fill at Atlantic City, NJ. *Proceedings of the 14th Coastal Engineering Conference, Amer. Soc. Civil Engr.,* pp. 1370–1388.

GALSTER, R. W. and M. L. SCHWARTZ (1990). Ediz Hook—A Case History of Coastal Erosion and Rehabilitation. *Journal of Coastal Research, Artificial Beaches,* M. L. Schwartz and E. C. F. Bird (editors). Special Issue 6, pp. 103–113.

GAUGHAN, M. K. and P. D. KOMAR (1977). Groin Length and the Generation of Edge Waves. *Proceedings of the 15th Coastal Engineering Conference, Amer. Soc. Civil Engrs.,* pp. 1459–1476.

GRIGGS, G. B. and K. FULTON-BENNETT (1988). Rip Rap Revetments and Seawalls and Their Effectiveness Along the Central California Coast. *Shore & Beach* 56(2): 3–11.

GRIGGS, G. B. and J. F. TAIT (1988). The Effects of Coastal Protection Structures on Beaches Along Northern Monterey Bay, California. *Journal of Coastal Research,* Special Issue No. 4, pp. 93–111.

GRIGGS, G. B., J. F. TAIT, and K. SCOTT (1991). The Impacts of Shoreline Protection Structures on Beaches Along Monterey Bay, California. *Proceedings of the 22nd Coastal Engineering Conference, Amer. Soc. Civil Engrs.,* pp. 2810–2823.

GRIGGS, G. B., J. F. TAIT, and W. CORONA (1994). The Interaction of Seawalls and Beaches: Seven Years of Monitoring, Monterey Bay, California. *Shore & Beach* 62(3): 21–28.

GROSSKOPF, W. G. And D. K. STAUBLE (1993). Atlantic Coast of Maryland (Ocean City) Shoreline Protection Project. *Shore & Beach* 61(1): 3–7.

GROSSKOPF, W. G. and D. L. BEHNKE (1993). An Emergency Remedial Beach Fill Design for Ocean City, Maryland. *Shore & Beach* 61(1): 8–12.

HALL, J. V. (1952). Artificially Nourished and Constructed Beaches. *Proceedings of the 3rd Coastal Engineering Conference, Amer. Soc. Civil Engrs.,* pp. 119–136.

HALL, J. V. and G. M. WATTS (1957). Beach Rehabilitation by Fill and Nourishment. *Transactions, Amer. Soc. Civil Engrs.,* pp. 155–177.

HALLEGOUET, B. and A. GUILCHER (1990). Moulin Blanc Artificial Beach, Brest, Western Brittany, France. *Journal of Coastal Research, Artificial Beaches,* M. L. Schwartz and E. C. F. Bird (editors). Special Issue 6, pp. 17–20.

HALLERMEIER, R. J. (1981). A Profile Zonation for Seasonal Sand Beaches from Wave Climate. *Coastal Engineering* 4: 253–277.

HANDS, E. B. and M. C. ALLISON (1991). Mound Migration in Deeper Water and Methods of Categorizing Active and Stable Berms. *Coastal Sediments '91, Amer. Soc. Civil Engrs.,* pp. 1985–1999.

HANSEN, M. and M. R. BYRNES (1991). Development of Optimum Beach Fill Design Cross-Section. *Coastal Sediments '91, Amer. Soc. Civil Engrs.,* pp. 2067–2080.

HANSON, H. and N. C. KRAUS (1990). Shoreline Response to a Single Transmissive Detached Breakwater. *Proceedings of the 22nd Coastal Engineering Conference, Amer. Soc. Civil Engrs.,* pp. 2034–2046.

HANSON, H., N. C. KRAUS, and L. D. NAKASHIMA (1989). Shoreline Change Behind Transmissive Detached Breakwaters. *Coastal Zone '89, Amer. Soc. Civil Engrs.,* pp. 568–582.

HEADLAND, J.R. (1992). Design of Protective Dunes at Dam Neck, Virginia. *Proceedings of Coastal Engineering Practice '92, Amer. Soc. Civil Engrs.,* pp. 251–267.

HEALY, T., R. M. KIRK, and W. P. DE LANGE (1990). Beach Renourishment in New Zealand. *Journal of Coastal Research, Artificial Beaches,* M. L. Schwartz and E. C. F. Bird (editors). Special Issue 6, pp. 77–90.

HEALY, T., C. HARMS, and W. DE LANGE (1991). Dredge Spoil and Inner Shelf Investigations off Tauranga Harbour, Bay of Plenty, New Zealand. *Coastal Sediments '91, Amer. Soc. Civil Engrs,* pp. 2037–2051.

HOUSTON, J. R. (1990). Discussion of: Pilkey, O.H., 1990. A Time to Look Back at Beach Replenishment (editorial); and Leonard, L., Clayton, T., and Pilkey, O., 1990. An Analysis of Replenished Beach Design Parameters on U.S. East Coast Barrier Islands: *Journal of Coastal Research* 6: 1023–1036.

HOUSTON, J. R. (1991a). Beachfill Performance. *Shore & Beach* 59: 15–24.

HOUSTON, J. R. (1991b). Rejoiner to: Discussion of Pilkey and Leonard, 1990; and Houston, 1990: *Journal of Coastal Research,* 7: 565–577.

HOUSTON, J. R. (1996). International Tourism and U.S. Beaches. *Shore & Beach* 64(2): 3–4.

HUDSON, R. Y. (1953). Wave Forces on Breakwaters. *Transactions of the American Society of Civil Engineers* 118: 653–674.

HUDSON, R. Y. (1959). Laboratory Investigations of Rubble-Mound Breakwaters. *Journal of Waterways and Harbors Division, Amer. Soc. Civil Engrs.* 85(WW3): 492–519.

HUDSON, R. Y. (1961). Laboratory Investigation of Rubble-Mound Breakwaters. *Transactions of the American Society of Civil Engineers* 126(part IV): 492–541.

JAMES, W. R. (1974). Beach Fill Stability and Borrow Material Texture. *Proceedings of the 14th Coastal Engineering Conference, Amer. Soc. Civil Engrs.,* pp. 1334–1349

JAMES, W. R. (1975). *Techniques in Evaluating Suitability of Borrow Material for Beach Nourishment.* U.S. Army Corps of Engineers, Coastal Engineering Research Center Technical Memo TM-60.

JOHNSON, C. N. (1987). Rubble Beaches Versus Rubble Revetments. *Coastal Sediments '87, Amer. Soc. Civil Engrs.,* pp. 1216–1231.

KAMPHUIS, J. W., K. A. RACHET, and J. JUI (1992). Hydraulic Model Experiments on Seawalls. *Proceedings of the 23rd Coastal Engineering Conference, Amer. Soc. Civil Engrs.,* pp. 1272–1284.

KIKNADZE, A. G., V. V. SAKVARELIDZE, V. M. PESKOV, and G. E. RUSSO (1990). Beach-Forming Process Management of the Georgian Black Sea Coast. *Journal of Coastal Research, Artificial Beaches,* M. L. Schwartz and E. C. F. Bird (editors). Special Issue 6, pp. 33–44.

KIRK, R. M. (1992). Artificial Beach Growth for Breakwater Protection at the Port of Timaru, East Coast, South Island, New Zealand. *Coastal Engineering* 17: 227–251.

KOIKE, K. (1990). Artificial Beach Construction on the Shores of Tokyo Bay, Japan. *Journal of Coastal Research, Artificial Beaches,* M. L. Schwartz and E. C. F. Bird (editors). Special Issue 6, pp. 45–54.

KOMAR, P. D., (1978). Wave Conditions on the Oregon Coast During the Winter of 1977–78 and the Resulting Erosion of Nestucca Spit. *Shore & Beach* 46(4): 3–8.

KRAUS, N. C. (1988). The Effects of Seawalls on the Beach: An Extended Literature Review. *Journal of Coastal Research, The Effects of Seawalls on the Beach,* N. C. Kraus and O. H. Pilkey (editors). Special Issue No. 4, pp. 1–29.

KRAUS, N. C. and R. A. WISE (1993). Simulation of January 4, 1992 Storm Erosion at Ocean City, Maryland. *Shore & Beach* 61(1): 34–40.

KRAUS, N. C. and W. G. MCDOUGAL (1996). The Effects of Seawalls on the Beach: Part I, An Updated Literature Review. *Journal of Coastal Research* 12: 691–701.

KRIEBEL, D. L. (1987). Beach Recovery Following Hurricane Elena. *Coastal Sediments '87, Amer. Soc. Civil Engrs.,* pp. 990–1005.

KRUMBEIN, W. C. and W. R. JAMES (1965). *A Lognormal Size Distribution Model for Estimating Stability of Beach Fill Material.* U.S. Army Corps of Engineers, Coastal Engineering Research Center Technical Report No. 16.

LARSON, M. and N. C. KRAUS (1991). Mathematical Modeling of the Fate of Beach Fill. *Coastal Engineering* 16: 83–114.

LEMÉHAUTÉ, E. M. and M. SOLDATE (1977). *Mathematical Modeling of Shoreline Evolution.* U.S. Army Corps of Engineers, Coastal Engineering Research Center Misc. Report No. 77–10.

LEONARD, L., T. CLAYTON, K. DIXON, and O. H. PILKEY (1989). U.S. Beach Repenishment Experience: A Comparison of the Atlantic, Pacific and Gulf Coasts. *Coastal Zone '89, Amer. Soc. Civil Engr.,* pp. 1994–2006.

LEONARD, L., T. CLAYTON, and O. H. PILKEY (1990). An Analysis of Replenished Beach Design Parameters on U.S. East Coast Barrier Islands. *Journal of Coastal Research* 6: 15–36.

LOUISSE, C. J. and F. VAN DER MEULEN (1991). Future Coastal Defense in The Netherlands: Strategies for Protection and Sustainable Development. *Journal of Coastal Research* 7: 1027–1041.

MAGOON, O. T. (1976). Offshore Breakwaters at Winthrop Beach, Massachusetts. *Shore & Beach* 44(3): 34.

MAY, V. (1990). Replenishment of Resort Beaches at Bournemouth and Christchurch, England. *Journal of Coastal Research, Artificial Beaches,* M. L. Schwartz and E. C. F. Bird (editors). Special Issue 6, pp. 11–15.

MAY, S. K., R. DOLAN, and B. P. HAYDEN (1983). Erosion of U.S. Shorelines. *EOS* 64: 551–553.

McDOUGAL, W. G, M. A. STURTEVANT, and P. D. KOMAR (1987). Laboratory and Field Investigations of Shoreline Stabilization Structures on Adjacent Properties. *Coastal Sediments '87, Amer. Soc. Civil Engrs.,* pp. 961–973.

McDOUGAL, W. G., N. C. KRAUS, and H. AJIWIBOWO (1996). The Effects of Seawalls on the Beach: Part II, Numerical Modeling of SUPERTANK Seawall Tests. *Journal of Coastal Research* 12: 702–713.

McLELLAN, T. N. (1990). Nearshore Mound Construction Using Dredged Material. *Journal of Coastal Research,* Special Issue No. 7, pp. 99–107.

McLELLAN, T. N. and N. C. KRAUS (1991). Design Guidance for Nearshore Berm Construction. *Coastal Sediments '91, Amer. Soc. Civil Engrs.,* pp. 2000–2011.

MIKKELSIN, S. C. (1977). The Effects of Groins on Beach Erosion and Channel Stability at the Limfjord Barriers, Denmark. *Coastal Sediments '77, Amer. Soc. Civil Engrs.,* pp. 17–32.

MUIR WOOD, A. M. (1970). Characteristics of Shingle Beaches: The Solution to Some Practical Problems. *Proceedings of the 12th Coastal Engineeering Conference, Amer. Soc. Civil Engrs.,* pp. 1059–1075.

NORDSTROM, K. F., J. R. ALLEN, D. J. SHERMAN, and N. P. PSUTY (1979). Management Considerations for Beach Nourishment at Sandy Hook, NJ. *Coastal Engineering* 2: 215–236.

PASKOFF, R. and R. PETOIT (1990). Coastal Progradation as a By-Product of Human Activity: An Example from Chañaral Bay, Atacama Desert, Chile. *Journal of Coastal Research, Artificial Beaches,* M. L. Schwartz and E. C. F. Bird (editors). Special Issue 6, pp. 91–102.

PEÑA, C., V. CARRION, and A. CASTAÑEDA (1992). Projects, Works and Monitoring at Barcelona Coast. *Proceedings of the 23rd Coastal Engineering Conference, Amer. Soc. Civil Engrs.,* pp. 3385–3398.

PILKEY, O. H. (1990). A Time to Look Back at Beach Nourishment. Editorial. *Journal of Coastal Research* vol. 6, pp. iii–vii.

PILKEY, O. H. and T. D. CLAYTON (1989). Summary of Beach Replenishment Experience on U.S. East Coast Barrier Islands. *Journal of Coastal Research* 5: 147–159.

PILKEY, O. H. and T. D. CLAYTON (1990). Reply To: Houston Discussion of Pilkey (1990) and Leonard et al. (1990). *Journal of Coastal Research* 6: 1047–1057.

PILKEY, O. H. and L. A. LEONARD (1991). Reply To: Houston, 1991, Re: Discussion of Pilkey and Leonard, 1990. *Journal of Coastal Research* 7: 879–894.

PILKEY, O. H. and H. L. WRIGHT (1988). Seawalls Versus Beaches. *Journal of Coastal Research, The Effects of Seawalls on Beaches,* N.C. Kraus and O.H. Pilkey (editors). Special Issue No. 4, pp. 41–67.

PLANT, N. G. and G. B. GRIGGS (1992). Interactions Between Nearshore Processes and Beach Morphology Near a Seawall. *Journal of Coastal Research* 8: 183–200.

POPE, J. and D. D. ROWEN (1983). Breakwaters for Beach Protection at Lorain, OH. *Proceedings of Coastal Structures '83, Amer. Soc. Civil Engrs.,* pp. 753–768.

POPE, J. and J. L. DEAN (1986). Development of Design Criteria for Segmented Breakwaters. *Proceedings of the 20th Coastal Engineering Conference, Amer. Soc. Civil Engrs.,* pp. 2144–2158.

PRESTEDGE, G. K. (1992). Sharp Rock Pier and Submerged Groyne: *Shore & Beach* 60: 6–14.

PRESTEDGE, G. K., I. G. STRICKLAND, and M. W. WATSON (1991). Artificial Headland and Submerged Groyne for a Beach Improvment Project. *Coastal Sediments '91, Amer. Soc. Civil Engrs.,* pp. 1901–1915.

PSUTY, N. P. and M. E. S. A. MOREIRA (1990). Nourishment of a Cliffed Coastline, Praia da Rocha, The Algarve, Portugal. *Journal of Coastal Research, Artificial Beaches,* M. L. Schwartz and E. C. F. Bird (editors). Special Issue 6, pp. 21–32.

ROELSE, P. (1990). Beach and Dune Nourishment in The Netherlands. *Proceedings of the 22nd Coastal Engineering Conference, Amer. Soc. Civil Engrs.,* pp. 1984–1997.

SEYMOUR, R. J. ET AL. (1995). *Beach Nourishment and Protection.* Committee on Beach Nourishment and Protection, Marine Board, National Research Council, National Academy Press, Washington, D.C.

SIMOEN, R., H. VERSLYPE, and D. VANDENBOSSCHE (1988). The Beach Rehabilitation Project at Ostend, Belgium. *Proceedings of the 21st Coastal Engineering Conference, Amer. Soc. Civil Engrs.,* pp. 2855–2866.

SMITH, A. W. S. (1990). Discussion of: Pilkey, O.H., 1990. A Time to Look Back at Beach Replenishment (editorial); and Leonard, L., Clayton, T., and Pilkey, O., 1990. An Analysis of Replenished Beach Design Parameters on U.S. East Coast Barrier Islands. *Journal of Coastal Research* 6: 1041–1045.

SMITH, A. W. S. and L. A. JACKSON (1990). The Siting of Beach Nourishment Placements. *Shore & Beach* 58: 17–24.

STAUBLE, D. K. (1990). *State-of-The-Art Report on Beachfill Design.* Technical Report No. 90, Coastal Engineer Research Center, U.S. Army Corps of Engineers., Waterways Experiment Station, Vicksburg, Mississippi.

STAUBLE, D. K. and W. G. GROSSKOPF (1993). Monitoring Project Response to Storms: Ocean City, Maryland, Beach Fill. *Shore & Beach* 61(1): 22–33.

SUH, K. and R. A. DALRYMPLE (1987). Offshore Breakwaters in Laboratory and Field. *Journal of Waterway, Port, Coastal, and Ocean Engineering, Amer. Soc. Civil Engrs.* 113(2): 105–121.

SWART, D. H. (1991). Beach Nourishment and Particle Size Effects. *Coastal Engineering* 16: 61–81.

TERICH, T. and T. LEVENSELLER (1986). The Severe Erosion of Cape Shoalwater, Washington. *Journal of Coastal Research* 2: 465–477.

TOUE, T. and H. WANG (1990). Three Dimensional Effects of Seawall on the Adjacent Beach. *Proceedings of the 22nd Coastal Engineering Conference, Amer. Soc. Civil Engrs.,* pp. 2782–2795.

VERA-CRUZ, D. (1972). Artificial Nourishment of Copacabana Beach. *Proceedings of the 13th Coastal Engineering Conference, Amer. Soc. Civil Engrs.,* pp. 1451–1463.

VERHAGEN, H. J. (1990). Coastal Protection and Dune Management in The Netherlands. *Journal of Coastal Research* 6: 169–179.

WALTON, T. L. (1994). Shoreline Solution for Tapered Beach Fill. *Journal of Waterway, Port, Coastal, and Ocean Engineering, Amer. Soc. Civil Engrs.* 120: 651–655.

WALTON, T. L. and J. S. PURPURA (1977). Beach Nourishment Along the Southeast Atlantic and Gulf Coasts. *Shore & Beach* 45: 10–18.

WALTON, T. L. and W. SENSABAUGH (1978). *Seawall Design on Sandy Beaches.* University of Florida Sea Grant Report No. 29.

WATSON, I. and C. W. FINKL (1990). State of the Art in Storm Surge protection: The Netherlands Delta Project. *Journal of Coastal Research* 6: 739–764.

WEGGEL, J. R. (1988). Seawalls: The Need for Research, Dimensional Considerations and a Suggested Classification. *Journal of Coastal Research, The Effects of Seawalls on the Beach,* N. C. Kraus and O. H. Pilkey (editors). Special Issue No. 4, pp. 29–39.

WIEGEL, R. L. (1992). Dade County, Florida, Beach Nourishment and Hurricane Surge Protection Project. *Shore & Beach* 60: 2–26.

ZWAMBORN, J. A., G. A. W. FROMME, and J. B. FITZPATRICK (1970). Underwater Mound for Protection of Durban's Beaches. *Proceedings of the 12th Coastal Engineeering Conference, Amer. Soc. Civil Engrs.,* pp. 975–994.

Index

Table of Symbols

C Velocity of wave propagation (phase velocity)

C_g Group velocity of waves ($= Cn$)

D Sediment grain diameter

d Horizontal major diameter of the elliptical water particle motion under waves

E Total wave-energy density

f Wave frequency ($= 1/T$)

F Fetch or distance over which winds blow in generating waves

g Acceleration of gravity

h Still-water depth

H Wave height

I_ℓ Total immersed-weight longshore sediment transport rate on a beach

k Wave number ($= 2\pi/L$)

L Wave length

L_e Longshore wave length of edge waves

n Ratio of wave-group velocity to wave-phase velocity

P Wave-energy flux (power)

P_ℓ Longshore-directed wave energy flux $[= (Ecn)_b \sin \alpha_b \cos \alpha_b]$

Q_ℓ Volumetric longshore sediment transport rate

R Wave run-up elevation on the beach

s Vertical major diameter of the elliptical water particle motions under waves

S Slope of the beach profile ($= \tan\beta$)

S_{xx} Component of the radiation stress of waves; the flux of x-directed momentum in the x-direction (onshore)

S_{yy} Component of the radiation stress of waves; the flux of y-directed momentum in the y-direction (alongshore)

S_{xy} The longshore component of the radiation stress of waves; the flux of y-directed (longshore) momentum advancing in the x-direction (onshore)

t time

T Wave period

T_e Period of edge waves

u Horizontal component of the water particle velocity under waves

U Wind velocity—important in the generation of waves

$\bar{v}_\ell$ Longshore current velocity

w Vertical component of the water particle velocity under waves

w_s Settling velocity of sediment grains

X_b Width of the surf zone